THE Environment AND THE People in American Cities, 1600s–1900s

Dorceta E. Taylor

The Environment and the People in American Cities, 1600s–1900s

Disorder, Inequality, and Social Change

Duke University Press
Durham and London
2009

Printed in the United States of America on acid-free paper ♾

Designed by Heather Hensley

Typeset in Adobe Garamond Pro by Tseng Information
Systems, Inc.

Library of Congress Cataloging-in-Publication Data appear
on the last printed page of this book.

Duke University Press gratefully acknowledges the support of
The Regents of the University of Michigan for providing funds
toward the production of this book.

Dedicated to

IAN

JUSTINE & SHAINA

CONTENTS

FIGURES, TABLES, AND BOXES

ACKNOWLEDGMENTS

I would like to thank my husband, Ian Robinson, and my twin daughters, Justine and Shaina, for their patience and understanding during the writing of this book. There were many times when the girls jostled for space on my lap amid the books and papers. Dinner was late too many times to recall, and playtime got overshadowed by research time. To all three of you, I simply say thanks.

I have worked with and taught many great students over the years at the University of Michigan. They helped to inspire this book. They asked probing questions about environmental history that led me to begin researching areas thinly covered. One thing led to another, and the trilogy of which this book is the first installment was born. Thanks to Latonia Phillips who helped to coordinate my programs and to all the students who have worked as my research assistants over the years; you have encouraged me, shown an interest in the work, and helped me to think about this topic more clearly.

Special thanks also go to the School of Natural Resources and Environment as well as to the Center for Afroamerican and African Studies for providing encouragment and support for this undertaking. The Office of the Vice Provost for Research also supported this project, and for this I am grateful.

I would also like to thank the staff at the University of Michigan's library. Thanks for assembling great resources for me to use and for your willingness in helping to find all the material I needed.

Thanks also go to the anonymous reviewers who read early

drafts of the manuscript and provided helpful feedback. I appreciate the time you spent reading the book and found your comments very useful. Finally I want to thank the wonderful editorial staff at Duke University Press. Thanks for your belief in the project and for your help and guidance through the publication process.

INTRODUCTION

Rivers of garbage and potent toxins burn in Guatemala. Deforestation of the Amazon is occurring at an alarming rate. Americans buy hamburgers made from cattle reared on denuded lands in Latin America. Imported agricultural produce returns pesticides to the United States that were banned years ago. Garbage barges travel the high seas searching for places to disgorge their hazardous cargo. Batteries tossed from American cars end up in Mexico or Taiwan, where workers process them barehanded. Famines affect vast swaths of the earth. The Arctic ice cap is melting. Massive tsunamis and powerful hurricanes devastate coastal communities.

In the United States cities are dotted with hazardous waste sites and the hulking, ghostlike remains of abandoned factories. Industrial accidents still make headlines as mines collapse, trapping workers underground, and deadly explosions occur in industrial complexes. Contaminated waterways abound. Some American cities have no space to put their garbage. Cancer rates soar in small towns in Cancer Alley, the stretch of the Mississippi River from Baton Rouge to New Orleans, and in urban communities such as the Toxic Doughnut on the Southside of Chicago. The term *affordable housing* is fast becoming an oxymoron as many Americans can't afford to buy homes and some have no hope even of finding an apartment they can afford to rent. Jobs that pay a living wage are becoming scarce. Because of globalization and economic restructuring many high-paying, unionized jobs that had health and safety protections have moved overseas. Such jobs have been replaced by low-paying, service sector jobs with few benefits.

Given all this it is not surprising that tensions run high in the United States over issues such as affirmative action, immigration, and welfare reform. Workers whose companies have closed, cut thousands of jobs, or moved all or parts of their operations overseas are more likely to vent their frustrations on minorities and immigrants for "taking their jobs" rather than confronting the corporate elites, policy makers, politicians, or the other architects of downsizing, corporate restructuring, free trade, and globalization. As increasing numbers of Americans have difficulty maintaining their standard of living and as social services disappear, people search for explanations. Some accept neoconservative explanations that use race as a wedge issue to distract people from focusing on the real sources of the problems. Some politicians seek convenient scapegoats, such as poor women and people of color, or attack social programs with weak political support. In this climate welfare and the other issues have been characterized as the source of all ills. The highly charged discussions have sometimes occluded rather than clarified matters for the public. These fault line's reveal rifts that have deep roots in American society.

Debates over urban environmental problems and social policies are not new. An analysis of urban environmental history shows that American cities have wrestled with poverty, health, sanitation, housing, open space, land uses, occupational safety, and pollution for centuries. Though many environmental history accounts begin and end with wilderness and wildlife, the city plays a critical role in our understanding of the relationship between people and the environment in America. This book focuses on the city because it provides important clues to understanding the evolution of American environmental activism. Cities provide contexts in which to understand the activists, the issues they fought, and how they perceived the environment.

The city is an understudied component of the environment, yet it is the place where human labor was exploited to transform forests into commodities, pavement, and buildings. Here intellectuals theorized about the relationship between health and open space, and reformers implemented policies and programs to improve quality of life for urban dwellers. To understand urban environmentalism one has to understand how cities grew and changed over time; class, gender, racial, and ethnic tensions; and the quest to impose order on the populace and the environment. Consequently I focus on the evolution of the city, the emergence of elite activists, and responses to the perceived breakdown in social order. I also draw attention to some of the environmental challenges faced by American cities over the past four centuries. The reader will notice some similarities between the problems facing older cities and cities of today; likewise there are some similarities in the ideology and policies of past and present urban reformers.

Overview of the Book

Seven major goals characterize urban environmental activism: (1) alleviation of poverty and improved quality of life; (2) sanitary reform and public health; (3) safe, affordable, and adequate housing; (4) the creation of parks, playgrounds, and open space; (5) occupational health and safety; (6) consumer protection, particularly food and product safety; and (7) land use and urban planning. The book has five parts; theoretical analyses appear at the end of parts I through IV to help synthesize the theoretical arguments embedded in each section. I introduce and discuss relevant sociological concepts and theories in this introduction. Three overarching conceptual frameworks are race relations, social movement, and organization theories. In part I (chapters 1 and 2) I focus on the condition of the city and the relationship between poverty and epidemics. I examine the social dynamics of cities, their evolution, and the transformation of the environment.

In part II (chapters 3–6) I examine attempts to reform the city. In particular I discuss elite responses to the chaos in the cities and the impetus to undertake social, political, and environmental reforms. I explain how elite perception and framing of issues such as poverty changed over time, and how elites developed and implemented programs to improve conditions. In part III (chapters 7–10) I focus on urban park development and social reform, looking at the evolving definition and use of open space, the urban parks campaigns, and the presumed role of parks in cultivating and maintaining social order. I also examine middle- and working-class conflicts over park use and consider the historical and contemporary models of financing urban parks.

Land use planning and the rise of comprehensive zoning are covered in part IV. Chapter 11 traces the rise in the use of restrictive covenants to control land uses. I analyze attempts by elites to separate different types of land uses, develop uniform residential and commercial districts, create aesthetically appealing neighborhoods, and bring coherence to the city planning process. Chapter 12 examines the rise of comprehensive zoning, the class and racial dimensions of zoning as well as legal challenges to different types of zoning ordinances. In the final section of the book, part V (chapters 13–14), I discuss occupational and community hazards arising from industrialization, specifically industrial accidents, unsafe working conditions, and pollution. I analyze the rise of industrial medicine and its effects on worker health and safety, attempts to reduce these hazards, and elite and working-class responses to these issues. The book ends with a short conclusion.

The Environment and the People in American Cities, 1600s–1900s provides a historical analysis of the way race, class, and gender shape environmental experiences, perceptions, activism, and the construction of environmental discourses. Not only do many historical accounts of environmental activism focus on wilderness and wildlife issues, but they tend to focus on elite white males.[1] This book departs from that tradition somewhat by including the experiences and activism of the poor and people of color in addition to the contributions of upper- and middle-class males and females to environmental change.

This book began as a project to write a comprehensive environmental history that examines the experiences of whites and people of color in urban and nonurban settings. Though my first impulse was to write one book, as the manuscript mushroomed in size and complexity I decided to split the book into three discrete but somewhat interrelated volumes. The current book, *The Environment and the People in American Cities*, is the first of three. The second book, *Outward Bound: Manliness, Wealth, Race, and the Rise of the Environmental Movement, 1830s–1930s*, will concentrate on conservation and preservationism and the role of the white middle class in the emergence of the mainstream environmental movement. It focuses on nineteenth- and twentieth-century environmental activism and ideology. *Outward Bound* will also show how some of the urban environmental activism discussed in this book influenced the conservation discourse. The third book, *People of Color and the Environment*, will examine the environmental experiences and activism of Native Americans, African Americans, Latinos, and Asians from the seventeenth century to the twentieth. It will also examine the rise of the environmental justice movement.

Conceptualizing the Environment and Social Inequality

Race is a lightning rod in American society; consequently the environmental discourse has avoided discussions of race for more than 150 years. In fact, some might wonder why I include racial terminology in an environmental book. What does race have to do with the environment? The answer is that local, national, and global environmental problems are intertwined with racial and ethnic relations, greed and overconsumption, resource and human exploitation, inequality, imperialism, and human and civil rights abuses. One cannot get a full understanding of environmental problems by analyzing only processes or policies. One has to examine the role of race, class, and gender relations to understand the forces at work in communities and work sites. Consider the fact that airplanes dust crops with deadly pesticides while workers and their school-age children either toil in the fields or are forced to reenter soon after spraying; studying only the effects of pesticides on food crops, the forest, and wildlife would ignore the human and social dimensions of the puzzle. Yet for most of the history of the environmental movement this was precisely how environmental research and analyses proceeded. In many cases, when the people affected were the poor or minorities, they received scant attention from environmental activists. This book will show that an understanding of social processes can enhance environmental analyses. The following sociological concepts will help us comprehend more clearly how labor market dynamics and social relations have influenced urban environmental affairs over time.

RACE RELATIONS CONCEPTS

Ethnicity

The term *ethnicity* comes from *ethnos*, the Greek word for "nation."[2] Ethnicity is commonly defined as membership in a collectivity that shares a common ancestry or country of origin, language, religion, historical past, and cultural traditions which are different from that of the dominant society, or a focus on symbolic elements defined as the essence of the group. The extent to which these features characterize any American group varies considerably with the size and history of the group and the length of time members of the group have been in the United States. Ethnic groups persist over long periods, but they also change, merge, emerge, and dissolve. Ethnicity is often situational; race may not be. An individual can choose whether and how to express his or her ethnicity, while many people have no choice about what race they display.[3]

Ethnicity plays a role in labor recruitment, the structure of the labor market, and the persistence of environmental inequalities. In addition to racial differentiation, people are also sorted and segregated along ethnic lines. In the historical analysis of environment and labor dynamics we will see that workers were first sorted by race, then by ethnic group. For example, during the nineteenth century and the twentieth, white workers were paid higher wages and given less dangerous jobs than nonwhites. However, among whites, those of

Anglo-Saxon ancestry were paid more and given better jobs than whites of Southern and Eastern European ancestry. Similarly in the early environmental movement organizational memberships were limited to whites of Northern European ancestry.[4]

Minority

Minority is a term used as a synonym for race, black, people of color, and nonwhites. The term carries a double meaning, the numerical and the political. In the United States a minority group is defined primarily in terms of disadvantage, less privilege, political oppression, economic exploitation, and social discrimination.[5] Wirth defined minority as "a group of people who, because of their physical or cultural characteristics, are singled out from others in the society in which they live for differential and unequal treatment, and who, therefore, regard themselves as objects of collective discrimination."[6] Wagley and Harris add that minorities are held in low esteem by the dominant segments of the society and that membership in a minority group is transmitted through kinship lines.[7]

Race and Racial Classification

Race is related to ethnicity and minority status. Race plays a key role in the discussion of environmental, labor, and social justice by helping us to understand how people are classified and assigned to groups and positions in society. Those positions and groups have differential access to power, privilege, and resources. Consequently race affects the experiences people have, how they perceive things, their identity, and the kinds of collective responses they organize. Although some researchers think of race simply as one aspect of ethnicity, others, such as Omi and Winant, believe it is a distinct concept and one worthy of greater discussion.[8] I share the view that though there is some overlap between race and ethnicity, ethnicity does not fully incorporate all the meanings of race and does not fully describe the experiences of people of color in the United States.

Since the word *race* first appeared in the English language in 1508 its meaning has changed several times.[9] During the seventeenth century the French used the phrase *especes-ou-races d'homme* to refer to "family" or "breed." Around 1700 Germans began using the word *Rasse* to mean "generation." In 1775 Kant made reference to the "races of mankind" to identify groups that could be distinguished from each other because of their physical characteristics. Anthropologists, naturalists, and botanists were at the forefront of efforts to develop racial classification schemes. In 1684 the naturalist François Bernier identified what he considered to be four "stocks" of human groups, and Linneaus identified four "varieties" (white, red, yellow, and black) of humans in 1734. Linneaus also attempted to correlate traits such as temperament and personality with the phenotypic features of skin color and body type. During the eighteenth century the natural scientist George Buffon also used skin color, stature, and body shape to delineate six different races.[10]

Hence early on, race was thought of as a biological category defined by physical characteristics such as skin or hair color.[11] As this line of thought developed, race was used to refer to subspecies, breeds, or zoological types to help explain perceived genetic variations in human groups. Researchers believed that distinct racial groups or biological subspecies developed when humans were isolated geographically for prolonged periods of time. Others thought that distinct breeds of people existed even in the absence of geographical isolation. Race has also been used to mean *species*, as in "the human race," and *nation*, as in the "French race" and the "German race."[12] According to the sociologist Robert Park people are bearers of a double inheritance.[13] That is, as members of a racial group people transmit a biological inheritance, but through their communication and interactions with others they also transmit a social inheritance.

By the mid-twentieth century many began to question the biological view of race and the notion of distinct human subspecies.[14] Consequently most contemporary scholars view race as a social construct rather than a system of biological types; Banton, for instance, argues that people use racial (phenotypical) differences to create and allocate roles.[15] Race is most commonly used today to describe a group of people who are socially defined as belonging together because of physical or phenotypic markers.[16]

According to Omi and Winant, in the American context the biological conceptualization of race has evolved during and since slavery to link racial inferiority and social inequality with phenotypic characteristics that are perceived to be part of a natural order; thus whites were considered to be the superior race while others were inferior, and white skin was presumed the norm while other skin colors were aberrations.[17] It followed that differences in intelligence, temperament, criminal behavior, and sexuality were viewed as racial in character. From this perspective, racial intermingling was unnatural because it could produce inferior beings or "biological throwbacks." Social Darwinists and eugenicists held these viewpoints.[18] Contemporary white supremacists still cling to these views.

Scholars began paying more attention to ethnic and cultural differences between groups during the twentieth century. Instead of looking for specific racial and genetic markers that distinguished one group of people from another, proponents of the ethnicity perspective saw race as a socially constructed category. Race was considered one of the components of ethnicity.[19] Sociologists from the Chicago school were in the vanguard of incorporating ethnicity and assimilationist perspectives in their work. The assimilationist (or melting pot) perspective argued that immigrants focused on overcoming barriers so that they could blend into the dominant culture. According to this perspective, immigrants followed a path that began with economic hardship, ethnic conflict, and discrimination followed by accommodation to American norms.[20] This eventually resulted in socioeconomic mobility as immigrants became more familiar with and accepted American culture.[21] Dissatisfied with this approach, some researchers modified the assimilationist model to reflect the cultural pluralism they witnessed in the society. Researchers such as Kallen, Greeley, and Glazer and Moynihan adopted a cultural pluralist approach that argued that immigrants didn't always "melt" into the mainstream American cultural milieu.[22] They contended that immigrants preserve important aspects of their ethnicity that make them distinct from others; rather than a melting pot, the image of cultural pluralism is a cultural mosaic or tossed salad.

Though a useful analytic tool, the ethnicity perspective is also problematic. First, ethnicity-based assimilationist perspectives focused on the experiences of white immigrants, not those of racial minorities. Second, researchers incorrectly assumed that the theories derived from studying white ethnics could be applied to minority experiences without modification.[23] Third, the ethnicity-based assimilationist approach tended to blame the victim for perceived lack of progress; that is, those who did not succeed failed because they did not follow the prescribed path. Thus when minorities did not fit into the white ethnic model of adaptation (accommodation, economic success, and absorption into mainstream society) they were viewed as uneducable and culturally deficient. As conceived, the ethnicity perspective was not well equipped to explain the condition of racial and ethnic minorities in American society. As later discussions show, resources and opportunities that were available to white ethnics to enhance their economic and social standing were unavailable to blacks and other minorities. Furthermore, rather than defining themselves in purely ethnic terms, many blacks, American Indians, and Chicanos define themselves in racial terms as well.[24] Hence minorities resisted simplified ethnic categorizations and modeling.

Though many now perceive race as a social construct, in many societies people's status and life chances are organized around racial classifications; thus, although scholars such as William Julius Wilson argue that the salience

Table 1 Merton's Typology of the Relationship between Prejudice and Discrimination

	Level of Discrimination	
Level of Prejudice	Does Not Discriminate	Does Discriminate
Unprejudiced	Unprejudiced nondiscriminator • All-weather liberal • Antiracist	Unprejudiced discriminator • Fair-weather or reluctant liberal • Strategic racist
Prejudiced	Prejudiced nondiscriminator • Timid bigot • Passive racist	Prejudiced discriminator • All-weather or active bigot • Active racist

Source: Adapted from Robert K. Merton, "Discrimination and the American Creed," in *Discrimination and National Welfare*, edited by R. H. MacIver (New York: Harper and Row, 1949), 99–126.

of race has declined, recent studies have shown that race continues to be a significant factor influencing life outcomes for blacks in America.[25]

Prejudice, Discrimination, and Racism

Though the terms *prejudice* and *discrimination* are often used synonymously in everyday language, their meanings differ. Prejudice is a set of beliefs and stereotypes that lead an individual or group of people to be biased for or against members of a particular group.[26] Discrimination refers to the unfavorable treatment of people assigned to a particular social group.[27] Though prejudice can lead to discrimination, this does not always happen. In distinguishing between prejudice and discrimination, Robert Merton identified four categories of people:

ALL-WEATHER LIBERALS: those who are not prejudiced and do not discriminate.

RELUCTANT LIBERALS: those who are not prejudiced but discriminate if it is in their interests to do so.

TIMID BIGOTS: those who are prejudiced but are afraid to show it.

ACTIVE BIGOTS: those who are prejudiced and are quite willing to discriminate. (See Table 1.)[28]

Although prejudice does not necessarily lead to discrimination, it can trigger discrimination. As Aguirre and Turner argue, prejudicial beliefs often highlight, unfairly and inaccurately, negative characteristics of ethnic groups. The negative imagery is then used to legitimize discrimination. Prejudice also instills emotions and feelings that can trigger intolerance, hostility, and acts of discrimination.[29] The reader will find ample evidence in this book of people and groups acting on their prejudices to stereotype and discriminate against the poor, immigrants, minorities, and women.

In an attempt to distinguish prejudice from discrimination, Tatum uses the analogy of a moving conveyor walkway like those seen at airports.[30] The active racists (such as the Aryan Nation) are the people who identify with white supremacist ideology; they are not afraid to express these views openly as they walk or run along the conveyor belt. Passive racists stand still on the walkway; they make no effort to move in either direction, but the conveyor belt pulls them in the direction of the active racists anyway. Some people on the moving walkway recognize the direction in which the active and passive racists are heading and do not want to follow them; these strategic racists might turn around and face the opposite direction, but unless they are walking actively they too will be carried along with the racists. The people on the moving walkway who take an antiracist stance walk vigorously in the direction opposite that of the racists.

Racism, Institutional Racism, and White Privilege

Until the 1960s racism was defined as a doctrine, dogma, ideology, or set of beliefs. It was believed that race determined culture. Since some cultures were deemed superior and others inferior, some races were also seen as superior and others inferior. During the 1960s racism was expanded to include the practices, attitudes, and beliefs that were rooted in the notion of racial superiority and inferiority.[31] In 1967 Stokely Carmichael and Charles Hamilton injected the concept of *institutional racism* into the discourse to refer to the institutional processes and apparatuses that support and maintain racial discrimination.[32]

Since then researchers have argued that to limit the understanding of racism to prejudicial and discriminatory behavior misses important aspects of racism. Racism is also a system of advantages or privileges based on race. In the American context many of the privileges and advantages that accrue to whites stem directly from racial discrimination directed at people of color. However, many whites do not think racism affects them; they do not consider whiteness to be a racial category that carries with it certain advantages and privileges. Omi and Winant refer to this as a transparent racial identity. That is, as a signifier of power, privilege, and dominance, whiteness often remains invisible to whites themselves.[33] Ergo whites benefit from racism and will defend their racial advantages, such as access to better schools, parks, housing, jobs, and higher wages, even if they do not espouse or endorse overtly prejudicial thinking.[34] McIntosh argues that because whites are not taught to recognize white privilege as an aspect of racism, white privilege is like an invisible package (or "weightless knapsack") of special provisions, unearned assets, and benefits that they can count on cashing in each day, but about which they remain oblivious.[35] Thus racism results from personal ideology and behavior, and those personal thoughts and actions are supported by a system of cultural messages and institutional policies and practices. Racism can be more fully understood as the execution of prejudice and discrimination coupled with power and privilege.[36]

Racism is buttressed and maintained by legal, penal, educational, religious, and business institutions, to name a few. Throughout American history racist ideology has been manifested through widespread and systematic racist actions that were supported by persistent, elaborate institutional infrastructures.

Environmental Racism and Environmental Justice

Though it was not labeled as such until the last decade of the twentieth century, environmental justice activism has been a part of the politics of communities of color for more than a century. During that time people of color tried to improve housing conditions among slaves; opposed the appropriation of land, erosion of treaty rights, and the share-cropping system; developed sophisticated farming techniques and irrigation systems; tried to acquire land; and fought for workers' rights. Throughout the twentieth century blacks in particular have opposed the segregation of housing, parks, beaches, and transportation systems. During the 1950s and 1960s blacks organized car pools to boycott segregated buses and developed extensive lead screening and education programs in several cities; Chicano and Filipino farmworkers opposed the use of pesticides on agricultural produce; and Native Americans fought for fishing rights.[37]

During the 1980s activists, scholars, and policy makers began investigating the link between race and exposure to environmental hazards. The term *environmental racism* is an important concept that emerged from these investigations to describe the environmental practices frequently observed in minority communities. In particular it linked racism with environmental actions, experiences, and outcomes.[38] Moreover the concept focused on the environmental inequality between whites and people of color as well as white environmental privilege.[39] The concept and the activism it spurred also linked past social justice activism in communities of color with past and present environmental experiences. In the broadest sense environmental racism and its corollary, environmental discrimination, is the process by which environmental decisions, actions, and policies result in racial discrimination or the creation of racial advantages. This book shows that environmental racism and discrimination have been pervasive throughout American environmental history.

Nativism

The term *nativism* refers to organized efforts by some members of a society to improve their quality of life or hold on to remnants of the past by attempting to exclude or eliminate from society groups of people or particular ideas, beliefs, customs, and objects associated with them. It also describes attempts by elites to limit immigration to homogeneous racial groups, for nativists reviled racial mixing. Nativist movements flourished in the United States during the nineteenth century and the twentieth, particularly during depressions and in tight labor markets. Members of these movements strongly asserted their Anglo-Saxon heritage to secure greater privileges. Nativism can arise from economic, cultural, or status conflicts sparked by increased immigration or the competition for jobs, commercial opportunities, housing, or outdoor space. It can also be triggered by one social group's discomfort with another's way of life or if one group perceives that it has lost status or power vis-à-vis another group. Nativism was rampant in New York and other large cities during the economic downturns of the 1840s. Vying for customers during the 1842 depression native-born merchants were angered by immigrant peddlers selling wares on the sidewalks of Lower Manhattan and in the streets in front of their stores. They saw the immigrants as invading what was once their exclusive space and expressed their anger by stereotyping particularly the Irish and organizing efforts to exclude immigrants from upscale commercial districts and to restrict immigration.[40]

New England Federalists had successfully pushed the Alien Acts through Congress in 1798; these extended the period of residency required for naturalization from five to fourteen years. But in Boston immigration skyrocketed during the 1840s, growing from around five thousand annually to thirty-seven thousand in 1847. Many of the immigrants were Irish peasants fleeing the potato famine. That wave of immigration prompted anti-Catholic marauders to burn the Ursuline Convent in Charlestown (now Somerville) and inspired Massachusetts and New England elites to form the Know-Nothing Party. When a new wave of Jewish and Greek immigrants moved to Boston nativists invoked social Darwinist principles to justify their claims that people of Anglo-Saxon descent were superior to the "inferior races" of Southern and

Eastern Europe. Though Anglo-Saxons portrayed themselves as a naturally democratic group, some were nativists who opposed immigration. To this end several Harvard graduates and lawyers founded the Immigration Restriction League in 1894 to push for tougher immigration legislation, including administering a literacy test to new immigrants; since many immigrants were illiterate they would have been denied entry visas. Though the League's bill was defeated in 1897, in 1917 the federal Immigration Act passed. Instead of the literacy test, the Immigration Act required immigrants to be able to read and write (in any language). The prominent Massachusetts senator Henry Cabot Lodge helped to push the bill through Congress. Further immigration restrictions occurred in 1924 with the passage of the National Origins Act, which specified immigration quotas for non-Anglo-Saxons.[41]

Split Labor Market

Race, ethnic, and gender relations are exploited in a split labor market, which exists when two or more groups of workers do the same work but their price of labor differs. The condition exists even if workers are not doing the same work but their price of labor would differ if they did the same work. The *price of labor* refers to the worker's total cost to the employer: wages, the cost of recruitment, transportation, room and board (if paid by the employer), health care, worker training, and the costs associated with labor disputes.[42] It is important to note that the presence of different ethnic or racial groups in the same labor market does not necessarily produce a split labor market. If several ethnic groups of the same racial background enter a labor market with similar resources or goals, a split labor market does not develop.[43]

Ethnic differences aren't the only factors giving rise to a split labor market. Such markets can also arise when a politically powerless part of the population competes directly with others in the labor force; for example, prison labor has been used to undercut the wages of nonincarcerated workers. Despite the fact that women gained significant political power in the twentieth century, gender has also been used to develop split labor markets because women are often paid less than men for doing the same work.[44]

According to Bonacich, a split labor market has three strata: the economic elites, higher paid laborers, and low-wage workers.[45] Conflict arises among these strata because their goals and motives are in opposition to each other: the economic elites desire an inexpensive, amenable workforce to gain a competitive edge, while workers try to improve their economic position by seeking higher wages and better working conditions; at the same time, high-wage laborers are fearful that low-wage laborers could drive down the price of labor.[46] Two additional strata also influence split labor market dynamics: the jobless and the incarcerated; both high- and low-wage workers compete with a reserve pool of unemployed job seekers and prisoners.

Dual Labor Market

A dual labor market is defined as one having two major sectors: a primary and a secondary labor market. In the primary labor market jobs are characterized by high wages, good working conditions and benefits, stability, security, promotions, and due process in the administration of work rules. The secondary labor market is made up of low-wage jobs characterized by poor working conditions, instability, harsh and arbitrary work rules, little or no due process in administering work rules, and limited promotions and benefits. In American society the poor, the uneducated, and many people of color and women are confined to menial jobs in the secondary labor market. Some people are confined to the secondary labor market because employers think they lack the traits (race, demeanor, accent, educational attainment, etc.) that would make them employable in the primary sector. Such discrimination plays a significant role in reducing the number of minorities employed in the primary sector.[47] This is compounded by pure wage discrimination, defined as the variations in wages paid to equivalent workers of different races (those with similar years and quality of schooling, skills, employment histories, age, health, and job attitudes). Studies find that blacks earn lower incomes than whites and that the difference can be partly explained by wage discrimination.[48]

Class

In the Marxist tradition class is defined in terms of the ownership of capital and the means of production. Individuals are propertied or propertyless, depending on their position in the system of production.[49] In contrast, the Weberian tradition sees class in terms of life chances; that is, a class is a group of people who share similar life chances or market positions.[50] Others define class in terms of common positions within status hierarchies, or as common positions within authority and power structures.[51] Bourdieu articulates class in terms of class habitus and cultural capital. He defines class habitus as a set of common conditions in daily life that lead to conditioned behaviors and dispositions. The dispositions include tastes and responses to certain kinds of appeals or calls to action. Class habitus develops in the family, the workplace, and the school and other community institutions that reinforce certain modes of thought and action.[52]

In the United States in the nineteenth century and early twentieth there were three general classes: the upper crust or better sort (upper class), the middling (middle class), and the inferior sort or lower orders (lower class). Class was defined as much by race as by lineage, education, occupation, wealth, and breeding. The upper class comprised the native-born, landed gentry of Northern European descent whose families had distinguished themselves in public life. Those of Dutch and English ancestry who were well educated, owned large plots of land or ran business enterprises, and who were political leaders or occupied powerful positions were among the upper class. They pursued prestigious, learned professions such as medicine and law. Even if one did not have great wealth, being born into a family whose ancestors were the founding fathers of American towns or cities accorded one great prestige and membership in the upper class. Upper-class Northern Europeans carried their class status with them when they migrated to the United States regardless of how much wealth they accumulated once there. The aristocrats occupied the upper echelon of the upper class; they were the patricians, families of old wealth, impeccable breeding and taste, and unparalleled power and prestige. The aristocrats defined themselves as such, but others recognized and defined them in these terms also. Though the nouveaux riches had wealth enough to buy their way into the upper class, they were not considered aristocrats.[53] Members of the upper class were elites, but not all elites belonged to the upper class; some elites were middle-class professionals and experts who developed and implemented policies.

The middling class consisted of newer arrivals: merchants of small and medium-size businesses, clerks, analysts, managers, artisans, and others of Northern European descent. Some in the middling class were in the process of generating family wealth, while others were already nouveaux riches aspiring to upper-class status. The middling class was also primarily native-born. The middle and upper classes were Protestants. The lower class comprised Southern and Eastern Europeans, foreign-born immigrants, and ethnic minorities. They were primarily renters, small shop owners, and unskilled laborers employed as factory and field hands or domestic servants or in other service jobs. Some were slaves or indentured servants. Many in the lower class were Catholic or members of other non-Protestant religions.[54]

A Social Constructionist Perspective

Concepts such as ethnicity, race, class, and environmental problems are socially constructed claims defined through collective processes; in other words, the problems, issues, concepts, and terminology are not static and are not always the product of readily identifiable, visible, or objective conditions.[55] That is, groups in a society perceive and define problems by developing shared meanings and interpretations of the issues; over time meanings change. A constructionist perspective is concerned with how people assign meanings to their social world.[56]

SOCIAL LOCATION AND THE CONSTRUCTION OF SOCIAL PROBLEMS

Social location, or positionality, also influences the construction of social problems. Social location refers to the position a person or group occupies in society. That position is influenced by factors such as gender, race, and class. Social location affects how people construct the meanings that define grievances, opportunities, and collective identities. In addition social location influences knowledge of collective action tactics and strategies. It also helps to determine the type and amount of resources available for social movement activities.[57] This book will highlight how social location influences participation; for instance, Figure 1 shows the interplay between race, class, gender, and environmental activism during the nineteenth century and early twentieth.

Not all white middle- and upper-class activists focused on wilderness and wildlife issues in the early environmental movement. During the nineteenth century two groups of white activists emerged. One group focused on wilderness and wildlife issues and the other concentrated on urban environmental reform. The white middle-class urban environmental activists interacted with the white working class and minorities in cities. Though the interaction was sometimes strained, there were times when elite urban activists collaborated with the poor. Though middle- and upper-class activists spearheaded many of the campaigns, the white working class and minorities also participated in environmental affairs. For the most part working-class whites and people of color focused on urban environmental issues. Though all groups had an interest in rural issues, this book focuses on urban environmental activism and the interrelationships between race, class, and gender. The forthcoming *Outward Bound* will focus on middle-class wilderness and wildlife activism and *People of Color and the Environment* will focus on minority activism.

Table 2 compares the two general types of white middle-class environmental movements that emerged in the nineteenth century. While the race and class backgrounds of the activists were similar, they differed on the issues they focused on, the location of their activism, the sphere of life they were concerned with, who they targeted or collaborated with, the role of social justice in their campaigns, and their general approach to environmentalism. Table 3

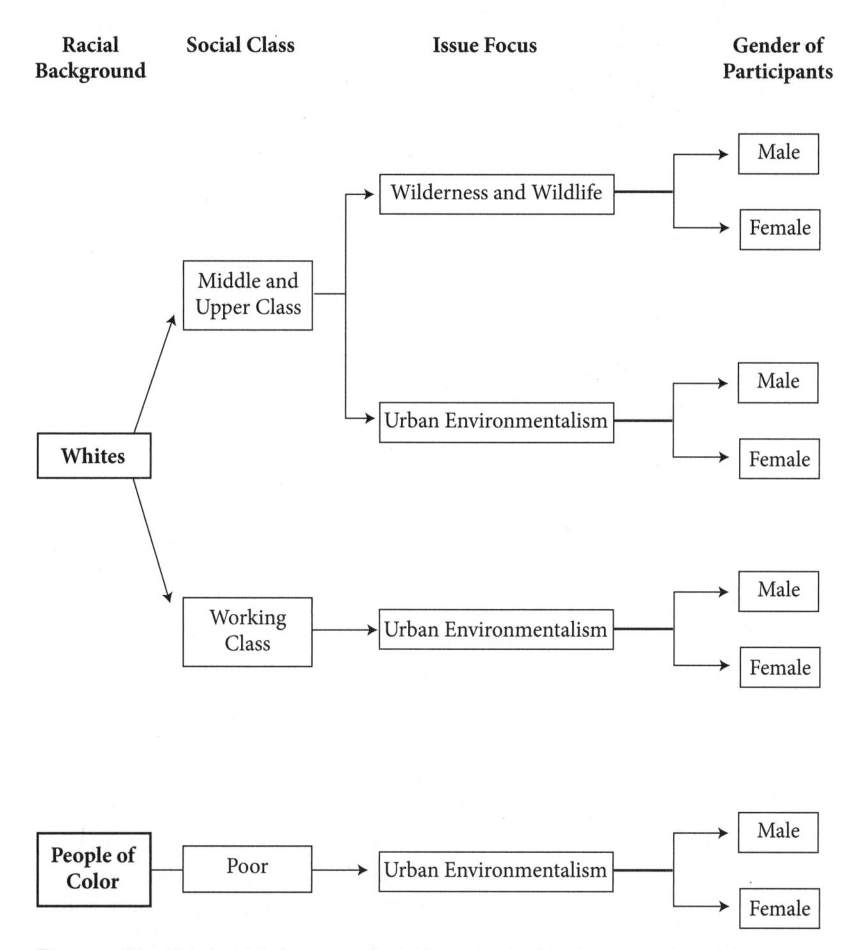

Figure 1 The Relationship between Social Location and Environmental Activism

shows the major issues that were on the urban environmental agenda during the nineteenth century: home and community, the workplace, and recreation or play.

Rhetoric and Claims Making

The social constructionist perspective is also very useful in helping to identify and analyze environmental claims making. Claims are complaints about social conditions that members of a society find offensive, undesirable, or unjust.[58] Table 4 shows a variety of urban environmental movements, the claims made, the frames used to articulate these claims, the placement of blame, and the targets of action. Best analyzes claims by concentrating on the rhetoric of the claims, the deliberate use of language to persuade others. According to Best,

Table 2 Urban Environmentalism and the Conservationist and Preservationist Movements Compared

Characteristics	Urban Environmentalism	Preservationism and Conservationism
Spatial focus	Urban	Rural, remote
Class	Upper and middle class Interaction with working class	Upper and middle class
Race, ethnicity	White, Northern European activists Serving poorer European immigrants and some minorities	White, Northern Europeans
Gender	Males and females Females dominate some areas	Predominantly male Some clubs exclusively male
Leadership	Many female leaders emerge	Few major female leaders Women in auxiliary or "helpmate" role
Spheres of activism	Recreation Home and community Work	Recreation
Orientation	Service provider to the poor Build, manipulate nature Community organizing	Conquest, domination of nature, exploration, preservation, conservation
Agenda	Broad environmental agenda Workplace, public health, parks, housing, zoning	Narrowly prescribed agenda Preservation Conservation
Human-Nature Focus	Humans and the environment	Wildlife, wilderness, wild nature
Political Orientation	Alterative, redemptive, reformative	Alterative, reformative
Movement Type	Consensus and conflict	Consensus
Power Relations	Shift power from rich to poor	Status quo
Social Justice	Advocates social justice	Status quo

Table 2 Continued

Characteristics	Urban Environmentalism	Preservationism and Conservationism
Business Ties	Weak in some areas, strong in others Where strong, interlocking directorates	Strong, interlocking directorates
Government Ties	Weak in some areas	Strong, activists hold key positions
Union Ties	Strong, collaborative efforts	None
Activist Networks	Loose, span several movements Dense in parks movement	Dense, interlocking relationships
Institutional Networks	Weak bridging ties to several movements and institutions	Strong intra-organizational ties, weak bridging ties to related movements

early on in the claims-making process activists usually employ a "rhetoric of rectitude," which calls on our values, morals, and desire to do good, to elevate the significance of an issue and motivate people to act. Over time activists shift to a "rhetoric of rationality," ratifying their claims through data collection and policy formulations.[59] Gamson and Meyer have identified a "rhetoric of change" that activists use to overcome malaise and inefficacy.[60]

Reliance on the rhetoric of rectitude is apparent among the evangelical and charity reformers who believed that immorality was the root cause of many urban ills. They called on slum dwellers to change their behavior. Settlement house activists evoked morality in a different way, by calling on industry and government to recognize corrupt behavior and make changes to improve people's lives. This book provides ample evidence of activists justifying their claims by collecting and analyzing data to support their position. Activists in the charity and settlement movements, housing reform, and occupational health and safety movements did this effectively.

All the movements and activists articulated a rhetoric of change; however, they differed on who should change (individuals, corporations, or the government) and how change should occur. Park and municipal housekeeping advocates evoked the loss of nature, culture, and civility to make their case. The City Beautiful movement activists and workers' rights, occupational safety,

Table 3 The Urban Environmental Agenda

Sphere	Agenda Item	Specific Issues
Community or Home	Public health	Infant health, infectious diseases, hygiene, nutrition
	Consumer health	Food and product safety, national standards
	Sanitation	Solid waste removal and disposal, environmental monitoring, pollution, housekeeping, clean water, street cleaning, sewers, sewage treatment
	Housing	Housing quality, availability, affordability, ventilation, safety, housing laws, overcrowding
	Land use planning	Zoning, building codes, congestion, aesthetics, restrictive covenants
Work	Occupational health and safety	Product substitution, workplace hazards, epidemiological studies
	Worker rights	Reduce the length of the workday and workweek, increase wages and levels of unionization
	Pay equity	Gender equity in wages
	Child labor	Reduce or eliminate exploitation of children, make school mandatory
Recreation	Open space	Access to parks, playgrounds, recreation equipment
	Recreation choice	Freedom to choose and engage in recreational pursuits
	Quality	Poor maintenance of recreational facilities, poor or inadequate facilities
	Supervision	Excessive middle-class supervision
	Type of recreation	Active and passive recreational opportunities

and child labor advocates relied on rhetoric evoking unreason, calamity, entitlement, fairness and justice, and endangerment to stake their claims. The zoning and urban planning movements expressed their claims in terms of a rhetoric of loss (of culture and civility).

Rafter posits that claims makers use archetypes or templates from which stereotypes are drawn to persuade others.[61] Ibarra and Kitsuse assert that claims are framed by using a variety of rhetorical idioms, images that imbue claims with moral meaning and significance.[62] Environmental discourse is filled with rhetorical idioms and motifs; for example, claims might have (a) a "rhetoric of loss" of nature, culture, and innocence; (b) a "rhetoric of unreason" invoking images of manipulation, discrimination, and conspiracy; (c) a "rhetoric of calamity" invoking images of environmental degradation or catastrophe; (d) a "rhetoric of entitlement" that demands justice or fair play; and (e) a "rhetoric of endangerment," which specifies intolerable risks or hazards. Claims makers also use rhetorical motifs; these are recurrent metaphors and figures of speech, such as "seat of Satan," "lungs of the city," and "swarming hordes," that amplify the problem and give added moral significance to the claim.[63]

Claims makers come from a variety of backgrounds and adopt different styles depending on the claim. The style, which can be scientific, comic, theatrical, civic, legal, adversarial, or cultural, must match the context in which the claim is being made and the audience at which the claim is being targeted.[64] For example, evangelical and charity reformers pioneered the use of spatial mapping techniques to track the people they served, and settlement house activists utilized surveys, maps, and pictures and legal and policy-making approaches.

Social Movement Theories

The environmental movement is one of the most influential social movements of our time.[65] A social movement is a type of collective behavior resulting from a set of opinions and beliefs aimed at changing some elements of the social structure of a society. As such, social movements can also be thought of as a sustained and self-conscious challenge to authorities or cultural norms by a collection of actors, some of whom might utilize extra-institutional means of influence to attain their goals.[66] Because collective behavior is emergent and action-oriented, it can appear, spread, contract, or change form suddenly and unexpectedly.[67] Emergent ideas, beliefs, and norms are in the process of being formulated; action-oriented behavior emphasizes recruiting individuals who want to and are able to do something about their grievances.[68]

One can analyze social movements in terms of the magnitude of change sought by activists and the level of change attempted. Regarding magnitude, movements can seek partial or total change. Moreover, movements can seek change at the individual or societal levels. Bearing this in mind, four general types of movements can be identified: alterative, reformative, redemptive, and transformative (see Table 5).

Alterative movements seek partial change in individuals. Such movements assume that individuals have some character flaws or tendencies that should be fixed or banished. Self-help movements such as those aimed at reducing or

Table 4 Urban Environmentalism: Movements, Framing, and Targets of Action

Movement	Problem Identified	Master Frame
Evangelical	Irreligiosity among poor	Immorality causes poverty and ill health
Charity	Increasing poverty	Poverty caused by sloth and immorality
Sanitary	Poor sanitation	Poor sanitation causes illness and death
Landscaped Parks	Lack of green space	Parks are breathing places that improve health, morality, civility, and culture
Settlement	Increasing poverty and environmental degradation	Ill health and poverty are caused by environmental and economic factors
Child Labor	Children being sent to work instead of school	Child labor causes illness. Children are being used to undercut the wages of adults
Worker Rights	Workers lacked rights	Eight-hour day, living wage, safety
Occupational Health	Excessive injuries and death on the job	Improve workplace safety, compensation for job-related injury or death
Pay Equity	Women were underpaid	Increase the wages of female workers
Housing	Poor, inadequate housing	People needed decent, affordable housing
Playground, Recreation	Lack of small parks and playgrounds	The poor need access to neighborhood parks. Children needed a place to play
City Beautiful	Drab, dirty cities and towns	Beautify the city and its surroundings
Municipal Housekeeping	Lack of sanitation around the home	Improving sanitation of food and beautify the home
Zoning, Urban Planning	Lack of planning in cities	Controlling land uses leads to increased property values and order in the cities

Table 4 Continued

Blamed	Targeted	Government Role	Activists	Made Policy
Poor	Poor	Minor	Mostly men	Elites
Poor	Poor	Minor	Mostly men	Elites
Industry, government	Industry, government	Major	Mostly men	Elites, government
Government	Government, park users	Major	Mostly men	Elites, government
Industry, government	The poor, industry, government	Major	Mostly women	Elites, government
Industry, government	Industry, government	Major	Men, women	Activists, government
Industry, government	Industry, government	Major	Men, women	Activists, government
Industry, government	Industry, government	Major	Men, women	Activists, government
Industry, government	Industry, government	Major	Mostly women	Activists, government
Builders, government	Slum lords, government	Major	Men, women	Elites, government
Government	Government, the poor	Major	Mostly women	Elites, government
City residents	City residents	Major	Women, men	Elites, government
Families	Families	Minor	Mostly women	Elites, government
Industry, government	The poor, industry, government	Major	Mostly men	Elites, government

Table 5 Level and Magnitude of Change Sought by Social Movements

	Level of Change Sought	
Magnitude of Change Sought	Individual Level	Societal Level
Partial	① Alterative	② Reformative
Total	③ Redemptive	④ Transformative

Source: Adapted from D. Aberle, *The Peyote Religion among the Navaho* (Chicago: Aldine, 1966); D. McAdam and D. Snow, "Social Movements: Conceptual and Theoretical Issues," in *Social Movements: Readings on Their Emergence, Mobilization, and Dynamics*, edited by D. McAdam and D. Snow (Los Angeles: Roxbury, 1997), xix–xx.

eliminating addiction fall into this category. The object of the change is the individual's weakness, deficits, or tendencies. Redemptive movements seek total change at the individual level. For example, evangelical reform movements perceive that social ills are rooted in an individual's behavior and beliefs; such movements seek to redeem individuals by advocating their total transformation. Religious movements and cults take this approach to social change. Reformative movements, such as the parks movement, seek to make limited or incremental changes in the system in which they are embedded. Such movements do not utter outright rejections of the system; instead they seek to work within the system to neutralize or amend wrongs or to reduce or eliminate threats. Transformative social movements (such as the environmental justice movement) seek broad or sweeping changes in the social structure and its ideological foundation.[69] Though all movements do not fit neatly into these idealized typologies, they tend toward one of these forms. For example, not all sectors of the contemporary environmental movement can be described as reformist; radical sectors such as the environmental justice movement tend more toward transformative environmental politics than reform environmentalism.

RESOURCE MOBILIZATION THEORIES

Resource mobilization theorists posit that social movements have institutional dimensions, are persistent and patterned, and are organized by rational actors and that resources and opportunities are more important than strain, grievance, and deprivation in causing the emergence of social movements. These theorists argue that this helps to explain why so many groups with grievances do not organize social movements: they simply don't have the resources and opportunity to organize. Resource mobilization theory posits that it is the most socially connected people, not the most alienated, who are most likely to be mobilized into movement participation.[70]

Resource mobilization arose in response to Mancur Olson's rational choice theory.[71] Olson claimed that rational, self-interested individuals will not participate in collective actions aimed at achieving collective benefits because public goods cannot be withheld from those who did not participate. The rational actor, therefore, is a "free rider." But this begs the question, If everyone behaves rationally and free-rides, how do mass mobilizations occur?[72] The social movement theorists—Oberschall, Gamson, McCarthy and Zald—and Tilly responded to Olson's arguments by developing alternative explanations that did not use grievance, deprivation, or strain to explain collective behavior.[73] Furthermore, they did not see free-ridership as a phenomenon that stymied collective behavior.

The resource mobilization model highlights the importance of resources while downplaying the role of grievances in collective behavior. Both a response to but also an outgrowth of the rational choice model, resource mobilization focuses on micromobilization. Theorists argue that social movement activists are socially located or embedded in networks that they identify with (some identities, such as nationality, race, ethnicity, class, gender, or religion, are of great significance to people). The social locations intersect and overlap to provide the cultural cues that activists draw on to identify and interpret social problems, resources, and opportunities.[74] In short, social movement activists have to understand how to organize and mobilize the resources necessary to initiate and sustain movements.

Resource mobilization research became heavily reliant on organizational studies while downplaying the role of psychological explanations (values, ideologies, identity, etc.) in social movement participation. While resource mobilization dominated social movement thinking during the 1970s and early 1980s, some social movement theorists began to identify problems with the theory, specifically that it minimizes the role of ideas and beliefs in social movements. Social movement theorists have sought to correct this oversight by examining the role of ideology, framing, collective identity, recruitment, micromobilization, and political opportunities and cognition in movement formation and maintenance.[75]

POLITICAL OPPORTUNITY THEORY

In contrast to the classical social movement theorists who view movement formation as a psychological phenomenon (i.e., a response to grievance, strain, or deprivation), political opportunity theorists believe that collective behavior is a political phenomenon. Eisinger defines political opportunity as the extent to which groups are able to gain access to power or manipulate the system.[76] Similarly McAdam believes that the factors influencing institutionalized political processes play significant roles in the emergence of social movements. Moreover, McAdam sees a movement as a continuous process from birth to decline rather than as a series of discrete developmental stages. His political process model posits that socioeconomic conditions lay the foundation for expanding political opportunities and the development of indigenous organizations. Political opportunities and indigenous organizational strength facilitate the process of cognitive liberation, which in turn enhances social movement formation.[77] Other proponents of the political process model, such as Rule and Tilly, Jenkins and Perrow, Tilly, and Tarrow, have also established a link between institutionalized politics and social protests or movement activism.[78]

In recent years scholars have sought to define the concept of political opportunity such that it is analytically distinct from social movement processes such as the mobilization of resources or identification and framing of issues. Though scholars operationalize political opportunity in a variety of ways, McAdam has identified four critical elements of these definitions: the relative openness or closure of the institutionalized political system, the stability or instability of that broad set of elite alignments that typically undergird a polity, the presence or absence of elite allies, and the state's capacity and propensity for repression.[79]

Cognitive Liberation

McAdam and Piven and Cloward argue that the emergence of a protest movement entails a transformation of both the consciousness and the behavior of individuals. McAdam uses the term *cognitive liberation* to describe the development of consciousness among potential movement participants that translates into collective action. Cognitive liberation occurs when (a) the system people once trusted loses legitimacy, (b) people who are ordinarily fatalistic begin to demand social change, and (c) people find and exercise a new sense of political efficacy. These three conditions cause significant shifts in the relations between the holders of power and those challenging the system.[80]

SOCIAL NETWORK THEORIES

Psychological or attitudinal accounts of social movement activism imply some psychological or attitudinal fit between the movement and supporters that impels activists to participate or makes them vulnerable to recruitment efforts. Despite numerous studies based on this assumption, the empirical evidence linking attitudes, predispositions, and activism is weak. Psychological attributes of individuals, such as frustration and alienation, are of little significance in explaining the occurrence of the high-risk collective actions of revolts, riots, and rebellions, for instance.

This is not to say that psychological and attitudinal factors are irrelevant to the study of activism. Both factors are important in identifying the "latitude of rejection" within which individuals are highly unlikely to get involved in a given movement.[81] This is important because in many movements the size of the pool of recruits positively disposed to the movement's message (the latitude of acceptance) is much larger than the number of people who will actually participate. That is, not everyone hearing a message and agreeing with or liking it is likely to become a movement participant. Research has shown that there is a disparity or gap between attitudinal affinity and actual movement participation. The weak link between psychological predispositions and attitudes and movement participation could be a function of this disparity.[82]

Microstructural network factors appear to be more helpful in explaining movement participation. The first aspect of this relates to interpersonal ties. Knowing someone already involved in a movement is a strong predictor of recruitment into a movement. Strong or dense interpersonal networks increase the likelihood that an individual will be asked to join a movement as such networks lessen concerns about movement participation.[83] The second network factor relates to organizational ties. Membership in an organization is an extension of the individual's interpersonal social ties; it increases the likelihood of meeting people and being drawn into a social movement. Movement organizers, recognizing the difficulty of recruiting single, isolated individuals, have long expended much of their recruiting efforts on getting support from existing organizations.[84] As later discussions will show, interpersonal ties and network connections were crucial to the formation of charities, the development of parks, and other causes.

Social Networks and Efficacy

There is a relationship between organizational membership and personal efficacy. Individuals belonging to several organizations have a stronger sense of personal efficacy than those who belong to few or no organizations.[85] Efficacy refers to a situation in which individuals believe they can assert themselves politically to make social and political changes. Political assertion can take place through citizens' organizations or through individual or group efforts.[86]

However, Sharp argues that political inefficacy, the perception that one's actions are unlikely to have an impact on governmental affairs because of unresponsive officials, is closely related to advocacy.[87] Inefficacious individuals or groups often do not recognize ways of advocating their needs. Later discussions of poverty and the condition of the urban poor will show how important the perception of efficacy is in the framing of discourses, the development of policies, and the design and implementation of programs.

To counter the pessimism and malaise encouraged by the rhetoric of reaction (inaction) social movement activists articulate an optimistic rhetoric of change. Instead of focusing on why actions should not be taken, they encourage supporters to act and to foster a culture of efficacy. To counter the negative themes of jeopardy, futility, and perversity social movement activists employ themes of urgency, agency, and possibility. Activists rebut the jeopardy theme by invoking a sense of urgency in supporters. They argue that if actions are not taken immediately things will get worse in the future; so, though action might be risky, inaction is riskier. Activists dispute the futility argument by stressing the opportunities of the moment; immediate action will open the window of opportunity wider, and inaction will lead to its closing. Finally, activists appeal to the promise of new possibilities, such as better policies and greater justice, as a counterclaim to the arguments of perverse effects and to motivate supporters.[88]

Social Networks and Identity Salience

In addition to the multiple network ties discussed earlier, identity salience is an important factor in explaining movement participation. According to Stryker, identities are organized into a hierarchy of salience, defined by the probability of the various identities being invoked or aroused in a given situation or a variety of situational contexts.[89] The salience (or significance) of any particular identity is a function of the individual's commitment to it. Commitment is affected by the individual's relationships with other movement activists. That is, the depth and importance of our relationship with others help to establish and sustain the salience of various identities.[90] Identity salience may be an important factor in determining the latitude of rejection within which individuals are highly unlikely to become active in a given movement.[91] Later discussion will show how settlement house activists developed a specialized and highly prized identity around settlements. Once they succeeded in shifting blame from the poor to other actors, the issues they organized campaigns around became more salient to the working class, who then participated in movement activities in greater numbers.

Framing

Though few scholars in the environmental field pay attention to framing, it is extremely important in the field. Environmental activists, policy makers, politicians, and those in government and business have long perceived, contextualized, and battled over environmental issues by establishing frames of reference. Framing is the process by which individuals and groups identify, interpret, and express grievances. It is a scheme of interpretation that guides the way ideological meanings and beliefs are communicated to would-be supporters. Beliefs are important because they can be defined as ideas that might support or retard action in pursuit of desired values, goals, or outcomes. Social movement collective action frames are injustice frames because they are developed in opposition to already existing, established, widely accepted frames. Such collective action frames serve to pinpoint, highlight, or define unjust social conditions. Activists trying to develop new frames have to overcome people's acceptance of the established or hegemonic frame as normal or tolerable. Collective action frames deny the immutability of undesirable conditions and promote the possibility of change through group action. Such framing defines social movement supporters as potential social change agents in charge of their own history.[92]

Frames organize experiences and guide the actions of the individual or the group. Collective action frames are emergent, action-oriented sets of beliefs and meanings developed to inspire and legitimate social movement activities designed to attract public support.[93] There are three components of collective action frames: injustice, agency, and identity. The injustice element refers to the moral outrage activists expound through their political consciousness. The indignation is more than a cognitive or intellectual judgment about equity or justice; it is a "hot cognition," one that is emotionally charged. Agency refers to individual and group efficacy, the sense of empowerment activists feel. Empowered activists or those exercising agency believe they can alter conditions. The identity component of collective action frames refers to the process of defining the "we" or "us," usually in opposition to "them."[94] Framing is also affected by suddenly imposed grievances: the spectacular, highly publicized, and unanticipated events, such as human-made disasters (factory fires, mine disasters, steamboat explosions, the burning of slums), that increase public awareness of and opposition to previously acceptable social conditions. Suddenly imposed grievances can provide a cognitive stimulus to the framing process.[95]

FRAME ALIGNMENT PROCESSES AND IDENTITY SALIENCE

Frame alignment is the process of linking the individual's interpretive framework with that of the social movement. Social movements also try to link activists' goals and identities with those of the general society. This is accomplished by expanding the personal identities of a constituency or group to include the collective identity of larger segments of society as one way they define themselves. Connecting the individual and societal levels is particularly important to movements that emphasize social change, target civil society rather than the state or economic institutions as the primary target of influence, and have a constituency that chooses to support the movement openly. Four kinds of frame alignment processes have been identified: frame bridging, frame amplification, frame extension, and frame transformation. Bridging personal and the collective identities can be a strategic part of the mobilization process. Frame bridging is the act of linking two ideologically compatible but structurally separate frames that refer to the same issue.[96] For instance, some consider the use of toxic products such as lead in the workplace an environmental issue, whereas others perceive it as a labor issue; bridging the frames necessitates focusing on the singular goal of eliminating lead in the workplace. There are also larger public health factors at play in the exposure of community residents to hazardous materials; for example, lax disposal practices can lead to contamination of wildlife. Effective framing could use one master frame to link the health of workers with that of the larger community and wildlife.

Frame amplification is the process whereby the meanings and interpretation of an issue are clarified to help people see how the issue is connected to their lives. Frame amplification also reduces the ambiguity and uncertainty that might prevent people from caring about and supporting an issue.[97] There are two aspects to frame amplification: value amplification and belief amplification. Individuals subscribe to a range of values that vary in the degree to which they are compatible or attainable. These values are ordered in a hierarchy according to their salience or significance to the individual. Value amplification refers to the identification and elevation of certain values to the top of the hierarchy to inspire movement participation.[98] Activists amplify people's beliefs by focusing on the seriousness of a given grievance or issue, identifying the causes and fixing blame for problems, stereotyping opponents, and emphasizing the likelihood of change and the efficacy of the movement and the urgent need to take a stand.[99]

Frame extension occurs when social movements broaden the frame of reference to make their message salient to pools of potential allies not normally targeted by the movement. In some cases social movements have to transform the framing of the issues. Frame transformation occurs when new ideas and values about movements or issues replace old ones; in addition, old meanings and symbols are discarded, erroneous beliefs and misframings corrected, and a general reframing of the issues occurs.[100] There is ample evidence of frame extension in this book as several groups of activists successfully broaden their message to reach new supporters. Several movements emerged to build on or challenge older ones by transforming the older framings. The transition from evangelical and charity reformation to settlements is an example of both frame extension and transformation. The discussion will show how the charities changed their framing of poverty to respond to changing times. There is evidence of frame transformation in the sanitary reform and worker rights movements also.

Activists do not typically fashion new collective identities from scratch; instead they redefine existing roles within established organizations and use these as templates for creating new identities. As McAdam and Friedman and McAdam argue, the civil rights movement grew as rapidly as it did because the movement appropriated a highly prized and salient role and identity in the black community, that of Christian or churchgoer, and used it as an effective means of creating a new identity.[101] As a result, for blacks to identify and retain their status as Christians or churchgoers they had to incorporate civil rights activism into their Christian identity. Similarly the role of the politically aware and active black college student was appropriated and expanded. Again, civil rights activism became one element of the identity that activists embraced. Thus as the movement matured one had little or no credibility as a black student activist or Christian if one didn't get involved in the civil rights movement. In contemporary working-class movements activists likewise appropriated salient identities (steelworker, community or labor organizer, churchgoer) and transformed them into valued and salient environmental activist identities. Environmental justice activists have adopted a similar approach by appropriating civil rights and community organizer identities and transforming them into environmental justice identities. Today community organizers have little credibility if they work in low-income minority communities and are not aware of or incorporate environmental justice issues into their repertoire of activities.

The environmental justice frame has emerged as a master frame used to mobilize activists desirous of linking racism, injustice, and environmentalism in one frame. Master frames serve the same functions as movement-specific collective action frames; however, their effects are exaggerated. That is, master frames are styles of punctuation, attribution, and articulation. They can be viewed as crucial ideological frameworks akin to paradigms. The master frame functions as a linguistic code, providing a language that connects experiences and events in the world around us to our own lives. Master frames play the crucial role of magnifying the attribution function of collective action frames; that is, they provide the interpretive medium through which activists identify problems and assign blame or causality as they make "causal attributions" or develop "vocabularies of motive."[102] In the case of environmental experiences of people of color, blame is externalized; that is, unjust outcomes in life circumstances are attributed to pervasive and persistent societal racism rather than the victim's imperfections.

As mentioned earlier, master frames serve important articulation functions, but not all master frames perform the same functions in the same manner. Some master frames are inflexible, while others are pliable. If the dialogues of master frames lie on a continuum ranging from restricted to elaborated, then the master frames can be characterized as restricted master frames or elaborated master frames, depending on where they lie on the continuum. The restricted master frames are considered closed; that is, they are developed from exclusive ideational systems that do not readily lend themselves to frame amplification or extension. As styles of expression they tend to organize a limited body of thought in a densely interlocking form; as modes of interpretation they are narrowly defined, thus allowing little leeway in how they are interpreted. In other words, restricted frames are syntactically rigid and lexically particularistic.[103]

Master frames are also developed in elaborated codes; such codes can express a wide range of ideas. Frames developed with elaborated codes tend to be more lexically universalistic; they offer more flexible means of interpretation and allow more comprehensive ideological amplification and extension than those created with restricted codes. Frames based on elaborated codes are also more inclusive: they are more accessible to aggrieved groups that can use them to express their complaints.[104] In the case of the environmental movement, the environmental justice discourse is framed in elaborated codes, while vegans

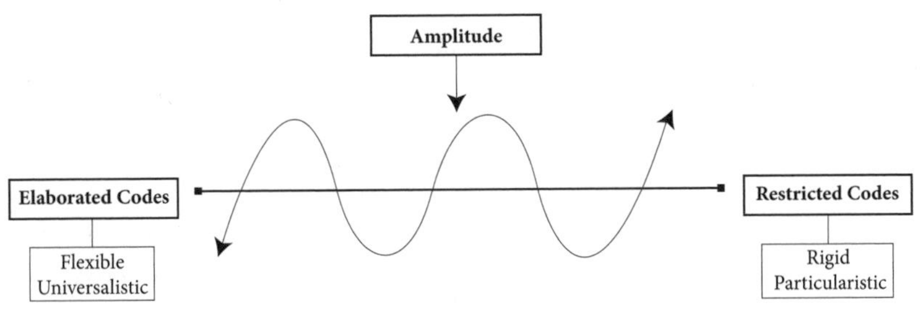

Figure 2 The Relationship between Amplitude and the Type of Codes Used to Construct Master Frames

and groups such as Earth First! tend to frame their discourse in restricted codes (see Figure 2).

Master frames are also potent. The potency of the frame is a function of where it rests on the restricted–elaborated continuum and the extent to which it resonates with its target audience. The amplitude of the resonance is affected by three interrelated factors: empirical credibility, experiential commensurability, and ideational centrality or narrative fidelity. There must be credible empirical evidence to support the master frame's claims; the target audience has to have some experience with the problem (direct and daily experience is more potent than infrequent, indirect, and inconsequential experience); and the issue must resonate with people or strike a chord in them. In short, the frame must have a high level of identity salience to potential recruits.[105] Movement activists must choose how to use the rhetoric of rectitude according to the stage the social movement is in. In addition activists use the rhetoric of change: rhetorical idioms and motifs to evoke moral responses, promote empirical credibility, and encourage supporters to be efficacious.[106]

The early environmental activism of people of color discussed in this book was often characterized by the absence of an explicit environmental master frame. However, careful analysis of early and later environmental actions by people of color and low-income communities indicates the existence of submerged environmental frames.[107] That is, though a master frame might have focused on the denial of rights or ending racism, prejudice, and discrimination, there were minor or submerged frames that were environmental in nature. Examples of this will be seen in the study of housing conditions in Philadelphia and African Americans' campaigns to gain access to parks and other recreational resources.

Today the master frames of environmental racism and environmental justice are used to amplify the framing used by minority and working-class activists. These concepts provide a label for the activism occurring in communities of color and low-income communities, linking racism or injustice with environmental actions, experiences, and outcomes. The concepts are also important because they bridge past social justice activism that focused on racial injustice and civil rights with past and present environmental experiences. That bridging elevates the environment from an implicit or submerged collective action frame to a master frame, one that is explicit, public, and potent. The terms also transform the environment into a salient frame for many minorities and low-income people.

Organizational Theory: The Organizational Field and Isomorphism

The works of organizational theorists can also help us to analyze the dynamics of urban environmental activism. The charities and the settlements involved in urban environmentalism constitute organizational fields. An organizational field is a group of organizations that comprise a recognized area of institutional life. In the initial stages of their life cycle organizations within a field tend to display considerable diversity in style and form. Once an organizational field becomes well established the level of homogenization increases.[108] This homogenization results in isomorphic organizations. Isomorphism is the process that forces one unit in a population (organizational field) to resemble other units that face the same set of environmental conditions. At the population level this process results in reconfiguration of organizational characteristics in the direction of increasing compatibility with environmental characteristics.[109] Isomorphism can result when organizational decision makers learn appropriate responses and adjust their behavior to suit the expected responses.[110] Organizations must consider other organizations because they compete for resources, members, political power, legitimacy, and financial support. There is evidence of increasing homogenization among institutions discussed in this book.

COERCIVE AND MIMETIC ISOMORPHISM

There are three processes through which institutional isomorphic changes occur: coercive, mimetic, and normative.[111] Coercive isomorphism results from political influence, the quest to establish legitimacy, and pressures placed on organizations by other organizations on which they are dependent. Pressure comes from such sources as government mandates and the cultural expectations of the society. Mimetic isomorphism results from the way organizations respond to uncertainty. When organizational technologies are poorly understood, goals are ambiguous, the environment is characterized by symbolic uncertainty, or the causes of problems and solutions are unclear, organizations tend to model themselves on other organizations. In the case of the environmental movement, larger, more well-established organizations led the way in defining the agenda.[112] Organizations also try to understand the patterns of interorganizational competition, influences, coordination, and flows of innovation that help to define the boundaries within which these processes operate.[113] In addition government agencies and other central actors have significant influence on that field.[114] Organizations cope by adopting successful strategies, all the time being keenly aware of their competitors, cooperators, and exchange partners.[115]

There is ample evidence of both coercive and mimetic isomorphism in the discussions that follow. Not only did charitable institutions copy or mimic each other's agenda and mode of operation, but leading charities coerced weaker ones to adopt certain policies and collectively streamline their operations. Mimetic isomorphism is evident in the settlement movement. Settlement houses developed similar styles of operation and agenda, collaborated with each other frequently, and swapped resident workers. In the City Beautiful movement local groups modeled themselves closely on successful groups in that movement.

Normative isomorphism arises from the level of professionalization of the field, the collective efforts of members of an occupation to control the methods and products of their work, provide a cognitive base for the occupation, and establish its legitimacy.[116] Normative isomorphism is accelerated through the process of filtering and recruitment of staff and members. Filtering occurs through the hiring of staff from the same firms or people trained in the same institutions, by employing similar promotion practices, or by specifying the skill requirements for particular jobs. Universities and professional associations socialize individuals, thereby creating a pool of interchangeable, almost indistinguishable workers who occupy similar positions in a range of organizations. These employees have similar orientations and dispositions that may override the variations in tradition and control that might otherwise result in diverse organizational behavior.[117]

Five key dimensions of professionalization accelerate the process of normative isomorphism in the environmental field: producing university-trained experts in the environmental field, creating an extensive body of knowledge, organizing new professional associations and strengthening existing ones, consolidating the professional elites (by working on collaborative projects), and increasing the organizational salience of professional expertise through expert testimony and the production of research reports.[118] Similar processes occurred in urban environmentalism with the reliance on graduates of elite northeastern colleges and universities to staff institutions and the development of programs to provide professional training for activists. This book discusses the professionalization of the charity and settlement movements, which eventually established a profession and helped to institute formal training for people in the field.

ORGANIZATIONAL FIELDS AND INSTITUTIONAL DEFINITION

Organizational fields exist only to the extent that they are institutionally defined. The process of institutional definition (sometimes referred to as structuration) is marked by four features: an increase in the level of interactions among organizations in the field, the emergence of clearly defined interorganizational structures of domination and patterns of coalition, an increase in the information load that organizations in the field must contend with, and development of mutual awareness of organizations working on common issues or having similar goals. Once disparate organizations in the same issue area are structured into the organizational field (by competition, the state, or professions) powerful forces lead them to become more similar to each other. Organizations may change their goals or develop new practices as new organizations enter the field, but the homogenizing forces lead organizations to develop similarities.[119]

Networks and Interlocking Directorates

Historical analyses of urban environmental activism indicate the existence of interlocking directorates among activists, businesses, and the government. This situation arises when leaders in one organization are board members in multiple environmental groups and businesses or hold government positions. Because organizations are dependent on each other, the most direct method of controlling dependence is to control the source of the dependence. Since organizations are not always in a position to acquire or merge with others, other kinds of interorganizational linkages are employed to coordinate interdependence, such as cooptation, trade associations, cartels, reciprocal trade agreements, coordinating councils, advisory boards, boards of directors, joint ventures, and social norms. Each represents a way of sharing power and a social agreement which stabilizes and coordinates mutual interdependence.[120] Readers will find evidence of interlocking directorates in several of the movements discussed in this book.

Pfeffer and Salancik claim that linkages to other organizations provide four primary benefits to organizations managing interdependence. First, interlocking directors among competitors can provide crucial information to each about strategic policies and plans. Second, a linkage provides a mechanism to transmit information from the focal organization to another organization. Third, a linkage and the exposure it provides is an important first step in obtaining commitments of support from important elements in the society. That is, board members exposed to the problems or viewpoint of the focal organization identify with the organization and over time become committed to and supportive of that organization. Fourth, linkages serve the function of legitimating the focal organization, such as when prestigious persons on the board of the focal organization help advertise the value and worth of the organization.[121] To this list I will add another benefit of linkages: interorganizational linkages can also serve a legitimating function for organizations other than the focal organization. For instance, businessmen and new philanthropists of the nineteenth century sought legitimacy and prestige by funding and becoming involved in social service institutions.

Linkages help to stabilize the organization's exchanges with other partners and reduce uncertainty. In addition to business interactions, organizations try to develop social bonds with each other. The more organizations become enmeshed in each other's social networks and the more extensive the overlap of friendship networks and business acquaintances, the more binding the relationship becomes and the more stable and predictable it is likely to be. Interpersonal links play a psychological role in reducing uncertainty because people prefer conducting business with others who are familiar to them.[122]

Power Elites

Power elite theory contends that environmental discourses and policies were conceptualized and orchestrated by elites in accordance with upper- and middle-class values and interests. In the context of this discussion, elites can be viewed as the people who run things, the key actors or inner circle of participants who play structured, functionally understandable roles in the formation and execution of environmental policies. Elites are those who get the most of what there is to get in the institutionalized sector of a society; that is, at every stage of any decision-making process, elites will inevitably accumulate disproportionate amounts of valued attributes, such as money, esteem, power, or resources which people desire and try to attain. Power is the ability to affect decisions; therefore powerful people are those who are able to realize their will even if others resist it. The truly powerful command major institutions, and it is by exercising institutional power that individuals enhance and consolidate their own power. Power elite theory helps us to identify six types of power elites who helped to shape events discussed in this book: ideological, innovative, planning, implementing, economic, and political. The theory also helps us to understand the role environmental advocates assume as the guardians of the nation's natural resources.[123]

Ethical Elites and the Party of Order

Edward A. Ross, writing about social order, urged the development of a new kind of "social religion" aimed at convincing the masses of "the conviction that there is a bond of ideal relationship between the members of a society." Ross argued that certain members of society, such as ministers, educators, jurists, and those with strong interests in economic stability such as merchants, have an interest in order out of professional necessity; in effect, they form a "party of order." However, other citizens in society, the "ethical elites," are also important components of the party of order. The ethical elites are a moral cadre, not drawn from any particular vocation or economic group per se, but people who have "at heart the general welfare and know what kinds of conduct will promote this welfare." According to Ross, the ethical elites "stand for an order that is right."[124]

Ross's conceptualization of elites taking on the role of leaders who define, monitor, and control issues related to ethics and morality can be applied to environmental elites. Many of the activists discussed in this book sought to create order out of chaotic situations. In so doing they controlled the discourse, identified right and wrong, created rules, and developed institutions to administer them. They had a sense of noblesse oblige; that is, they felt they had a higher calling to do certain tasks and an obligation to take them on.

Synthesizing Theoretical Approaches

Most historical accounts and analyses of environmental activism are not informed by sociological theory. In addition, much of the sociological research on the environmental movement does not adopt a social movement or organization theory approach in its analysis. While European social movement theorists have done extensive research on European environmental mobilizations, American social movement theorists have done limited research on the American environmental movement.[125] In the past social movement theorists tended to research a single theory; consequently we have bodies of research that analyze only resource mobilization, social psychological factors, or interpersonal networks and that are not integrated with each other. Recognizing

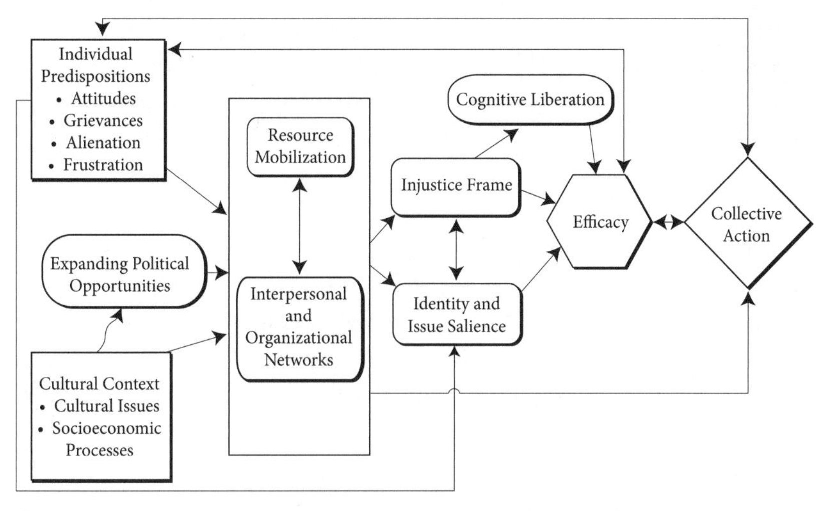

Figure 3 Factors Influencing Participation in Collective Action

the limits of the single-theory approach to explaining collective behavior, a number of social movement scholars have begun emphasizing and synthesizing three theoretical approaches in their work: political opportunities, mobilizing structures, and framing processes.[126] This book adopts this approach, but also incorporates organization theory in the analysis. While the social movement literature on networks focuses heavily on interpersonal networks,[127] less attention is paid to organizational networks and the interplay between these two types of networks. The marriage of these two traditions will enhance our understanding of environmental activism. Figure 3 displays a model synthesizing these theoretical approaches, showing the relationship between individual attitudes, framing, and collective action.

Figure 3 shows that the injustice frame can provide cognitive liberation that increases personal and group efficacy. Efficacy is also shaped directly by individual attitudes, framing, and collective identity. As efficacy increases, individual attitudes change to match the new sense of empowerment. Efficacious groups and individuals participate in collective action, which in turn reinforces efficacy, transforms individual attitudes, and increases access to resources. Collective action also strengthens both interpersonal network ties and organizational networks.

The model posits that social psychological factors such as grievances shape individual attitudes and predispositions. The cultural context and socioeconomic processes influence the political opportunity structure and elite alignments. According to McAdam, activists are likely to take advantage of political opportunities when the political system is open, they gain access to and manipulate the elite alignments, they have elite allies, and repression is unlikely.[128] Both the cultural context and expanding political opportunities shape the resources available for movement activities and the personal and organizational networks that are formed. Individual attitudes or dispositions also influence the kind of resources available to activists and how effectively they use these resources. Individual predispositions also affect the kinds of interpersonal and organizational networks that develop.

Resource mobilization and networks shape the collective action frames and identities that are forged. The injustice frame can provide cognitive liberation that increases personal and group efficacy. Efficacy is also shaped directly by individual attitudes, framing, and collective identity. As efficacy increases, individual attitudes change to match the new sense of empowerment. Efficacious groups and individuals participate in collective action, which reinforces efficacy, transforms individual attitudes, and increases access to resources. Collective action also strengthens both interpersonal network ties and organizational networks.

Part I

THE CONDITION OF THE CITY

This section of the book examines the rise of American cities. In chapter 1 I examine the environmental challenges that early towns and cities faced as they grew and analyze the ways city dwellers attempted to transform the environment. I discuss early attempts to incorporate green space in cities and how crowding, unsanitary conditions, and rising inequality resulted in growing unease and conflicts among the urban populace. While elites tried to exert control, they had limited success in stemming the tide of rising tensions and social unrest.

Chapter 2 focuses on the way epidemics exacerbated conditions, strained the ability of cities to function, and pushed elites to consider the role of the environment in public health disasters and develop environmental reforms to safeguard public health. Yellow fever, cholera, and bubonic plague tested civic leaders' will and forced them to implement sweeping reforms.

one

THE EVOLUTION OF AMERICAN CITIES

Early American Cities

During the seventeenth century and early eighteenth most of what we now know as major cities were trading posts or hamlets or had yet to be founded at all. Civic leaders, merchants, workers, and slaves lived in close proximity to their workplaces and to each other. The economic gap between the wealthiest and poorest citizens was not as great as it is now, and the economic and political elites were able to define and enforce social order. I focus on the development of five early American cities—New York, New Haven, Boston, Philadelphia, and Baltimore—particularly the challenges residents faced and the way they responded to environmental constraints. The issues that emerged in these cities were similar to issues that surfaced in other cities.

NEW YORK

Indian Farming and Dutch Settlement

The long history of farming on Manhattan begins with Native American women, who began growing maize on the island around 1100 and by 1600 were growing beans, squash, and other staples of the Indian diet. When the Dutch began settling the area in the early seventeenth century they copied the Indians' farming techniques, burning the brush, rotating crops, and cultivating hills. The Dutch West India Company promoted farming among settlers by providing tools, seed, and livestock if they agreed to farm the land. The settlers planted wheat and raised cattle, sheep, and pigs. Tensions between the Dutch and Indians rose as the livestock wandered about, trampling Indian crops.[1]

When Fort Manhattan was erected and a stockade built around 1614, it was a trading post.[2] Size notwithstanding, the desire to bestow order and morality on American cities and the emergence of elites willing to undertake the task dates back to the founding years of some of the earliest settlements. However, by the time Petrus Stuyvesant arrived in New Amsterdam (Manhattan) in August 1647 to take over as director-general, he found a town in which the inhabitants were wild and lacking in morality. He promptly pledged to reform the city and its inhabitants and told residents he would govern them the way a father governs his children. This same patrician attitude had gotten Stuyvesant expelled from Franeker University in the Netherlands at the age of twenty-two or twenty-three for seducing his landlord's daughter.[3]

Environmental Reforms

Stuyvesant, a military man and former governor of Curaçao, was the son of a Reformed clergyman. He wanted Manhattan to become a clean, well-regulated municipality with civic-minded citizens. Consequently he enacted a series of ordinances aimed at transforming the physical environment as well as the social and moral condition of the people. To impose some order to the haphazard lanes and footpaths, Stuyvesant commissioned three surveyors to establish property boundaries, lay out streets in a more orderly manner, and assign names to those streets. He also instituted a program of street paving. In 1658 residents of Brouwer (Brewer) Street became the first in the city to pave their lane with cobblestones. Six years later, in May 1684, the Common Council decreed that all the major streets of the city should be paved. The cost of paving was covered by a benefits assessment tax levied against those living on the paved streets. In 1787 the state legislature granted the Common Council the power to pave streets.[4]

Stuyvesant also began a sanitation program that cleared rubble from the streets and set a speed limit for wagons and carts. Prior to Stuyvesant's arrival residents had allowed their pigs, cows, goats, and horses to roam and forage at will in town. While animals feeding on the litter reduced the volume of garbage, they trampled gardens and orchards, and rooting pigs undermined the structural integrity of the fort. Commencing in 1648 stray animals were seized and placed in a pound and pigs headed for the fort or rooting around it were shot by soldiers. Residents had to clean the streets in front of their houses and were forbidden to throw rubbish, sewage, ashes, oyster shells, offal, or carcasses into the streets. Butchers were also prohibited from discarding offal into the streets. Toilets releasing excrement onto the ground were banned.[5]

The slaughterhouses were dirty, dimly lit wooden buildings scattered throughout residential neighborhoods, from which could be heard the sounds of dying animals throughout the night. Blood and other animal wastes drained into the streets, ponds, streams, and rivers until 1656, when measures regulating slaughterhouses were passed.[6]

Fire Prevention

Fire was a constant threat. To reduce the risk of fire, Stuyvesant banned the construction of wooden chimneys, thatched roofs, and haystacks, and residents were prohibited from using fireplaces on very windy days. Beginning in 1648 paid wardens inspected chimneys to see that they were swept regularly. A fire curfew was established; each evening all fires had to be extinguished and covered up. Around 1657 the city ordered 150 leather buckets and the following year acquired ladders and fire hooks. The "rattle watch" was also established; men patrolling the city sounded loud rattles whenever a fire was spotted. Each house and business establishment was ordered to have a fire bucket. At around this time brick began to replace wood as a building material, although most New Yorkers, too poor to afford brick, continued to build with wood.[7]

Stuyvesant also undertook a series of public works projects that hastened the transformation of the environment. He focused on projects that increased commerce, moral order, and security. In addition to the threat of invasion from other European nations, residents of New Amsterdam still frequently battled Native Americans; hence the fort was repaired. A pier was built on the East River and a canal, Heere Gracht (also known as "the Ditch"), was cut through marshy terrain.[8]

Feeding and Caring for City Residents

Cities wrestled with issues of health care, poverty, and destitution. New Amsterdam got its first hospital in 1658, but only after Stuyvesant convinced the Dutch West India Company that lack of proper health care impeded the recovery of sick slaves and soldiers. City governments also wanted to maintain an adequate and predictable supply of food for their residents. Stuyvesant issued an edict that a municipal market could be held every Monday along the East River; meat, bacon, butter, cheese, turnips, roots, and other produce were sold there.[9]

Poverty and Rising Income Inequality

Destitute children were another challenge for the city. As the number of orphans and vagrant children grew, pressure mounted to build an orphanage, but at first city leaders dealt with the matter by asking church deacons to look after destitute children. The first orphanage was not established until 1658. Poor relief too was deemed a responsibility of the church, not the city government. The Reformed Church opened the first almshouse in 1653 with funds obtained from collections made in church and at weddings. Eventually the municipality provided some funding to the poorhouses. Manhattan enacted its first poor laws in 1661 after deacons complained that poor people from outlying villages were drifting into the municipality for aid, outstripping the deacons' ability to meet the demand. The poor law required each community to collect donations to aid the poor and maintain a poor fund. Stuyvesant and his council also passed a law aimed at sorting out the "deserving" from the "undeserving" poor and forcing outlying villages to care for their own poor. The law sought to rebuke "the Lazy and Vagabond" so that "the really Poor" could be "assisted and cared for." In 1683 another poor law was passed that shifted the responsibility for the poor from the church to the county. During the early 1730s the municipal government provided the poor with food, firewood, shoes, clothing, medical care, funeral expenses, and small cash payments.[10]

But the attitudes of elites were changing: they began to perceive the poor not only as a nuisance, but as a threat to moral order and security, so much so that by the late 1730s the poor were criminalized and institutionalized. The first city-run almshouse was completed in 1736. The structure was located on the Common (also known as the Fields and later City Hall Park), right across from City Hall. At the time this location was on the outskirts of town. The almshouse functioned as a poorhouse, workhouse, and correctional facility all in one. The inmates ranged from "poor needy persons" and "idle wandering vagabonds" to "sturdy beggars," petty thieves, and "parents of bastard children." In addition "unruly and ungovernable servants and slaves" were sentenced to the facility to do hard labor. Inmates had to wear clothing bearing their initials. They were not fed if they did not attend prayer services, and they were put to work carding wool, shredding old rope that was being reused, or raising garden crops. Disobedient inmates were whipped. Once the almshouse began operating the city drastically cut its outdoor relief program. By 1747 outdoor relief was all but eliminated.[11]

Civic leaders articulated a framing and ideology about poverty that has guided poor relief and charitable aid for centuries. These sentiments are still recognizable in contemporary aid and welfare policies. By linking the causes of poverty to laziness and vagrancy civic leaders successfully implied that poverty was the fault of the poor. By this logic, aid should be given only to those who were *deserving* of assistance. Contemporary policy makers also seek ways of sorting out the deserving from the undeserving poor in an effort to help the former and withhold aid from the latter. Moreover some present-day social observers still try to link poverty to laziness and other behavioral characteristics, and they worry about the poor moving from one jurisdiction to another to take advantage of the most generous aid.

Life in many cities was becoming a life of extremes. As poverty increased for some, wealth increased for others. In 1664 10 percent of the merchants in New Amsterdam controlled 26 percent of the wealth; by 1676 the richest 10 percent of the city's 313 taxpayers (about thirty-one merchants) owned 51 percent of the wealth, the top 15 percent of the taxpayers owned 65 percent of the city's wealth, the five richest residents of the city owned 40 percent of the wealth, and the richest man owned 14 percent of the wealth. (It should be noted that at this time the city had a very limited tax base. In 1678 Manhattan contained only 384 houses.) This kind of wealth inequality continued into the eighteenth century: in 1703 the richest 10 percent of the population owned 47 percent of the wealth, while the bottom half of the population owned less than 20 percent of the wealth.[12]

Zoning and Public Works Projects

In 1664 the Dutch ceded control of New Amsterdam to the English after troops led by Richard Nicolls seized control of the territory. The name of the territory was changed to New York that year. Stuyvesant, the last Dutch governor, was succeeded by Nicolls, who served as governor till 1668. Francis Lovelace succeeded him and served from 1668 to 1674. During the 1670s the city leaders in New Amsterdam continued to manipulate the environment by undertaking more waterfront development projects. In 1675 the canal, Heere Gracht, now a foul inlet, was filled in and paved over to make Broad Street. This resulted in water shortages that hampered firefighting efforts, so the city dug six new wells. At that time residents slaughtered about four hundred cattle annually, and Governor Edmund Andros continued Nicolls's and Lovelace's policies of separating land uses such that noxious facilities were separated from residential and commercial establishments. Hence the slaughterhouses, tanneries, and breweries, known as the "stink factories," were relocated to the fringes of the city.[13]

Open Space

Lawn bowling was a popular pastime among the Dutch in New Amsterdam. Beginning in 1626 they held lawn bowling matches in the area that later became known as Bowling Green. When Fort James was being restored (after being ransacked by departing Dutch troops in 1664) "the Plaine afore the Forte" was designated as open space. There Bowling Green, the city's first park, began to take shape during the 1670s. This multiple-use open space was the site of parades and the annual fair where cattle, grain, and produce were sold. Bowling Green was officially laid out in 1733 and rented out to three residents for a nominal amount. The following year Abigail Franks described Bowling Green this way: "The Governor has made a Very Pretty bowling greens with a handsome Walk of trees Raild and Painted Just before the fort in t[ha]t Large Durty Place it Reaches three doers beyond." A statue erected in the park in 1770 was toppled by angry citizens in 1776. This was a harbinger of things to come, as later residents of New York and other cities vented their frustrations by rioting in parks and destroying structures in them.[14]

A Growing Population

By 1694 there were nine hundred and eighty-three houses in the city that had a population of five thousand, including eight hundred blacks, most of whom were slaves. One hundred and twenty-five vessels plied their trade in the harbor, and four thousand cattle were slaughtered annually, most for export.

In 1776 there were twenty-five thousand residents, four thousand of whom were black. Native Americans were also held as slaves in New York City.[15] The population reached thirty thousand in 1791 and had doubled by 1801. The East Side developed more rapidly than elsewhere, and "fashionable residences" for the wealthiest citizens began to appear along Nassau and upper Pearl Streets, lower Broadway, and the Battery.[16]

NEW HAVEN: A PLANNED CITY WITH OPEN SPACE

Whereas New York and Boston were settled before any town planning took place, New Haven and Philadelphia were planned cities. New Haven was one of the first planned cities in the country. Even more unusual, the town was planned with open space in mind. From the outset the eleven-square-mile town was built around a central open space called the Green. Some historians believe the town was laid out by Lion Gardiner, a close friend of John Davenport, who was one of the city's founding fathers. Gardiner was a military engineer, town planner, and architect. There is a striking resemblance between the town plans of New Haven and Saybrook, which he also laid out. Other historians have concluded that Robert Seeley was the original town planner; he also knew Davenport well, from their days at St. Stephens church in London, and he had the technical know-how to develop the plan. John Brockett was yet a third candidate given credit for planning the town. Though

Brockett, a surveyor, platted (mapped) the town in 1638, experts believe he wasn't the planner. The 1641 Brockett map, the oldest surviving one of the city, shows New Haven laid out in an orthogonal grid with nine squares and two suburban tracts.

The Green was used for common agricultural purposes, military training to prepare for attacks from the Dutch and Native Americans, and as a meeting place. The city's most important buildings, three prominent churches and the colony house, were built on the Green. Small burial grounds were attached to the churches.

Cambridge, Massachusetts, and other small towns along the Connecticut River were also laid out with squares, although it is unclear whether any of the squares were used for open space. Boston designated the Common in 1634, but it was situated on the Shawmut Peninsula, away from the busiest parts of the city. Though commons and greens dot New England towns today, most of these were not established until the nineteenth century.[17] The much older New Haven Green still survives as a park today.

At first New Haven's founders apportioned relatively small parcels of land to residents; the size of the lot corresponded to the size of the family and their wealth. Two years after its founding the city parceled out land for huge estates; for example, the Edgewoods had a 360-acre farm. Soon opulent estates sprang up on the choicest lots; Davenport's home, which had thirteen fireplaces, was one of these.[18]

BOSTON

As in New York, hordes of pigs roamed the streets of Boston, feeding off the waste. One of the earliest sanitation ordinances in America was passed in Boston in 1634, when civic leaders banned the dumping of garbage near the town dock. But it wasn't until 1652 that another ordinance forbade the disposal of animal entrails on the streets and banned the construction of outhouses within twenty feet of residences or roads unless the privies had an underlying vault at least six feet deep. Neither of these ordinances kept garbage and other wastes from being piled high in the streets, so in 1652 the city began hiring "scavengers" to remove dead animals and trash from the streets. Finally, four years later, the city prohibited residents from disposing of any kind of garbage on the streets. Believing that the waterways could solve waste disposal problems, civic leaders instructed residents to toss their trash into Mill Creek instead.

Early on Bostonians laid cobblestone streets. By 1715 a 795-feet-long common sewer had been laid along Prince Street extending to the point where it could drain into the Mill Pond. Property owners constructed sewers at their own expense that connected to the common sewer: for instance, Nathaniel Goodwin built 95 feet of sewer in 1715 from his residence to it. Other prominent citizens, James Tilleston, Robert Gutteridge, James and John Pecker, and John and James Clough, also built private sewers. By 1750 Boston residents had one of the best sewer systems in the country. In 1752 city ordinances mandated that residents sweep and clean the roadway outside their homes or risk heavy fines.[19]

Boston also wrestled with industrial pollution and malodorous air. In 1653 the city placed restrictions on where slaughterhouses, tanneries, and other nuisance industries could be located; thus it was ahead of other cities in dealing with this problem.[20]

Bostonians dealt with limited space and their peculiar geography by undertaking massive projects to expand the physical size of the city and transform the environment. By the time they finished they had more than tripled the size of the city. Beginning in 1711 residents began heaping rocks and wood in the harbor to make Long Wharf. By 1795 the area round the Boston Peninsula had increased dramatically through the process of "wharfing out": building a series of wharves outward from the shore and later filling in the slips between them. In 1855 they began to fill in Back Bay with garbage and the gravel obtained from leveling the hills.[21]

The "fountain society" emerged when Providence, Rhode Island, began distributing water through wooden pipes to its residents in 1772. Other cities undertook similar projects. By 1796 Boston was distributing water to residents through forty miles of wooden pipes from a source several miles south of the city. Soon the city began to tap more distant upland water sources. By 1848 residents got their water from an eleven-mile aqueduct that brought reservoir water to the city.[22]

In 1634 Boston became the first city in the country to set aside open space, the Boston Common, It was also the first to pass an ordinance to protect designated open space. In 1839 it became the first city to build a botanical garden. The Public Garden is situated in the lowlands on the edge of Back Bay adjoining the Common.[23]

Like other cities, Boston was plagued with fires. To combat the ever-present threat of fire, citizens were ordered to have water buckets and ladders and to participate in fire watches. The city had many "great" fires, the first of which occurred within twenty years of the establishment of the city. After each one, civic leaders instituted reforms to reduce the risk of fires.[24]

PHILADELPHIA: A GARDEN CITY ON THE DELAWARE

Green towns and garden cities have a long history in America. When William Penn got the charter to build a city on the Delaware River, he decided to build Philadelphia, a "greene Country Towne." Philadelphia began as a planned community. Penn envisioned a town of about ten thousand acres inhabited by gentlemen farmers. Plans called for all homes to be set back a uniform distance, at least six hundred feet, from the water's edge and spaced at least eight hundred feet apart. The homes would be surrounded by gardens, orchards, and fields. The city would have a small commercial center and a quay, warehouses, a market, and a state house. However, the main feature of the town would be the gentlemen's estates flanking the riverbank. Incorporated in the town plan were uniform streets, symmetrically placed houses, and open space along the river. In Penn's second city plan, developed in 1682, he offered eighty-acre lots to gentlemen farmers on the urban fringes of the city (in what is now North and West Philadelphia). Penn also laid out the city with wide streets, intersecting squares, and one-acre and half-acre lots. As Penn's cartographer described it, even the smallest lots had "room for House, Garden and small Orchard."[25]

At the time Penn planned Philadelphia, city planners in Europe were experimenting with new ideas as they sought to open up walled cities, reduce overcrowding, build wider streets, regulate façades, incorporate open spaces, and take advantage of dramatic vistas. Penn was influenced by these European ideas of city planning. Penn included five public squares in his city plan as a means of having an open feel to the city. Since each home was expected to have a yard and garden, Penn may have thought that this small amount of designated open space would be adequate for the city.[26]

Despite careful planning Penn's vision of a garden city did not come to fruition. The population of the city grew from a few hundred inhabitants in 1683 to over two thousand in 1700. By the late 1680s most residents and businesses crowded as close to the banks of the Delaware River as they could, rather than spread out from the Delaware to the Schuylkill, as Penn had conceived it. Owners of the original one- or half-acre lots subdivided them. In 1689, for instance, the original forty-three lots along Front Street were subdivided into seventy smaller lots; by 1703 they had been subdivided further into 102 lots, many of which were less than twenty feet wide. East of Front Street, along the Delaware River bank in an area Penn intended to be a public esplanade, the forty-one lots designated in 1689 had been subdivided into 110 lots by 1703. Because the original lots were about three hundred feet deep, the resulting long slivers of land were interlaced with a network of narrow alleys lined with rows of tenement housing. By 1698 the dream of a green town was no more, and Philadelphia took its place as one of the most congested cities in America.[27]

Inequality, Poverty, and Poor Relief

Income inequality was evident in Philadelphia soon after the city was founded. In 1693 the wealthiest 10 percent of the city's population owned 46 percent of the assessed property, while the poorest 30 percent owned about 2 percent. As poverty increased the city organized poor relief efforts. In 1713 an almshouse was opened; the facility was expanded in 1729. By 1793 Philadelphia was a bustling port handling almost a quarter of the country's export trade. The city's rapidly rising population topped fifty thousand in 1793, and the gap between rich and poor widened as rents skyrocketed. Immigration increased rapidly also, making it difficult for new arrivals to find affordable housing. At the time, more than half of the families owned no real property and about a third lived at subsistence level. Some poor residents of the city lived in caves dug out of the river bank.[28]

Environmental Planning and Open Space

Soon after the founding of Philadelphia city managers began instituting environmental reforms. In 1683 Penn designated a site near Coaquannock Creek, just north of the city, for a factory to build bricks from local clay. A year later the first brick house was built, and by 1690 there were four brick makers in the city. Because many of the homes were constructed of brick Philadelphia escaped the numerous and devastating fires that plagued other American cities. There were still fires, though, so in 1696 the Assembly decreed that each household acquire a leather bucket and a ladder.[29]

Philadelphians took steps early on to safeguard the integrity of their food supply. For example, civic leaders passed a meat inspection law in 1693. However, almost a century later, the city's inspector of salt provisions, arguing that spoiled and contaminated meat was being sold to consumers, warned merchants to sell only inspected meat.

Other environmental laws aimed at improving drainage and making it easier to move around the city were passed, but enforcement was lax. For instance, in 1727 residents were ordered to build sidewalks in front of their properties, but the ordinance was not enforced. That same year, an ordinance was passed requiring that the streets be pitched so that water could easily drain into the trenches in the center, but it too was not enforced. In 1739 a grand jury reported that many of the city's streets were unpaved and impassable; as a result, in the 1740s a street-paving program began in earnest.[30]

During the 1750s and 1760s some provisions were made for the public to have greater access to open space in the city. Some private pleasure gardens, such as the estate at Springettsbury, were opened to the public. Cherry Garden, a public garden, was extended from Front Street to the Delaware River. A six-acre square (now Washington Square) served as a potters' field for burying the indigent; Jacob Shoemaker, a farmer, pastured his cows there until complaints by other residents forced the city to charge him rent. In 1755 two small plots of land were set aside as a campground for Indians visiting the city. Just beyond the city limits was a forested area known as Governor's Woods.[31]

BALTIMORE

Laid out in 1730, Baltimore was a small hamlet with a collection of twenty-five houses in 1752. On the eve of the Revolution it had six thousand residents, ten churches, and a newspaper. The town grew rapidly; by 1774 there were 564 houses. There were also black and Indian slaves in the city. Like New Yorkers, Baltimoreans reshaped their physical environment to meet the needs of the burgeoning population. In 1784 the city began blasting rocks from the Susquehanna River channel to improve navigation. At the same time they began building a network of roads and bridges and public market places, and cutting down trees to provide building materials and firewood. By 1795 Philpot Hill had become a "respectable neighborhood," and in 1799 wealthy residents such as John Hollins and James Buchanan built the first "fine residences" on Calvert Street. Other merchants began building their homes on Gay Street.[32]

Rising Inequality and the Emergence of Slums

Increasingly the wealth of the cities became concentrated in the hands of a few. In Boston in 1771 15 percent of property owners owned 66 percent of the taxable wealth; the top 5 percent controlled more than 44 percent of the taxable wealth. At the same time 29 percent of the population did not own any property. Conditions were similar elsewhere. In 1770 in Portsmouth, New Hampshire, 10 percent of the residents owned half of the taxable wealth, while 10 percent of the population of Newburyport, Massachusetts, owned 57 percent.[33]

After 1790 the disparity between rich and poor increased even more rapidly and the cities became more racially and ethnically diverse. As immigrants flocked to the cities, longtime residents began moving out. Merchants ceased to live above their shops, and businessmen and their families moved away from the wharves and other congested downtown areas. As the wealthy and middle class fled, the poor became concentrated in neighborhoods that gradually became notorious urban slums. The economic decline caused by the Panic of 1819 accelerated this process. By the mid-nineteenth century the cities were associated with severe overcrowding, substandard housing, and homelessness; noise; disease, epidemics, and death; pollution, filth, and poor sanitation; crime, vagrancy, and corruption; riots and lack of control.[34] As the cities grew and became more crowded environmental inequalities became evident.

BLACKS IN NEW YORK

The Dutch West India Company began importing African males in 1626 to work as slaves in New Amsterdam to build the city's infrastructure and work in the fur trade. The first female slaves were imported in 1628.[35] Under Dutch rule slaves in New York City had more freedom than under British and American rule; they could be baptized, marry, own property, and work for themselves when their services were not required by the company or other masters. Slaves could bear arms during emergencies and live away from their master's home. They also had the right to legal recourse; they could sue whites and testify against them in court. Slaves were not allowed to have sexual relations with whites, however.

On February 25, 1644, the first eleven male slaves and their wives were granted "half freedom" after petitioning the Dutch West India Company. The Dutch, mired in a war with Native Americans and afraid blacks would take the Native Americans' side, granted blacks partial freedom. In exchange for half-freedom blacks were required to make annual payments to the company of thirty skepels of corn, wheat, peas, or beans and a pig. If they failed to pay their tributes, they could be re-enslaved. They received deeds to land some had farmed before their emancipation in what is now Greenwich Village. After being bonded for up to eighteen years slaves could petition for emancipation, although they still had to work for the Dutch West India Company. The children of half-free blacks were enslaved by the Dutch West India Company.[36]

Blacks acquired property and established "Negroes' farms" on the fringes of the city, on marginal lands that were not in high demand. This suited the Dutch because it meant that whites and blacks did not have to live in the increasingly cramped confines of the city, and blacks could act as an early-warning system of impending raids by Native Americans.[37] It also meant that blacks would have land on which to grow crops and raise animals in order to pay their annual tribute. Blacks with land were self-sufficient; they fed and housed themselves, thereby absolving their master of these responsibilities.

The conditions grew much harsher for slaves after the English took control of Manhattan in 1664, at the same time that the slave population increased dramatically, accounting for about 20 percent of the city's population of eleven thousand in 1741. Any public assembly of more than three blacks or Indians was banned, curfews for blacks and Indians were established, and free blacks were prosecuted if they were suspected of harboring slaves. Soon after a 1712 slave uprising that culminated in the deaths of several whites and the execution of nineteen blacks, a law was passed prohibiting blacks from owning land. Blacks could not vote or hold public office, and a 1785 law prohibited them from testifying in court against whites. Whites dealt with the competition from Native Americans for land and resources by expelling most of them from the city; some remained to serve as laborers.

The lot of blacks began to improve in 1799; that year a Gradual Emancipation Act was passed in which every male slave born after July 4, 1799, could gain full freedom at age twenty-eight and every female at age twenty-four. So instead of petitioning one's master for partial freedom and relying on the master's discretion, after 1799 slaves could be guaranteed their freedom on reaching a specific age. In 1821 the state legislature ruled that blacks could vote if they owned property worth at least $250 ($3,300 in 2005 dollars); at the same time the legislature eliminated the property qualifications for white male voters. Only sixteen blacks qualified to vote in 1825 and sixty-eight in 1835.[38]

Five Points and the Freshwater Pond

In 1825 most blacks in Manhattan lived in Five Points and nearby neighborhoods. Close by was a body of water known as the Collect or Fresh Water Pond (located in what is now Foley Square), also known by the Dutch names Kolch (a small body of water) and Kalch-hook (lime shell point). The spring-fed pond was sixty feet deep and covered forty-eight acres. Before being driven from the area Indians had collected oysters from the pond and deposited the shells on its banks. Just northeast of the Collect a promontory called Bunker Hill rose to a height of more than a hundred feet. From Bunker Hill picnickers could watch the wildlife around the pond and have a view of the city to the south. Blacks established a community on the southern marshy shores of the Collect. A second black community was established north of Houston Street between Lafayette Street and the Bowery, and a third was established in what is now Greenwich Village. Blacks also lived on the southwest corner of what is now Washington Square Park and along the marshy banks of Minetta Creek.[39]

During the seventeenth century and early eighteenth Native Americans and the Dutch and English settlers drew their water from the ponds, streams, and wells on Manhattan Island. One of the earliest sources of the city's water supply was the Collect. However, by the 1780s the water quality in the Collect had declined noticeably because city residents dumped garbage, sewage, and the carcasses of dead animals in the pond. The city also drew water from a well close to the Collect in the early 1800s, but water quality was so poor that the Manhattan Water Company, a privately owned firm that tried to sell the water to customers, abandoned the project shortly after it began operations. The Tea Water Spring was also located close to the Collect, just east of present-day Chinatown. For several decades it had provided the best quality water in Manhattan, and New Yorkers who could afford it purchased their water from this spring. However, by the 1780s water from this spring too had declined in quality, although it was still the best water in the city.[40]

New York followed Boston's lead in segregating land use. When Mayor Andros expelled the stink factories and other noxious facilities from the central city, they relocated in the Collect and other areas. Private slaughterhouses clustered along the southern and eastern edges of the pond. Mulberry Street became known as "slaughterhouse row." The tanneries processing the hides were built near the slaughterhouses. Soon the Collect stank from the pungent odor of the breweries, curing carcasses, and chemicals used to tan the hides. By the end of the eighteenth century the Collect, which was still the source of drinking water for the Chatham Street neighborhood, had been transformed into a putrid, semiaqueous eyesore. Around this time Pierre L'Enfant proposed cleaning up the pond, encircling it with a park and making it a focal recreational amenity around which the city could grow, but the plan was abandoned when landowners refused to sell. In 1802 the Common Council decided to fill in the Collect with earth obtained from leveling Bunker Hill, which it accomplished between 1803 and 1811. A notorious slum neighborhood soon enveloped the site of the former pond.[41]

Hence early on in the evolution of the city several environmental inequalities converged to place Native Americans and blacks at a disadvantage. Noxious facilities were sited in or adjacent to black communities after such facilities were expelled from white communities, and practices that were banned in white residential neighborhoods such as slaughtering animals and disposing of their remains were allowed to go unchecked in black neighborhoods and sparsely populated outlying communities. These patterns still exist today and are major issues on the agenda of environmental justice activists.

SLUMLORDS

The Schermerhorns, Astors, and other wealthy New York families owned property along the edges of the former Collect and played an active role in degrading the pond and surrounding communities. These entrepreneurs saw the potential for real estate development in the area once they relocated some of their tanneries and factories to the northern rural parts of Manhattan and elsewhere. By the time the Collect was filled in, the population of the city had exploded and there was great demand for cheap dwellings to house the immigrants who were arriving daily. The factory owners became real estate developers and converted some of the former factories into cramped, shoddy, multi-unit dwellings or built tenements in the neighborhood. Their properties became the source of great wealth for New York's slumlords. Because the ground under the Collect remained damp the houses shifted and tilted as the ground settled. The area flooded when it rained or snow melted, so basements that housed families (even though they were not designed for that purpose) were almost always flooded with water and sewage. As developers built dark, dank, two-and-a-half-story tenements over an entire lot, leaving little room for air or light to penetrate, the demographic characteristics of Five Points changed. By 1825 immigrants accounted for about 25 percent of the population and blacks made up 15 percent. The per capita income of the neighborhood was 40 percent lower than the rest of the city. The population of the area grew increasingly dense; while an average of thirteen people occupied the buildings in the 1830s, anywhere from thirty-five to several hundred people were crammed into single tenements by the 1850s.[42]

Residential and industrial land uses coexisted in this neighborhood long after the Collect was filled in. In 1825 Daniel Drew drove an estimated two hundred thousand cattle, pigs, horses, and sheep through the neighborhood to the Astors' slaughterhouse. Drew's cattle drives increased in size until 1845, when, unable to compete with the railroads, the practice was discontinued.[43]

Some slumlords, including the Astors, opposed housing reform. They objected to improving the tenements because improvements would result in higher tax assessments against their properties. The Astors also opposed the development of the upper part of Manhattan, reasoning that the more congested Lower Manhattan was, the fewer housing options the poor had and the more price gouging could occur.[44]

SLAVERY IN NEW ENGLAND

New York wasn't the only northern city organized around and buttressed by slavery. Many New England towns and cities had slaves. The ruling elites in these municipalities held blacks, Native Americans, and white slaves. New Haven, founded in 1638, had slaves from the start; its settlers made heavy use of indentured servants, who served a specific term of bondage. By the end of the seventeenth century blacks and Native Americans were enslaved for life and their children were automatically enslaved. Life changed quickly for the Quinnipiacks after 150 Europeans settled the area. The Quinnipiacks gave up all their lands around New Haven harbor for a small amount of cloth, spoons, knives, hatchets, hoes, porringers, and scissors. They were then placed on the nation's first Indian reservation, a thirty-acre plot of land on the eastern side of the harbor.

At first Europeans looked askance at Indian farming techniques, whereby women usually did the farming and where mixed cropping prevailed. Yet more than sixty of the New England towns established before 1650 were settled on lands that Indian tribes had already cultivated.

While black slavery was not outlawed in New England until the Emancipation Act of 1848, Indian slavery was outlawed in 1715. It was very difficult for the colonists to hold Indians as slaves and there was a growing fear of Indian slave insurrection. By 1774, when Connecticut prohibited the importation of slaves, there were sixty-five hundred slaves in the New Haven colony alone, accounting for 3.4 percent of the population. Slaves in Connecticut could own property and sue their masters in court.[45]

Northeastern slavery was very much an urban phenomenon. By 1700 blacks made up 16 percent of the population of Boston. Thirty-three percent of the blacks in New Hampshire lived in Portsmouth and 50 percent of those in Rhode Island lived in Newport; by the 1750s slaves constituted more than 19 percent of Newport's population. Slaves also made up 6 percent of the population of New Bedford, Massachusetts.[46]

THE SOUTH

New Orleans

By the 1870s one could find well-defined elite residential neighborhoods in New Orleans. The founding families had built their homes in the uptown Garden District along Levee (now Decatur) and Chartres Streets, where they could take advantage of the cooling breezes blowing in from the river. The homes of the well-to-do contrasted with the crowded, filthy shacks occupied by black and Irish residents between Camp Street and the river in the district known as the Irish Channel. Garbage was so plentiful that contractors used it as street filling. A report by the Louisiana Board of Health in 1879 cited the privies as a major health hazard, and an 1880 report called attention to the urgent need for improved drainage in the city. At the time New Orleans residents deposited over two million pounds of excrement in the soil annually. The 1880 health survey found that the city had 39,932 buildings with more than 44,000 toilets. Inspectors declared that about half the privies were either "foul" or "defective." As a result in the 1890s the city embarked on a plan to improve the drainage system and reduce the health problems associated with the open canals. The plan was to connect the entire city to sewers and pumping stations. The ambitious drainage program opened new areas of the city to settlement as lands were drained. However, real estate practices controlled access to the best neighborhoods, reserving them for whites.[47]

Grave inequities existed in the delivery of urban services in New Orleans and other parts of the South. Black neighborhoods were not provided with sewage systems, clean water, paved streets, or garbage removal as early or as frequently as white neighborhoods. In some cases black communities were provided with these services only to prevent diseases from spilling over into white neighborhoods. As was the case in New York, blacks were relegated to the swampy lowlands that served as the sites of garbage dumps and other noxious facilities or the drainage pools for sewage. For instance, while blacks lived in Congo Square adjacent to the turning basin, whites lived on higher ground in the better drained parts of town. One result of these inequities was the differential rate of death from malaria in the early twentieth century. While the death rate fell from 104 per 100,000 to 8 per 100,000 between 1900 and 1912 in the well-drained parts of town, the death rate in the swampy areas remained high. Because blacks inhabited the low-lying, water-logged parts of town and could not afford the $25 or $50 deposit to get sewer connections, the death rate among blacks remained much higher than among whites. For instance, in 1918 the death rate from typhoid was 20.4 per 100,000 for whites and 31 per 100,000 for blacks. The typhoid death rate dropped for whites but rose for blacks; in 1926 the death rate among whites was 13 per 100,000 but had climbed to 42 per 100,000 for blacks, primarily because black neighborhoods still had poorly developed sewage and drainage systems.[48]

Blacks were effectively corralled into the swampy, disease-ridden lowlands of the city. In 1912 the state of Louisiana passed an act that effectively segregated the city of New Orleans. "In the interest of peace and welfare" the city sought to "foster the separation of white and negro residential communities." As a result of Act 117 of 1912 and Act 118 of 1924 the city engineer was authorized to deny building permits to whites wanting to build in black neighborhoods or blacks wanting to build in white neighborhoods. Violators could be fined up to $25 and imprisoned for up to thirty days. These acts were challenged in court by the Land Development Company of Louisiana in 1926. The company argued that it was being denied the right to rent its property to blacks and that this constituted a taking of the property without due process. The district court in which the case was heard dismissed the case, arguing that the Land Development Company hadn't shown how they were being harmed in a direct and clear way. The court argued that if a black person who had been denied the right to occupy a building had brought the suit, then the court would be in a better position to decide whether harm was inflicted on such a plaintiff. The ordinances stood until 1927, when the case went to the Fifth Circuit Court of Appeals, which reversed the district court's decision and struck down the prohibition on racially mixed neighborhoods.[49] The court of appeals was swayed by the U.S. Supreme Court's decision in *Buchanan v. Warley*, which struck down racial zoning in Louisville, Kentucky.[50]

Atlanta

Great wealth and poverty coexisted in Atlanta by the second half of the nineteenth century. The rich lived on Peachtree Street, while poor blacks inhabited a low-lying area amid the garbage dumps, where contaminated water, milk, and food were prevalent. Local dairies, expelled from the city in 1878, built facilities on its edge. The milk became contaminated as diseased cattle drank sewage-tainted water. By the 1870s well-to-do Atlantans bought drinking water from Ponce de Leon and other springs. However, the majority of city residents, unable to afford the spring water, continued to drink the contaminated water. In 1878 there were twenty-three public pumps and seven drinking fountains from which city residents got their water. An 1883 study showed that many of the wells were polluted.[51]

Blacks bore the brunt of the pollution. Their homes and the wells they drew water from were located in the sewage runoff areas. The extensions of the sewer outfalls also terminated in the black neighborhoods. As a result the death rate among blacks was two and a half times higher than that for whites in 1885. In 1888 the city began to intensify its inspection program in an effort to improve the quality of the food and milk that reached consumers.[52]

There were elite residential communities in Memphis by the 1870s, but the Irish lived in the Pinch settlement, where they established a crowded shantytown called Happy Hollow. The huts there, which housed people as well as their animals, were constructed of disintegrated flatboats and old sheet iron and other materials. In the poorest section of Memphis fifteen to twenty blacks were typically crowded into small dilapidated rooms.[53]

Horrid conditions prevailed in poor neighborhoods in other cities as well. In Baltimore as early as 1816 the drinking water supply was polluted with excrement from the privies.[54] The worst environmental conditions existed in black communities. In Charleston, South Carolina, blacks were forced to live in the Neck, a low-lying dumpsite.[55] Birmingham, Alabama, financed its sewage system through special assessments that effectively kept black homes from connecting to the system. This happened despite pleas from health administrators in the first decade of the twentieth century urging the city to get rid of the large number of privies in the black neighborhood. Health officials thought the privies were a health hazard.[56]

Urban Riots

Though riots occurred in American cities during the eighteenth century, they tended to grow out of protests against the British or government edicts such as new taxes and treaties or town-and-gown tensions. For instance, the 1765 Stamp Act triggered a series of riots in Boston, Newport, and New Haven. In 1788 a doctors' riot occurred when thousands of people, believing that doctors and medical students had stolen cadavers from the cemeteries, marched to New York Hospital, ransacked it, and briefly held some medical students hostage until protestors collected and reburied all the cadavers they found.[57]

During the nineteenth century tensions rose as the population of cities grew larger and more racially and ethnically diverse and the gap between rich and poor widened. In addition to protests against the government, riots arose from interethnic (white-on-white) conflicts, class tensions between the rich and poor, religious differences, political rivalries, and racial animosities between whites and blacks.

NEW YORK RIOTS

On Christmas Day 1806 Irish Catholics and mostly native-born Protestants battled each other in the Sixth Ward. Labor strikes occurring between 1825 and 1828 also turned violent.[58] Class warfare touched well-to-do New Yorkers directly when they became the target of working-class rage on New Year's Eve 1827. That night, several thousand men and boys marched through downtown streets. The crowd headed down Pearl Street, breaking crates and barrels. When they got to the Battery they broke the windows of surrounding homes and tried to tear down the iron railing around the park. When they left the Battery they marched up Broadway just as revelers were leaving a fancy-dress ball at the City Hotel. However, the watchmen were able to avert an attack on the rich partygoers.[59]

Riots broke out in New York again in 1832 and 1834. Political rivalries led to the fight between Whigs and Democrats during the election of April 9–10, 1834.[60] One of the most destructive riots was the anti-abolitionist riot that broke out in Five Points on June 12, 1834. The events triggering the riot began with a simple gesture of courtesy, when Lewis Tappan, a silk importer and the former editor of an African American newspaper, invited Samuel Cornish, a black preacher whom he knew from his work in the abolition movement, to sit with him in his front-row pew at the Laight Street Presbyterian Church. Tappan's gesture caused an uproar among members of what was considered to be a relatively liberal congregation, who requested that Tappan not embarrass them that way again. The minister of the church, Samuel Cox, noting the commotion, condemned the intolerance of his parishioners in his sermon that day. The *Commercial Advertiser*, the *Courier and Enquirer*, and other conservative newspapers printed accounts of the events at the church, calling Cox an "amalgamationist" who promoted the intermarriage and social integration of blacks and whites. The papers advocated violence to suppress the abolitionist movement and the threat it posed to national unity. Tensions between abolitionists, members of the colonization movement, and anti-abolitionists rose, and on July 4 a mob disrupted an abolitionist lecture under way at the Chatam Street Chapel. The abolitionists quickly canceled the lecture, but the mob marched over to City Hall Park, where they assaulted blacks and ordered them out of the park. The police arrested six of the rioters and dispersed the rest. Three days later blacks who gathered at the Chatam Street Chapel to celebrate the anniversary of the abolition of slavery in New York were assaulted. When the *Courier and Enquirer* announced erroneously on July 9 that an abolitionist meeting would be held at the chapel that evening, the building was mobbed. Finding the chapel empty, rioters broke in and held their own meeting. Later that evening the crowd broke into Lewis Tappan's house, burned the furniture, and destroyed the house. They also broke into the Bowery Theater. The

Box 1 New York's Abolition Movement

During the 1830s the movement to abolish slavery was considered radical in New York. Most New Yorkers who opposed slavery did not support its immediate abolition; instead they supported the colonization movement, which sought to send voluntarily emancipated slaves back to Africa. Lewis Tappan and his brother were active in the colonization movement around 1830, but later changed their position to support the abolitionist cause. In his attempt to discredit the colonization movement Tappan publicly interrogated a man who had recently returned from Liberia. The man reported that drunkenness, sexual debauchery, and licentiousness characterized the country. This embarrassed the supporters of colonization, particularly the editors of the *Commercial Advertiser* and the *Courier and Enquirer*, who responded with scathing attacks on the Tappans whenever the opportunity arose.

Source: Tyler Anbinder, *Five Points: The 19th-Century New York City Neighborhood That Invented Tap Dance, Stole Elections, and Became the World's Most Notorious Slum* (New York: Free Press, 2001), 7–12.

following night mobs returned to Tappan's house; they also attacked his store on Pearl Street and Reverend Cox's house.[61]

On the night of July 11 the rioting turned into a full-scale assault on blacks. Mobs attacked African American homes, churches, businesses, and the African American Relief Hall and other institutions that served the black community. As the violence spread, so did word that whites should leave a light on in their windows so that the mob would spare their homes. The crowds rampaged through the black neighborhood until a street inspector addressed the crowd. He announced that all blacks should flee the neighborhood by noon the following day. The crowd cheered, shook hands with the officer, and moved farther down the block, where the rampage began again. When they attacked Thomas Mooney's barbershop at 87 Orange Street, Mooney, who had armed himself, fired three shots at the mob; one person was injured. Mooney's business was spared, but the rioters continued to destroy other black businesses, churches, and homes. The rioters dispersed only after attacking and destroying virtually all African American homes and institutions they came across. Many blacks were seriously injured; more would have been injured had they not fled the city when the rioting began.[62]

Riots broke out again in Five Points in 1835, and the Panic of 1837 precipitated the Flour Riot. On February 12, after a large crowd held a meeting in which they decried the plight of the poor, a portion of the crowd broke into Eli

Box 2 The 1863 Draft Riots: Recollections of Maritcha Lyons, an African American Woman, 1928

One July afternoon, a rabble attacked our house breaking window panes, smashing shutters, and practically demolishing the main front door. Lights having been extinguished, a lonely vigil of hours passed in mingled darkness, indignation, uncertainty, and dread. Just after midnight, a yell announced that a second mob was gathering to attempt assault. As one of the foremost of the rioters attempted to ascend the front steps, father advanced into the doorway and fired point blank into the crowd. Not knowing what might be concealed in the darkened interior, the fickle mob more disorganized than reckless, retreated out of sight hastily and no further demonstration was made that night. . . . The next day a third and successful attempt at entrance was effected. This sent father over the back fence to the Oak Street station, while mother took refuge on the premises of a neighbor. This was a friendly German who in the morning had loosened boards of intervening fences in anticipation of an emergency. This charitable man, some weeks after, was waylaid and severely beaten by "parties unknown." In one short hour the police had cleared the premises and both parents were at home after the ravages. . . . [The interior of the home] was dismantled, furniture was missing or broken. . . . A fire, kindled in one of the upper rooms was discovered in time to prevent a conflagration.

The dismayed parents had to submit to the indignity of taking refuge in the police station house. . . . During this state of anarchy, many were rendered homeless, maltreated, outraged, even put to death by hanging, burning, and similar barbarous acts. Under the cover of darkness the police conveyed our parents to the Williamsburg ferry; the steamboats were kept in readiness to either transport fugitives or to outwit rioters by pulling out to midstream.

Source: "The Draft Riots of 1863," Main Collection, 1928, Brooklyn Historical Museum. Reprinted in Kenneth T. Jackson and David S. Dunbar, *Empire City: New York through the Centuries* (New York: Columbia University Press, 2002), 265–66.

Hart's warehouse, seized flour, and gave it away; two other warehouses were also attacked.[63] Between 1834 and 1844 there were over two hundred major gang wars in New York City alone. The 1849 Astor Place Riot between supporters of two rival entertainers was quelled only after the militia was called in. Twenty-two people were killed and nine more may have died of injuries sustained in the melee; forty-eight were wounded. The Five Points gang war occurred in 1857, as did the Kleindeutschland Riot in the German neighborhood. The Draft Riot took place in 1863. It started as a class riot when poor Irish residents protested a law that exempted anyone who paid $300 ($5,500 in 2005 dollars) from being drafted into the Civil War. At first the mob targeted the symbols of power and privilege by burning police stations, arsenals, and the homes of the wealthy. A man who resembled Horace Greeley, the newspaper publisher and pro-war advocate, was almost beaten to death, and a friend's house where Greeley had stayed during a recent illness was ransacked. Then the rioters began lynching, mutilating, and assaulting African Americans and destroying black institutions and homes. Estimates of the death toll from this riot range from 105 to 1,000.[64]

More class riots occurred in the economic depressions of the 1870s. The Tompkins Square Riot occurred in 1874, when seven thousand workers tried to hold a rally in the park. When the organizers were denied a permit at the last minute the workers refused to leave. Sixteen hundred police charged the crowd, assaulting protesters and arresting some. When news of the disturbance spread, hundreds more people mobbed police headquarters, demanding the release of those who had been arrested.[65]

RIOTS IN NEW ENGLAND

A town-and-gown riot occurred in New Haven in 1812, when city residents and Yale students armed with clubs and knives clashed with each other. In 1824 a doctors' riot broke out when town residents became suspicious that medical students had removed the body of a woman from a West Haven cemetery. As rumors of the alleged body-snatching spread, mobs attacked the medical school, founded the year before, for five days.[66]

A riot in 1831 is noteworthy for its shifting of race relations in New Haven. In 1811 the Artisan Street School opened in the city, the first school founded for the purpose of teaching "colored children." A similar school opened in 1825. In 1820 Simeon Jocelyn founded the United African Society, the first black church in the city. Given the city's tolerance for emerging black institutions and the preeminence of Yale University, Jocelyn went to the 1831 Philadelphia Colored Convention and proposed that a "Negro college" be founded in New Haven. Residents of the city responded to the proposal with outrage. The newspapers condemned the idea and city officials convened an emergency meeting at which they passed resolutions to prohibit the establishment of the college. Residents voted 700 to 4 against the establishment of the college. In the meantime blacks were assaulted in New Haven. The mayor, Yale professors, and students rioted in front of the homes of supporters of the proposal. Things were calm for a few weeks, then a mob attacked the black community, nicknamed "New Liberia." Fourteen white men and four white women suspected of befriending and fraternizing with blacks were assaulted. Jocelyn was forced to stop overseeing the church he had founded in 1837 when mobs attacked his residence.[67]

Boston Common was frequently the site of rioting as marauders regularly set the trees on fire. The burning of the Catholic Ursuline Convent by Protestants in the greater Boston area resulted from an interethnic conflict with religious overtones.[68] The Hard-Scrabble and Snow Hill Riots that occurred in Providence in 1831 arose from racial conflicts between white and black laborers. During the nineteenth century there were strikes that led to rioting in Pawtucket, Rhode Island; Dover, Delaware; and Manchester, New Hampshire.[69]

PHILADELPHIA RIOTS

Two of the earliest riots in Philadelphia were religious riots triggered by the actions of the rector of Christ Church, Reverend Francis Philips, who had affairs with female parishioners and proceeded to brag about his conquests. In 1726 poor residents of the city rioted spontaneously, destroying the pillories and stocks beside the High Street market. In 1738 a law against fish weirs and racks in the Schuylkill River touched off a riot by working-class residents who thought the law would cut off their access to fish. The high price of bread triggered riots in 1741 and 1742.[70]

Riots grew more violent during the nineteenth century. In 1830 Philadelphia had a population of 755,000 whites and 15,624 blacks. The black population had increased by 40 percent between 1810 and 1830, and with that increase came heightened racial tensions. In 1829 a series of antiblack riots wracked the city; similar riots occurred sporadically until the 1840s. In 1834 several hundred club-wielding white males marched down Seventh Avenue to nightspots frequented by blacks; there four hundred to five hundred people engaged in an all-out brawl. Buildings were torn down and people assaulted. The police gained control of the crowd, but only temporarily. The following night a black church and more than twenty black residences were burned. The riot was not completely spontaneous. As was the case in New York, lights left on to shine through windows indicated which homes were occupied by whites; those homes were spared. On the third night of rioting blacks barricaded themselves in a building as a white mob surrounded it; thirty-one homes and three black churches were burned that night. Before the riot was over two blacks were killed and several people were severely injured.[71]

A year later violence broke out as a mob of whites attacked blacks living close to the intersection of Seventh and Lombard Streets. The mob set fire to a row of houses. Many black women and children fled the city. On the second night of the riot armed blacks barricaded themselves in a building and had to be persuaded by the mayor to come out. Pennsylvania Hall was burned to the ground on May 17, 1838, after white and black antislavery activists held a meeting there and marched arm in arm.[72]

A race riot in 1842 began when a mob attacked a parade of the Negro Young Men's Vigilant Association as they marched to commemorate the abolition of slavery in the West Indies. Blacks were beaten and their homes looted. The Smith Beneficial Hall, an abolitionist meeting place, and the Second Colored Presbyterian Church were burned to the ground. Race riots broke out again in Philadelphia in 1849. One began on October 9, when whites attacked an interracial couple and the hotel they operated for blacks. Violence escalated into a two-day race riot in which blacks in the area were attacked. Other riots arose from labor strikes or religious conflicts. In 1842 striking weavers ransacked the homes and destroyed the looms of weavers who did not go on strike. Armed skirmishes between Irish Catholics and Protestants broke out in 1844; the homes of thirty Irish families were burned, as well as St. Michael's Catholic Church, the Female Seminary of the Sisters of Charity, and St. Augustine's Church.[73]

RIOTS IN THE SOUTH

Social conflicts manifested themselves in riots that swept across Baltimore from 1829 to 1835. The increasing violence coincided with worsening economic times. Riots sometimes broke out over wage disputes or when contractors tried to pay in scrip, postpone wages, or otherwise manipulate the sum workers were owed. In 1829 one person was killed and several injured when rioting broke out among railroad workers. Conflicts between black and Irish workers escalated into a riot in 1831. Five people were killed in 1834 when rival gangs of canal workers fought each other. President Jackson sent in army units from Fort McHenry to help keep the peace, but rioting continued as white laborers from different ethnic groups attacked each other and black workers. A banking scandal triggered riots in 1834 and 1835. Beginning in February 1835 a three-month wave of arson targeted factories, law offices, a library, the Athenaeum, a church, a female orphan asylum, and an engine house.

Quite frequently the wealthy found themselves the target of the arson. During the Monument Square Riot, which began on a Saturday night in August, a mob faced off against six hundred recently deputized, stick-wielding citizens acting as law enforcement officers. The deputies were no match for the crowd, which succeeded in destroying Reverdy Johnson's home, furniture, and ten-thousand-dollar law library; John Morris of South Street lost 171 dozen bottles of wine from his cellar; and John Glenn in North Charles lost a twelve-thousand-dollar law library and 4,000 bottles of wine. The homes and furniture of other well-to-do citizens were turned into bonfires. By Sunday morning between five and eight people had been killed and the mayor was forced to resign.[74]

A riot broke out in Memphis on May 1, 1866, when Irish policemen began fighting with black Union soldiers. The three-day riot turned into an orgy of antiblack violence, during which time a city magistrate urged white rioters to "clean every Negro son-of-a-bitch out of town."[75] A similar incident occurred in New Orleans on July 30 of that year; white rioters and the police killed 34 blacks and injured 119.[76]

two

EPIDEMICS, CITIES, AND ENVIRONMENTAL REFORM

Illness has always been a part of life for the poor, but during the eighteenth century and the nineteenth virulent epidemics swept through American cities, forcing urbanites to search for the causes and cures of these plagues. An analysis of the responses to these diseases shows how people's thinking about diseases evolved and how epidemics forced municipalities to confront environmental problems. Yellow fever, cholera, and the bubonic plague were three diseases that forced Americans to adopt environmental reforms to safeguard public health.

These three diseases occurred at a time when elites were trying to regain control of the morality and social order of the cities. The epidemics made this goal more fleeting because the plagues disrupted life in the city, hampered the delivery of social services, and added to the chaos evident in urban centers. Unfortunately the patterns of infection and mortality arising from these diseases reinforced some of the beliefs that social reformers held about poverty.

Yellow Fever

Yellow fever devastated cities and towns, causing mass hysteria and the curtailment of commercial activities and severely limiting municipalities' capacity to function. There were sporadic yellow fever outbreaks in the country during the seventeenth century and the eighteenth. The disease broke out in Philadelphia in 1699, 1747, 1762, 1793, 1794, and 1798, one of the worst occurring in

1793. During the "yellow fever era" (1793–1806) there were outbreaks in New York, Baltimore, and New Haven as well. The first outbreak of yellow fever in New Orleans occurred in 1796. Yellow fever is thought to have killed about forty thousand troops in Santo Domingo from 1802 to 1803 and about one hundred thousand Spanish soldiers during the colonial wars with Cuba.[1]

In its mild form yellow fever causes flu-like symptoms that last about a week. The more severe form of the disease is characterized by the abrupt onset of shaking chills, fever, muscle aches, chilly fits, quick tense pulse, hot skin, headache, inflamed eyes, flushed countenance, moist tongue, sore stomach, especially with applied pressure, and dark stools. This is followed by jaundice, creating a yellowish-purple tinge to body, as the virus invades the liver, liver failure, and delirium. Hepatic congestion combined with a failure of the blood to clot results in hemorrhaging from the gums, nose, and stomach lining. This causes victims to vomit black blood (the disease is known as *vomito negro* in Spanish). Victims stop defecating and urinating and the kidneys fail. Death ensues about a day or two after renal failure, within a week of the onset of the symptoms. Depending on the demographic group, between 10 and 60 percent of those infected with the disease die from it.[2]

Yellow fever tends to affect mature adults more frequently than young children or the very old. Very slender individuals seemed to be less susceptible than obese individuals and women are less vulnerable than men. Those who recover from the mild form of yellow fever seem to have immunity afterward.[3]

THE PHILADELPHIA EXPERIENCE

In 1793 Philadelphia was a city of 55,000 inhabitants, with about 2,500 births and 1,400 deaths in the city annually. The gap between rich and poor was widening. As new immigrants poured into the city they were met with escalating rents. Poor residents settled near the congested dock areas. More than half of the families in the city did not own any property, and about a third lived at subsistence level. That year yellow fever struck the city with such ferocity that by the end of the outbreak, which lasted from early August to mid-November, nearly 5,000 people died and 200 children were orphaned. The city was so overwhelmed by the number of deaths that the deceased were stacked ten to fifteen deep in cemetery plots.[4]

Philadelphia was an utterly dirty city in 1793. The wharves jutted out into the Delaware River, disrupting the flow of the water. High tide deposited rotting carcasses and entrails on the muddy banks. Most of the streets were unpaved. There was no water system; holes were dug to catch runoff from the gutters, but these reeked from the putrefying vegetable matter and animal remains hurled into them. There was only one sewer, running under Dock Street. When the sewer malfunctioned the garbage was piled high in the streets, in the markets, and on the banks of the river. In 1739 a group of citizens tried unsuccessfully to get the Assembly to expel noxious industrial facilities from the heart of the city. Some doctors concluded that the disease was caused by foul odors or miasma and the filthy conditions of the city and so was not contagious. They believed that when thousands of people evacuated infected areas the disease did not follow them.[5]

Though yellow fever infected both wealthy and poor residents it hit hardest in poor neighborhoods. A large number of red flags, ordered by the Board of Health to mark where infected people lived, hung from the houses in the poor sections of the city. The poor were particularly vulnerable because they had no money for doctors or medicines and often had no one to care for them. As early as the 1750s Dr. Thomas Bond linked pollution and what he saw as "imbalances" in the environment to illness in humans. He began campaigning to clean up the city. Bond renewed his efforts to improve the sanitation of the city after the 1762 yellow fever epidemic reinforced his belief that there was an environmental connection to the disease. This resulted in the Corporation and Assembly overseeing the paving and cleaning of streets, removal of solid wastes, and extension of the city's drainage system. The upper third of Dock Creek was turned into a drainage sewer.[6]

Many of the doctors, policy makers, and planners of the time were miasmatists; because of their limited understanding of disease transmission, they believed that scourges like yellow fever were caused by obnoxious odors. They also thought that the poor were disproportionately afflicted because immorality, debauchery, and irreligiosity weakened their constitution, making them more susceptible to the harmful miasma. This view led medical experts and civic leaders to focus on ameliorating the conditions that produced malodorous environments and immoral behavior.[7] Among the causes they targeted was city residents' habit of placing offal, consisting of bones with flesh on them along with entrails and other rubbish, in barrels in their yards and cellars. The air reeked from the smell of the putrefying wastes. Consequently during the 1793 epidemic Dr. Benjamin Rush, who had apprenticed with Dr. Bond, advised the mayor that the disease arose from filth, impurities in the air, the open sewer, decaying animal and vegetable matter, and corruption in the city, and that it could be combated by cleaning up the city. The mayor immediately ordered workers to clean the streets and gutters. The Assembly passed an ordinance ordering people to remove the trash from their yards and pile it in the streets for the garbage collectors.[8] In a letter to the city commissioners, the mayor wrote:

> As there is great reason to apprehend that a dangerous, infectious disorder now prevails in this city, it is the duty of every department of authority to take the most effectual precautions to prevent its spreading: And as the keeping the streets at this time as clean as possible may conduce to that desirable object, I require that you will immediately upon the receipt of this letter, employ the scavengers in making the streets and gutters in every part of the town as clean as possible, and that as fast as the filth be laid together, that it be *immediately hauled away*. I recommend that they begin to clean first in Water Street, and all the alleys and passages from thence into Front Street, and then proceed to clean the other more airy streets. I expect that the inhabitants will have the satisfaction of feeling this business going on, this afternoon or tomorrow morning; any delay on your part will reasonably be considered as an improper attention to a very essential duty.[9]

Rush also hypothesized that there was a relationship between deforestation and the yellow fever epidemic. He observed that when the British army cut the trees between the Schuylkill River and the city, an epidemic broke out, which ended when the land was ploughed. He contended that the fever was originally confined to the shores of rivers in Pennsylvania, but once the trees were felled, it spread inland about eight to ten miles.[10] Rush was not alone in thinking that trees reduced the incidence of yellow fever. In 1812 a burial ground was opened just outside the wall of Colonial Park Cemetery in Savannah, Georgia, to accommodate the large number of interments during yellow fever epidemics that engulfed the city periodically. Residents of the city thought "poisonous effluvia" arose from this burial ground. Believing that miasma caused yellow fever, Savannah residents planted trees and shrubs in the burial ground in 1846 to counter its effects.[11]

Miasmatists believed that diseases, caused by foul odors from swamps and dumps, were local and environmental in origin. However, not all doctors believed in the miasma theory. Some were importationists; they believed that the disease was of foreign origin, imported into the city by immigrants and spread by contagion. Their belief was fueled by the fact that during the months of July and August 1793 over two thousand exiles fleeing the revolution in Santo Domingo settled in Philadelphia. The refugees were black, white, and of mixed race. Because the vessels on which they traveled to Philadelphia and Baltimore were frequently robbed at sea, many arrived penniless. Some of these refugees might have been exposed to yellow fever as the West Indian islands were frequently ravaged by the disease. The refugees were at first welcomed into the city, but as newspapers began publishing notices in French and the new immigrants began opening their own businesses and expressing their unique culture (cockfights, street games and songs, rope dancing, gambling, etc.), suspicion and resentment of the new immigrants grew. When yellow fever broke out importationists identified the immigrants as the source of the disease.[12]

Blacks and the Philadelphia Yellow Fever Epidemic of 1793

In 1793 about thirty-three hundred of Philadelphia's residents were black. During the epidemic more than seventeen thousand people fled the city, including the governor, Thomas Mifflin, and many city council members and other elites. A greater proportion of whites fled the city than blacks. Though 750 blacks lived in one of the areas hardest hit by the epidemic, 14 percent of the blacks fled the area compared to 40 percent of the whites. Even Benjamin Rush's wife and children remained in New Jersey throughout the epidemic. Blacks were thought to be immune to the disease, so many, untrained for the tasks they were persuaded to assume, remained to help the sick. During an earlier epidemic in Charleston, South Carolina, it was reported that "the black people [had] in no one instance been infected" with the disease. Dr. Rush, who was aware of this report and who had worked with black Philadelphians to raise funds to build a church, convinced the leading African American ministers and former slaves Absalom Jones and Richard Allen that blacks should help during the epidemic. Rush thought it would be an opportunity for blacks to show their gratitude for emancipation and that their civic participation might foster racial harmony and reduce antiblack prejudice. (Though Rush was seen as an ally of blacks, he kept a slave as his personal servant.) Jones, Allen, and Rush convinced the mayor that the idea was sound. Black leaders also persuaded the mayor to release black prisoners to help care for the infirm; two-thirds of those freed were blacks volunteering to be nurses. An estimated three hundred blacks participated in the relief effort, functioning as nurses (there were about twenty black nurses for each white nurse), digging the graves, and burying the dead. Poor whites were also called on to be nurses, but some shied away from contact with the sick and dead.[13]

According to Jones and Allen, blacks worked with yellow fever victims because "it was very uncommon, at this time, to find anyone that would go near, much more, handle a sick or dead person."[14] Jones and Allen argued that some blacks worked without compensation, while being subjected to the fear and hostility of the whites whom they tried to help. Jones and Allen also arranged for blacks to transport the indigent to Bush Hill Hospital. Dr. Rush trained blacks in the controversial techniques of purging and radical bloodletting; in addition, Jones and Allen kept a rigorous schedule of visiting and caring for sick patients. Doctors who opposed Rush's treatment of bloodletting vented their ire on blacks practicing the techniques.[15]

Given the critical role blacks played in keeping the city functioning, when some blacks succumbed to the disease doctors were reluctant to say that they died from yellow fever. Hence doctors helped perpetuate the myth that blacks were immune to the disease. But they weren't; estimates indicate that 198 blacks died of the fever. When black nurses caring for ill white patients contracted the disease they were chased from the homes in which they worked.[16] When blacks began to die from the disease some felt that the report of their immunity was a ruse to get them to do the work abandoned by others. It should be noted that blacks did not begin to succumb to the disease until it had raged for a full month and had taken a staggering toll on whites. Only then did doctors begin revising their earlier assumption about black immunity to yellow fever.[17]

Despite the role of blacks in helping to maintain order in the city during the crisis, rumors published in the newspapers and reports of the epidemic circulated that blacks pillaged the homes of whites who had fled the city or who hired them to care for the sick, charged exorbitant fees for their services, did not do the work they were hired to do, and were the source of the contagion (contagionists believed that infected people passed the disease directly to others). A twenty-three-year-old Quaker, Elizabeth Drinker remained in the city during the epidemic with her husband, Henry, a merchant. Drinker records in her diary, rumors persisted that the "Negroes may have poisoned the wells" to spread the disease. The mayor summoned Jones and Allen to discuss the allegations. In one of the earliest environmental health and social justice documents written by blacks, Jones and Allen wrote an essay describing the black community's response to the public health emergency. They provided data showing that nurses were paid six dollars per day for their service and that whites drove up the price by trying to outbid each other. Jones and Allen, who organized work crews to bury the dead, received £23 10s for their services and were £110 in debt for burying several hundred people without being paid. The Free African Society, which was founded by Jones and Allen and served as the institutional base of their activities, ran a deficit of £177 9s 8p that year.[18]

Economic Impacts of the Epidemic

The epidemic disrupted economic activities throughout the city. Businesses closed and some businessmen fled the city, leaving many workers without a means of livelihood. So many seamen became ill that ships lay idle at the docks, making it impossible for other, incoming vessels to dock. However, not all businesses suffered, and the stores that remained open profited. For instance, those who made or rented carriages did very well; so did coffin makers, apothecaries, bleeders, and doctors. Owners of country farms and guest houses also profited.[19]

YELLOW FEVER IN NEW YORK

Yellow fever first broke out in New York in 1703 and appeared several more times during the eighteenth century and early nineteenth. About seven hundred people died in the 1795 epidemic, and more than two thousand died in 1798. Despite the fact that blacks succumbed to yellow fever in the later stages of the 1793 Philadelphia epidemic, when yellow fever broke out in New York in 1801, 1803, and 1807 the myth persisted that blacks were immune to the disease.[20]

During each outbreak, merchants quickly fled the areas close to the docks:

> As soon as this dreadful scourge makes its appearance in New York the inhabitants shut up their shops and fly from their houses into the country. Those who cannot go far, on account of business, remove to Greenwich, a small village situate[d] on the border of the Hudson River about two or three miles from town. Here the merchants and others have their offices, and carry on their concerns with little danger from the fever, which does not seem to be contagious beyond a certain distance. The banks and other public offices also remove their business to this place; and markets are regularly established for the supply of the inhabitants. Very few are left in the confined parts of the town except the poor classes and the negroes. The latter [are] not . . . affected by the fever.[21]

Similar events unfolded during the 1819 outbreak. Shockwaves spread through upper-class communities when a wealthy merchant of a prominent family, Gilbert Aspinwall, died of the disease.[22] John Pintard, the manager of The Bank for Savings, wrote, "Stores are shutting up & goods removing to the upper parts of Greenwich & Bdway. The Banks have not yet started, but several of the Insurance Offices remove this day, on the principle that they must follow their customers. I presume that the same rule must guide us, altho' I absolutely declare that I think the city is panick struck. . . . We are like a city in a siege, the inhabitants fleeing they knew not wither."[23]

As was the case in Philadelphia, blacks kept the city functioning during epidemics. Among other acts, they cared for the sick and buried the dead.[24]

NEW ENGLAND'S RESPONSE

City officials in Boston responded to the 1799 yellow fever outbreak by creating the Boston Board of Health. The board comprised miasmatists, who ordered the cleanup of dumpsites such as the Frog Pond at Boston Common and the town dock. Other Massachusetts communities followed Boston's lead and created their own boards of health: in Salem in 1799, in Marblehead in 1802, in Plymouth in 1810, in Lynn in 1821, and in Cambridge in 1828.[25]

YELLOW FEVER IN THE SOUTH

The site of major yellow fever epidemics shifted to the southern part of the Atlantic seaboard and the Gulf Coast states, where outbreaks occurred between 1840 and 1905. The disease took a serious toll on New Orleans, Mobile, Norfolk, Charleston, Galveston, Savannah, Memphis, Holly Springs, Mississippi, and Jacksonville.[26]

During the 1878 epidemic in Memphis 25,000 people fled the city. Of the 25,000 remaining, 14,000 were black and most of the remainder were Irish. About 17,000 of those remaining in the city contracted the disease; 4,204 of the 6,000 infected whites died, as well as 946 of the 11,000 infected blacks.[27] The city was on the brink of chaos as public drunkenness, looting, and violence erupted. In an effort to restore social order the Citizens' Relief Committee appointed blacks to its membership, enlisted the services of black organizations, and appointed black men to police the city. As was the case in Philadelphia, blacks also took care of the sick and buried the dead.[28]

At the beginning of the 1878 outbreak in New Orleans many of the early cases affected children living in a relatively clean part of the city. Because residents had believed that children born in the city were immune to yellow fever, soon after the children's deaths were made public mass panic ensued, and 40,000 people out of a population of 211,000 fled the city.[29] It is estimated that yellow fever epidemics cost New Orleans about $10.5 million annually in lost labor and capital from 1846 to 1851.[30]

A massive fire devastated a large section of Savannah in January 1820. Soon after the fire torrential spring showers produced conditions favorable for mosquitoes to breed. In August of that year a ship from the West Indies carrying yellow fever aboard docked in the harbor. An epidemic broke out and six hundred and sixty-six people died. The death toll was low because many people who had left the city to escape the fire had not returned. Furthermore, by the time of this outbreak many people routinely spent the summers inland to escape the disease; others fled the city as soon as word of the outbreak spread. Only about fifteen hundred people remained in the city throughout the epidemic. It is estimated that nineteen of every twenty people who contracted the disease died. Two later epidemics broke out in the city: in 1854 more than one thousand people died of yellow fever; by the 1876 epidemic improved sanitary practices, better public health policies, and more stringent quarantine practices limited the spread of the disease.[31]

Preventive Measures and Cures

Yellow fever was known as the "stranger's disease," a label which reflected the growing nativist and anti-immigrant sentiments of the time. The nickname also referred to the fact that childhood cases were often mild and conferred immunity upon the victim. Hence people who grew up in areas where there had been earlier epidemics were not very susceptible to the disease. Blacks also seemed to have a higher level of immunity to yellow fever; in contrast, Irish and Italian immigrants were very susceptible to the disease. Though there was a tendency to blame the victims of the disease for becoming infected, there was less personal blame attached to yellow fever than other diseases, such as cholera. This is the case because catching yellow fever seems less related to personal behavior such as hygiene and morality and more correlated with prior contact with the disease, where a person lived, the ability to flee infested areas, and how long one had lived in a particular area. Because lack of proper sanitation seemed to be linked to the spread of the disease, government officials shared the blame for the epidemics.[32] Nonetheless, some of the preventive measures that residents were urged to adopt focused on behavior modification. For instance, residents of Philadelphia were urged to bathe, avoid intemperance, and drink alcohol sparingly as a means of protecting themselves from yellow fever.[33]

Residents of cities plagued by yellow fever were besieged by doctors, public figures, neighbors, and friends offering a bewildering array of preventive measures and cures for the disease. Suggestions included diffusing tobacco smoke and sprinkling vinegar throughout one's house, placing tarred rope in a room or carrying it in one's pocket, hanging camphor in a bag around one's neck, lighting bonfires in the streets, and pouring oil in containers of water.[34] During the 1793 epidemic, the College of Physicians published a list of suggestions residents should follow:

> Avoid every infected person, as much as possible.
>
> Avoid fatigue of body and mind.
>
> Don't stand or sit in a draft, or in the sun, or in the evening air.
>
> Dress according to the weather. Avoid intemperance.
>
> Drink sparingly of wine, beer, or cider.
>
> When visiting the sick, use vinegar or camphor on your handkerchief, carry it in smelling bottles, use it frequently.
>
> Somehow mark every house with sickness in it, on the door or window.
>
> Place your patients in the center of your biggest, airiest room, in beds with curtains. Change their clothes and bed linen often. Remove all offensive matter as quickly as possible.
>
> Stop the tolling of the bells at once.
>
> Bury the dead in closed carriages, as privately as possible.
>
> Clean the streets, and keep them clean.
>
> Stop building fires in your houses, or on the streets. They have no useful effect. But burn gunpowder. It clears the air. And use vinegar and camphor generally.
>
> Most important of all, let a large and airy hospital be provided near the city, to receive poor people stricken with the disease who cannot otherwise be cared for.[35]

In 1793 the *Federal Gazette* carried the following "Preventative against the Raging Yellow Fever":

> IT has been suggested . . . that, to avoid being infected with the epidemic malady now prevailing in this metropolis, it is necessary to breakfast early, and that without those appendages of the tables commonly called *Relishes,* whether in fish or flesh. To avoid lassitude and fatigue, as much as may be; and to dine moderately, on fresh animals, vegetable food, about one o'clock in the day; drinking Beer, Cyder, or good

> Brandy, respectively diluted with water, as the wholesomest beveridge at meals. In the evening, Tea or Coffee may be drank, with simple Bread and Butter, as in the morning; but suppers are to be avoided. Dram drinking, (which some persons practice in the morning, and indeed at other times of the day,) is at all times an evil and destructive habit; but at present, is doubly pernicious in its effects.[36]

People responded by staying indoors; scrubbing, whitewashing, and otherwise "purifying" their residences; burning tobacco, niter, and gunpowder; and sprinkling vinegar everywhere. Those who ventured outdoors armed themselves with tarred rope and camphor bags, chewed garlic constantly, doused themselves in vinegar, and carried smelling salts and tobacco with them. People stopped shaking hands, socializing, or visiting with friends. Everyone walked in the middle of the street to avoid contact with contaminated houses. They also tried to avoid contact with anyone else in the street.[37]

THE SPREAD OF THE DISEASE AND FLEEING YELLOW FEVER

Despite the beliefs of eighteenth-century physicians, the disease did follow refugees fleeing infected areas. And expanding rail transportation meant that yellow fever could reach hitherto isolated inland communities. In 1878, for instance, yellow fever traveled up the Mississippi River and along the rail lines to cities in Tennessee, Kentucky, Illinois, Indiana, and Ohio. Hysteria about the disease spread and residents fled their communities. Some vowed to tear up the train tracks in order to keep the "yellow jack" out of their towns; in some communities armed bands of vigilantes met the trains and forced them to keep moving. Those fleeing on foot were often denied shelter; some died of exposure or starvation.[38]

Yellow fever struck such fear that vigilantes tried to rid communities of yellow fever victims. In Norfolk, Virginia, angry mobs burned Irish neighborhoods because they blamed the yellow fever epidemic on the condition of the "dirty Irish."[39] Quarantine riots erupted in New York in the 1850s as residents of Tompkinsville and Castleton, near the quarantine station at Seguine's Point on Staten Island, established vigilance committees aimed at keeping sailors and immigrants sequestered in the station. When several yellow fever victims were placed at the quarantine station, a mob of rioters attacked in 1857. The marines and police from Manhattan quelled the riot.[40] Nonetheless, whenever a yellow fever outbreak occurred almost everyone who had the means to leave the area did. Wealthy southerners preempted the outbreaks by regularly scheduling vacations in the North or elsewhere during what became known as the "yellow fever season." Once the exodus from infected areas began, economic and social activities dwindled. Quarantines were placed around cities and institutions were shuttered for months.[41] Southern slaveholders often fled their plantations during epidemics, leaving slaves under minimal supervision. In some instances slaves planned insurrections while their masters were away.[42]

YELLOW FEVER, IMMORALITY, AND POOR RELIEF

Prior to Reed's discovery of the real cause of yellow fever, not everyone believed the epidemic was a purely medical event. Some, including physicians, thought the disease was connected to irreligiosity and immorality. They thought that the wrath of the Deity manifested in the fever. In fact advice issued by Philadelphia's College of Physicians suggested a link between intemperance and lack of cleanliness and yellow fever. Indeed Benjamin Rush, one of the city's most prominent miasmatists, was adamantly opposed to alcohol consumption and began advocating temperance as early as the 1770s. Philadelphia's pious were more inclined to see yellow fever as communal punishment for the evils of the city rather than vengeance against individual wrongdoers. In the quest to identify the communal wrongs being avenged, advocates of the divine retribution thesis identified the completion of the "Synagogue for Satan" (the lavish new bordello-like Chestnut Street Theatre), the pride and vanity of communal leaders, French immorality, the adoration of citizen Edmond Genêt, and the wild parties held in Philadelphia in the summer of 1793 as likely reasons for the punishment. From this perspective the epidemic was a calamity sent from heaven to purify the city. To reinforce the notion of divine reprisal, sixteen of the city's leading clergymen and Quakers petitioned the state to close the new theater.[43]

The framing of epidemics as retribution visited on the poor became even more pronounced during cholera epidemics. However, in the case of cholera retribution was not seen as a collective punishment that included elites, but as individualized punishment visited on the poor for deviant behavior. The Philadelphia yellow fever epidemic occurred at a time when elites were becoming more concerned with poverty and social disorder in cities but had not yet fully articulated the problem as one of lawlessness, irreligiosity, and immorality for which diseases were divine vengeance. This can be seen in Philadelphia's approach to poor relief during the epidemic. The city organized relief efforts to help the poor and get them medical care. As the disease raged a committee of forty-six male volunteers was formed to canvass the city and collect information on victims, investigate people's needs, and dispense relief certificates redeemable at City Hall. As the printer Mathew Carey noted at the time, never before had so many indigent people been helped with so little probing into their personal affairs. Poor people who refused charity were given weekly loans, but no one really tracked their debt or ensured that the loans were repaid. The committee assisted about 1,200 people. In one day 118 families were given money and provisions; on another day, $100 in cash, more than twelve cords of wood, and bread, meat, and vegetables were dispensed to 112 families.[44] During the nineteenth century, when diseases were framed as the scourge of the poor, aid was restricted, charities and relief committees avoided dispensing cash, and there was much more oversight of the poor before and after aid was dispensed.

YELLOW FEVER AND ENVIRONMENTAL REFORMS

Even before the cause of yellow fever was known, the 1793 epidemic indicated a link to sanitation as the disease took the worst toll on the poorest, most polluted parts of the city. Consequently Philadelphia created committees to oversee cleanup efforts. The city also approved plans in 1799 to build a centralized system to deliver clean water. Prior to the outbreak Philadelphia drew its water from wells and cisterns; these were located in low-lying areas, easily contaminated with surface runoff from the streets and from the underground flow of cesspools and toilets. The new waterworks began operation in 1801.[45]

Southerners copied Philadelphia, New York, and other northeastern cities in cleaning up their cities. Thus by the 1850s Charleston, South Carolina, had improved its drainage and the city streets were paved and cleaned. In New Orleans the street-cleaning contract was taken away from a corrupt operation and placed under municipal control. Sanitarians also pushed city leaders to clean up Savannah and Mobile.[46]

CAUSES OF YELLOW FEVER

It was not until 1901 that Walter Reed and his colleagues discovered the cause of yellow fever and how it was transmitted. Yellow fever is an acute infectious disease introduced to the Western Hemisphere by the African slave trade. It is caused by an arbovirus transmitted in a human-vector-human cycle from one victim to another by the blood-sucking female *Aedes aegypti* mosquito. The mosquito acquires the virus after feeding on the blood of an infected human in the first three or four days after infection. After an incubation period of about twelve days, in which time the virus travels from the mosquito's stomach to its salivary glands, the mosquito can transmit the disease for the remainder of its life. Because frost kills the mosquito, the disease is indigenous to tropical areas. Though Philadelphia, Baltimore, New York, and New Haven are north of the natural range of the mosquito, *Aedes aegypti* breeds prolifically, and once introduced into the population can survive for a season. In the United States the disease tended to last from about July to the first autumn frost. The *Aedes aegypti* mosquito breeds in still bodies of fresh water, such as cisterns, puddles, buckets, and flowerpots. The disease is most rampant in hot, crowded waterfront areas of cities. Because of the mosquito's limited range, the disease does not generally appear far inland. Hence mosquitoes aboard ships docking in Philadelphia's harbor would have found ample breeding grounds in Dock Creek or in the barrels that Philadelphians left beside their houses to collect rainwater.[47]

THE CONTEMPORARY PICTURE

Yellow fever is still a potent disease. Between 1960 and 1962 as many as 30,000 people were killed by the disease. In 1985 805 cases of yellow fever were reported in Nigeria; of those infected, 416 died. Five people died in Brazil from the disease in 2008. Today yellow fever is still a disease linked to poverty. It is found in the poorest parts of developing countries.[48]

Cholera

Cholera can be identified in Sanskrit writings from around 500 B.C., but the disease remained confined to Asia for at least two thousand years. As global commerce increased, the disease spread along trade routes, eventually making landfall in Europe in the early nineteenth century. The British first encountered the disease when five hundred British soldiers stationed in Ganjam, India, fell ill in 1781. In 1873 news filtered out that twenty thousand pilgrims in Haridwar had died from an outbreak of the disease. Cholera first appeared in Britain in 1831. During the 1832–1833 epidemic, twenty thousand died. Another epidemic in 1848–1849 claimed the lives of fifty thousand in Britain. At the time of these outbreaks most British leaders were miasmatists who explained the deaths and disease affliction on rank odors and depravity.[49]

By the time cholera began infecting residents in American cities a wave of evangelicalism was sweeping urban areas and the disease was firmly linked to immorality and analyzed and framed in religious imagery more than previous epidemics had been. Evangelicals proffered religious explanations and sought religious cures for the disease. Cholera first appeared in the United States in

1832, but before the century drew to a close it had infected people in virtually every part of the country.

The onset of cholera is marked by diarrhea, vomiting, convulsions, and cramps. This causes dehydration, cyanosis, and cold and darkened extremities. As much as 25 percent of the body's fluids and up to 30 percent of body weight may be lost in a matter of hours. The symptoms appear without warning and victims die within hours of the appearance of the first symptoms.

The disease ravaged New York, Chicago, Cincinnati, and New Orleans. Of the major cities, only Boston and Charleston escaped. The epidemic spread to small communities along road, rail, river, and canal routes. There were outbreaks of cholera in 1832–1834, 1848–1854, 1866, 1873, and 1892. Like yellow fever, cholera epidemics forced municipalities to pay more attention to the environment and public health.[50]

CHOLERA, IMMORALITY, AND SANITATION

On June 15, 1832, when word reached New York City that cholera had breached the Atlantic and broken out in Montreal and other parts of Quebec, New Yorkers went on high alert. Based on the lessons learned from the yellow fever epidemics, the city's Medical Society formulated a program for public hygiene. They suggested that the streets be swept and kept clean; streets, yards, sinks, and cesspools should be disinfected with quicklime; water should be flushed from the hydrants several times a week; and individuals should wash frequently and remain calm and temperate in drinking and dining. A new system of street cleaning was instituted, and householders were asked to clean and purify their buildings and yards. However, within a week the discarded garbage was piled so high on the streets that the cleanup program was halted. The water quality was so poor that the well-to-do had spring water delivered to their residences; the poor, on the other hand, relied on water from the city pumps and other contaminated water sources.[51]

The first case of cholera was diagnosed in New York on June 26. In July, when cholera broke out at the almshouse, the city released all prisoners who had been locked up on misdemeanor charges. The belief in the retribution thesis was entrenched. Groups like the Special Medical Council reinforced this notion by making daily announcements such as the one appearing on July 10 claiming that forty-five people had died and that "the disease in the city is confined to the imprudent, the intemperate, and to those who injure themselves by taking improper medicines."[52] The Special Medical Council also drew up recommendations which were published in the newspapers and distributed on flyers:

Notice

Be temperate in eating and drinking,

Avoid crude *vegetables* and *fruits*;

Abstain from *cold water*, when heated;

And above all from *ardent spirits* and

if habit have rendered it indispensable, take much less than usual.

Sleep and clothe warm

Avoid labor in the heat of the day.

Do not sleep or sit in a draught of air when heated.

Avoid getting wet

Take no medicines without advice.[53]

At first cholera was seen as an affliction of the poor and sinful whose behavior and lifestyle predisposed them to being infected. These assumptions were reinforced when the disease hit slum dwellers, particularly the Irish and blacks, very hard. As evangelicals, the press, and some medical experts saw it, the disease was caused by sinfulness; it was the rod used to smite transgressors in retribution for their sins. In the 1830s no one recognized that it was the poor sanitary conditions of cities that hastened the transmission of diseases.[54]

Even before cholera hit American cities the disease was linked to those with "barbarous habits," the weak, immoral, and imprudent in Europe. In Paris, where 18,000 people perished in the 1832 outbreak, it was reported that 1,300 of 1,400 prostitutes had died from the disease. In other European cities the disease killed many drunkards. In America the large numbers of deaths in the slums of New York and other cities confirmed the disease's reputation of afflicting the poor and immoral.[55] As one newspaper reported, "The disease is now, more than before rioting in the haunts of infamy and pollution. A prostitute . . . who was decking herself before the glass at 1 o'clock yesterday, was carried away in a hearse at half past three o'clock. The broken down constitutions of these miserable creatures, perish almost instantly on the attack. . . . But the business part of our population, in general, appear to be in perfect health and security."[56]

In letters to his daughter John Pintard, a diarist, prominent merchant, banker, founder of the New York Historical Society, and secretary of the American Bible Society, noted that the city would be very alarmed if cholera attacked "regular householders." He was thankful that the disease was "almost exclusively confined to the lower classes of intemperate dissolute & filthy people huddled together like swine in their polluted habitations." In another letter Pintard declared that the disease attacked "chiefly . . . the very scum of the city."[57]

Eventually the city's Board of Health created five makeshift cholera hospitals, but before the epidemic abated it had killed 3,513 people. Even as cartloads of coffins were hauled to the cemeteries, dead bodies lay in the gutters. At the same time the air was filled with acrid smoke as residents burned the clothing and bedding of the dead.[58] Other cities were also hit hard. The disease killed 5,000 people in New Orleans, and 853 died in Baltimore.[59] The 1849 outbreak took a heavy toll also; 4,822 died in New Orleans; and 400 died in Memphis. Another 2,509 New Orleans residents died in the 1854 outbreak, and 1,875 died in 1866. The last cholera outbreak occurred in New York City in 1892; fewer than 120 people died from it.[60]

By 1866, even before the cause of the disease was clearly understood, many had concluded that the disease resulted from poor sanitation. While evangelists in 1832 had urged people to repent and behave morally as a means of escaping or surviving the disease, by the 1850s it was understood that prayer and religious conversion were not the answer, and sanitarians urged cities to adopt stricter environmental reforms to prevent the spread of the disease.[61]

FLEEING FROM CHOLERA

As news of the 1832 cholera outbreak in Montreal reached New York City, fifty thousand New Yorkers packed hastily and fled in steamboats and stage coaches or piled their belongings on carts or wheelbarrows or on their back and left on foot. Many physicians joined the exodus; those remaining charged exorbitant fees for their services.[62] As the *Evening Post* reported, "The roads, in all directions, were lined with well-filled stage coaches, livery coaches, private vehicles and equestrians, all panic struck, fleeing from the city."[63] By the end of July just about anyone who could afford to leave the city had done so. The wealthy fled to seaside resorts such as Marine Pavilion in Far Rockaway in the southeastern section of Queens. All accommodations within a thirty-mile radius of the city were filled. Many of those who left their houses unguarded would come back to properties that had been looted and vandalized.[64]

By 1833 small communities outside of the major cities were reluctant to host urban refugees fleeing the scourge of cholera. Several people suspected of being infected with cholera and the man who sheltered them were murdered in Chester, Pennsylvania. Armed Rhode Islanders prevented New Yorkers fleeing across the Long Island Sound from coming ashore. Armed residents of Ypsilanti, Michigan, fired on the mail stage from cholera-infested Detroit.[65]

The poor couldn't afford to flee. As waves of epidemics hit the slums the New York City Board of Health decided that reducing the population of the slums would at least slow the spread of the disease. Slumlords interested in collecting rents from overcrowded hovels opposed the Board of Health, even filing lawsuits to prevent the evacuation of the poor. The Board of Health formed the Committee to Provide Suitable Accommodations for the Destitute Poor to find affordable housing. The committee rented several buildings and supervised the construction of housing in six locations in the city. Blacks were housed in poorly constructed shanties that leaked when it rained.[66]

CAUSES AND ROUTES OF EXPOSURE

As early as 1849 John Snow, a London physician, theorized that cholera was spread through contaminated water supplies. Snow had treated victims of a cholera outbreak among miners in Newcastle in 1831. When he toured the Killingworth colliery he found that the conditions were deplorable: workers were eating and defecating in the same areas of small, stifling caverns. He was intrigued by cholera and began looking for clues to its origins and routes of transmission. During the 1848–1849 London epidemic Snow observed that the disease tended to concentrate in the poorest areas and to break out in the late summer, when people drank large quantities of unboiled water. He began taking samples to study the water quality in these areas; he suspected that the water was the source of the illness and that it was being contaminated with something, he wasn't sure what, that caused the disease.[67]

Snow's belief that cholera was caused by waterborne contaminants was not typical at the time, when the leading doctors and civic leaders were ardent miasmatists. They were convinced that diseases such as cholera attacked the poor because of their moral failings. Consequently they were dismissive of other theories of disease transmission, including Snow's.

As cholera ravaged London Snow noted that different water companies served different sections of London. Water, filtered through sand, was piped to homes in upscale West End homes in Chelsea by the Chelsea Company. The Southwark and Vauxhall companies delivered unfiltered water drawn from the Thames from the tidal reach to communal cisterns and pumps in Lambeth and Southwark. The tidal reach was contaminated with London's sewage, so much so that Parliament passed legislation in the early 1850s ordering water companies to draw their water above the tidewater mark by 1855. Southwark and Vauxhall were the two companies that were not in compliance with the law. As a result, Snow was able to study the rates of cholera in groups of people drinking from different water supplies. He found that those drinking the water on the south side of the Thames River (in neighborhoods like Lambeth and Southwark) were twice as likely to get cholera than people on the north side of the river.[68]

In 1854 an outbreak of cholera swept through the densely populated and impoverished Soho section of London. The outbreak, which killed almost seven hundred people living within 250 yards of the Broad Street pump in less than two weeks, provided Snow, whose practice was only six blocks from the epicenter of the outbreak, with the *experimentum crucis* he had been waiting for to furnish convincing evidence that cholera was caused by pathogens that contaminated water sources. The outbreak was confined to one area of town, and deaths seemed to radiate from a central location around a single pump. Snow needed to prove that those who died had consumed water from that particular pump, while others living in the same area, inhaling the same foul miasma, but who did not consume the suspect well water did not fall ill. Though Snow was able to amass an impressive amount of data from door-to-door interviews, mortality statistics, and maps indicating that the outbreak was caused by contamination of the Broad Street water pump, he probably would not have been able to sway the miasmatists without the work of a clergyman, Henry Whitehead. Though Whitehead was a miasmatist who was initially skeptical of Snow's theory of the waterborne transmission of cholera, he was open to exploring the idea. Like Snow, he had been investigating the outbreak by visiting the sick and interviewing neighborhood residents. The two met when they began working on a commission to investigate the cause of the outbreak. It was Whitehead whose detective work provided the crucial piece of evidence that Snow needed to solidify his case. Whitehead identified the first individual, a baby, to succumb to the disease in 1854, and after talking with the baby's mother found out that she had discarded excrement and bodily fluids from the sick infant in a cesspool located only a few feet from the Broad Street pump. An investigation of the cesspool showed that its contents had leaked into the water at the pump. Together Snow and Whitehead were able to establish a definitive link between sewage contamination of water sources and the transmission of cholera. The publicity surrounding Snow's investigation led to the passage of a series of bills that forced the city to clean up its water supply and sewage system. Once this occurred cholera rapidly disappeared from London.[69]

Unbeknown to Snow, an Italian anatomist, Filippo Pacini, who sought to find the organism responsible for the 1854 outbreak in Florence, had discovered the cholera bacteria, which he named *Vibrio*. Instead of examining water samples as Snow did, Pacini examined histology slides of the intestinal mucosa of patients on whom he conducted autopsies. Pacini detected the cholera bacteria and published his findings in a paper titled "Microscopical Observations and Pathological Deductions on Cholera." Like Snow's, Pacini's results were dismissed because the miasma theory was the dominant paradigm in Italy too. Consequently his findings were not widely circulated,[70] and his discovery lay dormant for decades. It wasn't until 1883 that Robert Koch, a German scientist working in Egypt, isolated the organism, *Vibrio cholorae* (formerly *Vibrio comma*), that caused cholera. Once the bacterium found its way to the intestine, it produced the disease that killed about 50 percent of those infected.

Cholera is spread along any pathway leading to the human digestive tract. Unwashed hands or fruits and vegetables or uncooked food can cause infection. However, ingestion of sewage-contaminated water was responsible for the most widespread and severe cholera outbreaks.[71]

Water and salts normally traverse the intestinal wall in both directions—enough to keep the fecal matter in the bowels moist without causing dehydration. However, the toxin secreted by the cholera bacillus disturbs this balance, resulting in the expulsion of copious quantities of water into the gut's interior. As a result victims pass large quantities of fluids and experience violent convulsions.[72] Ingestion of one or a few cholera bacteria is not enough to cause infections. Since a glass of water may contain up to two hundred million cholera bacteria, all it took was the consumption of a small amount of contaminated water to become ill.[73]

Yellow Fever, Cholera, and Public Health Reforms

Epidemics created significant public health problems for municipalities. In New York, for instance, the piles of garbage removed from the streets were dumped into the harbor in places where the current moved slowly. During the epidemics the bone-boiling facilities that normally disposed of thousands of carcasses each day were ordered closed because they were considered a menace to public health. Consequently animal remains were also dumped into the harbor. In addition the city had difficulty coping with the large quantities of dead bodies. After the dead were removed from the streets or their homes the bodies were taken to the cemetery on Randall's Island, where they were buried in long, deep trenches, one body stacked on top of another, to within a foot or two of the surface. Rats swarmed over the island, unearthing the contents of the trenches.[74]

IRISH AND BLACK WOMEN AND THE CAMPAIGNS AGAINST PIGS

The city also launched a campaign against pigs. Though physicians warned that it was unhealthy to have pigs sharing living quarters with people, living in cellars and yards overflowing with sewage, or wandering the streets, poor families ignored these warnings. They did so because the pig was a vital part of the working-class diet. At the time poor families ate primarily potatoes, corn, peas, beans, cabbage, and milk from cows fed on swill, a by-product of the breweries. Small amounts of salted meat, cheese, butter, sugar, coffee, and tea supplemented this diet in times of plenty. Many working-class women kept their own animals to provide meat for the family; pigs fattened on garbage were a source of cheap food for such families. Thus whenever the police tried to round up the pigs to remove them from within the city limits they encountered determined women who hid their pigs in every conceivable spot.[75]

Civic leaders in New York began a concerted effort to remove the pigs from the slums in 1818. In that year Mayor Cadwallader Colden argued, "Our wives and daughters [upper-class women] cannot walk abroad through the streets of the city without encountering the most disgusting spectacles of these animals indulging the propensities of nature [defecating, urinating, and rooting through litter]." He empanelled a grand jury which indicted a butcher, Christian Harriet, for keeping his pigs on the streets. At the grand jury hearing Harriet's lawyer argued that though the pigs might offend upper-class ladies and dandies, without their hogs many poor families would have to resort to begging or risk being sent to the almshouse. Mayor Colden ruled that it was irrelevant to consider the subsistence value of the pig and Harriet was convicted. The Common Council ordered a roundup of pigs in 1821, but when officials began seizing hogs hundreds of Irish and black women quickly mobilized and freed the animals. There were other pig roundups in 1825, 1826, 1830, and 1832. In each case the women succeeded in retrieving their pigs.[76]

During the 1849 cholera epidemic the city got much tougher on pig owners. Subterfuge, riots, and protests notwithstanding, in June of that year up to six thousand pigs were removed from the slums by midmonth. By the end of the summer an estimated twenty thousand pigs had been relocated to the northern boundaries of the city. Bounties were also placed on stray dogs; as a result 3,520 dogs were killed in the streets, most by club-wielding boys. Then, between 1851 and 1852, all the bone-boiling facilities were expelled from Lower Manhattan.[77]

CATTLE IN THE CITY

While civic leaders expelled pigs from the city and exterminated dogs, cows got a reprieve. Since cattle and slaughterhouses were owned by merchants, the city was more reluctant to round up and expel the cows in the same manner in which they removed the pigs. Moreover cattle were not sharing living quarters with people. However, removing cattle would have reduced wastes in the city by significant amounts. Finally, in 1853, the city passed an ordinance that banned cattle drives south of Forty-second Street. At the time New York was the largest center of beef production in the country; Manhattan's 206 slaughterhouses butchered 375,000 animals annually. In August 1853 alone, a city scavenger reported clearing away the carcasses of 2,281 animals, 1,303 tons of offal, and 62 tons of bones from the slaughterhouses. In 1870 the Board of Health banned slaughterhouses between Second and Tenth Avenues. The slaughterhouses were replaced by abattoirs. In 1877 the city designated an offal pier on West Thirty-eighth Street to service the four major abattoirs in the area. Nonetheless, 52 small, private slaughterhouses continued to operate in the city. Between 1884 and 1885 the Ladies' Health Protective Association organized a campaign against the slaughterhouses. In 1898 slaughtering in Manhattan was further restricted to a few areas of the city.[78]

THE IMPETUS FOR CHANGE

While yellow fever affected southern cities to a greater extent than cities in the North, cholera had almost the opposite effect. Consequently yellow fever became the impetus for sanitation and public health reforms in southern cities, while both diseases provided that stimulus in the Northeast. Though there were cholera outbreaks in New Orleans, Richmond, Nashville, and Memphis (which stimulated reform in all but the first) it was yellow fever that pushed civic leaders in New Orleans, Mobile, Pensacola, Savannah, Charleston, and Norfolk to adopt sanitary reforms.[79]

The Bubonic Plague

The plague struck terror wherever it appeared. Between A.D. 541 and 542 it killed seventy thousand people in Constantinople alone and millions of others elsewhere in Europe. It killed more than a fourth of the world's population during the middle of the fourteenth century; approximately 40 percent of the English population died from the plague in the 1348–1349 epidemic. From the late fourteenth century until the end of the eighteenth recurring outbreaks killed fifty million people in Europe, including 30 to 60 percent of the population of Genoa, Milan, Padua, Venice, and Lyons. The last pandemic of the bubonic plague occurred during the 1890s, but unlike earlier outbreaks, this time the disease made landfall in the United States. This latest epidemic broke out in south-central China before 1870 and by the end of the decade had ravaged the region. By the early 1890s the disease spread along the trade routes and river valleys toward the coast, killing thousands of people in Canton in 1893 and in Hong Kong in 1894. Identified as *Yersinia pestis* or black death, the bubonic plague was quite virulent; more than 75 percent of those infected with it died; in Hong Kong the rate was 95 percent. In all, the pandemic of the 1890s killed an estimated twenty million people. It spread through India from 1898 to 1948, killing thirteen million.[80]

SYMPTOMS AND MODE OF TRANSMISSION

Bubonic plague is caused by a bacterium known as *pestis*. The first symptoms are high fever, intense pain, chills, headache, and nausea. The symptoms often receded after several hours and patients are lulled into thinking they are recovering. However, after a day or two the symptoms return suddenly and with greater intensity. When victims relapse, their body temperature skyrockets and they suffer excruciating pain and almost complete loss of energy. Additional symptoms are rapid and progressive tumescence and hardening of the lymph glands (buboes) in the groin or neck, under the armpits, or in the inner thighs; diarrhea and vomiting; lethargy, mental stupor, speech disorders, delirium, and insomnia; rapid pulse, seizures, and ruptured blood vessels; liver and spleen disorders; and skin lesions. Some patients die quickly from toxic shock; others linger for up to five days in semi-consciousness or a coma before dying. Internal bleeding often leaves black marks on the bodies of the dead; hence the nickname black death.[81]

Though advances in bacteriology allowed scientists to isolate the bacteria causing the plague by the time of the latest pandemic, they still did not know how the disease infected humans. It wasn't until the early twentieth century that scientists determined that the disease is transmitted when infected rats are bitten by fleas. The flea sucks the blood containing the *pestis* bacteria, which multiply rapidly, forming a clot that causes constriction and blockage of the foregut. The flea is forced to expel the clot in order to suck fresh blood. In so doing the flea expectorates the infected clot into the animals or humans it bites, thereby spreading the disease. The flea desiccates quickly in hot dry weather if it is away from a host. It thrives when the humidity is above 65 percent and temperatures are between 26 and 37 Celsius; under ideal conditions it can survive for about six months without feeding.[82]

Rats were suspected long before it became accepted wisdom that they were involved in the transmission of the bubonic plague. In 1894 Dr. Mary Miles in Canton, China, reported that large numbers of rats died during plague epidemics. Her warning was ignored because it was thought that the rats caught the disease from humans. That same year Dr. Shibasaburo Kitasato and Dr. Alexandre Yersin published findings showing that they had independently identified the bacillus that caused the plague. In 1897 a Japanese doctor, Masaonori Ogata, argued that when rat fleas leave their dead hosts they can transmit the disease to humans. Ogata was ignored; so was Paul Louis Simmond who proved that rat fleas bit humans. This body of thought was not taken seriously until a British commission published a report in 1908.[83]

RACIALIZATION OF THE DISEASE

Though by the 1890s reformers knew enough about epidemics to recognize that they weren't a form of divine retribution, the plague was still framed in racial and class terms and was linked to the lifestyle of poor nonwhites. The pandemic of the 1890s was seen as an Asian disease incubated and transmitted in filthy conditions; whites seemed to forget that the plague had ravaged Europe in earlier outbreaks and believed they were immune to the disease. With the 1890s outbreaks observers noted that the disease seemed to strike hardest in areas with poor sanitation: for instance, in China, India, and the Middle East the filthiest slums in the dirtiest cities suffered the highest death toll. There also seemed to be a racial pattern to the disease: while many people living in China and Southeast Asia died of the disease, Europeans (colonial officers and their families) living in the same region did not succumb. Because observers were ignorant of the way the disease spread and did not consider that the lifestyles and social locations of colonialists and poor Asians differed dramatically, they concluded that Asians were more susceptible to the disease than Europeans. There was also a general conclusion that Asians might be carriers of the disease.

Now we know that colonial officers and their families were not infected because they did not live in filthy, rat-infested slums. Moreover they wore long pants tucked into their boots, making it more difficult for fleas to bite them if they did come in contact with the insect. In contrast poor slum dwellers often walked around with their legs exposed and feet bare, making it easy for fleas to bite and transmit the disease to them.[84]

THE HONOLULU EXPERIENCE

The pattern of infection and death from the plague dovetailed with the racial theories prevalent in the United States. The plague strengthened the arguments of white supremacists who believed that Europeans were superior to other races and had somehow developed immunity to the disease. Thus when it was first reported that someone died from the plague in Honolulu's Chinatown in December 1899, white residents of the city were confident that they were safe from the disease. The print media helped to frame the disease in racial terms. *Austin's Hawaiian Weekly* reported that Europeans and Americans living in Asia did not get infected, and other papers assured residents that the disease did not attack whites. However, for about four decades elites were concerned about the filthiness of the city and worried that the plague might break out.[85]

The plague first appeared in Honolulu in the form of a dead Chinese passenger transported aboard the *Nippon Maru* in June 1899. Suspecting a plague death, city officials took immediate action, holding the ship and its passengers and cargo at Quarantine Island for seven days. However, ignorance of the mode of transmission of the disease led to critical errors that might have been responsible for the outbreak in Honolulu. While health officials screened and quarantined people, they ignored the rats. Chinatown residents, however, noticed a large number of dead rats, particularly in areas close to the wharves. The *America Maru* docked on October 20, and by the first week of November dockworkers at the Pacific Mail pier noticed large numbers of rats darting in and out of the light and writhing in pain as they died. Workers swept hundreds of dead rats into the water. Though the number of dead rats increased steadily, no one thought to examine their tissue for signs of the disease or to link the sudden increase in rat mortality to the plague. Rat fleas prefer to jump from one rat to another, but as Honolulu's rat population dwindled the fleas began to seek out other warm-blooded hosts, including humans.[86]

By the late fall rumors of a new illness were circulating through Chinatown, though residents kept news of the scourge to themselves and turned to traditional healers in the community for answers. The memory of how shabbily they were treated during the 1895 cholera outbreak was still fresh in their minds: as soon as physicians suspected they had cholera on their hands, the Board of Health sealed off the affected parts of Chinatown, closed schools, banned all public meetings, and instituted twice-daily medical inspections of everyone living in the quarantine zone. Many of the affected Chinese residents did not speak English and so had no idea what was happening and why. The government also instituted an aggressive campaign to clean up Chinatown. They transported tons of human waste, animal remains, and garbage out to sea for disposal; toilets were cleaned and disinfected. The Board of Health instituted new standards for handling and storing food. Charities provided meals for workers who couldn't leave the quarantine zone to go to work, while the United Chinese Society issued vouchers for people to get medicines from drugstores and herbal shops. Cholera killed eighty-five people (seventy-three of whom were Hawaiians) before it petered out. While the new government of Hawaii congratulated itself for its swift and effective response to the disease, Chinatown residents were angered and frustrated by the invasion of their privacy, the financial devastation wrought by the strict quarantine, and the shoddy treatment they received at the hands of white volunteers. The episode left them with a deep mistrust of white physicians and the Board of Health.[87]

As a result of the treatment they had received during the cholera epidemic, Chinatown residents kept quiet about this new disease; some even resorted to burying the dead in secret. Though two Chinese doctors, Li Khai Fai and Kong Tai Heong, who had trained at the Canton Medical School but who were not living in Honolulu at the time of the 1895 cholera epidemic, urged Chinatown residents to report the illnesses and deaths to the Board of Health, residents ignored them. All of this changed on December 11. When traditional Chinese medicines failed to bring about any improvement in You Chong, a bookkeeper for the Wing Wo Tai Company, his employer sent for an English physician. When You Chong died the physician ordered an autopsy, which revealed bubonic plague, Chinatown residents were unable to contain their secret any longer. Within a day four more deaths were reported.[88]

Newspaper headlines on December 13 blaring the arrival of the plague in Honolulu horrified the city's white community. Whites did not fear that there would be widespread outbreaks in their own community; they were distressed that outsiders might link their city with other poor Asian cities affected by the plague. As the *Hawaiian Gazette* noted, "The city promises to put itself on a par with other cities, having among its people a large number of highly intelligent and progressive Anglo-Saxons."[89] Honolulu's white residents were trying to put behind them descriptions like the following:

> Honolulu in 1820 was an ugly, barren, hot, and feculent town, a straggle of grass huts and a very few adobe stone houses. Its residents sweltered amid clouds of dust or skidded along in mud often ankle-deep. . . . Foreigners . . . [were] . . . too impatient for an early escape from this dismal port to care about beauty, or comfort. Natives, knowing nothing about

> towns in other parts of the world, thought that this was how all seaports must be. . . . Honolulu's residents seemed determined to earn for their town the name by which eventually it was known to sailors throughout the seven seas: The Cesspit of the Pacific.[90]

Whites believed that the disease hit hardest where filth and chaos reigned and among the debased; they didn't want Honolulu to be seen in this light. They also feared that an outbreak of the epidemic in Honolulu might jeopardize Hawaii's chance at statehood.[91]

It was in this context that Honolulu policy makers, the Board of Health in particular, took bold steps to assert their authority, create a perception of order in the city, and project that perception to the rest of the world. Elites have long understood the need for order in times of crises that are likely to accompany the outbreaks of virulent diseases like the plague. Sheldon Watts argues that the Italians established an "ideology of order" in Italy during the 1448 plague epidemic in which civic leaders and other elites exercised control over the lower class. This model spread to other European countries.[92]

Once the plague deaths were confirmed the Board of Health moved quickly to establish quarantines and place bans on the movement of people and goods. Fearing that ships might carry the disease to other islands or that there might be a stampede of folks trying to leave Honolulu, the Board put an immediate halt to all inter-island shipping and placed travel restrictions on anyone trying to leave or enter Honolulu. Believing that Asians were carriers of the disease, they imposed an absolute ban on travel by any Chinese or Japanese traveling to or from Honolulu (non-Asians could get a travel permit if they underwent an examination by a physician appointed by the Board). The Board of Health also established a quarantine zone encompassing most of Chinatown. No one could enter or leave the zone without permission from agents appointed by the Board. The national guard, composed primarily of Native Hawaiians, was called in to patrol the perimeter of the quarantine zone to see to it that no one breached it. The highest concentration of national guardsmen was stationed along Nuuanu Street, marking the border between Chinatown and the city's bustling commercial center, which was dominated by white-owned businesses. Once the quarantine was announced on December 12 no one residing inside the district could leave until further notice. Lines were painted down the middle of boundary streets to prevent people from wandering in and out of the zone. Schools and restaurants were closed. Even work on the sewer system that was being built ground to a halt as the workers—who were Chinese and Japanese—could not get to work.[93]

In 1900 Honolulu had a population of 44,252 residents: 12,820 were Hawaiians; 10,741 were Chinese; 7,298 were Japanese; 5,466 were Portuguese; and 7,927 were foreigners from elsewhere. Chinatown was also home to 72 of the city's 153 Chinese stores. Almost a third, or fifteen hundred, of Chinatown's residents were placed under quarantine in December. Of those quarantined, about a third were Japanese. Most of the Chinese in the quarantined district lived in the low-lying areas close to the wharves, while most of the Japanese lived in the newer developments in the upper part of the district.[94]

Chinatown was the city's dirtiest slum. In 1886 a devastating restaurant fire had burned most of Chinatown. In the aftermath the government decreed that all new buildings be constructed of brick or stone and built to sanitary code; however, these orders were ignored. Small, poorly constructed wooden one- and two-story buildings, shacks, storage sheds, livestock pens, and chicken coops were crammed into the area. The many rooming houses sheltered scores of single men in overcrowded conditions. Very little fresh water from the hills drained into the area; nonetheless low-lying parts of the neighborhood were covered with muck and stagnant water year round. Overflowing outhouses and cesspools as well as putrefying garbage added to the noxious effluvia and unsanitary conditions that pervaded the community. The *Hawaiian Gazette* described Chinatown as a "pesthole . . . where rats, the agents which spread the plague, are abundant." Despite the problems of overcrowding and lack of sanitation in Chinatown, not all of the residents or the merchants owning businesses in the community were poor. A small group of powerful merchants, manufacturers, and financiers owned businesses and lived in Chinatown. These entrepreneurs wielded considerable influence there.[95]

The effects of the quarantine were felt almost immediately by Chinatown residents. Workers couldn't leave the quarantine zone to go to work, not even on the docks down the street; fishermen couldn't go fishing, and stores closed because they couldn't get goods in or out. Families could not replenish meat and other goods after they exhausted their supplies. Price gouging was common, but officials paid no attention. Though charitable organizations got food to people inside the quarantine zone, there were still intense hardships. To add to their misery, quarantined residents had to submit to twice-daily inspections by medical personnel of the Board of Health. These intimate bodily inspections, conducted by complete strangers, were intrusive and sometimes carried out in plain sight of others. When residents locked their doors the inspectors kicked the doors in to conduct the examinations. Chinatown residents were also ordered to clean up their neighborhood; they worked alongside volunteers to remove the garbage and sewage from streets and yards.[96]

Soon the Chinese in Chinatown began to make accusations of racial discrimination. They accused the Chinese consulate of ignoring their needs and failing to protect their rights. The Chinatown Chinese community also questioned why the quarantine line had been gerrymandered to include Chung Kun Ai's new waterfront establishment, the City Mill, and other Chinese-owned facilities but had excluded the white-owned Honolulu Iron Works, which was surrounded by Chinese-owned facilities. Rumors were rampant in Chinatown that workers at Iron Works were being allowed to slip in and out of the quarantine zone to go to work. Chinese and Japanese merchants decried the devastating economic effects the quarantine was having on their businesses. Noting that businesses elsewhere in the city were allowed to operate freely they saw the restrictions as a way for white businesses to limit the economic clout of Asian businesses.[97]

In the meantime the Board of Health took steps to expeditiously dispose of the bodies of those who died of the plague. Because it was believed that the plague bacteria could survive in the soil or groundwater and spread infection that way, the corpses were immediately cremated. At first they were burned in a spare furnace at Iron Works, but after a few days Iron Works laborers built a crematory on Quarantine Island to burn the bodies. These cremations violated a widely practiced cultural tradition of sending the dead to China for their final resting place. If the family did not have enough money to ship the body to China, it was usually buried temporarily until enough money could be found. Cremation made this impossible. Not surprisingly, Chinese residents of Chinatown began hiding their relatives when they became ill and surreptitiously buried them without reporting that the plague had killed them.[98]

After the first five deaths occurred in a twenty-four-hour period, no other deaths were reported for several days. A sense of relief spread throughout Honolulu, and some even questioned whether the plague had caused any deaths at all. Some believed that the vigorous cleanup efforts in Chinatown had prevented the plague's spread. During this lull newspapers, doctors, and others took the opportunity to question why the unsanitary conditions in Chinatown had been allowed to get so bad. Some blamed the Dole government in Hawaii for disregarding Chinatown; others blamed the three doctors who ran the Board of Health (Nathaniel B. Emerson, Clifford B. Wood, and Francis R. Day) for paying too much attention to the inspection of imported goods while ignoring the enforcement of the city's health and housing codes; still others blamed slumlords who profited from unsanitary dwellings.[99]

On December 17, after no deaths had occurred for five days, the Board of Health loosened quarantine restrictions. Trams resumed running through

Chinatown with the proviso that they not stop in the quarantine district. The next day the Board voted to lift the quarantine completely. Then news filtered out that a woman in the district had died of the plague. Doctors broke down the door of the house in which she lived and immediately examined the body, relieved to discover that the woman had died from injuries sustained in a fall. Then a genuine case of the plague was reported. This time the victim was a white teenager, Ethel Johnson, who lived in a respectable neighborhood outside of Chinatown. Doctors speculated that she had contracted the disease from the garbage being hauled past her house to a nearby seaside dump. They prohibited the transportation of garbage in residential areas, and on December 20 finally lifted the quarantine ban on Chinatown.[100]

There were rumors that traditional healers and morticians were helping people to disguise plague deaths. Some Chinese, fearing that their home would be quarantined if someone died in it, took dead plague victims elsewhere in the city. Then, on December 24, word leaked out of Chinatown that a Chinese resident, Ah Fong, had died of the plague. Another Chinese, Chong Mon Dow, died of the plague on Christmas day. The Board of Health members abandoned their Christmas dinners to meet and discuss the resurgence of the plague. Fearing the reinstatement of a quarantine, Chinese and Japanese residents of Chinatown fled to relatives and friends in other parts of the city.[101]

Heeding the advice of fellow doctors who suspected that rats were transmitting the plague, the Board of Health announced a bounty of ten cents apiece for dead rats. On December 27 the newspapers printed a flyer circulated by the Board listing "Precautions against Bubonic Plague" that Honolulu residents should heed. The flyer warned that "plague germs flourish in filth, in garbage and in damp, dark or foul places." Residents of the city were told to wash frequently, to cover up all cuts and scrapes, and to cook food properly. They were warned that the plague can reside in the soil and can attack from the ground up: "It is dangerous to go barefoot in times like this; wear shoes." Residents were urged to "destroy all the rats and vermin on [their] premises" so that "the danger of the plague" would decrease.

In the early hours of December 28, without warning, the Board of Health called in the national guard and reinstated the quarantine in Chinatown. Chinatown residents awoke to national guardsmen with bayonets fixed patrolling the perimeter of the quarantine zone. As the newspaper *Ke Aloha Aina* noted, "While people . . . were sleeping in their homes, they suddenly became surrounded by government troops who were sent to guard the area with their rifles projecting everywhere. Residents were panic-stricken, but helpless to defend themselves from the will of the government." Though the newspaper was sympathetic to the plight of Chinatown residents, it repeated the Board's line that the quarantine was reinstated because the Chinese were sending plague victims outside of Chinatown to die elsewhere.[102]

The three doctors who sat on the Board of Health did not believe that the quarantine by itself could stem the spread of the plague, so they looked for other means of stopping the disease. They delegated the task of developing a plan for improving sanitary conditions in Chinatown to an ad hoc committee, which came up with a ten-point plan that called for the construction of a garbage incinerator, improved site grading and sewage treatment, new zoning laws and more vigilant oversight of the laws, and other improvements. The Board approved the plan and allotted money to execute it. At the same time the Board tried fumigating buildings by burning sulfur or disinfecting them with formalin and other acid washes. They then came up with two new measures for controlling the disease: quarantining people who had plague victims in their homes, then burning the belongings and property of the evacuees. When questions were raised about the legality of seizing and torching property, the attorney general of Hawaii suggested that the Board of Health had two choices: they could give landowners and residents time to make necessary improvements to their property, or they could destroy the property and deal with the legal consequences later. The Board of Health pursued the second option.[103]

The deliberate use of fire was a new tactic in the fight against the plague. In earlier epidemics the belongings of victims were sometimes burned because it was thought such items spread contagion. Some accidental fires, such as one in London in 1666, did seem to help halt the spread of the plague. Doctors supported the systematic use of fire because they believed the *pestis* bacteria survived inside buildings, behind walls, and beneath the floor, so burning the structure and its contents was an effective way of killing the germs. The Honolulu doctors weren't the first to use fire. In fact, they were following the lead of the British, who had burned 6.25 acres (containing 384 houses) of the Taipingshan district of Hong Kong in 1894 as way of fighting the plague. To reduce the likelihood of fire spreading beyond the intended target, the British demolished the homes first, then burned the rubble.[104]

In Honolulu the Board of Health supervised the first burn in Chinatown on New Year's Eve, 1899. The three doctors inspected the building, in which a plague victim had died, condemned it, and at 11 a.m. ordered it to be incinerated. Occupants were forced to evacuate the building immediately. The fire department burned the building at 3 p.m. that day, leaving eighty-five

Chinese, Japanese, and Native Hawaiians homeless. All the evacuees were taken to a quarantine station. This exercise was repeated several more times on New Year's Day and in the days following. The *Pacific Commercial Advertiser* heartily supported the prescribed burns and what it saw as a sound plan for "fighting the devil with fire" because "destruction was the only certain disinfectant." The *Hawaiian Gazette* also argued, "The cost of burning down whole blocks of houses, if necessary, for the public good and to save life will not be objected to by thoughtful people."[105]

Chinatown property owners, some of whom were wealthy white landowners who leased their property to residents, were outraged and immediately began legal proceedings. Chinese and Japanese landowners also protested and sought legal advice. Asian doctors inside Chinatown continued to cooperate with the Board of Health and refer patients suspected of suffering from the plague to the Board for transference to the plague hospital. Chinatown residents, hostile to their own doctors for cooperating with the Board during the first quarantine, vented even more of their anger on Japanese and Chinese doctors, who were seen as pawns of the Board. One Japanese physician was attacked by a mob of Japanese and barely escaped with his life. Flyers appeared all over Chinatown threatening death to the "agents" of the Board of Health and to Chinatown doctors who cooperated with the Board. One doctor was mobbed in quarantine camp. Native Hawaiians also protested. They argued that once a site was condemned, "men were taken as if prisoners of war by lines of military guards and women were taken away by wagon under guard to be placed under quarantine." Native Hawaiians also objected to being driven to quarantine camps in the same wagons used to transport plague-infested corpses to the crematory.[106]

Though the death toll from the plague was relatively low, the Board of Health went forward with the policy of burning the homes of victims. By January 4, 1900, there had been nineteen known plague deaths. However, reports of more plague deaths continued to filter in daily, as many as five on some days. Most of the victims were Chinese and Japanese; a few were Native Hawaiian. Some of the inspectors working for the Board of Health, members of the Citizens' Sanitary Commission, and white merchants pushed for more aggressive burning; they wanted all unsanitary structures burned whether or not plague deaths had occurred in them. Some wanted to expand the burning, arguing that after the first quarantine was lifted many Asians had moved away from Chinatown to live in other Asian neighborhoods; these neighborhoods were now so densely populated that they were in danger of becoming public health menaces—in short, miniature Chinatowns.[107]

In the meantime *Ke Aloha Aina* decried the treatment of Native Hawaiians and other Chinatown residents. It framed the whole operation in military terms as it exposed the inhumane treatment of quarantine victims: "Those whose homes were taken and burned are like prisoners of war." The paper reported that Native Hawaiians were being taken by armed agents of the Citizens' Sanitary Commission and hustled off to armed quarantine camps. The Japanese, Chinese, Gilbertese, and Indians were treated similarly. Some of the people forced to evacuate had "tears streaming down their faces" as they walked the gantlet of "white men and women . . . taunting those being led off by the government guards."[108]

The Citizens' Sanitary Commission, charged with the task of monitoring the health of Chinatown residents, was authorized to enter residents' homes at will to conduct twice-daily examinations. Commission agents were not popular in Chinatown; they were accused of stealing, racial harassment, invasion of privacy, and sexual assault (there was a charge of attempted rape after an agent entered the home of a young white woman). All evacuees were processed at a disinfection station before being sent to quarantine camps. At the disinfection station people were stripped of all their belongings, examined by medical officers, and given a disinfecting shower. Men, women (some menstruating), and children were given intimate physical examinations in plain view. After the shower evacuees were provided with government-issued fumigated clothing. White guards, who did not return valuables taken from evacuees, stood by and watched as hundreds of people, upwards of 250 in a single morning, were processed in this manner. *Ke Aloha Aina* campaigned to expose the dehumanizing conditions of the disinfection stations and quarantine camps: "The wife of the president of the Board of Health would be treated better had she been among these Hawaiian women who are subjected to abuse. What are his policies regarding these lustful inspectors? We imagine he would quickly order a halt if such things were happening to his wife. Can he not just as quickly put an end to the glaring of the inspectors at the privates of men and women and the seizure of their belongings?"[109]

The Japanese consul, Saito Miki, also complained about the shoddy treatment of the Japanese. He described evacuees being "driven along the street by the guards who had a stick in their hands as though they were driving sheep." He also objected to the practice of putting naked men and women together in the fumigation line and complained to the Board of Health about the extensive losses that merchants were suffering. Their goods were confiscated with the promise that they would be fumigated, stored in warehouses, and returned to the owners after the quarantine was lifted. However, the goods of Chinese and Japanese merchants were stolen while in storage, all the perishable items spoiled, and some merchandise was damaged. Dr. Wood of the Board of Health denied that Japanese goods were being intentionally destroyed or that there had been "any ill treatment of the Japanese of any kind." Nevertheless, the *Pacific Commercial Advertiser* ran an article detailing the special privileges a French woman who was in quarantine received and the harsh treatment meted out to the Japanese.[110]

On January 14 the newspapers reported that forty-six-year-old Sarah Boardman, an art director living in an upscale section of the Nuuanu Valley of Honolulu, had the plague. It was believed that Boardman contracted the plague at a dry goods store on Fort Street that had recently been infested with rats. The news that the plague had infected a white victim in a prosperous part of town was unsettling for white Honolulu residents. All those present in the Boardman household were sent to a quarantine site, their show dogs were chloroformed, and a quarantine was imposed on the lower part of the Nuuanu Valley. Whites placed under quarantine were housed at the Queen Hotel. Wealthy residents living in the Boardmans' neighborhood were outraged that the family had to leave their home and that part of their neighborhood was under quarantine. They registered their protests with the Board of Health.

Sarah Boardman died on January 16. In response the Board of Health closed all schools and public amusement sites in the city. Prior to Boardman's death the torching of homes had occurred only in Chinatown. Her death forced the Board of Health to consider burning homes in a wealthy white neighborhood too. In the meantime the white-owned media was in a frenzy. Certain that the rat that was responsible for Boardman's death was "a Chinatown refugee," they urged the wholesale burning of Chinatown as a means of controlling the spread of the plague. Editorialists opined, "The plague does not always respect the white skin," hence Chinatown should be burned to protect those living outside of that neighborhood. The *Pacific Commercial Advertiser* demanded, "Abolish the plague spot. . . . The rotting spot must be burned out."[111]

Since the burning began in Chinatown on New Year's eve, the fire department had grown confident that they could conduct controlled burns with surgical precision. Thus they weren't worried when they set out to burn the buildings on block 15 on Saturday morning, January 20. All of the city's firefighters were on hand and all four pumping engines were on standby when the fire was lit sometime between eight and nine o'clock. At around ten o'clock the buildings were burning vigorously and everything seemed to be proceeding smoothly when the winds shifted suddenly. Instead of gentle zephyrs wafting in from the ocean, as the weather charts had predicted, strong gusts came down the *pali* slopes that rose abruptly from the edge of the city. Though the reports of the events claim that pali winds were unexpected at this time of the day and year, an employee of the Hawaii Weather Bureau interviewed in 1999 claimed that the prevailing winds in January flow down from the hills overlooking Honolulu and are funneled into the valleys leading to the city. The wind gusts became stronger as they reached the city. Within minutes the strong winds reversed the path of the fire, and the ensuing conflagration burned out of control. Flying embers lit nearby buildings not targeted for the burn. Events changed so quickly that firemen had no chance of preventing the fire from spreading. Realizing that they could not save nearby structures, they set their sights on trying to stop the fire from burning the Kaumakapili Church. The church, which had escaped destruction in the 1886 Chinatown fire, was of great religious and cultural significance to Native Hawaiians. But the firemen's efforts were fruitless. Though they sprayed the building with a steady stream of water, embers caught the steeple on fire at a point that was out of reach of the hoses. Fire enveloped the building, fed by a huge pile of clothing from condemned plague sites that had been brought to the church for fumigation and storage.[112]

Firefighters kept redeploying themselves and their equipment to new locations in Chinatown to stop the blaze, but their efforts were futile. Residents were frantic as an ever-widening swath of the fire raced through Chinatown. They tore down buildings and tried to remove flammable material in order to create firebreaks. The fire department even began blowing up buildings to create firebreaks, but the fire, propelled by vigorous pali winds, simply leaped from one building to the next and from one street to another. Soon the fire burned in all directions. One leading edge burned toward the ocean. At the same time the fire resumed its original intended course and recommenced burning toward the slopes. Fires also began burning to the east and west of the original blaze as the winds kept gusting into the afternoon. Chinatown burned until sunset.[113]

The people in Chinatown tried to rescue the young, elderly, and infirm and save any valuables they could, but the fire spread so quickly and burned so intensely that most simply ran for their lives. To their horror, they discovered that even in the face of a massive conflagration they were not allowed to flee Chinatown. As soon as the fire broke out, guards manning the quarantine zone requested backup support to keep people from escaping. Armed police officers were deployed to help keep people inside the cordon. In addition, citizens—mostly white—armed with firearms, baseball bats, pickaxes, iron bars, meat cleavers, and fence posts, swarmed the perimeter of the quarantine zone to prevent anyone from escaping. As the conflagration spread the mob on the perimeter grew larger and more frenzied. By the time the fire was spent thirty-eight acres of the fifty-acre Chinatown had burned to the ground.

The Board of Health responded to the crisis of so many homeless by releasing those in the quarantine camps who had not been exposed to the plague for nine or more days. This freed up a thousand spots. Chinatown residents made homeless by the blaze were forced to march through a gantlet of armed guards, police, and vigilantes to detention camps established for quarantine victims and to new camps that were being hastily constructed. Approximately 1,780 Chinese, 1,025 Japanese, and 1,000 Native Hawaiians were relocated to the grounds of the Kawaiahao Church. About five hundred women and children were housed inside the church. An armed posse surrounded the church until U.S. Army troops arrived at 5 P.M. Most of the Native Hawaiian fire victims were sent to the empress dowager's property or to the Boys' Brigade headquarters. About 1,200 refugees, most of whom were Japanese, were sent to the Drill Shed grounds; 3,500 people were sent to the Kalihi Beach detention camp, where they joined 1,500 people already detained there. About 1,000 people were housed at the Kerosene Warehouse site. Even as Chinatown residents were being corralled into detention camps, their stores were being pillaged by those who were not quarantined.

A day after the fire the Board of Health decided to continue the twice-daily medical inspection of Chinatown residents. Two days after the great fire, the fire department resumed burning buildings in Chinatown.[114]

After the fire wealthy whites donated liberally to relief efforts and volunteers organized clothing and food drives for the fire victims; the newspapers praised the volunteer efforts. However, many inside the camps were not as enthusiastic. As one person saw it, "This is like taking an arrow and wounding a man from a distance and then secretly go[ing] up to [him to] dress his wounds and soothe his pain in order to receive his thanks." Outraged detainees complained about the destructive fire and the humiliating disinfecting process they were subjected to. In March more than 200 destitute Hawaiians crowded into the Executive Building to seek aid from the government; they were turned away and told to return to the relief camp in the Punchbowl.[115]

Fifteen people had died from the plague during the week leading up to the fire; only one died of the disease on the day of the fire. After the fire the plague death rate leveled off and then began to decline. On January 31 the Board of Health reported that altogether fifty-four people had died of the plague up to that time: twenty-six Chinese, seventeen Native Hawaiians, nine Japanese, one German, and one American. Reports filtered out that whites in upscale neighborhoods were surreptitiously burning dead and dying rats and neglecting to report these occurrences to the Board of Health. Whites feared that they would be quarantined or have their property condemned and burned. When Ng Gee, a servant working in one of these homes, died the Board quarantined the family and ordered the property condemned and placed on the list of places to be burned. A third white victim was reported to have contracted the plague on February 3. J. Weir Robertson was a grocery store owner who had organized and participated in a rat-catching contest, which he won, with other downtown grocers a few days earlier. Robertson died of the plague soon after his illness was reported. In the meantime Asians living outside of Chinatown grew fearful as threats were made against their property and person. There were reports that Citizens' Sanitary Commission members or people pretending to be Commission members extorted gifts from Asians in exchange for excluding their businesses from the list of plague sites. There were continued complaints about the medical inspection of women conducted by the Sanitary Commission, which was composed solely of male inspectors. Though there weren't any reported cases of plague in the detainee camps, tensions rose around the city when refugees at the camps were released on February 8. Since many had no homes to return to they stayed in the camps.[116]

Given its location close to the heart of Honolulu's central business district, there was a rush to redevelop Chinatown. Board of Health officials, fearing that *pestis* might still be in the ground, wanted to douse the area with gasoline and burn it thoroughly, then seal off the area for a year before rebuilding on the site, as the British had done in Hong Kong. However, developers pressured the Board and they consented to collect soil samples and test them before proceeding with any further burns. The soil samples did not contain any plague bacteria, so redevelopment plans moved ahead. White developers pressured the city to turn over Chinatown land to them and resettle Chinatown residents elsewhere. The city resisted this pressure and set about upgrading the sewer system and other public infrastructure in the area. The original Chinatown residents were allowed to return and rebuild, this time with brick. They also began the process of making claims for their property. Approximately 6,750 claims were filed. The merchants, who filed claims first, were compensated roughly the amount they requested. However, as the process dragged on claims administrators, suspecting that some claims were fraudulent, developed more elaborate filing procedures. Asian voluntary associations such as the Japanese Victims Representation Committee and the United Chinese Society stepped in to help fire victims make their claims. In an effort to appear reasonable these organizations urged claimants to reduce the amount of their requests. In turn, the claims administrators reimbursed only about half of what was requested. Most claimants did not get reimbursed until 1903; then the amount was not enough to compensate for their losses. This further devastated Chinatown residents and left many feeling victimized several times over: by the government, which had promised full compensation at the time of the fire; by claims administrators; by the Board of Health; and by their own ethnic institutions and elites. Some Chinatown residents were so disenchanted by the whole affair that soon after the fire, rumor quickly spread among them that the Board Health had deliberately chosen a windy day to set the fire in the hopes that it would spread. Prominent Chinese citizens also publicly blamed the Board of Health for the plague. The Chinese and Japanese also attended a mass rally in April to air their grievances. In all, 71 people were infected with the disease and 61 died. On May first the *Hawaiian Gazette* published a list of all the victims. Among the dead were 25 Chinese, 13 Japanese, 15 Hawaiians, 7 whites, and 1 South Sea Islander. Men were more susceptible to the disease than women—58 men and 13 women were infected. Chinatown residents did not escape scrutiny after the epidemic waned. As the *Hawaiian Gazette* article identifying plague victims noted, "If there is one lesson that the plague has taught more forcibly than any other, it is that the dwellings of Orientals must be constantly watched. . . . The Chinese are too eager after the almighty dollar to devote either time or money to the task of keeping their shacks clean. . . . Constant watchfulness is the only possible way to keep Chinese quarters from becoming foul and dangerous, and the need of inspecting these will never cease."[117]

BURNING MAUI, SAN FRANCISCO, AND BEYOND

The Honolulu Board of Health was convinced that fire was an effective weapon in fighting the plague and urged others to adopt their techniques. They sent advice to the Board of Health in Sydney, and when seven people died of the plague in Maui's Chinatown the three-acre area was ordered condemned and burned.[118] Fire was used again in Honolulu to fight a yellow fever outbreak in 1911. A passenger aboard the *Hong Kong Maru* was found ill with yellow fever when the ship, traveling from South America to Japan, stopped for coal and supplies at Honolulu on October 21. On October 28 a Hawaiian man, one of the quarantine guards who boarded the ship to fumigate it, contracted yellow fever. The man lived at Kalihi Camp, which was no longer used as a detention center. At the time Kalihi Camp was a cluster of forty shacks built on nine acres. Its residents were immediately sent to a quarantine station, and trees within a hundred yards of the perimeter of the camp were felled. Residents' clothing was fumigated and the shacks were demolished and burned. The man infected with yellow fever recovered quickly, and no new cases were reported.[119]

During the summer of 1899 a ship from Hong Kong entered San Francisco harbor with two cases of plague on board. The ship was quarantined at Angel Island, but as was the case in Honolulu, no attention was paid to the rats that might have been aboard the ship. Two weeks after the great prescribed burn in Honolulu, the San Francisco Board of Health, fearing an outbreak of the plague, sent a representative to study events in Honolulu. Dr. Ichitaro Katsuki, born in Japan and educated at the University of California, was very impressed by the Honolulu model. Chinese living in San Francisco's Chinatown, fully aware of the fate that befell Honolulu's Chinatown, began to prepare for the worst, moving out for fear of quarantine, medical inspections, fumigation, and prescribed burns. Those who remained had packed their bags in case they needed to make a speedy exit. Plague did strike the neighborhood: on March 6, 1900, a man's plague-ridden corpse was found in a basement in Chinatown. The San Francisco Board of Health as well as the California State Board of Health immediately called for the burning of Chinatown. The entire area was quarantined and armed police officers conducted a house-to-house search for other victims. The Chinese responded by concealing the sick and the dead.[120]

Fearing that a quarantine would hurt commerce, business leaders strongly opposed the imposition of one in San Francisco. Though additional cases of plague were found, officials, bowing to pressure, denied that there was any plague in the city. The Board of Health lifted the quarantine three days after it was imposed. The city wanted to quarantine Chinatown but it also wanted to portray the image of a plague-free city. The second impulse won out: the governor, business leaders, the U.S. attorney general, and the press conspired to conceal the existence of plague in the city. The impasse was broken only after several states threatened to quarantine vessels and merchandise from California and a new governor was in office. Officials debated how to handle the plague for two months, during which time they tried different strategies to fight the plague without resorting to burning. However, when more plague cases were discovered President McKinley stepped in and quarantined all Chinese and Japanese persons living in San Francisco. The railroads were forbidden to transport out of the city any Asians or members of "races liable to the plague" if they didn't have a government-issued health certificate indicating that they were free of the plague. Asians hired top American law firms to defend them. They challenged McKinley's order in court, arguing that it violated their equal protection rights guaranteed by the Fourteenth Amendment. They won, and the quarantine was removed.[121]

The city launched an extensive cleanup of Chinatown in 1901. Between 1900 and 1904 plague killed 118 people in San Francisco (Chase reports that 113 died in this outbreak). There was another outbreak of the plague following the 1906 earthquake and fire. The two catastrophic events resulted in hundreds of thousands of people living in refugee camps, where rats and fleas thrived in the unsanitary conditions. Plague killed another 78 people during the 1907–8 epidemic.[122]

Though San Francisco's Chinatown was not burned, prescribed burning became a standard method of fighting the plague in Kobe, Japan, in 1901; Mazatlan, Mexico, in 1903; and Manchuria in 1907. Furthermore the international public health protocols developed in Berlin in 1902 adopted the Honolulu model. Though the use of fire to fight the plague was discontinued after the disease was better understood, prescribed burns continued well into the twentieth century. Los Angeles was probably the last city in the United States to employ this method, when buildings in a Mexican neighborhood were burned when the plague broke out in 1924.[123]

There were also outbreaks of the plague in New York City and New Orleans in 1924 and 1926 and in Washington State. A small outbreak occurred in Hawaii in 1943, but this time rat-control measures were used to prevent the spread of the disease. Between 1900 and 1925 there were 432 reported cases of plague in the continental United States, resulting in 284 deaths. Since the 1990s there have been fewer than 10 cases reported per year. Because there is effective antibiotic treatment the disease is not fatal if treated early.[124]

RACE, CLASS, AND THE PLAGUE

Fire is an extreme response, and in the case of Honolulu its full force was unleashed to fight the plague. Though the homes of four whites were burned and some white merchants lost property in Chinatown, thousands of Chinese, Japanese, and Native Hawaiians were burned out of Chinatown in fires deliberately set by government officials. Though Honolulu officials showed that they were willing to burn some white-owned property, they deliberated long and hard before doing so. Questions remain: If the plague had broken out in upscale white neighborhoods the way it had in Chinatown, would fire have been seen as such an effective tool for fighting the epidemic, and would it have been used on such a massive scale? Would the doctors of the Board of Health have ordered their own families and wealthy white neighbors to strip and take fumigating showers in public view while guards ogled them and pilfered their belongings? Would naked white men and women be shunted through the same lines? After white families were deliberately burned out, would they have to march through a gantlet of armed and screaming vigilantes as they made their way to detention centers?

The fact that the plague mapped onto a geographically defined area inhabited by nonwhite immigrants and indigenous people made it easier to impose draconian sanctions. The residents of Chinatown, many of whom did not speak English, were not a powerful political force in Honolulu. They lived in a slum. Not only were the squalid conditions of the slum and environmental degradation therein blamed on them, but they were considered to be the reservoirs and transmitters of the disease. As the disease was framed in racial and cultural terms that implicated the Asians and Chinatown, the media, merchants, and policy elites saw the burning of the neighborhood as a logical and efficient means of eradicating the disease. As soon as plague victims were found in Chinatown a broad quarantine zone was established. In comparison, a very delimited quarantine zone was placed around the homes of white victims. Chinatown residents were sent to crowded quarantine camps, while whites were sent to hotels. All Chinese and Japanese were prohibited from leaving Honolulu right after the first plague victims were identified, while no such general ban was placed on whites when white victims came down with the plague. Race might not have been the overriding factor driving all the decisions in Honolulu's handling of the plague outbreak, but it definitely played a role in how events unfolded.

Theoretical Analysis

Cities were hampered by their inability to deliver basic social and environmental services to residents. Though environmental policies and legislation appeared on the books early on, these were for the most part paper tigers. There was lax or no enforcement mechanism, so environmental conditions continued to deteriorate. Ineffectual city leaders delegated the responsibility of providing social services to religious and fledgling community organizations. In the absence of governmental will or ability to deal with the problems plaguing cities a host of religious and social institutions emerged to develop social and environmental policies and take on the task of improving urban conditions.

Around the time elites, the party of order, were beginning to exert more social control over the lower classes they also began to separate themselves spatially from the poor by segregating themselves around the few existing open spaces in cities. As poverty increased and city life became more tumultuous elites changed their perception of the poor. Increasingly they came to view the poor less favorably and grew less sympathetic to their plight. These evolving attitudes were reflected in reduced aid and more draconian sanctions being levied against the poor.

Poverty was linked to sloth and immorality, as were diseases. Noticing the link between poverty and the intensity of disease outbreaks, elites propagated the idea that epidemics were caused by poverty and immorality. This perception led elites to adopt extreme measures in their attempts to rid cities of diseases.

In the face of epidemics and the cataclysms they spawn, public officials are willing to go to extreme lengths to maintain order. The spread of the plague during the Middle Ages provided the impetus for developing quarantine protocols and laws.[125] Since then quarantines have been widely used to fight disease outbreaks, but as Edelson argues, these are not imposed in the same way on all sectors of the population. Echenberg also critiques the differential treatment meted out to wealthy whites and poor minorities in the Honolulu plague episode.[126] Although some wealthy people have been quarantined, by and large quarantines have been imposed more frequently and extensively on the poor and marginalized, such as ethnic minorities and colonial subjects. During epidemics the wealthy who contracted the diseases were usually allowed to remain in their home or sent to hotels, while the poor were carted off to cholera hospitals or crowded quarantine camps. Even under routine conditions poor immigrants have been subject to stricter quarantine procedures than wealthy immigrants. At Ellis Island, for instance, first- and second-class passengers were either not screened or given cursory screening in private and allowed to enter New York City, while passengers traveling third and fourth class and in steerage were subjected to extensive medical examinations conducted in public view. If they were suspected of mental illness or of harboring any contagious diseases, they were quarantined on the island before being allowed to enter the city.

Preexisting stereotypes influenced how public health officials and other policy elites responded to disease. In the case of yellow fever the preconceived notion that blacks were immune to the disease predetermined their role even in the face of evidence to the contrary. It took some time after blacks began dying from the disease for policies to reflect that reality. In the case of the plague the stereotype that Asians were carriers and whites were immune greatly influenced policy decisions. Blanket restrictions on travel and movement were placed on Asians as soon as the plague struck. During the plague epidemic in Honolulu, for instance, Asian doctors in Chinatown who collaborated with white colleagues were quarantined, burned out, and sent to detention camps; however, the white doctors and inspectors who went in and out of the plague-infested areas, conducted autopsies on victims, and even injected themselves with plague serum were not subjected to quarantine, fumigation showers, or detention centers. White doctors did not order their own property condemned or burned, yet they had extensive contact with the *pestis* bacteria. In the case of the Asian doctors, their educational attainments, research skills, knowledge of the plague, help in identifying early victims, and medical degree were not enough to put them on an equal footing with the white doctors or exempt them from the degrading quarantine process.

The spread of diseases was often blamed on the lifestyle and culture of particular groups. Poor and marginal people were blamed most often for the occurrence of epidemics. As disease transmission became better understood effective interventions and environmental reforms helped to improve conditions in cities and reduced the risk of contracting and spreading diseases.

Part II

REFORMING THE CITY

In this section I examine efforts to transform the physical environment of cities and reform city dwellers. In chapter 3 I examine how the rich responded to overcrowding, the mixing of the races and classes, unregulated land uses, and lax enforcement of ordinances to devise collective strategies to protect the integrity of upscale neighborhoods. In chapter 4 I analyze how elites perceived poverty and disorder and the policies promulgated and implemented to ameliorate conditions. I include an analysis of the evolution of elite and working-class ideologies.

In chapter 5 I examine the role of data collection and analysis in helping civic leaders to understand the social dynamics of the city and formulate and implement policies and future plans. In particular I examine the use of spatial and statistical data to impute blame, help ratify claims, and devise solutions. Chapter 6 focuses on the sanitation and housing reform efforts of activists. I examine the interplay between social reformers and political elites as they wrestled with a number of questions, among them What should be the role of the government in safeguarding public health and in improving the quality of life for citizens? Should the government play a role in providing safe and affordable housing in adequate quantities? This chapter illustrates how reformers changed their perceptions and tactics to place less blame on the poor and more on the government and major corporations.

three

WEALTHY URBANITES

Fleeing Downtown and Privatizing Green Space

Wealthy urbanites had several reactions to the turmoil and environmental degradation in the cities, among them acquiring or privatizing green space, forming charitable institutions and instituting workfare programs, establishing settlement houses, instituting sanitary reforms, and spearheading housing reforms. The wealthy made great efforts to separate themselves from the poor, live in close proximity to each other, and interact with their compatriots. They also emulated European aristocratic tastes and lifestyles.

Country Estates and Summer Homes for the New York Gentry

Early on the wealthy responded to urban disorder by moving to the fringes of the city or the countryside to escape the mayhem or cloistering themselves in exclusive residential enclaves within the city. Homberger argues that aristocrats have a thirst for exclusivity not only in choosing where they live, but in deciding who lives next to them, the institutions they build or belong to, and their recreation and schooling. American aristocrats sought to be more exclusive than their European counterparts. During the 1870s, when British elites were said to consist of five hundred families, New York aristocrats recognized four hundred families as belonging to the society set. Though aristocrats in Boston, Philadelphia, Savannah, and Charleston were influential locally and regionally, the sphere of influence of the New York gentry was widespread, so they were the arbiters of who belonged to elite society.[1]

Though the process whereby the rich cocooned themselves in exclusive communities accelerated rapidly during the nineteenth century, it really began in the seventeenth. Again New York aristocrats led the way. Even when New Amsterdam was a mere hamlet, economic and political elites began to separate their residential quarters from those of the masses and from their place of work. Petrus Stuyvesant, for example, lived downtown in the governor's house in Fort Amsterdam when he arrived in 1647, but he bought a farm on the East River at Seventeenth Street in 1651. In 1655 he built an elegant home, named White Hall, at the intersection of what is now Whitehall and State Streets and made it his official residence.[2]

The image of gentlemen farmers living on lavish estates relatively close to cities was one the upper class embraced. To this end they hired the leading architects to re-create European mansions and stately homes for their downtown abodes and country estates. While American gentlemen farmers tried to emulate longstanding European traditions, there were important differences between the lifestyles of American gentlemen farmers and their European counterparts. European gentlemen of leisure inherited estates that had been in their families for generations; the American upper class acquired property that had to be transformed into farms from scratch. Furthermore, though American gentlemen farmers had tremendous land holdings, these were not the primary basis of their wealth; they usually derived their wealth from business enterprises.[3]

In the early- to mid-eighteenth century other wealthy New Yorkers followed Stuyvesant's lead and began building country estates on the fringes of the city. In some instances the city helped to subsidize these homes. For instance, in 1730 Colonel Anthony Rutgers, whose family made their money in trade and brewing, approached the city with a proposal to drain the Lispenard wetlands if the city gave him the seventy acres of swampland. Rutgers, who already owned a farm adjacent to the wetlands, wanted Lispenard Swamp free of charge. He knew his proposal would be taken seriously because the Common Council viewed the marsh as a nuisance. Cattle and other animals got lost and trapped in the smelly swamp, and people living nearby believed they got fevers and chills from it. Moreover the swamp was a physical barrier extending across about two-thirds of Manhattan that limited the northward expansion of the city as development in the vicinity was confined to a narrow strip of land along the eastern side of the island.[4] The Trinity Church historian William Berrian described Lispenard as a "wild and marshy spot of no inconsiderable extent, surrounded with bushes and bulrushes, which in winter was a favorite place for skaters, and at certain seasons for gunners."[5] Rutgers, a miasmatist, described the swamp in bleaker terms: "The said swamp is constantly filled with standing water for there is no natural vent, and being covered with bushes and small trees is, by the stagnation and rottenness of it, become exceedingly dangerous and of fatal consequence to all the inhabitants of the north part of the city bordering near the same, they being subject to very many diseases and distempers which, by all physicians and by long experience, are imputed to the unwholesome vapours arising thereby."[6]

Rutgers got Dr. Moses Buchanan to provide an affidavit swearing that several people who lived close to the swamp were under his care for fevers and agues that he thought was caused from the miasma of the swamp. Persuaded by these arguments the Common Council granted Rutgers the land on condition that he should pay a "moderate quit-rent" for it and drain it within a year. In 1733 Rutgers built a drain and a bridge across the swamp at what later became Greenwich Street.[7]

A decade after Rutgers drained the Lispenard swamp wealthy New Yorkers began looking for suitable places to build summer and country homes. At the time the city did not extend much more than a mile beyond the southern tip of Manhattan. Greenwich Village, a tiny community separated by more than a mile of open space from the city proper, became a popular site for these homes. It was a marshy area dotted with hillocks known as Sapokanican to the Canarsee Indians, who used to camp and fish in the area. The Dutch named the area Bossen Bouwerie, meaning "farm in the woods." The marshes fed a meandering trout stream called Manetta, an Indian word meaning "devil water." The stream was later renamed Minetta Brook. Bossen Bouwerie was one of the four designated farm areas reserved for the Dutch West India Company by the first governor, Peter Minuit. Free blacks also farmed parcels in the area. However, soon after the second governor, Wouter Van Twiller, arrived in Manhattan in April 1633 he appropriated a large parcel of land in the area for his own private tobacco farm. Van Twiller's farmhouse was probably the first house erected by Europeans on Manhattan outside of Fort Amsterdam. After the English conquest the country hamlet was designated Grin'wich in the 1713 Common Council records.[8]

Chief Justice and later Lieutenant Governor James De Lancey purchased the 340-acre Dominie Farm that comprised about 120 blocks of the Lower East Side. Oliver De Lancey also owned an estate in the area.[9] Sir Peter Warren, vice admiral of the British navy and commander of its New York fleet, became a trendsetter when he acquired several hundred acres of land at Greenwich. Warren, who married one of the James De Lancey's daughters, built a Georgian country house around 1745 on the property (the city later gave

Warren additional land in recognition of his accomplishments at Louisburg in 1745). Warren lived on Lower Broadway close to Bowling Green, but he and his family used their country home during the summers to avoid the heat, plagues, and diseases that ran rampant in the city. The summer homes provided fresh air, ample green space, and access to the water. They also allowed the gentry to pursue a refined lifestyle both in and outside of the city.[10]

The Warren estate was the nucleus around which other country estates clustered. Robert Richard Randall, a sea captain, and the Brevoorts, an aristocratic Dutch family, also owned estates in Greenwich. Free blacks, who had established an independent community in the sparsely populated hamlet, had farmed parcels of land in Greenwich before wealthy New Yorkers began building their country estates there. In 1750 Thomas Clarke, a retired British army captain, built a country house on a tract of land he named Chelsea, just north of Fourteenth Street. In 1767 Major Abraham Mortier, commissary to His Majesty's forces and a member of Trinity's vestry, secured a ninety-nine-year lease on the Richmond Hill estate from Trinity Church. The estate consisted of 220 lots between Spring and Christopher Streets. Trinity Church owned fifty-seven blocks containing almost one thousand city lots between Fulton Street and Greenwich Village. The church acquired the original thirty-two acres by securing a patent for the land from Queen Anne in 1705. The church enhanced its holdings by acquiring more than a hundred acres of farmland and by gifts and grants from the city for shoreline lots. Trinity Church, the Astors, and the City Corporation of New York were among the largest landowners in the city. Because the original grant stipulated a limit on the profit the church could earn from the land, Trinity leased out its property to provide a steady and reasonably high income stream.[11]

Soon a string of large estates snaked their way up the west and east sides of Manhattan. Trinity Church granted Anthony Lispenard an eighty-three-year lease on eighty-one lots north of Chambers Street and granted ninety-nine-year leases on Broadway property to wealthy merchants and church vestrymen Walter Rutherford, William Antell, and Alexander de la Montagne, among others.[12] By 1770 the opulent merchants William Bayard and James Jauncey and the lawyer and military officer John Morrin Scott had built country homes on the West Side. On the East Side, members of the Rutgers family built their country home on a hundred-acre estate south of where Seward Park is currently located. Petrus Stuyvesant's grandson still kept the original White Hall estate as newer mansions sprung up around him. Two of Petrus's sons built mansions, Petersfield and the Bowery, in the area. Other East Side estates included John Watts's Rose Hill, John Murray's Inclenberg (located in what is now Murray Hill), and James Beekman's Mount Pleasant. The mansions of the Schermerhorns, Rhinelanders, and Lawrences were located along the East River north of Fiftieth Street. John Watts was a lawyer and son-in-law of Governor DeLancey. The Scotsman John Murray was a merchant and friend of Alexander Hamilton's. James Beekman was also a merchant descended from one of the founding families of New York City. His lavish coach dating from 1771 is on display at the New-York Historical Society. The Schermerhorns, related to the Beekmans by marriage, made their fortune in the lumber and shipping business. The Rhinelanders were also shipping magnates; a part of their fortune also came from their sugar refining business. The Lawrences belonged to a wealthy English family. The British colonel Roger Morris built his country home—Mount Morris—in Harlem Heights in 1765. Located in Roger Morris Park, the mansion—once occupied by George Washington—is currently the oldest surviving house in Manhattan. Others went farther afield to Westchester, Long Island, and the Bronx. For instance Frederick Van Cortlandt, whose family made their fortune from milling operations and a large grain plantation, built his mansion in the Bronx in 1748. Born in Ireland of Scottish parents, Cadwallader Colden, who was trained as a minister, became a merchant and a doctor after migrating to Pennsylvania. By the 1760s Colden—who built his country home in Flushing—was one of the most powerful men in New York. The merchant and civic leader, Philip Livingston, built his country estate in Brooklyn Heights, while Colonel William Axtell purchased Melrose Hall in Flatbush to serve as his country seat.[13]

Country Estates, Gentlemen Farmers, and the New England Gentry

By 1760 the countryside around Boston was dotted with the country estates of the wealthy, in the nearby hamlets of Danvers, Medford, Middleborough, and Brookline. Such influential families as the Cabots, Lowells, and Lawrences built their estates in Brookline and nearby villages. The Cabots grew wealthy from their shipping, banking, and privateering ventures. They built lavish homes in Beverly and Brookline, Massachusetts. The Lowells, originally merchants, gained wealth and fame from their textile mills in Lowell, and were united by marriage with the Cabots, another merchant family. The Lawrences served in the military and founded the Groton Academy, but the family is best known for their textile mills in Lawrence. The country seats were designed to show off the wealth, status, and refinement of the owners. After the Revolutionary War many gentlemen farmers in New England became interested in experimenting with new crops, livestock, tools, and techniques. Their foray into this new type of agriculture helped to transform the perception and identity of gentlemen farmers as individuals undertaking a gentlemanly pursuit for the public good, especially when they began studying crop diseases. They encouraged experimentation and innovation by founding agricultural societies, publishing farmers' journals, and organizing agricultural fairs. However, by the late nineteenth century gentlemen farmers from New York who built their country estates in New England shifted away from this more utilitarian type of farming to planting and developing ornamental trees and elaborate formal gardens.[14]

During the nineteenth century Newport emerged as a major hot spot for building country estates. Newport's salubrious summer climate beckoned southerners from Charleston and Savannah who were fleeing epidemics. As talk about the abolition of slavery heated up, southern aristocrats found Newport more welcoming than older resort towns such as Saratoga, New York. Because so many southern plantation owners (George Noble Jones, Hugh Ball, Henry Middleton, and Ralph Izard) built their mansions in Newport, it was more comfortable for them there. In addition, some of the rich northeasterners who built their estates in Newport, such as Ward McAllister, had southern family ties and could mediate the tensions between the northern and southern gentry. Jones owned two plantations in Florida, while Ball was a member of a wealthy Charleston plantation family. Middleton was a U.S. Senator and a rich plantation owner from South Carolina whose family had early built a mansion in Newport. Izard was also a South Carolina plantation owner. McAllister, who came from a family of prominent Savannah lawyers, was also a lawyer. He married a New York heiress and became a prominent member of high society who defined good taste and guarded membership in the society set. The relaxed atmosphere of Newport was facilitated by a ban on business, politics, and religion as topics of polite conversation. This ban was strictly enforced. When Saratoga aristocrats pushed for and succeeded in getting Jews excluded from hotels and other resort areas, Newport gained even more cachet by attracting wealthy Jews and others who felt snubbed at Saratoga. By the late nineteenth century New York's wealthiest families had built mansions in Newport. The resort was easily accessible because from 1847 on daily steamship service from New York and Boston became available.[15] The mansions of the Astors (Beaulieu and Beechwood), Vanderbilts (The Breakers and Marble House), and many others lined Bellevue Avenue. A visit to the Bellevue mansions reveals that millionaires such as Philadelphia's Edward Berwind spared no expense in building his French-style chateau, The Elms.

The summer season spent in Newport was a period of excess for the rich.

Women typically took a hundred or more gowns to Newport for the eight-week season; they changed into eight to ten different outfits per day as they flitted from one party to another. A typical mansion, complete with French chef, was staffed by more than forty servants.[16]

Exclusive Residential Enclaves

Since wealthy families still owned homes in cities, by the early eighteenth century there were sporadic attempts to develop organized, upscale residential neighborhoods in the city as well. In 1718 real estate developers announced a new residential development at Barton's Point in Boston. They described the neighborhood as consisting of house lots with a "very fine prospect upon the River and Charlestown" in a "great part of Boston."[17]

NEW YORK

From the late eighteenth century on, elites made more deliberate attempts to organize exclusive urban residential neighborhoods. In 1784, when New Yorkers thought the federal capital was going to be located in their city, they made plans to turn City Hall into Federal Hall. They tore down the remains of the old Dutch fort (Fort George) in 1788 to build a mansion to house the president. The streets surrounding Fort George, the old colonial seat of government, was known as the "Court-end" of town. The homes of Dutch colonial families such as the Van Dams, De Lanceys, Livingstons, Bayards, Morrisses, Crugers, and De Peysters were found in this section of town. The Van Dams were prominent merchants and political leaders. The Bayards, who were related to Petrus Stuyvesant, were also wealthy merchants, large landowners, and prominent political leaders. The Crugers' wealth came from shipping, privateering, and trading in slaves. They were influential political leaders, too. The Morrises were a wealthy mercantile family that held important political offices. The De Peysters were rich merchants. Several members of the family obtained high ranks in the British and American military, and they commanded regiments on the American frontier. The most prestigious streets were those facing the site of Fort George and those encircling Bowling Green. The wealthiest and most powerful residents of New York City lived in Lower Manhattan along Broadway (from Bowling Green up to Chambers Street) and around Battery Park. This location was convenient for a number of reasons. First, the rich built their homes around the existing green space (Bowling Green and the Battery). Second, the neighborhood was close enough to the commercial center of Manhattan that executives had an easy walk to work along tree-lined avenues. The neighborhood was also within walking distance or a short carriage ride to churches frequented by the elites and to places of entertainment. Third, for a brief period the neighborhood was considered to be the future home of the president. So many wealthy residents lived around the Battery, Bowling Green, and adjacent blocks that Bowling Green was nicknamed "Nob's Row."[18]

The Battery was a popular site for courtship and other upper-class rituals. As William E. Dodge, one of the "merchant princes" of Wall Street and a renowned public speaker, recalled, "In the summer and early fall a band of music in the evening enlivened the scene, and the grounds were crowded with the elite of the city; it was as polite and marked a compliment for a young lady to be invited by a gentleman to take a walk on the Battery as now to be invited to drive in the park."[19] However, it wasn't only the rich who wanted to enjoy the music and walkways of the Battery. Large crowds of working-class people began to flock to the park. The cultural diversity of the crowds bothered the elites; they feared that the park was no longer a place for upper-class single women to stroll in unmolested.[20] In 1802 the writer and politician James Kirke Paulding, who was a friend of Washington Irving's, commented on the changing nature of the crowds in the Battery: "I was in the midst of a crowd assembled on the Battery, to hear a Band of Music which played from the Flag Staff. Gentlemen, Boys, Negroes, wenches, Ladies, all jumbled together in the true spirit of sublime Equality, presented a Scene of confusion which Baffles the powers of my Pen, and I was ready to exclaim, as I surveyed the Scene at a distance, 'Chaos is come again'!—I did not dare to mix with this moving Sea of Living Animals."[21] Another writer, Jeremy Cockloft, echoed Paulding's sentiments: "[The] Battery [was] a very pleasant place to walk on a Sunday evening—not quite genteel though—every body walks there, and a pleasure, however genuine, is spoiled by general participation—the fashionable ladies of New-York turn up their noses if you ask them to walk on the Battery on Sunday—quere, have they scruples of conscience, or scruples of delicacy?—neither—they have only scruples of gentility, which are quite different things."[22] The German-born journalist, writer, and scholar of American aristocracy Francis Grund lamented, "I am in the habit of taking a stroll here [in the Battery] every evening; but have not for the space of two months, met with a single individual known in the higher circles. Foreigners are the only persons who enjoy this spot [now]."[23]

To the distress of the upper class the Battery became even more popular after the Southwest Battery (later Castle Clinton and then Castle Garden) was constructed between 1807 and 1811. Crowds were drawn to the twenty-eight

32-pound cannons and the firing of blank cartridges from the site, as well as the soldiers parading and drilling.[24] The Battery, renovated in the 1820s, had sculptured lawns, shade trees, and ornamental iron railings. It was the city's premier park and promenade. It grew even more popular when the city acquired Castle Clinton from the federal government in 1824. Castle Clinton, which was joined to the Battery by a wooden bridge, was refurbished as an upscale theater and renamed Castle Garden.[25]

Broadway was another aristocratic address, so much so that in 1828 more than one hundred of the city's five hundred richest men lived there. The Roosevelts, Grinnells (children from both these families would later play important roles in the environmental movement), Astors, Aspinwalls, and Hones all lived on Broadway. The Aspinwalls were merchants whose wealth was derived from their steamship and railroad businesses. The Hones were in the auctioneering business; some members of the family held important political offices. Such notables as Alexander Hamilton, George Washington, and John Jay also lived on Broadway, as well as aspiring aristocrats such as Herman Melville's parents, Allan and Maria.[26] The Melvills (before they added an "e" to the end of their name) had moved to Lower Manhattan from Albany. When they settled on Broadway Allan commented, "I have at last gained my point which has always been a House on Broadway—My spirits are better & I am a more agreeable companion than I have been for some time past."[27]

The author Frances Trollope described "splendid" Broadway as that "noble street [that] may vie with any I ever saw, for its length and breadth, its handsome shops, neat awnings, excellent *trottoir* [sidewalk], and well-dressed pedestrians."[28] Trollope also noted, "The dwelling houses of the higher classes are extremely handsome and very richly furnished. Silk or satin furniture is as often, or oftener, seen than chintz; the mirrors are as handsome as in London; the cheffoniers, slabs, and marble tables as elegant. . . . Every part of their houses is well carpeted, and the exterior finishings, such as steps, railings, and door frames are very superior."[29]

By the end of the 1820s the wealthy became uneasy with Broadway as it lost its exclusivity. It wasn't only the well-to-do businessmen who walked along the avenue and elegantly dressed men and women who promenaded in their carriages anymore; shopkeepers, laborers, vendors, and clerks also traversed the thoroughfare on their way to and from work. Domestic animals and dogs from nearby slums such as Five Points wandered over to Broadway and other ritzy streets. As the commercial district expanded and older private homes were converted to boardinghouses for young apprentices and clerks, the mixing of the classes and land uses led wealthy citizens to search for more exclusive neighborhoods elsewhere.[30] Thus in 1826 John Pintard noted, "Our principal merchants are all resorting up Bdway."[31]

Mary Enos Pinchot (mother of Gifford Pinchot, the renowned conservationist and first director of the U.S. Forest Service), who grew up in a prominent family living in one of the houses surrounding the Battery, had a memorable run-in with one of the stray dogs when she was a little girl. As she walked through the park with her nursemaid she saw a small dog lying on the grass asleep and reached out to pet it. The dog bit her on the hand. A gentleman walking in the park at the time swooped up the child and took her home as the nurse, who was holding Mary's baby sister in her arms, led the way. A doctor attended to the wounds. In the late 1840s the Enoses joined other wealthy families in their flight uptown when they moved to Washington Square.[32]

Yellow fever also prompted rich New Yorkers to move away from the tip of Manhattan. In 1801 rich families such as the Bleeckers moved to rented houses up Broadway to escape the disease. The 1803 epidemic forced merchants to move their families and businesses inland. The disease first broke in the low-lying marshy area around the Old Slip. Within a few days of the first outbreak on Rector Street all the printers, as well as banks, insurance companies, the post office, and the Customs House, relocated to Greenwich Village or Upper Broadway. When yellow fever broke out again in 1819 and 1822 merchants temporarily moved their businesses away from the southernmost portion of Manhattan.[33]

St. John's Park (Hudson Square) was one of the uptown destinations of choice. Hudson Square, bounded by Laight, Varick, Beach, and Hudson Streets, was developed around St. John's Chapel by Trinity Church. Lavish homes and open space weren't the only prerequisites of aristocratic neighborhoods. The most desirable enclaves were anchored by prominent churches, exclusive restaurants such as Delmonico's, and cultural institutions such as the New-York Historical Society, theaters, and social clubs. In their heyday the Bowling Green, Battery, Lower Broadway, and Bond Street neighborhoods were home to the St. Nicholas Society (1935), the Union Club (1836), and Stuyvesant Institute (1837), as well as dining societies. The St. Nicholas Society had five hundred members of Dutch ancestry, all of whom had to prove that their families migrated to New York before 1785 to obtain membership in the club.[34]

The institutions of the rich had to keep abreast of the demands of their members or risk becoming irrelevant; the churches were no exception. Sometimes the churches followed their members to uptown locations; at other times the churches led the way in anchoring the development of new enclaves. Wealthy families guaranteed their spot in church by buying or renting pews. Like almost everything else, aristocrats vied with each other to see who could buy the most prominent pews in the most elite churches. The elite downtown churches such as Trinity and Grace Church were oversubscribed, their pews sold out. Those who were waitlisted were growing impatient, as pews rarely opened up because families who owned pews passed them on in their wills. It was in this context that Trinity sought to build an uptown church, St. John's Chapel, to meet the demands of its parishioners. Trinity acquired the Hudson Street property and adjoining burial ground in 1803 and completed St. John's Chapel in 1807.[35]

Other factors also drove Trinity to develop Hudson Square. Because the original land grant to the church capped the profits that could be generated, in 1807 when church leaders wanted to develop fashionable residences like the villa parks in England the church decided to lease the house lots. There were few takers in 1807. However, in 1827, when the vestry decided to sell the lots rather than lease them and deeded Hudson Square itself to the purchasers, wealthy buyers flocked to the neighborhood. It was also in 1827 that the homes of the rich that flanked the Battery were vandalized by rioters. The new landowners built a high iron fence around St. John's Park and landscaped it with catalpas, cottonwoods, horse chestnuts, silver birches, flowerbeds, and gravel walkways. The square was soon surrounded for blocks with the homes of the most prominent families and was deemed "the fairest portion of the city."[36]

Critics of Trinity Church, by now one of the largest slumlords in New York, and of St. John's Park voiced their opposition to the development of this private park. Mike Walsh, a politician, argued, "A more exclusive concern than this park does not exist on earth. Its gates are all locked, and keys for it are sold by the church which *claims* it, for ten dollars a year, to none however, but the upper ten thousand, who reside in the surrounding palaces.—Can anything be more insultingly aristocratic than this?" Walsh protested by climbing over the railing and taking a stroll in the park.[37] The author Lydia Maria Child was also offended by the restricted access to the park. She wrote, "St. John's Park, though not without pretensions of beauty, never strikes my eye agreeably because it is shut up from the people, the key being kept by a few genteel families in the vicinity."[38]

Hudson Square got a boost when the Richmond Hill estate was converted into the opulent Richmond Hill Theater in 1831. Finally the neighborhood, which was located away from the main traffic and commercial arteries, had a cultural institution nearby. However, when the opera house closed a few years later, the St. John's neighborhood, with its weak cultural and social institutional infrastructure, became even more vulnerable to the flight of restless elites and commercial incursions.[39]

Other wealthy families began settling even farther uptown at the northern edge of the city where Broadway intersected with Bleecker, Bond, Great Jones, and East Fourth Streets. Here the wealthy bought larger lots to accommodate their mansions, ample gardens, and stables. It was said of these homes, "[They] may vie, for beauty and taste, with European palaces." The Minturns, who were merchants, made this neighborhood trendy when they built the first house on Bond Street, the most fashionable street in the neighborhood, in 1820. Their house, which had a white marble façade, set the standard for the three- and four-story homes that would soon line the block.[40] The rich built mansions large enough to entertain hundreds of people at a time. After Henry Brevoort threw a fancy-dress ball at his Lower Fifth Avenue home in 1840 patricians and parvenus competed with each other to throw lavish balls and soirées.[41]

As the elites fled the crowded portions of the city some of their downtown churches were demolished or left to languish. Even Hudson Square became a victim of its own success. As the desire to move far away from the working class became an obsession of the upper class, well-to-do families began moving away from St. John's Park in the 1840s and 1850s in search of greater exclusivity and higher status neighborhoods. Industrial facilities and commercial facilities began to move into the neighborhood around the same time. Workmen traversed the streets surrounding the park at all hours of the day and night. The residential character of the neighborhood declined even further after Cornelius Vanderbilt ran his Hudson River Railroad tracks down the west side of the neighborhood in 1851, and other industrial facilities, such as the American Express stables and warehouses, grocers, iron works, refineries, and factories, began operating nearby. The horse-drawn carriages of the Hudson River Railroad clanged and clattered incessantly along Hudson Street, the western border of the park. Vanderbilt bought out the area and covered most of it with a railroad freight yard in 1866. He paid Trinity Church $1 million ($12.3 million in 2005 dollars) for St. John's Park in 1867. Trinity got $400,000 of the payment, and the remaining $600,000 was split among the lot owners. By the time Vanderbilt purchased the site the disheveled park was surrounded by run-down tenements and rum shops. The two hundred trees adorning the park were felled and a three-story freight depot built on the site where the park once stood. The area around the park had degenerated into a slum by the 1890s. The congregation that once sauntered to St. John's Chapel on Sunday mornings in their expensive apparel dwindled steadily, so in 1894 Bishop Potter decided to close the church. Angry Episcopalians filed suit to keep the chapel open. Despite strong opposition from aristocrats such as John Pierpont Morgan and Theodore Roosevelt, the chapel was demolished in 1919. The former St. John's Park is now the entrance to the Holland Tunnel.[42]

By the 1850s the most fashionable residential enclaves were found around Union Square and Gramercy Park. Both of these projects, begun in the 1830s, were inspired by the success of St. John's Park; however, they stalled during the economic depression and were not built until Hudson Square and similar neighborhoods fell into disrepute. The Grinnells, Lenoxes, Belmonts, Schermerhorns, and Astors made Union Square the most fashionable neighborhood in the 1850s. The Belmonts were in financing. Members of the family were business partners with J. P. Morgan, the banking and railroad magnate. As the wealthy moved their city residences farther uptown in search of exclusivity, they also moved their country estates to more sparsely populated areas. The original estates, subdivided and sold, were engulfed by development.[43]

The parks and squares around which the aristocrats built their homes served as places to glimpse nature, resorts for genteel recreation, and informal resource centers. In Gramercy Park the well-to-do screened household servants for each other by providing recommendations for honest and industrious household help and warned their neighbors of dishonest workers.[44]

Trinity Church wasn't the only church that built sparkling new Gothic edifices "above Bleecker." Grace Church built a fashionable church at 800 Broadway in 1846. St. Thomas's Episcopal Church (built in 1826), St. Bartholomew's (1835–36), and the Collegiate Dutch Reform Church (1839) served the elites as they marched uptown. Becoming a pew-holding member in these churches was an expensive proposition and another form of conspicuous consumption that enabled the rich to display their wealth as they laid claim to the most highly coveted pews. The churches used the sale of pews to build and maintain their structures. When Grace Church was located downtown, John Pintard, who bought a pew for $300, sold it for $400 in 1927. The Church of the Messiah raised $70,000 from the sale of its pews in 1839. A pew at Calvary Church cost $550 in 1848. Lorenzo Delmonico's pew at St. Ann's cost him $575. When the new Grace Church was built 206 pews were auctioned; prices ranged from $400 to $900. The price of the pews escalated with demand; when the new

St. Bartholomew's opened in 1872 at Madison Avenue and Forty-fourth Street Cornelius Vanderbilt's pew cost him $1,509. The church raised $321,900 from the sale of two hundred pews that year. The richest, most powerful families held the most expensive pews with the best sight lines.[45]

Some social critics of the time called attention to the economic divide between the rich and the poor. The housing reformer Samuel Halliday noted that while 400 or so families lived in the 2-mile stretch of Fifth Avenue between Washington Square and Forty-second Street, 700 families numbering 3,500 individuals lived in one block in lower Manhattan. Around that time a square mile of London's East End, the neighborhood made famous in Charles Dickens's novels, had about 175,000 dwellers; New York's Fourth Ward had a population density of 290,000 people.[46]

BOSTON: RESIDENTIAL DEVELOPMENT FOR "THE SIFTED FEW"

Toward the end of the eighteenth century Mount Vernon Proprietors began developing homes on the westernmost hill of the "Trimountain" peaks in Boston. John Hancock had already built an estate near the peak of Beacon Hill, the central Trimountain peak. In 1795 construction began on the State House, which was located in Hancock's old pasture. Soon after that Mount Vernon Proprietors acquired land surrounding the State House and Mount Vernon in the hope of building elegant homes with ample gardens. To accomplish this they flattened the peaks of Beacon Hill and Mount Vernon and used the dirt to fill in and create Charles Street. By the early nineteenth century elites in Boston flocked to Beacon Hill. In addition to John Hancock, Beacon Hill residents included Governor James Bowdoin and Reverend William Emerson, father of Ralph Waldo Emerson.[47]

Pemberton Square, built after a hill was leveled in 1835 to provide landfill material, was designed with a small park surrounded by the elegant homes of Boston's wealthy citizens. Captains and shipowners settled in homes surrounding the park on Bunker Hill after the Quincy granite obelisk was completed in 1843. From Bunker Hill residents had a commanding view of the park and the rest of the city. The residences of the wealthy overlooked other green spaces too. Park Street, once home to a poorhouse, asylum, and jail on the fringe of the city, became a fashionable residence after Charles Bulfinch built the State House at one end of the street and elegant homes overlooking the Boston Common in the early nineteenth century. The fashionable residences along Park Street became known as Bulfinch Row, and the street itself became one of the favorite promenades of the city's elites. Beacon Street, which had an unobstructed view of the Common, was another prestigious address (cattle grazing was outlawed on the Common in 1830). As Oliver Wendell Holmes characterized it, Beacon Street was "the sunny street that holds the sifted few."[48]

Built in the 1830s–1840s in the Greek Revival style, the Louisburg Square neighborhood was another elite residential enclave, with homes overlooking a private park; notable residents were Louisa May Alcott and Henry James. Other well-to-do residents built their homes flanking the Public Garden located on the "flat" of Beacon Hill. The Common is the oldest publicly held open space and the Public Garden the oldest botanical garden in America. Wealthy Bostonians who could not find room on Beacon Hill and other fashionable hillocks and squares settled along Boylston Street and in Back Bay once the area was filled in during the 1850s.[49]

PHILADELPHIA'S ELITES

Philadelphia's wealthy lived in elegant mansions and kept extensive libraries and art collections. Though city residents deviated from Penn's original plan and subdivided their lots, the wealthy lived in mansions surrounded by formal gardens and expansive grounds. In 1747 the wealthy Quaker patron of several of the city's public institutions Israel Pemberton Jr. purchased Clarke Hall on Chestnut Street. The prominent Quaker merchant Charles Norris bought a three-story Georgian mansion on Chestnut Street; it had a staircase made of wild cherry and outside a greenhouse and a hothouse with pineapples. The writer Charles Stedman built an imposing mansion that he later sold to the mayor Samuel Powel. The wealthy also built country estates on the banks of the Schuylkill and near what is now Germantown. Their estates were a combination of mansions and farms. The mansions, maintained by indentured servants, sported granite and marble façades. In addition to orchards and nurseries, the estates had fishponds, racehorses, sheep, and cattle. Between the 1820s and 1840s the wealthy moved away from the downtown areas in droves and their homes were subdivided and turned into tenements for the poor.[50]

Elites also formed literary, artistic, and recreational clubs. Anglers formed the Schuylkill Fishing Company in 1732 and the Fort St. David's Fishing Company. The Jockey Club was founded in 1766 to "encourage the breeding of good horses and to promote the pleasures of the turf." A fox hunt was established in 1766. During the early nineteenth century the United Bowmen held annual shoots in front of large galleries. In the 1830s, when trotting races were all the rage, the Nicetown Hunting Park held races. The racing enthusiast

John Craig built his own race track and had a large collection of race horses at Carlton, his Germantown estate. General Callender Irvine owned thirty-eight thoroughbreds at the time of his death in 1841.[51]

BALTIMORE'S ELITES ON THE MOVE

In Baltimore elites began developing residential enclaves in 1799 on hilltops on the edge of the city. General Samuel Smith who made his fortune from brewing bought a five-hundred-acre estate he named Montebello, while the merchant Robert Oliver acquired the sixty-eight-acre Green Mount estate. The merchant and privateer John Donnell inherited the twenty-six-acre Willow Brook when his uncle Thorowgood Smith, the second mayor of Baltimore, died. The Scottish-born merchant Robert Gilmor developed the ten-acre estate named Beech Hill.

By the 1840s, the old homes and mansions of the elites were torn down or converted to offices and warehouse space as the wealthy moved into large dwellings flanking Waverly Terrace, Franklin Square, Mount Vernon, Charles Street, Madison Avenue, and Bolton, Hoffman, Preston, Lexington, and Pearl Streets. Decorated with Italian marble and Maryland soapstone, the buildings were three stories tall. Evergreen House, the mansion of the president of the Baltimore and Ohio Railroad John Garrett on North Charles Street, occupied twenty-six acres of wooded grounds. The forty-eight-room mansion was a Gilded Age marvel, with a twenty-three-karat gold-plated bathroom and a library of thirty thousand books. It had formal gardens adorned with classical sculptures.[52]

Baltimore followed New York's lead in building elegant homes around squares. In the 1830s the Canton Company proposed the development of four squares "on the Commanding Eminences" that would rival the squares of Manhattan while preserving the trees. When Mount Vernon and Washington Squares were being developed Frederick Law Olmsted was commissioned to help. By 1860 Baltimore had developed six new squares on hilltops.[53]

four

SOCIAL INEQUALITY AND THE QUEST FOR ORDER IN THE CITY

The wealthy realized that moving from one neighborhood to the next did not solve the major problems facing cities. Ergo they attempted to restore order to the cities, reduce social conflicts, and eliminate poverty. To this end, they undertook a series of initiatives to increase social control. Not surprisingly, the working class attempted to influence the process.

Social Control and the Party of Order

According to Edward Alsworth Ross, the ethical elites of the early twentieth century formed a party of order dedicated to ensuring that moral order prevailed in the city. In 1901 he advised that under the benevolent surveillance of the ethical elites, social control in urban areas should not involve the application of "rude force." Instead control should rely on "sweet seduction" through "inobvious . . . social suggestions." Ross contended that if society were to gain control of the "unpleasant and slimy things lurking in the . . . undergrowth of the human soul," it had to "supplement control by sanctions with control through the feelings."[1]

Talcott Parsons refined Ross's conceptualization of order and social control, claiming that social control is concerned with the processes intended to counteract deviant tendencies.[2] Every social system rewards conformity and punishes deviant behavior. Social systems also contain complex unplanned and largely unconscious mechanisms to counteract deviant tendencies. Generally speaking, social control is an attempt to manipulate the behavior of others by means other than a chain of command or requests.[3] Efforts to restore civility and social order to the city involved punitive sanctions as well as manipulative social control techniques and strategies. Social control and order were two dominant themes in urban environmental activism as well. In short, activists relied on a variety of rules, sanctions, seductions, and social suggestions to accomplish their goals.

Though most of the discussion of social control in this book focuses on attempts to control the masses, it should be noted that the upper class expended great effort to control their own behavior. Leading aristocratic figures such as Caroline Astor developed and enforced elaborate codes of conduct that covered just about every aspect of private and personal life. Aristocrats developed a very complex system of formal rules and rituals as well as informal controls and constraints to govern their daily lives. They decided who remained in good standing and what kinds of indiscretions made a person or family fall out of favor. The upper class created and enforced order within their class as a means of distinguishing themselves from the lower class. Order was important within the class so that each member understood how the society functioned. Creating and enforcing order among themselves was also a way for elites to gain a sense of stability amid the turmoil.[4] Though internal conflicts and rifts divided the upper class at times, for the most part they appeared to the outside world to be in solidarity and cohesive. While many of the controls and techniques used to enforce order in the upper class went unnoticed by most Americans, aristocrats were as concerned with controlling who entered and remained in the upper class and on what terms as they were with seeing greater order instituted among the masses.

Emergent Social Movements

As late as the nineteenth century a laissez-faire approach dominated the thinking of industrialists. That is, owners and managers of commercial enterprises felt that they had no responsibility for the welfare of their workers outside of the immediate confines of their facilities. They believed that the life outcomes of workers were a reflection of their moral fiber and standing in the eyes of the Divine. However, as the century wore on some industrialists adopted the position that owners should play a paternalistic role in their workers' lives.[5]

Several urban movements emerged to deal with the problems elites identified, such as immorality, irreligiosity, disorder, poverty, ill health, and environmental degradation. Evangelical reform was the first movement to arise; charity reform and company towns developed simultaneously. These were followed by the settlement house movement. Each movement grew out of the preceding one and emerged partly in response to perceived failures of earlier movements. Right or wrong, the movements served the important functions of identifying and amplifying urban problems, framing them, attributing blame, identifying and implementing solutions, and developing policies.

URBAN EVANGELICAL REFORM

Two of the earliest evangelical reform groups to operate in American cities were British missionary groups aimed at Christianizing Native Americans and slaves and converting to Anglicanism colonists and unchurched people living in the outlying areas of Boston. The Society for the Propagation of the Gospel in New England, better known as the New England Company, was formed in 1649 for these purposes. A rival organization, the Society for the Propagation of the Gospel in Foreign Parts, was formed in 1701.[6]

Settlers launched their own homespun missionary and revival efforts. The Great Awakening was a series of religious revivals that swept through New England towns from 1720 to 1750. The first Great Awakening did not have the same impact as the Second Great Awakening, which spread well beyond New England and focused on creating order in American cities. In the Second Great Awakening the evangelists were more deliberate in their attempts to use religion to stimulate reform. Many of the evangelical institutions and activities described in this chapter emerged out of the Second Great Awakening.[7]

Urban Missionaries: No Place for a Woman

Though evangelicalism had limited success in restoring order, the evangelical impulse played a crucial role in urban social control strategies. Toward the end of the eighteenth century evangelicals began to make a connection between poverty and lack of religious affiliation and participation: many of the poor did not attend church because there weren't many churches in their neighborhoods and they couldn't afford to dress up to attend churches in wealthy neighborhoods. Furthermore, as cities became more culturally diverse, people desired a broader range of religious experiences than those available in existing Protestant churches. However, the rich did not recognize these barriers to participation in religious activities. Consequently they began to question charitable aid to irreligious people; they also began linking poverty to irreligiosity.[8]

Initially American missionary outreach was not an urban phenomenon. The New-York Missionary Society, formed in 1796, focused on ministering to Indians and settlers on the frontier. The following year Scottish immigrants

Isabella Graham and her daughter and son-in-law Joanna and Divie Bethune (who grew wealthy in the mercantile business) founded the Society for the Relief of Poor Widows with Small Children. This shifted the focus of religious outreach to the cities. Joanna Bethune founded the Orphan Asylum Society in 1806. Six years later the Society for Supporting the Gospel among the Poor of the City of New-York was formed to preach to people in hospitals, prisons, and almshouses. Joanna Bethune combined charitable aid with work and education when she established the Society for the Promotion of Industry among the Poor in 1814. This institution had a House of Industry in which more than five hundred women earned wages from sewing while they were being taught basic literacy skills. Efforts to enhance religious outreach in the cities occurred around the time cities began getting more crowded and disorderly; income inequality was rising, and the rich were cloistering in elite neighborhoods. In 1815 more female evangelicals began to work among the poor in cities and formed the Female Missionary Society for the Poor of the City of New-York. Three years after its founding, the Society established a free church in the "seat of Satan" in the African American neighborhood of Bancker Street. A year later a chapel was built near Corlear's Hook, an area with a large number of prostitutes (the term *hooker* is said to come from the name of this area). The female missionaries visited the poor in their homes and tried to coax destitute mothers to attend church. Ministers and other male church members were alarmed by the bold, hands-on approach of the female missionaries, and in 1821 Reverend William Gray told the women that such was not the "appropriate sphere" of work for women to be engaged in. He transferred to men the work the women had been doing. Soon after that the Presbyterian Young Men's Missionary Society of New York built churches near Corlear's Hook and on Bancker. Young men took over the task of conducting family visits and holding prayer meetings with families.[9]

Female urban missionaries operated more freely in Philadelphia. In 1847 prominent church women formed the Rosine Association, whose goal was "to rescue from vice and degradation a class of women who have forfeited their claim to the respect of the virtuous." The Rosine Association operated a fourteen-room building housing 30 women at a time. By 1854 they were serving 325 women, most of whom were young women who had recently moved to the city. Rosine Association missionaries also made friendly visits to brothels.[10]

In New England, too, female-led organizations were among the first to recognize and confront the growing problem of poverty. The Female Benevolent Society was founded in 1814 in Lynn, Massachusetts, a manufacturing town. Two years later the Cambridge Female Humane Society was founded. A similar society, the Chesterfield Female Benevolent Society, was formed in Hopkinton, New Hampshire.[11]

Bible and Tract Societies

From the 1820s onward religious leaders developed a series of voluntary organizations and activities aimed at helping individuals to recognize their personal failings and reforming deviant behavior through repentance. Thus the tract and Bible societies emerged to teach morality and order and to spread religion.[12] In 1816 a group of reformers—Henry Rutgers, David L. Dodge, Divie Bethune, Gardiner Spring, Richard Varick, Governor De Witt Clinton, and others—formed the American Bible Society to print and distribute Bibles in the United States. Within a few years the American Bible Society press was producing tens of thousands of Bibles each year. Because Society members believed that poverty could be alleviated by preaching the gospel, they helped to organize a network of Bible societies in 1816, including the New-York Female Auxiliary Bible Society, the Female Juvenile Auxiliary Bible Society, and the New-York Union Bible Society. They formed the New-York African Bible Society and the New-York Marine Bible Society in 1817. Volunteers from these organizations distributed Bibles to people in slums, brothels, grog shops, gambling dens, hospitals, almshouses, and prisons.[13]

However, the Bible was too large, long, and complex to be a convenient instrument of evangelical outreach that targeted poor and uneducated people. This realization led to an expanded role for the tract societies. The tracts were short religious stories or biblical texts, written in simplified language and illustrated to make them interesting. The New York Religious Tract Society was founded in 1807 by many of the same men who had founded the American Bible Society. Initially the Tract Society imported and distributed British inspirational tales to frontier missionaries, but in 1815 it joined in the urban evangelical campaign. In 1825 the New York Tract Society joined forces with the New England Tract Society to form the American Tract Society. The Tract House installed the first steam-powered press four years before any of the commercial publishers had this new technology. With sixteen presses at their disposal, the Tract Society produced six million tracts (about 61 million pages) and three thousand Bibles in 1829 alone. The evangelicals distributed the tracts and Bibles by organizing corps of agents and hundreds of branches in each state. The agents went door to door to distribute the tracts and Bibles. They also founded lending libraries, placed advertisements in local newspapers, and made special deliveries to sailors and boatmen. In New York the Tract Society saw the distribution network as "the lever which shall move the foundation of Satan's empire in this city."

The Tract Society's goal was to deliver a different tract to every city resident each month. To this end, each ward had a committee and a chairperson and was subdivided into districts of sixty families each. There were over five hundred districts, each with a team of distributors. Each team reported their distribution efforts to a central committee monthly. In 1829 the Tract Society reported that of 28,771 families visited, only 388 declined to accept a tract.[14]

Sunday Schools

The Sunday school movement began in 1871 in Gloucester, England. Founded by Robert Raikes, the movement did not provide religious instruction at first; the goal was to improve the lives of poor children. Many churches initially opposed Sunday schools because they thought that teaching on Sunday was blasphemous. Despite this, Sunday schools spread quickly across America. The first American Sunday school for adults, organized by Isabella Graham, was held in 1792 in New York, on Mulberry Street in Five Points. The Methodist Gilbert Coutant opened a school in the Bowery Village in 1793 or 1794. In 1794 Katy Ferguson's School for the Poor in New York City opened Sunday schools for adults and children; organized by Catherine Ferguson, a former slave whose freedom had been purchased by the Grahams and Bethunes, the school was open to both blacks and whites (Graham's school also taught blacks and whites). By 1813 there were fifty Sunday schools in New York with six thousand students.[15]

The Sunday school movement promoted literacy, the Protestant ethic, thrift, self-reliance, temperance, and order. Using the Bible the schools taught reading and writing as well as religious lessons. The target of reform was lower-class children, who were taught the values of their middle-class teachers. The schools aimed "to arrest the progress of vice and to promote the moral and religious instruction of the depraved and uneducated part of the community." These "free schools," as they were called, were organized by the church. Organizers were driven by the principle of noblesse oblige, whereby the fortunate or wealthy have a responsibility to guide the children of the poor.[16] In 1820 activists proposed that a national society be formed; thus the American Sunday School Union was founded in Philadelphia in 1824. At the time there were over 700 Sunday schools teaching 50,000 children in Pennsylvania. A year later, 33 percent of the children in Philadelphia were enrolled in Sunday school; in 1829 so were 41 percent of the children in New York.[17] In 1825 4,000 children were enrolled in Sunday school classes in Baltimore. By 1830 350,000 children were enrolled in Sunday schools nationwide and about 150 new Sunday schools were being formed annually.[18]

CHARITIES AND SOCIAL ORDER

By the middle of the nineteenth century elites had grown frustrated with the inability of religious reform to change urbanites. It was clear that simply urging religious conversion was not enough to change society; riots remained commonplace, and business leaders became increasingly concerned about social instability as their establishments and dwellings were targeted during riots. Consequently there emerged a new brand of charitable organization aimed at eliminating poverty and lifting up the poor. Wealthy merchants, who had also funded the tract societies, joined forces with social reformers to form charitable organizations to help lift slum dwellers out of poverty. These merchant-led antipauperism societies firmly believed that depravity and wickedness caused poverty, and only greater moral and social control by elites could restore order.

The Influence of Malthus

Though the impetus to help only the deserving poor preceded the introduction of Malthusian ideology to American reformers, that ideology played a very influential role in how American elites understood and dealt with poverty. The reverend Malthus decried the fact that the ability of the eater to reproduce was considered "of superior order" to the ability of the individual to produce food. Malthus contended that humans had a tendency to degenerate toward sloth and savagery; the poor in particular lacked moral restraint. He thought the threat of hunger prompted people to exert their full capacities and become more civilized and that labor and self-discipline would elevate humans to divine grace. He proposed that England abolish the Poor Laws that offered minimal help to the destitute, for by easing misery they removed one of the natural checks on the population as well as the impetus to work. Malthus believed that teaching the poor self-control and responsibility would limit family size and reduce population size.[19]

Malthus wrote his important essay in 1798, when there was a concern with savagery and civilization as extremes of the human condition. Civilization was seen as the nobler, more desirable state, while savagery was perceived as a degraded condition to be exterminated.[20] The concern over the level of savagery or civility present in a given society grew in Europe and America throughout the nineteenth century. In America it infused the debates about charity, urban environmental reform, wilderness, and wildlife conservation.

Malthus had his critics. One Baltimorean, Daniel Raymond, an outspoken opponent of slavery, argued that poverty was not caused by idleness but by the economic structures that allowed a few people to accumulate enormous wealth by exploiting the labor of others. Writing in 1819 he argued that while the rich appropriated the wealth generated by the labor of the poor, the wealthy blamed the poor for lack of thrift and aversion to hard work, thus blaming the victim: "If a man were to plant his field with trees, and then complain of the corn for not growing under them, it would not be more unreasonable. . . . The laws of justice, as well as the laws of the land, require the rich either to furnish the poor with labour, or support them without labour." Raymond argued that the alleviation of poverty would require "such modification of the laws of a country, as shall produce a more equal division of property."[21] He contended that the unequal distribution of resources contributed to poverty and environmental calamities.[22]

Malthus's framing resonated more than Raymond's with American charity reformers. Malthus deftly aligned several ideas related to poverty and bridged them in a way that formed a very effective master frame that linked poverty, laziness, religiosity, morality, and civility. Charity reformers found Malthus's arguments compelling on several fronts. They were drawn to the religious undercurrents and prescriptions of the arguments and believed in the Malthusian message of morality and civility. Malthus's arguments also pointed to a plan of action that did not challenge the status quo in any way: Malthusian ideology encouraged charities to focus on alterative and redemptive behavior-modification programs rather than reforming or transforming the social structures giving rise to poverty. Hence charities believed that refusing to help most of the poor was justifiable because they were allowing natural processes to run their course. When aid was given it came with instructions on how to improve perceived character flaws; following the model pioneered by the missionary, tract, and Bible societies, charity workers devised a scheme of "friendly visits" to poor people's homes to dispense advice and scrutinize them. However, the charities transformed the Bible societies' friendly visitor model by developing an elaborate data-collection infrastructure around the visits. In essence, the friendly visitors acted as a surveillance system and a benevolent guardian in one.

The Society for the Prevention of Pauperism

In 1817 a New York charity announced that fifteen thousand people (about one-seventh of the city's population) had been helped by charities. However, the rapid increase in the number of people seeking help impelled charities to begin sorting out the "deserving" from the "undeserving" poor. Following the lead of British thinkers such as Malthus, David Ricardo, and Jeremy Bentham, American reformers began to distinguish between poverty and pauperism; this helped to amplify and clarify the framing around poverty and aid. Since poverty and pauperism were two different conditions, they required different responses. Poverty occurred when the individual was poor through no fault of his or her own. Thus the sick, handicapped, aged, orphaned, widowed, deserted, victims of epidemics, and casualties of war were poor because of circumstances beyond their control; they were granted minimal charity. Pauperism, in contrast, stemmed from laziness, fraud, immorality. and other sins or character flaws; thus the remedy was punishment and behavioral modification, not charity.[23]

As was the case in New York, local municipalities in New England were responsible for the poor within their borders. Civic leaders were also concerned about the poor moving around to take advantage of charitable aid. Consequently town residents took to "warning out" anyone suspected of being a drifter; that is, nonresidents were harassed into leaving town before they could establish themselves.[24]

American reformers were well aware of the methods by which the British dealt with poverty through organizations such as the London Society for Bettering the Condition of the Poor. In December 1817 Thomas Eddy, warden of Newgate Prison (New York's first prison), John Pintard, and the doctor John Griscom called a meeting in New York, which resulted in the formation of the Society for the Prevention of Pauperism, which drew a number of prominent merchants, lawyers, and clergymen to its membership.[25] The following year the Society outlined what it saw as the causes of pauperism: idleness, intemperance, gambling, extravagance, early and imprudent marriage, prostitution, lack of education, pawnshops, and indiscriminate almsgiving. Members argued that giving charity to the undeserving poor undermined their independence and increased taxes while sapping the wealth of the entire community. They recommended that all paupers be denied all forms of public assistance immediately.[26]

Not all elites espoused this framing or supported the policies of the Society for the Prevention of Pauperism, instead arguing in favor of a gentler approach to dealing with the poor. For instance, Mayor Colden argued that unemployment triggered by economic depressions and the introduction of labor-saving technology increased poverty, and so the city government should play a role in alleviating poverty. An almshouse chaplain named Stanford argued that since the poor labored to create the wealth of the rich, the wealthy should help to improve the lives of the poor.[27] But because these critics did not coordinate their responses or organize any sustained effort to counter the framing and activities of the Society for the Prevention of Pauperism and other charities, their arguments were largely ignored. The organized voices and actions of the charities held sway. At the time of these discussions, most cities had a limited tax base; therefore they had little or no funds earmarked for the delivery of social services. As a result most cities followed the lead of the Society for the Prevention of Pauperism and other charities that emerged to provide social services and create social policies in the face of government inaction.

The Society worked in concert with the municipal and state governments to implement its policies: give minimal material aid to the poor because such aid encouraged dependency on charity, and end pauperism by inculcating the poor with middle-class values such as frugality, sobriety, good manners, punctuality, industriousness, cleanliness, religion, and education. With this in mind, the Society advocated a ban on street begging and pushed for the establishment of workhouses to institutionalize the poor. The poor were encouraged to open savings accounts in banks and attend Sunday school. Municipal and state governments banned outdoor aid, except in New York City, where the large numbers of poor made it difficult to completely get rid of this form of aid. In 1824 the city shut down the soup kitchens and halted the distribution of food and firewood in the winter. The state government also took the Society's advice and began institutionalizing the poor in almshouses, jails, workhouses, juvenile detention facilities, orphanages, and hospitals. Work became a prerequisite for aid; even the frailest inmates had to work in order to receive food and shelter. Brooklyn followed New York's lead and closed its soup house.[28]

The Society for the Prevention of Pauperism also advocated the moral superintendence of neighborhoods. Like the tract societies, it divided the city into districts, to each of which was assigned two or three middle-class volunteers who visited poor families to advise them on domestic issues, child rearing, and personal conduct. These visitors also collected data and kept files on the background and character of the people they visited. However, by 1821 the visitor system had foundered.[29]

Some female reformers, such as those who organized the Society for the Relief of Poor Widows with Small Children, argued that it was impossible for a widow to support children on her own. The Society for the Prevention of Pauperism attacked the work of this organization by arguing that giving charity to widows with small children encouraged them to reproduce more, which was immoral and un-Christian and only increased the number of paupers.[30]

Political Leaders and Workhouses

New York City's political leaders largely echoed the Society's thinking on poverty and aid. Leaders of both political parties accepted the position that healthy, able-bodied men and women were not victims of economic depression, and therefore not eligible for aid. In 1838 a Whig Common Council argued that charity increased pauperism while promoting dependency. In 1840 the *Evening Post* attacked the charities by arguing that they were grossly lavish in giving aid and careless of the consequences of their actions. The following year the mayor of New York proclaimed that the able-bodied poor should be isolated and put to work in a separate workhouse on Blackwell's Island; the Common Council supported the mayor's proposal and contended that such an institution would discourage the dissolute and idle population from seeking aid in the first place. Some citizens of New York went further by calling for the complete elimination of any form of public aid. One letter writer to the *Tribune* who took this position argued, "They who will marry and beget children in dirty cellars are a curse to the world." Such sentiments persisted, despite Horace Greeley's insistence that it was the depressions that caused economic hardship. Greeley's view was not widely held among elites.[31]

The Association for Improving the Condition of the Poor

Though charities reduced the level of religious instructions and sought other ways of aiding and reforming the poor, striking the right balance between religious outreach and other forms of aid remained a challenge for urban reformers. Thus in 1839, when the American Female Moral Reform Society began blending religious missionary work with aid (they gave out food and clothing and helped people find housing and jobs), the purely missionary societies took notice. In 1843 the American Tract Society spun off a separate organization, the Association for Improving the Condition of the Poor, to focus on charitable work.[32]

The Association for Improving the Condition of the Poor was a major player in the field of charitable giving. Firmly rooted in the ideology that poverty was caused by irresponsibility, the Association elaborated on past systems of identifying, sorting, and tracking the deserving poor to ensure that the "right" people got aid. According to the director, Robert Hartley, the organization had to be vigilant about how and to whom it dispensed aid because prior to the Association's establishment "the most clamorous and worthless" beggars got the largest amount of charity, while "the most modest and deserving" got none.[33]

The Association applied the social work model of tracking and observing the poor to justify dispensing aid and ensuring that aid recipients behaved according to the rules prescribed by the charities. Using middle-class male agents, charity workers, and volunteers, reform organizations divided the cities into districts, and representatives from the charities collected information and compiled dossiers on families in each district. For instance, in Philadelphia the Union of Benevolent Associations used five thousand volunteers to canvass the city. They collected detailed information on prostitution, gambling, drinking, crime, and other behavior they claimed resulted in poverty. Those applying for aid had to agree to have their home inspected; neighbors were interviewed to confirm the applicant's work history and drinking and sexual habits. This information was used to help identify and aid the "legitimately" poor. It was also used to prevent families from receiving aid from more than one charity.[34]

In 1844, the year after it was established, the Association for Improving the Condition of the Poor aided 28,062 families in New York City. Four years later it claimed that its 298 "philanthropic laborers" had made more than 135,000 visits of sympathy and aid to people who were poor through no fault of their own.[35] By the 1850s the Association had 400 friendly visitors and about four thousand contributing members, including some of the most prominent bankers, merchants, and industrialists in the city. With its focus on centralized control of the aid-giving process, the organization emerged as the most influential charitable institution in the mid-nineteenth century.[36]

The Association for Improving the Condition of the Poor strongly supported the idea of putting the able-bodied poor in workhouses. Blackwell's Island in New York quickly became swamped with inmates, 60 percent of whom were Irish. Between 1853 and 1856 7,000 people were sent to Blackwell's Island. The sick and immigrant poor were sent to Bellevue Hospital, which admitted 3,728 individuals in 1850; 70 percent were Irish. These institutions were ravaged by diseases and epidemics.[37]

The Association encouraged New Yorkers not to give anything to street beggars, but instead to donate the money to the organization, which would redistribute the money to the deserving poor.[38] Very strict aid-giving policies were observed, even during the Panic of 1857, when a severe winter combined with a sharp economic downturn. By its own calculations the Association claimed that forty-one thousand people had sought shelter in police stations during the winter and thousands were evicted. Nonetheless, the Association strictly controlled the distribution of aid, blocked the formation of independent ward relief committees, and refused to establish branch offices across the city. This forced the poor to travel to its central office to be interviewed by trained personnel, who applied rigorous means-testing screens before aid was dispersed; three-quarters of those applying for aid were denied help. Even as economic conditions worsened fewer people were given aid. In October 1857, 25 percent fewer people were granted aid than in the prior, more prosperous year. The organization also pressured other charities to follow its lead.[39]

The Association for Improving the Condition of the Poor also tackled the issues of public health and housing. Though bathhouses had been built in the city since 1792, they served only the elites paying to use them. The Association built the first public bathhouse in 1849; it was open only in the summer and charged a fee for its swimming pool, laundry facilities, and baths. The bath served sixty thousand people annually at first, but it closed in 1861 for lack of patronage.[40]

The Association worked to improve tenement conditions in the slums by encouraging businessmen to build model tenements. John Griscom, one of the founders of the Society for the Prevention of Pauperism, promoted the idea of developing well-managed tenements that returned a 6 percent profit to the investor. At the time, investment in upper-class housing brought a 10 percent return and slumlords could reap a 50 to 75 percent return by cramming people into decrepit buildings. Not surprisingly, most businessmen spurned the model tenement idea. Consequently the Association decided to build its own tenement, and in 1855 it opened what was then the largest apartment building in New York, a six-story, eighty-seven-unit tenement just north of Five Points. Known as the Workingmen's House or the "Big Flat," it had indoor plumbing, toilets for each family, and gas lights. The apartments were rented to African Americans, who at the time were forced to live in the worst housing in the city. The venture did not succeed; the rooms were small and cramped and the rents were too high for many occupants to afford but too low to ensure a 6 percent profit. After twelve years of operation the property was sold to the Five Points House of Industry to become a residence for working women.[41]

Though it was a secondary aim, the Association for Improving the Condition of the Poor and other charities did help institute environmental health reforms such as the pure milk campaigns and the development of medical dispensaries.[42]

New Englanders developed their own method of dealing with the poor. Instead of investing in poorhouses and jails many small New England communities resorted to auctioning off the poor in their midst. In what became known as the "New England method," individuals and whole families were put on auction blocks and "sold" to the lowest bidder; that is, auctioned individuals would be sent to live with someone who agreed to feed, clothe, and house them for a year for the lowest cost to the town. In return, auctioned individuals had to work for their keep.[43]

The Children's Aid Society: Westward Ho!

Some charities instituted the "placing-out" system, removing the poor from the cities and relocating them to rural areas. The Association for Improving the Condition of the Poor largely failed in this endeavor because adults were not easily coerced into moving once they had established themselves. But the Children's Aid Society made rural outplacement a major part of its agenda—with mixed results. The Children's Aid Society was founded in 1853 by Charles Loring Brace, who grew up in Hartford, Connecticut. Brace got his first taste of working with New York's poor when he was a friendly visitor at the Five Points Mission. He also worked with the poor on Blackwell's Island. A graduate of the Yale Divinity School and Union Theological Seminary, Brace was a social Darwinist, though he rejected the claim that survival of the fittest was a predetermined outcome. He believed that institutional structures and interventions could change the direction of the lives of the poor. He believed in the benefits of self-help and opposed charitable aid, which he thought encouraged dependence. Brace was also influenced by his minister, Horace Bushnell, who argued that infants were not damned or depraved but could be influenced and changed by unconscious mechanisms.[44]

Before the formation of the Children's Aid Society few groups in the city focused on the plight of poor children; the Society was formed to fill that lacuna. The organization operated six lodging houses (five of them for boys), as well as twenty-one industrial schools that taught carpentry, woodworking, printing, dressmaking, laundry work, and typewriting. The industrial schools were free, but the Society charged a small fee for room and meals at the lodging houses.[45] A Society annual report explained, "Our charges are six cents each for supper, lodging, and breakfast. We make these charges in order to create a feeling of independence amongst our boys. The poor and needy are welcomed without money, and many such have been received . . . 33 percent of our lodgers are received gratuitously."[46] Though the lodging houses generated cash, the Society was also funded by wealthy and influential New Yorkers such as the Roosevelts, Grinnells, Astors, Aspinwalls, William Cullen Bryant, and the mayor. In fact Brace succeeded in getting funding by preying on the fears of the upper class, framing the children as dangerous and threatening. Hence his plan to institutionalize, educate, and remove the children from the city was endorsed by the rich. Aristocrats were so enamored by the plan that Charlotte Augusta Gibbes, the wife of John Jacob Astor III, paid to send seven hundred children west during the 1870s.[47]

The Children's Aid Society is best known for its "emigration plan," or placing-out system. Though the Society is often credited with pioneering the system in America the Boston's Children's Mission (founded in 1849), the New York Foundling Hospital, the Philadelphia Women's Industrial Aid Association, the Association for Improving the Condition of the Poor, and other organizations advocated or developed emigration programs before the Children's Aid Society did. For instance, the Boston's Children's Mission placed out seventy boys and girls in 1850. The placing-out system was loosely modeled on the British "transportation" program, begun in the 1700s, whereby Britain's less desirable citizens were shipped to America, Australia, and South Africa. Initially only convicts were transported out of Britain, but as the program expanded women and children were included. The French had a similar program; they rid themselves of troublesome adolescents and emptied their crowded jails by booking youths and convicts on a one-way trip to Louisiana.[48]

The Children's Aid Society began its emigration program in 1854 by placing 207 children in rural homes. The number of emigrants grew steadily until more than thirty-three hundred were being placed annually by the 1870s. Brace was a firm believer that slum children were redeemable; they could be reformed by hard work, religious guidance from the families adopting them, and the innocence and pastoral beauty of the rural lifestyle he idealized. Brace also saw the placing-out system as a means of emptying the jails and city streets of juvenile offenders and vagrants while meeting the demand for labor in the countryside. In 1860 New York's population was 814,224; of that number 5,880 girls and 2,708 boys were incarcerated. By 1871 the number of children incarcerated had fallen dramatically, to 548 girls and 994 boys. The emigration program also relieved the city of the responsibility of aiding those who were placed out.[49] Brace's logic for instituting the emigration program was clearly articulated in an 1859 essay in which he argued that the Children's Aid Society should "connect the supply of juvenile labor of the city with the demand from the country, and . . . place unfortunate, destitute, vagrant, and abandoned children at once in good families in the country. . . . Young vagrants and petty offenders . . . might with ease be placed in good religious homes in our rural districts where every influence exerted upon them would be far healthier and better."[50]

Brace felt compelled to reform children because they were more malleable and easier to reform than adults. He wrote that the children formed a "perilous element in large cities" which he called the "dangerous class," who grew up to become "ignorant voters."[51] He contended that neither the rich nor rural inhabitants were aware of this danger:

> Beneath the busy and brilliant surface of a city like New York . . . there is a most inflammable and explosive material. It only needs a sufficient cause or opportunity, such as the accidental absence of the guardians of public peace and some supposed wrong inflicted by the rich on the poor . . . and we might experience an outbreak . . . which would leave large portions of the city in ashes and blood, and destroy, as it did in Paris, the accumulation of wealth and treasured art of generations. A "dangerous class" has always for its worse element those who are simply and solely criminals, and who come forth mainly in the night to prey upon and plunder the community. . . . These latter are the neglected youth of a city. They are the children of the poor . . . who have been cast out on the street or left to themselves, or have abandoned their disagreeable tenement homes: the *enfants perdus* become men. . . . This great class are always discontented, and stirred up by designing leaders, they are frequently ready to revolt against the existing order of things. . . . The only really and permanently dangerous element in this city, and one the perils from which cannot be exaggerated, are the neglected youth. They swell the criminal class; or, more dangerous still, they help to form the vast body of ignorant voters.[52]

The Society's friendly visitors scoured the streets to find children for the lodging houses, industrial schools, and emigration program: "The first duty of the 'Visitors' is to pick up the little waifs of the streets; its 'Industrial Schools' teach those poor children, whose parents cannot or do not care for them . . . its 'Lodging-Houses' shelter the orphan and homeless boys and girls . . . and . . . the crown of all, its Emigration System . . . [takes] . . . these children before they are too much corrupted by the streets, transfers them to honest and kind homes and an industrious life in the far West."[53]

After being taught remedial skills, the children were bathed, issued clean clothes, and sent on "orphan trains" to live with families in the Midwest, the West, and to a lesser extent the South. On the first leg of their journey from New York City the first group of forty-five children spent the night on a riverboat traveling up the Hudson. Most of the boys were between the ages of ten and twelve; one was about six years old, and a few were teenagers. They slept on mattresses under clean blankets because the boat's captain took pity on them after hearing their stories. They weren't as fortunate for the remainder of their trip. In Albany they boarded a train, sleeping on the floor in a dark freight car. After leaving Buffalo they crossed Lake Erie in a steamer, from which they emerged soiled with vomit from the motion sickness they suffered and the excreta of the animals on the deck above them. Finally they took a train from Detroit to Dowagiac, Michigan, arriving at 3 a.m. on October 1, 1854. They huddled together at the meeting house until daybreak, when families arrived to pick which boy they wanted.[54]

Brace's description of the process was much cheerier:

> On a given day in New York the ragged and dirty little ones are gathered to a central office from the streets and lanes . . . [and] are cleaned and dressed, and sent away, under the charge of an experienced agent. . . . When they arrive in the village a great public meeting is held. . . . Farmers come in from twenty to twenty-five miles, looking for the "model boy" who shall do the light work of the farm . . . ; childless mothers seek children to replace those that are lost; housekeepers look for girls to train up; mechanics seek . . . boys for their trades. . . . Thus in a few hours the little colony is placed in comfortable homes.[55]

Though critics argued that the Children's Aid Society supplied cheap labor for families to exploit, had lax placement procedures, exercised very little oversight and supervision of the children once they were placed, and deliberately placed Irish Catholic children in Protestant homes, the organization sent thousands of children to the country each year. Dubbed "orphans," about 40 percent of the children had one or both parents living, and 13 percent did not know of their parents' whereabouts. By the time the last orphan train left New York with its emigrants in 1929 the Society, with branches in Baltimore, Boston, Philadelphia, Chicago, Washington, St. Louis, Cleveland, San Francisco, and Brooklyn, had sent 250,000 children to the countryside.[56]

Though the emphasis of the emigration program was on children (boys more so than girls), adults, especially poor women, were placed out too. In 1871 the New York branch placed out 1,856 boys, 887 girls, 303 men, and 340 women. Between 1854 and 1893 approximately 39 percent of those placed were girls.[57]

The Massachusetts Board of State Charities embraced the emigration system because it diffused rather than aggregated the poor. The organization sought to restrain "from the career of crime orphaned children, and [train] them up to virtue; and has done this quite as much in order to prevent future as well as to cure present ills. . . . It has striven to show that the best reformatory is a good family. . . . Into such natural reformatories the Board could fain place as many of this class of children and youth as is possible, without even consigning them to a public reformatory."[58]

The Boston branch of the Children's Aid Society, founded in 1865, developed a modified version of the placing-out system called the "cottage plan." The chaplain of Suffolk County, Rufus Cook, routinely attended police court and secured the release of male juvenile offenders to his custody. Cook then had the boys' parents sign a release granting him custody of the children for a certain number of years. The boys were sent to the Society's Pine Farm in West Newton, Massachusetts, for training; after that they were sent to live with and work for a local farmer. In exchange the farmers were expected to school, feed, and clothe the boys. The Boston branch developed a program for girls in 1866.[59]

The Fresh Air Fund: A Variation on the Placing-out System

In June 1877 Willard Parsons developed the Fresh Air Fund, a variant of the placing-out system in which poor children were sent to the countryside for two weeks of summer vacation. Parsons, a clergyman from Sherman, Pennsylvania, asked his congregation to host children from New York during the summer. In its first year of operation sixty city kids went to Sherman. Later that year the *New York Tribune* took over operation of the program, which it ran until 1962, when the Fund faced bankruptcy. It became an independent, nonprofit organization in 1967. In 1976 the Fund introduced a farm-training program for teenagers. By the mid-1990s it had provided vacations for more than 1.7 million poor children from New York between the ages of five and sixteen. Today about six thousand children vacation with families in thirteen states and Canada annually; an additional three thousand spend time at Fresh Air Fund camps.[60]

The Charity Organization Society: Streamlining Charitable Giving

Charities became even more systematic in dispensing aid during the Progressive Era (1880s–1920s). Josephine Shaw Lowell was one of the driving forces behind making charitable institutions more efficient. Born in West Roxbury, Massachusetts, in 1843 into a wealthy family, Lowell moved to Staten Island in 1847. She began working with charities during the Civil War. After losing her brother and husband during the war she immersed herself in charitable work. Convinced that the jails and almshouses failed to rehabilitate inmates, in 1875 she conducted a statewide study of the able-bodied poor. The following year she was appointed to the New York State Board of Charities. Lowell, the first woman appointed as its commissioner, analyzed the duplication of work of the city's charities and found them to be "wasteful" and "encouraging pauperism and imposture." From her perspective the charities were insufficiently coordinated, despite decades of effort by the Association for Improving the Condition of the Poor to streamline the city's almsgiving efforts. She thought that the state should apply a business model and the principles of scientific management to charities. She believed that the role of state-sponsored charities was not to relieve immediate suffering but to reform the individual. She also believed that the intermingling of the poor in asylums and poorhouses led to hereditary pauperism and created a culture of perpetual dependence and negative attitudes toward work. Hence in 1882 Lowell founded the Charity Organization Society, whose goal was to organize and streamline charitable giving and reduce the likelihood of families receiving aid from multiple sources and of families failing to live up to the standards necessary to receive ongoing aid. The Charity Organization Society was modeled after the London Society for Organizing Charitable Relief and Repressing Mendicity, founded in 1869 to curb "indiscriminate" almsgiving by West End elites. Through the Society Lowell oversaw a range of charitable organizations, wealthy families, and institutions contributing to those charities. All told, the Charity Organization Society pooled and managed the resources of more than five hundred churches and societies and nearly a thousand families.[61]

Lowell made it more difficult for the poor to get aid, arguing that gratuitous charity caused more evil than good. She claimed that helping a widow with infants made the woman's life easier, but because aid relieved the woman's anxiety for the children she lost her love for them. Giving aid to the unemployed helped them through tough times but taught them that it was easier to make a living without working for it.[62] The following statistics give us some indication of how the Charity Organization Society approached almsgiving. In one year the organization investigated 5,169 applicants seeking aid; of those, only 327 (6.3 percent) got continuous relief, 1,269 (24.6 percent) were eligible for temporary relief, 2,698 (52.2 percent) were urged to work rather than given aid, and 875 (16.9 percent) were rejected outright.[63]

To reduce the likelihood that aid recipients received help from multiple charities, the Charity Organization Society developed a centralized Registry Bureau to which all relief agencies, asylums, churches, and the city's Department of Public Charities and Correction sent data on their clientele. By 1887 the Society had collected data on nearly 90,000 families, plus 27,400 houses that were "occupied by the dependent and disreputable classes." By the mid-1890s the organization had dossiers on 170,000 families. In effect, the Society acted as a clearinghouse of information on the poor, providing that information to any person or group that was "charitably interested," such as charities, prospective employers, landlords, banks, and police departments.[64]

The Charity Organization Society was much more stringent than the Department of Charities and Correction. It used volunteers to probe into aid seekers' past and put them through a rigorous screening process. The Society relied on volunteers because Lowell thought the Association for Improving the Condition of the Poor had lost touch with the poor after it switched to using salaried friendly visitors in 1879. The Charity Organization Society also switched to salaried friendly visitors, but only when volunteers became too unreliable. Other cities followed New York's lead. For example, Seth Low, the mayor of Brooklyn, set up a similar program at the Brooklyn Bureau of Charities, and the Cleveland Charity Organization Society coordinated information on thirty thousand families compiled from data gathered by thirty different agencies.[65]

The Charity Organization Society managed to keep aid giving to a bare minimum even during the economic depressions of the 1880s. To reduce panhandling the Society established a Committee on Mendicancy that hired "special agents" to arrest beggars; in 1885 they arrested 700 people for street begging. At the Society's instigation, 2,594 street beggars and 1,474 "persistent offenders" were arrested over a five-year period. In 1889, 2,633 people were arrested for vagrancy, 36 percent of whom were women. Furthermore, the police notified the Society when beggars were released from Blackwell's Island. In all, 138,332 people were placed by the police and social service organizations that year in jails, asylums, poorhouses, workhouses, hospitals, and other institutions dealing with the poor. The Society also opposed the practice of allowing the homeless to sleep in the basements of police stations; though it couldn't stop the practice altogether, it got the legislature to establish a Municipal Lodging House that gave food and a night's lodging (no more than three times per month) in return for labor in the Society's wood yard. In 1893 John M. Kennedy donated the United Charities Building to the Society. In turn the Society persuaded other leading charities, such as the Association for Improving the Condition of the Poor, the Children's Aid Society, and the Mission and Tract Society to move into the building. In 1939 the Charity Organization Society merged with Association for Improving the Condition of the Poor to form the Community Service Society.[66]

Boston's and Baltimore's charities also responded to the changing environment that charities faced. In 1879 the Cooperative Society of Visitors was absorbed by the Boston Associated Charities. Under the guidance of Zilpha

Smith, the Boston Associated Charities began to provide formal training for its charity workers; the program became a model for the rest of the country. In 1897 Mary Richmond of the Baltimore Charity Organization Society called for the establishment of professional training schools in applied philanthropy for those engaged in charitable work. Richmond's idea was acted upon in Boston. The Boston School of Social Work was founded in 1904 by Zilpha Smith and Jeffrey Brackett; it affiliated with Harvard University and Simmons College (males registered at Harvard and females at Simmons).[67]

The YMCA

Between 1860 and 1900 the urban population of the United States grew almost 500 percent, from 6.2 million to 30 million,[68] and disorder remained commonplace. In response some social control agencies, such as the Young Men's Christian Association (YMCA) and Young Women's Christian Association, began focusing on the opposite end of the social spectrum. They provided a safe, Christian environment and recreational opportunities to young men and women recently settled in cities before they became corrupted.[69]

The YMCA was founded in England in 1844 by George Williams, and the first U.S. branch was opened in Boston in 1851 by Thomas V. Sullivan. Two years later, a freed slave, Anthony Bowen, founded a branch for black men in Washington, D.C. The YMCA model caught on quickly, and soon there were branches in Worcester, Massachusetts, and New York City. By 1860 there were over two hundred local branches serving twenty-five thousand members; by the end of the nineteenth century there were fifteen hundred local associations and more than a quarter-million members. Today the YMCA serves about 18.3 million people in the United States.[70]

CHARITIES AND THE CHANGING DEFINITION OF POVERTY

Despite all these reform efforts poverty remained widespread in the late nineteenth century and riots were still commonplace. Ergo groups like the City Homes Association began to question the basic ideological assumptions of the charities movement. Increasingly activists expressed doubts about the assertions that poverty was caused by a person's character flaws and that it could be alleviated by the social and religious uplift efforts of friendly visitors. Even such traditional charities as the Charity Organization Society began to question these assumptions. As Edward T. Divine of the Society asserted, "It is possible that in the analysis of the causes of poverty, emphasis has been placed unduly upon personal causes, such as intemperance, shiftlessness, and inefficiency, as compared with causes that lie in the environment, such as accident, disease resulting from insanitary surroundings, and death of bread-winner due to undermined vitality." The City Homes Association placed the blame for poverty squarely on environmental factors such as bad housing and illness arising from unsanitary conditions. Hence as the nineteenth century ended the definition of poverty evolved to take environmental and economic factors into consideration.[71]

By and large the charities and emigration programs discussed earlier bypassed blacks; their primary purpose was to help poor whites. In response black reformers established charities and orphanages that aided African Americans.

Corporate Paternalism and Benevolent Control

COMPANY TOWNS

Industrialists too tried their hand at social engineering. Textile manufacturers had to place their facilities in the countryside, where they had access to ample water power. There they experimented with developing industrial company towns and selective hiring. New England industrialists pioneered this form of social and industrial reform, and Samuel Slater was at the forefront of these efforts. When Slater landed in New York from Belper, England, in 1789 he was twenty-one years old and had completed an apprenticeship as a mechanic in a textile mill. At the time almost all the yarn manufactured in America was home-spun; there was no mass production of cotton. Slater moved to Pawtucket, Rhode Island, and began working with artisans there to develop water-powered spinning machinery. In 1790, with the backing of the wealthy investors William Almy and Moses Brown, Slater opened a small water-powered textile mill. He hired his workforce selectively and outsourced much of the work, and instead of hiring individual workers he hired entire families, including children as young as eight. Under the Slater (or Rhode Island) system, he established a dawn-to-dusk workday and standardized the yarn production.

Slater and other New England mill owners were appalled by the squalid neighborhoods that typically surrounded factories in industrial cities, so they strove to reconcile the agrarian character of New England's countryside with the facilities they were constructing. Eventually Slater created mill villages that included cottages, churches, stores, and artisan shops for the workers. He also instituted mandatory Sunday schools for the children as well as banks to encourage thrift and savings.[72]

Slater's strategy of hiring women and children into the workforce was endorsed at the highest levels of government. In his 1791 report to Congress on the status of manufacturing in the United States, Secretary of the Treasury Alexander Hamilton urged the acceleration of the process of industrialization of the country. He identified increased division of labor and the use of untapped pools of labor as mechanisms for intensifying industrialization. In particular Hamilton encouraged industrialists to hire whole families because "the husbandman himself experiences a new source of profit and support from the increased industry of his wife and daughters." He continued, "In general, women and children are rendered more useful, and the latter more early useful, by manufacturing establishments, than they would otherwise be." Hamilton noted that the British had already incorporated this sector of the population into their workforce; more than 50 percent of the cotton manufacturing workforce in Britain was women and children, "of whom the greatest proportion are children and many of them of a very tender age."[73]

Like many of the other social reform efforts discussed in this chapter, the development of industrial company towns was influenced by events occurring in Britain. One of the first true company towns was developed by Robert Owen, a former textile worker turned industrialist and social reformer. In the early nineteenth century Owen developed New Lanark, Scotland, as a company town that had high-quality workers' housing, schools, and stores. Francis Cabot Lowell, a Boston Brahmin and a 1793 graduate of Harvard College, was impressed with Lanark when he visited it on his 1810–12 trip to Britain. On his return to America the thirty-seven-year-old Lowell entered into a partnership with his brother-in-law Patrick Tracy Jackson and Nathan Appleton to found the Boston Manufacturing Company (later called the Boston Associates). He also joined forces with an inventor, Paul Moody, to develop an efficient power loom and a spinning apparatus. They built their first mill on the Charles River in Waltham, Massachusetts, in 1813–14. Lowell borrowed ideas from Slater and Owen but also introduced several innovations to his mill. In a pioneering move, he financed the construction and operation of the mill by selling stock in it. Lowell took Slater's innovations one step further by building the first integrated textile mill; that is, the Waltham mill integrated and mechanized all the steps involved in the production of textiles—carding, spinning, and weaving—under one roof.[74]

Lowell and his partners also deviated from the Rhode Island system in the type of workforce they hired. Under the Lowell system the mill hired a segment of the workforce that had hitherto been ignored by industries: unmarried, native-born, young farm women and girls from northern New England, primarily New Hampshire and Vermont. Concerned with maintaining order and alleviating rising social tensions, Lowell and his partners sought to avoid creating a permanent working class around the mill by hiring young women and building housing for them. These women were expected to be in the workforce only temporarily, until they got married, earned enough money to send a brother to college, or helped their family in some other way. (An investigation into factory conditions found that women worked an average of three to four years in the mills.) Moreover, farm women had a home to return to when work was slack or if they couldn't find work in town. To convince parents to let their young daughters, some younger than ten, work in the factories, the Boston Associates developed a rigid system of surveillance and guardianship of their workers. They also created a wholesome atmosphere to prevent youthful workers from falling from grace while toiling in the factories. The company constructed redbrick boardinghouses in which the young women lived under the exacting supervision of matrons. They had a strict schedule that they could not deviate from (except in unusual circumstances) that included a lights-out bell at 10 p.m., when the doors to the boardinghouses were locked and no one entered (except under unusual circumstances). The "mill girls," as Lowell's and other female mill workers were called, had to adhere to a strict dress code, shared meals, and developed a community spirit in the boardinghouses. The company also provided libraries, lecture series, and recreational opportunities for the workers.[75]

Lowell pushed for the expansion of their operations, but he died in 1817, before this came to fruition. To complete his vision, in 1823 his surviving partners in the Boston Associates began investing in textile manufacturing at the confluence of the Concord and Merrimack Rivers, about twenty-five miles northwest of Boston. They established the town of Lowell in 1826; it covered about fifteen square miles and had a population of twenty-five hundred. During the 1820s and 1830s the Associates raised enough to build a string of factories along the Merrimack River. The Lowell system was hailed as a success, and soon other company towns sprang up.

Not only did the Boston Associates invest in charities in Lowell, but they gave generously to Greater Boston's most venerable institutions: Harvard University and Massachusetts General Hospital. They encouraged their workers to open bank accounts and save as well as donate to charitable causes.[76]

As in the Waltham mill, company officials kept strict control over the daily lives of the workers. The days were ruled by the bell. Bells rang to wake workers at four-thirty (a little later in the winter months), to start work at five, and to announce breakfast, the half-hour lunch break (forty-five minutes during the summer), dinner, the end of the workday at seven or seven-thirty, and lights out. The workday was between 11.5 and 13.5 hours per day, depending on the season, and workers got off early on Saturdays and had Sundays off. They had four holidays and no additional vacation. The women had little free time, and privacy was almost nonexistent. Though the boardinghouses were clean, they were crowded and the rooms were poorly ventilated. Each room had three beds, and six or seven women shared each room. Work at the mill was demanding, and as the company faced increasing competition, there were frequent speedups, when each worker attended three or four looms at a time (it required great skill and concentration to attend two looms at a time).

Roughly seven to eight thousand women lived in Lowell in 1846. Church attendance was compulsory; factory rules stated that anyone who did not go to church regularly or was "known to be guilty of immorality" would not be employed by the company. In addition, anyone caught stealing could be prosecuted. The workers got free visits from a physician.[77]

Like Slater, Lowell and his partners hired children to work in the mills. Lucy Larcom was twelve and Harriet Hanson Robinson was ten when they started working at Lowell. Some of the girls employed at the mill were even younger.[78] Seth Luther, a children's advocate who spoke out against child labor, published accounts of the brutal treatment of children in the mills:

> In some of the prisons in New England called cotton mills, instead of rosy cheeks, the pale, sickly, haggard countenance of the ragged child—haggard from the worse than slavish confinement in the cotton mill [is apparent. An observer] might see that child driven up to the "clock-work" by the cowskin [whip], in some cases . . . the child [is] taken from his bed at four in the morning, and plunged into cold water to drive away his slumbers and prepare him for the labors of the mill. . . . The female child too, [is] driven up to the "clockwork" with the cowhide, or well-seasoned strap of American manufacture. We could show [an observer] many females who have had corporeal punishment inflicted upon them; one girl eleven years of age who had her leg broken with a billet of wood; another who had a board split over her head by a heartless monster in the shape of an overseer of a cotton mill "paradise."[79]

The Lowell mill workers were encouraged to write and publish in their spare time. During the 1840s the company sponsored a newsletter, the *Lowell Offering*. Though the essays and poems paint a rosy picture of life in Lowell, one can discern hints of the rigors and dangers of factory life. One author wrote about the omnipresent bells: "Soon the breakfast bell rings; in a moment the whirling wheels are stopped, and she hastens to join the throng which is pouring through the open gate. . . . The short half-hour [for lunch] is soon over; the bell rings again; and now our factory girl feels she has commenced her day's work in earnest." The essayist also wrote, "Some part of the work becomes deranged and stops; the constant friction causes a belt of leather to burst into a flame."[80] Another wrote, "Between them, swings the ponderous gate that shuts the mills in from the world without. But, stop; we must get 'a pass,' ere we go through, or the 'the watchman will be after us. . . .' Look into the first room. It is used for cleaning cloth. . . . There are, occasionally, *fogs and clouds; and* not only fogs and clouds, but sometimes plentiful showers." The author continued, "And we have the 'carding room.' . . . I beg of you to be careful as we go amongst them, or you will get caught in the machinery."[81]

Around the time the Boston Associates were developing mills in Waltham and Lowell, mills that employed a variant of the Lowell system were being developed at Amoskeag Falls on the Merrimack River in Manchester, New Hampshire, located about fifty-eight miles from Boston. In 1809 a small cotton mill, the Harvey Mill, was constructed at this location. Though the proprietors knew little about processing cotton, they expanded the company in 1810 to form the Amoskeag Cotton and Wool Manufacturing Company. The name of the town was changed from Deerfield to Manchester that same year. The Amoskeag workers did not work at the mill; instead the work was "let out" to women in the community. While Slater hired whole families, Amoskeag owners hired primarily women. On average women wove ten to twelve yards per day for two to seven cents per yard. Even so, the operation was not profitable, and as late as 1822 little manufacturing was actually done at the mill. It was sold more than once and eventually ended up in Slater's hands (as part owner). Slater and his partner sold three-fifths of the company to a group of Boston investors (Oliver Dean, Lyman Tiffany, and Willard Sayles) around 1826. The new management purchased the water power rights of the river and fifteen thousand acres of land and set about building a new mill and developing a town with stores, shops, and boardinghouses. The Amoskeag Manufacturing Company was chartered in 1831.[82] By 1905 Amoskeag was the largest textile mill in the world. At its pinnacle in 1915 it employed seventeen thousand workers, occupied eight million square feet of space, and produced 164,000 miles of cloth in seventy-four different departments.[83]

Mill owners exerted a mix of corporate paternalism and benevolent control over their workers and towns. They saw the workers as the corporation's children and treated them as such. Desirous of having a stable community around the mill and a pliant workforce within it, the owners of Amoskeag funded the development of community services, local churches, and charities, and provided their workers with good benefits compared to workers in large urban areas. They provided housing, mortgages, medical care, and recreational facilities as well. As was the case in Lowell, Amoskeag developed a company town. In it about 15 percent of the workers lived in low-rent company-built brick tenements, and the remainder lived in single-family company houses clustered around churches in ethnic neighborhoods. The boardinghouses for the "mill girls" were rented at submarket rates, about one dollar per month for a room. Company housing was well maintained. Amoskeag controlled the growth of the town by placing deed restrictions on prime industrial land. They were also very selective in their hiring, employing only native-born New England women at first. The boardinghouses were locked at 10 p.m., church attendance was mandatory, and alcohol consumption banned. The company's stranglehold over the workers' lives began to loosen after the Civil War, when immigrants began to replace native-born women in the mills.

Company towns flourished well into the twentieth century. George M. Pullman built the town of Pullman (now part of Chicago) around his Pullman Palace Car Company.[84] Gary, Indiana, started as a carefully planned company town supporting a steel mill in which workers' housing was sorted by status and class. The best housing and jobs were reserved for native-born managers, while the smallest, shabbiest housing and worst jobs went to Eastern and Southern Europeans. When the workforce became racially mixed the sorting took race into consideration. Once U.S. Steel began to hire blacks and Hispanics as strikebreakers, the worst jobs and housing went to those two groups.[85] Other steel mills, such as Sparrows Point in Baltimore, also developed company towns where ethnicity, race, and class determined the kind of job one got and the location and quality of house one lived in.[86]

LABOR UNREST

After the introduction of Slater's and Lowell's innovations the textile industry grew rapidly in the United States, and New England was the epicenter. In 1860 there were 81,000 cotton mill workers and 25,583 wool mill workers in New England. By 1905 there were 23 million cotton spindles operating in the United States, 14 million of which were in New England. Of the 310,458 cotton mill workers in the country in 1905, 155,981 lived in New England. Most of the wool mill workers in the country also lived in New England.[87] Therefore it is not surprising that New England was at the center of labor conflicts in the textile industry.

Women were an integral part of the labor unrest that hit the textile industry. In 1824 male and female textile workers in Pawtucket walked off the job to protest wage cuts. The following year the women of the United Tailoresses of New York City struck for higher wages. In 1828, less than two years after the mill was incorporated, about four hundred female workers struck at the Cocheco mills in Dover, New Hampshire, to protest "obnoxious regulations." This was the first strike organized by female textile workers.[88] Cocheco mill workers had to sign a draconian contract before they were hired by the company, and the workers resented this. The contract stipulated that workers had to work for wages "as the company may see fit to pay, and be subject to the fines" administered by the company. Workers had to "agree to allow two cents each week to be deducted from [their] wages for the benefit of the sick fund," and they could not "be engaged in any combination [union activity] whereby the work may be impeded or the Company's interest in any work injured."[89]

The first walkout occurred at Lowell in 1834. This was an ephemeral strike in which about eight hundred workers walked off the job on February 20 to protest a 15 percent pay cut. Young female activists organized the strike, led a procession around town, made public speeches (unusual at the time), and advised their coworkers to make a run on the Lowell Bank as well as the Savings Bank. About one-sixth of the workers joined the walkout. Strike instigators were promptly fired. Though the all-male National Trades Union supported the strikers, the women did not get any help in organizing or money to support the strike. The strikers used the master frame of slavery to make their point, likening work in the factories to the enslavement of white New England women, and they focused on the immorality of the situation. They also juxtaposed the condition of bondage with their fondness and longing for liberty. The strike began on Saturday, and on Monday morning all the women were back on the job without getting any concessions from management.[90]

Throughout the 1830s and 1840s Lowell workers objected to pay cuts, rent increases, production speedups, longer working hours, and the regulation of their behavior. In 1835 the Lowell mills were producing about 39.2 million yards of cloth annually. That year the average wage for women was $1.90 per week after room and board was paid for, while males earned about $4.80 per week after expenses.[91] Reiterating the master frame developed in the 1834 strike, that Yankee women would not be treated as slaves, about two thousand workers, one-fourth of the workforce, went on strike in 1836, and the Factory Girls' Association was formed, with a membership of twenty-five hundred. In 1844 the Lowell mill worker Sarah Bagley transformed the Factory Girls' Association into the Lowell Female Labor Reform Association. The group formed an alliance with the New England Workingmen's Association, and together they petitioned the Massachusetts Legislature to get the workday reduced to ten hours. The petition had more than two thousand signatures, but was unsuccessful. The Lowell workers struck again in 1846, but with limited success. They were not able to secure the ten-hour day until 1874.

In general, labor activism among native-born women peaked in the 1840s. As they left the mills during the second half of the nineteenth century, they were replaced by immigrants; as early as 1842 the mills began hiring mostly Irish foreign-born workers. The influx of Irish immigrants into New England mill towns led to ethnic tensions and conflicts. In 1854 a riot broke out between the Irish and Yankees in Manchester, New Hampshire.[92]

White slavery was a salient theme that mill activists used in organizing. Northern whites, concerned with the condition of white women in the workforce, could identify with the image. An account of the recruiting practices of the mills described the situation:

> We were not aware, until within a few days, of the modus operandi of the factory powers in this village of forcing poor girls from their quiet homes to become their tools and, like Southern slaves, to give up their life and liberty to the heartless tyrants and taskmasters. . . . [They drive a] "long, low, black wagon" . . . that is what we term a "slaver." She makes regular trips to the north of the state [Massachusetts], cruising around in Vermont and New Hampshire, with a "commander" whose heart must be as black as his craft, who is paid a dollar a head for all he brings to the market, and more in proportion to the distance—if they can bring them from such a distance they cannot easily get back. . . . This is done by . . . representing to the girls that they can tend more machinery than is possible, and that the work is so very neat, and the wages such that they can dress in silks and spend half their time reading. . . . They are stowed in the wagon, which may find a similarity only in the manner in which slaves are fastened to the hold of a vessel. . . . Philanthropists may talk of Negro slavery, but it would be well first to endeavor to emancipate the slaves at home. Let us not stretch our ears to catch the sound of the lash on the flesh of the oppressed black while the oppressed in our very midst are crying out in thunder tones, and calling upon us for assistance.[93]

The theme of white slavery also made its way into the writings of Herman Melville. After describing female mill workers who were "pale with work, and blue with cold" in the "The Paradise of Bachelors and the Tartarus of Maids" Melville wrote, "Machinery—that vaunted slave of humanity—here stood menially served by human beings, who served mutely and cringingly as the slave serves the Sultan."[94]

By the 1840s activists had broadened their framing beyond wages and white slavery to include hazardous work conditions, ill health, and substandard living conditions. Their petition before the Massachusetts Legislature in 1845 to reduce the length of the workday shows this evolution and the way activists bridged several ideas to create new master frames. As the workers put it, they toiled "from 13 to 14 hours per day, [remained] confined in unhealthy apartments, [were] exposed to the poisonous contagion of air . . . [were] debarred from proper Physical Exercise, Mental discipline, and Mastication cruelly limited, and thereby hastening us on through pain, disease and privation, down to a premature grave."[95] In addition to more than eleven hundred mill operatives at Lowell, most of whom were women, who signed the petition for the ten-hour day, hundreds of male mill workers in Andover and Fall River signed similar petitions that were sent to the legislature.

The petitions prompted the legislature to form a special committee to investigate conditions in the mills. Though the committee was skeptical of the petitioners' claims, their investigation corroborated reports of the grueling work schedule. Some women reported to the committee about getting sick on the job. Olive J. Clark of the Lawrence Corporation believed the long hours made her ill; she reported that almost every week someone in her unit missed work because of illness. She thought the air was polluted with small, suspended particles of cotton. When committee members inquired about the number of workers who simply left the factories and returned home when they fell ill, mill supervisors indicated that this was a minuscule number. However, the committee suspected that the numbers were higher than reported.[96]

Committee members interviewed the town physician, who reported that there was less illness among employees of the mill than among townspeople who did not work at the mill. In 1844 Lowell had a population of 25,163; of that number 15,637 were females. The special committee examined the rates of diseases from 1840 to 1844 and found that respiratory diseases were common among inhabitants: 314 cases of consumption (tuberculosis), 115 cases of inflammation of the lungs, and 46 cases of croup were reported. There were also 134 cases of cholera, 91 cases of scarlet fever, 26 cases of measles, 95 cases of dysentery, and 36 cases of inflammation. The death rate had been climbing steadily from 1840 to 1842 but fell dramatically in 1843 and 1844, after the town built a sewer system and a hospital. The committee warned that the number of illnesses and deaths might be misleading because some female workers left town when they got sick. The agent for the Merrimack Corporation responded to this by saying that sick women returning home from the mill was not a problem because their sisters or friends often showed up to replace them.[97]

The special committee also examined an 1841 survey of 203 women who worked at Boott Mill No. 2 and found a relationship between the women's age and where they worked in the mill. The average age of the women in the weaving room, where 55 percent of the women surveyed worked, was the lowest (22.27 years), and they worked in the mill for about 3.23 years. The women in the spinning room tended to be over 28 years old. While 14.3 percent of the women reported that their health had improved since working in the mill, 27.1 percent reported that their health had deteriorated. Though the committee acknowledged that factory conditions could be improved, they concluded, "The factory system . . . is not more injurious to health than other kinds of indoor labor." They also concluded that it would be impossible to reduce the length of the workday without affecting wages and that therefore wages should be negotiated between workers and factory owners, not legislated.[98]

Despite the "Amoskeag spirit" that the company worked hard to create among its laborers, workers began to unionize in 1917 and struck a few years later. Though the workers got some help from national unions, after a nine-month strike in 1922 to protest pay cuts and longer working hours they returned to work without winning any concessions. There were strikes in 1934 and 1935 also. Amoskeag responded to the latest round of strikes by shutting down the company on Christmas Eve 1935, while declaring its intention to reopen again. In the meantime the company filed for bankruptcy protection; its antiquated equipment made it difficult to compete with the lower operating costs of the southern mills. Before the company could reopen in 1936 a devastating flood inundated several of the mills, ruining the machinery. This was the final straw. A court ordered the company to liquidate its assets, and Amoskeag ceased to exist.[99]

INDUSTRIAL UTOPIANISM

During the middle of the nineteenth century a small number of industrialists experimented with industrial utopianism. South Manchester, Connecticut, was established in 1838; St. Johnsbury, Vermont, in the 1830s; Hopedale, Massachusetts, in 1842; Peace Dale, Rhode Island, in 1848; and Ludlow, Massachusetts, in 1868. Unlike the absentee mill owners who pioneered the development of company towns, industrial utopians planned small towns in which they lived side by side with their workers. These industrial utopian communities sprang up in small New England mill towns of two thousand to three thousand residents. Mill owners controlled the size and construction materials of the homes and the layout of the towns. The communities were landscaped; streams and reservoirs used to operate steam engines and turbines were protected and made into amenities for the community. The homes were usually set apart from the factories, and trees were sometimes planted to buffer the noise from the factories. Though skilled and unskilled workers lived in close quarters and intermingled, class distinctions were quite apparent. The largest homes belonged to the mill owners and were set apart from those of the workers. Single-family and duplex cottages were constructed for the workers and rented at affordable rates. Some of these model communities won prizes at World's Fairs.[100]

The utopian villages instituted programs that acculturated immigrant workers into the American lifestyle. Workers' children were educated, and the libraries had free lending programs. There was entertainment or town hall meetings during the evenings. The companies encouraged workers to participate in village improvement projects and compete against each other for prizes for gardening and yard maintenance. Yet despite the veneer of egalitarianism that characterized these towns, corporate paternalism and social control was deeply embedded within them. To achieve an atmosphere in which business prospered while workers happily tended their tasks, the companies exerted significant control over the workers' lives. By the late nineteenth century communities like these were falling out of favor as economic pressures supplanted the utopian visions and as workers grew more resistant to companies having total control over their lives.[101]

TRANSCENDENTALISM AND INDUSTRIALISM

Transcendentalism arose, in part, as a response to the growing industrialization of America. Transcendentalists were concerned with the loss of nature in this process and the resulting lowered quality of life in the cities. They were also concerned with the relationship between humans and the industrial world. Living in Concord, Massachusetts, and similar towns, the Transcendentalists had ample opportunity to observe the growth of industrial towns all around them, and they provided social and literary commentary on what they saw. Some Transcendentalists were attracted to attempts to blend industrialism with the development of utopian industrial communities in places like Hopedale. However, Transcendentalists like Henry David Thoreau were critical of the economic policies that drove the industrial revolution so evident around him. Unlike the urban reformers discussed in this chapter, the Transcendentalists did not involve themselves directly with reforming the lives of poor urbanites, although two well-known Transcendentalists tried to link Transcendental ideas with social change. Bronson Alcott, who developed a utopian community named Fruitlands, encouraged the children enrolled in his Temple School to learn through open discussion and respond freely to the Bible; George Ripley established a utopian community at Brook Farm. However, for the most part Transcendentalists focused on the ideas that emerged from their discussions, writing, thinking, and lecturing. This approach earned them a reputation for having their heads in the clouds. Even some Transcendentalists criticized this approach. For instance, in "The Celestial Railroad" Nathaniel Hawthorne, who spent time at Concord and at Brook Farm, wrote about the "Giant Transcendentalist": "We caught a hasty glimpse of him, looking somewhat like an ill-proportioned figure, but considerably more like a heap of fog and duskiness. He shouted after us, but in so strange a phraseology that we knew not what he meant, nor whether to be encouraged or affrighted."[102] Nonetheless, the contributions of Transcendentalists should not be underestimated. Their concern with nature and their writings about relations between humans and nature led to greater public environmental awareness by the end of the nineteenth century. In addition, Thoreau's argument in *Civil Disobedience*, that individuals have a right to oppose the state when the state is morally wrong, has been very influential.[103]

Though they were sometimes critical of Transcendentalism, detractors of industrialism such as Hawthorne and Melville sided with Transcendentalists when they critiqued industrialism. Hawthorne spoofed this in "The Celestial Railroad," while Melville depicted the leisurely life of the bachelors and the degradation of the female factory workers in "The Paradise of Bachelors and the Tartarus of Maids" in 1855.[104]

Settlement Houses and Social Change

In the late 1880s, around the time charity workers began questioning the causes of poverty and approaches to alleviating the problem, a new kind of social reform organization, the settlement house, emerged in American cities. Settlement houses resembled the traditional charities in some ways, but were radically different in others. Staffed by middle-class college graduates, they took the friendly visitor model further: settlement house workers actually lived in slum neighborhoods. By establishing model middle-class homes and by living and interacting with the poor on a daily basis, settlement workers could demonstrate the lifestyle they wanted the poor to emulate.[105]

The settlement house movement had strong disagreements with the

Charity Organization Society and other traditional charities. The movement rejected the idea that poverty was caused by character flaws and that religious conversions and punitive sanctions would lift people out of poverty. Though some settlement workers believed that personal flaws were linked to poverty, for the most part they believed that poverty was caused by structural factors in the society and that social change required government intervention. Thus they focused attention on government, business, and environmental reforms. Settlement workers believed that improving environmental conditions would improve people's lives. Like the older charities, the settlement houses collected data on slum dwellers, but instead of using the data to dispense aid and sanctions to the poor they used neighborhood-level data to push for reforms and help restructure the urban environment.[106]

THE FIVE POINTS HOUSE OF INDUSTRY: A PROTOTYPE

The "House of Industry" was an early prototype that had some of the elements that characterized settlement houses. The earliest House of Industry was founded by Joanna Bethune in 1814 in New York. Decades later the Five Points House of Industry elaborated on Bethune's model and in so doing, became a transitional form between the early houses of industry and the settlement houses. The Five Points House of Industry arose out of a conflict between female missionaries and a brash young resident minister they hired to staff their mission in the slums. In 1848 the New-York Ladies' Home Missionary Society of the Methodist Episcopal Church established a missionary program with a resident staff member in the New York slum called Five Points. The Missionary Society, run by middle- and upper-class married women, wanted to convert slum dwellers into Methodists. The group established the Five Points Mission in a former brewery in the center of the neighborhood in 1850, with a chapel, schoolrooms, bathrooms, wardrobes, and apartments. One female administrator of the mission claimed that the children served were "a more vivid representation of hell than [she] had ever imagined." A year after its opening the Five Points Mission announced that so much progress had been made that churchgoers could be kept orderly without the aid of police.[107]

The resident missionary was Lewis M. Pease, a thirty-year-old minister. Within a year of accepting the post Pease and the women who ran the mission began feuding bitterly. While the Missionary Society concentrated on turning Five Points residents into Methodists and wanted Pease to visit potential converts and preach to them, Pease quickly realized from his interactions with Five Pointers that his efforts would be fruitless unless he could help them find jobs. Because the mission's board refused to support secular activities, Pease arranged jobs on his own. He made an agreement with a clothing manufacturer to provide cloth; he then hired destitute Five Points women to manufacture shirts. At first the women showed up to work drunk and dirty, so Pease established strict work rules: workers had to be sober, come to work on time, eschew alcohol, and attend church regularly. Though the first batch of shirts were ruined and Pease had to pay for them himself, he continued the project. Within a few weeks he had hired one hundred women.[108]

Pease's employees had great difficulty shunning alcohol, so he decided to insulate his workers from neighborhood vices by renting an entire building to house them. When he couldn't find a tenement large enough, he convinced the police and district attorney to evict and arrest brothel keepers in two houses on Little Water Street, on which the Five Points Mission was located, and chose 30 of his workers to live there. The women were charged $1.25 per week ($28 in 2005 dollars) for room and board. By the spring of 1851 Pease had rented additional houses to lodge 120 of his workers. Though he was accumulating debt for these endeavors, he continued to expand the services offered to Five Pointers. When the Ladies' Missionary Society balked at funding a secular day school he obtained funding elsewhere and opened the school in 1851.

By midyear it was obvious that Pease and the Missionary Society had incompatible agendas, so he left. He announced that he wanted to form an Industrial House for the Friendless, the Inebriate, and the Outcast, and the National Temperance Society funded the operation. The name Five Points Temperance House was chosen for Pease's operation. Within ten months the Temperance Society stopped the funding because of a change in leadership at the society; luckily his work in Five Points had caught the attention of the Episcopal Church, philanthropic organizations, and businessmen willing to fund the organization. In May 1852 the organization was renamed the Five Points House of Industry.[109] By 1853 the House of Industry had gained control of all seven buildings on Little Water Street facing Paradise Park. The headquarters of the organization was on Worth Street, directly across from the park and the Missionary Society's new building. This complex was a dominating presence in Five Points.

Pease was not the first to articulate the idea of a House of Industry; neither was he the first to open one since Bethune's. In the early 1840s Horace Greeley suggested that the city start a house of industry where the poor would live and learn job skills and thrift. A House of Industry and Home for the Friendless was opened in 1848 a few miles north of Five Points by an Episcopal minister,

Stephen H. Tyng, and John Griscom. But this facility operated mainly as a shelter for homeless girls. Thus Pease was really the first to fully implement Greeley's idea.[110]

Pease expanded the type of job training offered, adding baking, shoemaking, corset making, basket weaving, and hat making. In 1854 he employed more than 500 people. Top seamstresses earned $2.50 per week ($50 in 2005 dollars). In 1857 the House of Industry helped to find jobs for 630 of its trainees.[111]

When Pease realized that children came to the school hungry he opened a "soup room" so they could get a hot lunch. The House of Industry extended its charitable acts to others in the community as well. While other charities tightened aid during severe winters and economic depressions, the House of Industry expanded its efforts to help the poor. Consequently during the 1854–55 winter the House of Industry served 39,267 free meals in a four-month period.

The Five Points Missionary Society didn't serve as many meals as the House of Industry, but focused on other kinds of aid. The Missionary Society gave away 17,569 pieces of clothing, 922 pairs of shoes, 355 quilts, 250 caps, 150 bonnets, and 25 tons of coal during the twelve-month period ending in early 1856. Neither the Missionary Society nor the House of Industry gave any aid to children who did not attend school, nor did they aid adults who did not pledge sobriety. Both organizations held meetings in Paradise Park to preach temperance and made friendly visits to check up on the people they aided.[112]

Like Charles Loring Brace of the Children's Aid Society, Pease believed in the placing-out system. In 1853 the House of Industry purchased a sixty-four-acre farm sixteen miles north of Manhattan in Eastchester. Pease thought the fresh air, bucolic rural setting, fresh food, and invigorating labor would be effective in reforming Five Pointers. However, the farm could accommodate only a few children and adults at a time. The House of Industry and the Missionary Society expanded their placing-out activities by putting up large numbers of Five Points children for adoption around this time.[113]

Though the House of Industry is not considered a settlement house in the classic sense of the term, a careful analysis leads one to conclude that the organization could be considered an early form of a settlement house. The Progressive Era reformers would popularize the settlement houses in the 1890s, but Pease gave life to Greeley's idea decades earlier in the slums of Five Points.[114]

THE EMERGENCE OF THE SETTLEMENT HOUSES

Toward the end of the nineteenth century the settlement houses emerged as the latest iteration of social reform organizations trying to improve the lives of the poor, alleviate poverty, and ameliorate environmental conditions in the city. Like many other ideas about charity and social reform, the settlement house idea came partly from Britain, specifically London and the work of Oxford University professors and their students. In 1884 Canon Samuel Barnett, with the help of some Oxford students, purchased and renovated an old boys' industrial school building in the Whitechapel area of London's East End. The building, Toynbee Hall, served both as a residence for reformers and a community center for neighborhood residents. Toynbee Hall was established under the auspices of the University Settlement Association, a collaborative effort between a coalition of Oxford professors John Ruskin, Sidney Ball, and others, their students, and Parliament; some Cambridge students also participated. Members of the University Settlement Association investigated the condition of the poor and developed plans to improve their well-being. Barnett, the vicar of St. Jude's Parish in Whitechapel, was selected as director of Toynbee Hall. The reformers saw themselves as settlers in the neighborhood; hence the term *settlement house* was used to describe organizations of this nature.[115]

Settlements brought together recent college graduates and residents of poor neighborhoods for mutually beneficial collaboration. The organizers of University Settlement believed that if educated, well-to-do young men lived among the poor, those young men could establish a sense of community and assert some social control over and transmit culture and education to the slum dwellers. The settlement workers lobbied politicians to provide better housing, cleaner streets, and recreational facilities for poor neighborhoods. Today Toynbee Hall still operates as a community center in which volunteers live in residence.[116]

In America the Neighborhood Guild was an early prototype of the settlement house. After earning a doctorate at Humboldt University in Berlin in December 1885 Stanton Coit spent three months volunteering at Toynbee Hall. Upon his return to New York in 1886 he founded the Lily Pleasure Club, a boys' club, at 146 Forsyth Street on the Lower East Side. Coit, who was born in Columbus, Ohio, and studied at Amherst College and Columbia University, before going to Germany also participated in the Ethical Culture Society and was influenced by the Social Gospel movement, an attempt by Protestant ministers to apply the teachings and doctrines of Christianity to the social, economic, and political problems they encountered in their parishes. Coit was also influenced by the writings of Emerson, Kant, and Coleridge. The Ethical Culture Society stressed the importance of expressing personal virtue and commitment through service. Coit, having studied German culture and society, was committed to the idea of the positivist state that should use its power to advance the general welfare of people. He was convinced that he could best express his beliefs and commitment through social reform work. He changed the name of the Lily Pleasure Club to the Neighborhood Guild in 1887 and invited a small group of male reformers, among them Charles B. Stover and Edward King, to participate in the Guild. Unlike other reformers of the time Coit believed in the agency of slum dwellers. He believed that grassroots leaders could be found in the community and advocated a system in which residents were organized into guilds of about one hundred families in size. These groups operated as cooperatives to help residents develop their skills and abilities. The Guild joined forces with local labor unions, organized lectures and theater performances, and established boys' and girls' clubs, a kindergarten, and a gymnasium. When Coit moved away in 1887 to take up a ministerial position in London, the Guild collapsed.[117] But Charles Stover and Edward King were committed to the idea of developing a settlement. In May 1891 they reorganized the Guild as University Settlement, and Stover became the head resident.

That same year Vida Scudder and other Smith College graduates who met at an alumni gathering began making plans to develop a settlement. Julia Davida Scudder was born in 1861 in India, where her father, David Coit Scudder, was a Congregationalist minister and missionary. She and her mother returned to Boston in 1862 after the accidental drowning of her twenty-eight-year-old father. Scudder had had a privileged upbringing in a wealthy Boston family that traced its lineage to Governor John Winthrop and the earliest settlers of Boston. She was a member of the first graduating class of Boston Girl's Latin School in 1880. After graduating from Smith College in 1884 she and Clara French became the first two American women to be admitted for graduate study in literature at Oxford, where she studied under Ruskin and was introduced to the settlement idea. Scudder, Jean Fine, and Helen Rand had seen the settlements in London and were enthusiastic about the idea. In 1887 they founded the College Settlements Association to help the poor while providing college-educated women a chance to broaden their horizons by engaging in meaningful work. Katherine Coman, Katherine Lee Bates, Cornelia Warren, Jane E. Robbins, and Helena S. Dudley joined the College Settlements Association, which established its first settlement on September 1,

1889, on Rivington Street on New York's Lower East Side. It was called College Settlement New York or the Rivington Street Settlement. The first residents were seven graduates of Smith, Wellesley, and Vassar. Jean Fine was the head resident. In the first year of operation eighty women applied for residency in the settlement house. In 1899 the Association opened College Settlement in Philadelphia (reorganized as the Starr Center in 1900), and in 1902 it opened Denison House in Boston. In addition College Settlements Association established chapters at Wellesley, Vassar, Bryn Mawr, Radcliffe, and other colleges, as well as in several girls' finishing schools. The Association offered fellowships to men and women ranging from $300 to $500 per year to cover the cost of room and board at the settlements.[118]

The Rivington Street Settlement, located not far from the Neighborhood Guild, opened a week before Jane Addams and Ellen Starr opened Hull House in Chicago.[119] Addams and Starr, friends since the time they attended Rockford Seminary in Illinois, toured Europe in 1888 and visited Toynbee Hall. They were attracted to the idea of creating a social settlement in Chicago. Addams, who would become the first American woman to win the Nobel Peace Prize for her work at Hull House, was also motivated after seeing the Armour Mission in Chicago. In December 1886 the millionaire Phillip Armour, a meat-processing magnate, opened a mission on Thirty-third Street in memory of his brother. The Armour Mission, which was surrounded by middle-income housing (also built by Armour), had a kindergarten, day nursery, kitchen, library, classrooms, auditoriums, and dispensary. Addams and Starr opened Hull House in September 1889. William Kent donated an old mansion that activists converted into Hull House. Kent's family gave up their law practice in Connecticut to rent a pork house in Chicago in 1858. In 1860 they bought a packinghouse and later made millions from the business.[120]

The settlement workers were quite young, on average twenty-five. Addams was twenty-nine when she founded Hull House. Though some activists spent a lifetime working with the settlements, most spent an average of three years living in poor neighborhoods and working with particular settlement houses. Ninety percent of the settlement workers went to college; 80 percent had a bachelor's degree, and 50 percent had done some graduate work. Most were unmarried, born in the Northeast and Midwest, and grew up in middle- and upper-class homes in cities.[121]

The settlement house idea spread rapidly in the United States. In 1897 there were seventy-four settlement houses; by 1900 more than one hundred had been established, and in 1910 there were more than four hundred. Most were in major cities such as Boston, Chicago, and New York. New England alone had fifty-eight settlement houses, thirty-five in the Boston area. Many small cities and rural communities had settlement houses also.[122]

Hull House is one of the most renowned settlement houses in the United States. Located on South Halsted Street in Chicago, it housed the settlement workers Julia Lathrop, Grace and Edith Abbott, Florence Kelly, Alice Hamilton, and Sophonisba Breckinridge. Addams and Starr were the first two residents to move in. Hull House had once belonged to a wealthy businessman, Charles J. Hull; by the time the settlement was established the old mansion had been surrounded by factories, slaughterhouses, and a teeming tenement district.[123]

Settlement houses required their workers to live among the poor for extended periods of time because reformers thought that living "within the limits of one's class belittles manhood." They saw it as their duty to improve poor neighborhoods while at the same time broadening their horizons by moving out of the comfort zone of their own cozy upbringing. Moreover female settlement workers saw themselves as examples of the New Women, which required them, the first generation of college-educated women, to engage in social reform as a means of liberation. By liberating others, they would liberate themselves. According to Vida Scudder, "It is not enough for woman to be free, she must be useful."[124] Other wealthy white women of the time took to the mountains and wilderness to express their feminism and liberation; however, the female mountaineers and wilderness and wildlife activists expressed their feminism and undertook activist environmental agendas within the milieu of their social class. Working-class white women unionized and flocked to dance halls, circuses, and amusement parks as a means of defying convention, expressing their womanhood, and demonstrating their independence.

The activists developed agendas that linked access to recreation with improved health and working and living conditions. The settlement houses were places where the playground, the occupational safety and public health movements, urban environmental improvement, human rights, and social justice activism converged. Hull House provided services to poor immigrants to help them adjust to the city; it also contained a model playground and comprehensive recreation facilities. Hull House and other settlement house activists tried to understand the connection between the environment, occupational hazards, disease, and political action.[125]

It would be unfair to portray slum neighborhoods as social lacunae or a tabula rasa where settlement house activists created and maintained the community institutions. Some slum neighborhoods, such as Chicago's Packingtown (surrounding the stockyards), had an elaborate infrastructure of churches and voluntary associations that helped to meet the needs of the residents before settlement houses were established there. An 1889 directory lists 215 voluntary associations in the district; 59 of them were churches and 91 were secret societies, among them the Masons, Odd Fellows, and the Ancient Order of Hibernians. The YMCA, Salvation Army, and Knights of Labor had branches there, as well as several friendly and benevolent societies.[126] Similarly New York's Five Points neighborhood had a large number of churches, banks, businesses, missions, and other charitable organizations by the time settlement houses were established.[127]

SETTLEMENT WORKERS AND BLACKS

Settlement workers had a complex and sometimes strained relationship with African Americans. Though they worked in centers of social justice and progressive politics, some expressed the racism and nativism common in the nineteenth century. Hence the question of service to blacks was a thorny issue for the settlements. Some established segregated facilities for blacks, though in 1910 there were only ten settlement houses nationwide designated for blacks. A few settlements had integrated houses, but by and large they served white immigrants. For instance, Hull House served primarily the Italian and Slavic immigrants living in the Halsted Street area. Similarly Northwestern University Settlement (opened in 1891), the Maxwell Street Settlement (opened in 1893), the University of Chicago Settlement (opened in 1894 by Mary McDowell in the Packingtown district), Graham Taylor's Chicago Commons (also opened in 1894), and Association House (opened in 1899) all focused their attention on Chicago's white immigrants. Like many of the charities before them, some settlement workers believed that the social reforms were intended to enhance the progress and assimilation of whites. They also ignored blacks because they believed that whites would not use the settlements if they were integrated. Since the raison d'être of the settlements was to improve the condition of whites, even when settlement workers worked with blacks it was not their core mission.[128]

Some settlement workers were at best ambivalent and at worst hostile toward blacks. As one worker in a black settlement house in Philadelphia put it, "Our settlement has its unique problem for it deals not with a race that is intellectually hungry." A committee of settlement workers in Boston concluded that the city was not a "healthy or suitable home for the Negro race." Consequently the committee urged settlement workers to dissuade blacks from leaving the South, where they "naturally" belonged, although the committee decided that they would still work with blacks "foolish" enough to stay in the North. In a more extreme case Frederick Bushée, an avowed nativist settlement worker, claimed that the typical black Bostonian was "low and coarse, revealing much more of the animal qualities than the spiritual."[129]

THE STATUS OF BLACKS

Motivations aside, some settlement workers sponsored or helped to conduct some of the earliest studies of northern urban black communities. In 1899 the Philadelphia College Settlement sponsored W. E. B. Du Bois, who produced the classic study, *The Philadelphia Negro: A Social Study*.[130] Du Bois demonstrated that the plight of urban blacks was severe, and that as bad as it was for poor whites conditions were worse for blacks. He reiterated this point when he published a series of articles on the condition of blacks in three northern cities, New York, Philadelphia, and Boston, in the *New York Times* in 1900.[131] In 1909 Du Bois and some progressive settlement workers helped to form the National Association for the Advancement of Colored People (NAACP).

When Du Bois conducted his study he was a young African American scholar who had just completed a doctorate at Harvard. While other settlement workers and sanitarians before them had conducted studies that examined the plight of the poor, these studies were conducted mainly among white ethnics. In many instances those studies ignored the racial inequalities that were evident or gave them cursory treatment; they did not explore how race and class interacted to produce even more severe inequities. Du Bois broke new ground and racialized the discourse around environmental inequality in a way that had not been done before. His Philadelphia study was done with a submerged environmental justice frame. It is the earliest systematic neighborhood study of its kind that employs an analysis recognizable in contemporary environmental justice framing, though the term *environmental justice* does not appear in the book. Du Bois links the study of housing and living conditions to racial inequality, differential access, low quality of life, discrimination, crowding, high cost of living, ill health, and wage inequality.

Du Bois found that roughly 750,000 blacks lived north of the Mason-Dixon line in 1900; around 400,000 of those were in the New England and mid-Atlantic states. Sixty thousand blacks lived in the greater New York City area alone. Despite the fact that 90 percent of the black men in the city worked, black families lagged behind whites because they earned one-third less in weekly wages. The study found that one-fourth of blacks enjoyed a good living, about half could make ends meet, and the remainder lived in poverty as "God's poor," the "devil's poor," or "poor devils." About 6,000 blacks lived in the poorest districts, the Eleventh and Thirteenth Wards. Du Bois found that 19 percent of the families lived in one- or two-room apartments; four hundred of the rooms had no access to outdoor air. Yet blacks paid one to two dollars more per month for rent than whites living in similar quarters. This amounted to a black rental penalty of a quarter-million dollars throughout the city. About 40 percent of black families took in boarders to help pay the rent. Roughly 50 percent of the lodgers were not related to the families they boarded with. This resulted in overcrowding and disruption of family life in black households. The black death rate exceeded that of whites; in 1901 the death rate for blacks was 28 per thousand, while it was 20 per thousand for whites.[132]

Philadelphia had just over 1.29 million people in 1900; 62,613 were black. Most blacks were still confined to the Seventh Ward. Blacks had a higher death rate than whites partly because they lived in the unhealthiest part of the city with the worst housing. During the 1830s, when the death rate was 32.5 per thousand for blacks, it was 24 per thousand for whites. During the 1890s the death rate fell to 28.02 for blacks. However, in the Fifth Ward the death rate among blacks was 48.5 per thousand. In the Thirteenth Ward, where blacks generally lived in better conditions, the rate was 21 per thousand. The death rate for whites in the Thirteenth Ward was similar to that of blacks. Twenty-two percent of black families paid less than five dollars for their monthly rent. Most (57 percent) paid between ten and twenty dollars per month. Sixty-six percent of blacks earned between one and ten dollars per week in wages.[133]

There were twelve thousand blacks in Boston in 1900. About 63 percent were employed. The lower rate of employment is partly due to the fact that a large number of black children were in school and a large number of black women were homemakers. Though the death rate among blacks was lower in Boston than in New York and Philadelphia, it was still higher than that of whites; in 1900 the death rate among blacks was 25 per thousand and among whites was 20 per thousand. Blacks traditionally lived on the north side of Beacon Hill in the West End, and as late as 1880 it was difficult for blacks to buy or rent outside that crowded district. However, the study also found that a sizable number of blacks owned homes worth between $2,000 and $10,000, and several thousand blacks had bought homes in the suburbs of Roxbury, Cambridge, Dorchester, and Chelsea. Suburban blacks did not cluster in black enclaves but were scattered throughout the white neighborhoods.[134]

A similar pattern was found in Chicago, where mortality rates were 9 per thousand for wealthy whites, 12 per thousand for poor whites, and 28 per thousand for blacks. The tenuous status of blacks was exacerbated because they had difficulty finding jobs, were excluded from unions, and had few support services. During the Civil War, when large numbers of blacks moved to the North, some settlement workers served them in segregated facilities.[135]

Because white settlement workers were not serving the black communities, African Americans formed their own settlements. In Chicago Reverend Reverdy Cassius Ransom, a spokesperson for militant blacks and an ardent proponent of the Social Gospel movement, collaborated with Jane Addams, Mary McDowell, Clarence Darrow, and Graham Taylor to establish the Institutional Church and Social Settlement on South Dearborn Street in 1901. This was the first attempt in Chicago to provide the services of a settlement house to blacks. The settlement offered a wide range of social services, including a day nursery, a kindergarten, a mothers' club, and an employment bureau, as well as classes in cooking, sewing, and music. They also organized public lectures by leading white and black speakers and built a fully equipped gymnasium. The Institutional Church attracted both poor and middle-class blacks (middle-class blacks were drawn to the teachings of the Social Gospel). The church was bombed in 1904 during the stockyard strike after Ransom tried to mediate between black strikebreakers and white strikers.[136]

Noting that social service organizations bypassed blacks, Chicago's black leaders began raising funds in 1910 to open a black YMCA. Black women also became active in the Colored Women's Club movement. Fannie Barrier Williams and Ida B. Wells-Barnett were two leaders in the women's clubs.[137] Wells-Barnett also collaborated with Jane Addams to develop a settlement house for blacks. The National Association of Colored Women was organized in 1896. However, the Ida Wells Club predated the national organization. Several of the women's clubs in Chicago banded together to form the Colored Women's Conference of Chicago. These clubs operated kindergartens, parent-teacher associations, mothers' clubs, day nurseries, sewing schools, employment bureaus, and a penny-savings bank. The women's clubs also sponsored the Phyllis Wheatley Home for Girls, opened in 1896. This organization provided living accommodations, social activities, and employment assistance to young black women who migrated to Chicago but were excluded from the Young Women's Christian Association and similar organizations. The Colored Women's Conference of Chicago also operated a Home for Aged and Infirmed Colored People.[138]

In 1905 a white Unitarian minister, Celia Parker Woolley, founded a full-fledged settlement for blacks in Chicago. The Frederick Douglass Center was built on the South Side, in Chicago's growing Black Belt. Prominent black leaders such as Wells-Barnett, George Hall, and S. Laing Williams were involved in the center. Despite its location on the South Side, the Frederick Douglass Center was built in a middle-class black community on the edge of the Black Belt and was still some distance from the poorest black neighborhoods. The Center ran into problems because, while Woolley was interested in fostering interracial harmony, many of the destitute blacks visiting the center were not particularly interested in the interracial teas and discussion groups. Moreover, Woolley offended blacks by addressing them as "you people," and she was reluctant to appoint blacks to leadership positions in the Center. She was also offended when blacks raised questions about the operation of the Center.[139]

Twenty blacks got together in 1907 to establish a West Side settlement that catered to black men. The following year they opened the Wendell Phillips Settlement (it took this name in 1911). The settlement was staffed entirely by blacks, but it had an interracial board of directors. The Wendell Phillips Settlement operated till the 1920s.[140]

Blacks developed settlement houses in other cities as well as in rural areas. In 1897 the Tuskegee Women's Club, led by Margaret Washington (the third wife of Booker T. Washington), established the Elizabeth Russell Settlement near the Russell Plantation in Macon County, Georgia. The Phyllis Wheatley Community Center was established in Minneapolis in 1924, and in 1927 Mary E. Williams organized a settlement house in New Orleans.[141]

THE CHANGING ROLE OF THE SETTLEMENTS

Despite their best efforts, not all the settlement houses emerged as major centers of reform; only about 10 percent achieved this status. After 1900 the most influential settlements, such as South End House, Greenwich House, and the College Settlement Association, offered free room and board and a small stipend to entice college graduates to become settlement workers. Thus within the settlement community, *settlement* had a double meaning. On the one hand it represented a place in poor neighborhoods where young, often well-to-do, idealistic reformers took up residence in an effort to bring about social change; on the other hand it represented reformers' attempts to help immigrants, and to a lesser extent blacks, establish themselves and become assimilated in the cities.[142]

Despite the fact that settlement workers lived among the poor, it was difficult for them to "belong" to the neighborhood they lived and worked in. The social distance was not effectively bridged. For instance, though settlement workers worked and lived in ethnically diverse neighborhoods where many languages were spoken, few settlement workers spoke any foreign languages;

the communication difficulties that ensued caused some immigrants to view the workers with distrust. Community residents sometimes felt they were under a microscope, given the numerous surveys, community studies, investigations, books, and committee meetings being undertaken by the settlement houses. Though men were an integral part of the immigrant family, the settlement houses tended to attract primarily women and children to its programs (particularly those dealing with health and nutrition). Men sometimes showered at the settlement houses but otherwise they stayed away. So, despite their best efforts, these reformers didn't completely fit into the neighborhoods where settlement houses were located. Over time this lack of connectedness contributed to a diminished role for the reformers.[143]

The role of settlement workers and other reformers also changed because many of the issues they had worked on for years finally gained acceptance during and after the First World War. Labor made gains during the war in terms of collective bargaining, the minimum wage, acceptance of the eight-hour day, improved working conditions, reduced exploitation of workers, and workers' compensation. The government began building public housing, spurred in part by the need to house war workers. Improvements in health care enhanced the lives of the poor. In addition, women gained employment in occupations closed to them before the war. Though there was some retrenchment in some of these gains after the war, conditions improved overall for the poor.[144]

Other factors worked against the settlement workers, however. Despite the fact that labor activists and settlement workers had collaborated prior to the war, the relationship between the two deteriorated during the 1920s as the labor movement grew more confident yet conservative after the war. Meanwhile the reformers were not as bold as they used to be, and they lost some of their credibility with the working class because of their ambivalence regarding the passage of restrictive immigration laws sponsored by the nativist Know-Nothing Party. Though some settlement workers opposing the anti-immigrant hysteria formed the Immigrant Protective League in 1908, in 1917, when the restrictive immigration bill was passed, settlement workers did not muster an effective opposition to it. After the bill passed a few settlement workers responded by trying to soften the language in the legislation.[145]

The pool of settlement workers dwindled during World War One as some left the settlements to work overseas. In addition, settlement workers became increasingly vulnerable to the criticism that they were too radical, and they responded by tempering their activism even further. The changing demographics in the cities also altered their work. By the 1920s many of the immigrant families the settlement workers served had moved out of the slums to better neighborhoods in the cities or to the suburbs. White immigrant families were replaced by blacks and Latinos. Consequently settlement workers had to change their approach or face obsolescence. Rising racial tensions between white ethnics, blacks, and Latinos made settlement work far more complicated than before. The settlements had always operated under severe fiscal constraints, and now the war bond drives siphoned off potential funding. Furthermore now that their clientele had changed from whites to blacks and Latinos, the philanthropic organizations that once funded them were reluctant to provide funding for work directed at the new clients. By this time some of the individuals who had supported settlement and other reform work for decades were dying off. Others believed the problem of poverty was solved. Settlements found themselves in dire straits. Some closed, others transformed themselves into cultural arts centers, and the remainder became mainstream charities stressing community service rather than social reform. The Great Depression forced settlements to focus on relieving suffering rather than engaging in social reform.[146]

After the war the settlements had great difficulty recruiting new workers at a time when they really needed them, but they could not afford to offer fellowships and stipends anymore. As many of the workers aged and burned out, they left the settlements. Tensions escalated between the older generation of settlement workers and the newer ones. Some of the pioneers who stayed with the movement were very set in their ways. At the same time, charities and settlement workers had pushed for the professionalization of the field and in so doing created the new field of social work. This change affected the settlements in ways they may not have anticipated. New professional settlement workers identified themselves and thought about the work very differently from the way the old guard did. The new professionals saw themselves as social workers, not social reformers. They were trained as caseworkers, recreation specialists, and the like, who saw the neighborhood people as clients, not neighbors or friends. Moreover the social workers did not believe they had to live in slum neighborhoods to be effective; they wanted to go home at night, not spend it organizing the community. They also rejected another core settlement value: they did not believe that they needed a settlement house to anchor them in a community.[147]

Though sociologists such as Robert Park and Louis Wirth criticized the settlement house community studies as sloppy and emotional, they, like the social workers, built their work on the foundation laid by the settlement workers. The settlement movement declined further as cities and the federal government began providing services that the settlement houses traditionally provided. As public libraries, museums, community centers, and parks and playgrounds were built in neighborhoods across the country and housing quality improved dramatically, the settlement house libraries, picture exhibits, playgrounds, and baths became obsolete.[148]

Another important factor in the reduced efficacy of the reformers was the emergence of industrial medicine, which replaced some of the work done by some settlement houses and unions. As community-based public health gathered adherents, corporations came under increasing pressure and scrutiny from community activists, workers, and unions. In response, companies began to provide their own medical services for their workers. Rather than allowing workers to visit independent doctors or have their health care needs met in local clinics, companies took on the task of diagnosing and treating workers' health problems and keeping health records. This shift made it more difficult to detect occupational health hazards, protect workers from being dismissed when they were injured on the job, and assess workers' health histories.[149]

SETTLEMENTS TODAY

Today there are more than nine hundred settlement houses in the United States and about forty-five hundred worldwide. Some, like Hull House, have endured through time. Today the original Hull House is a museum at the Chicago campus of the University of Illinois, but the Jane Addams Hull House Association of Chicago has an annual budget of over forty-four million dollars and serves more than two hundred and twenty-eight thousand people in one hundred programs in five community centers. It employs about seven hundred people.[150]

In 1911 Jane Addams founded the National Federation of Settlements and Neighborhood Centers; the name was changed in 1979 to the United Neighborhood Centers of America. Of the nine hundred U.S. settlement houses, 195 are members of the organization. These centers are located in fifty-seven cities in twenty-two states. There is also an International Federation of Settlement Houses and Neighborhood Centers, with member institutions in over thirty countries.[151]

Working-Class Reformers and the Discourse of Poverty

The language of poverty and pauperism was not the sole domain of middle-class reformers. It filtered into the discourse of working-class reform organizations too. In 1807 the General Society of Mechanics of New Haven was formed for the purposes of advancing and regulating the mechanical arts and making loans and donations to "worthy paupers." Other mutual aid groups were the Mutual Benevolent Society of Cordwainers of New Haven (formed in 1821) and the New Haven Mutual Aid Association (formed in 1833).[152]

Working-class people did not sit back and wait for reformers to resolve all their problems for them. A number of working-class groups and institutions tackled social problems as well. Workers campaigned against hazardous work conditions, occupational injuries and diseases, poor housing conditions, lack of free time, and the dangers of child labor. Unions took steps to frame the issues and object to some of the actions of the charities. In 1833 the first industrial union in the United States held a convention in Boston. The New England Association of Farmers, Mechanics and Other Workingmen brought together delegates from all over the region. At the convention delegates called for radical reforms on a variety of issues affecting workers' lives. They argued, "[The] laws [of the country] are made to operate partially and unequally for the benefit of the accumulator and the degradation of the producer." The working class lacked opportunities to participate in government even though the government controlled their property. More than half a century before settlement workers campaigned to end child labor, union members at the convention called for an end to child labor. They argued, "If children must be doomed to these deadly prisons, let the laws at least protect them against excessive toil and shed a few rays of light upon their darkened intellects." They also contended that because of heavy labor, children were "sinking under oppression, consumption and decrepitude into an early grave." Union activists linked the end of child labor with increased educational opportunities for children; settlement workers used a similar kind of frame-bridging technique in later years as they campaigned to eliminate child labor. Convention delegates also called for the abolition of debtors' jails. Noting the inequities in the tax system, whereby much of the wealth of major corporations went untaxed, delegates called for tax reform. Three decades before antidraft riots erupted in cities across the country, union members also called for an end to the system whereby rich men could buy their way out of military service so that only poor men served.[153]

Working-class activists took steps to make it easier for poor families to buy food. During the Panic of 1873 an organization called the Sovereigns of Industry was founded in New Haven. This group did not focus on unionism per se but on organizing cooperatives to buy in bulk to reduce costs. The organization charged an initiation fee of two dollars for men and one dollar for women. Within a few months of its founding eight hundred and fifty members had signed up in New Haven. The co-op idea caught on quickly in other

Connecticut towns as well; one thousand members signed up in Hartford, five hundred in Meriden, and four hundred in Waterbury. The Sovereigns of Industry estimated that by buying in bulk they could reduce the cost of coal by about 20 percent and flour from eleven dollars per barrel to eight dollars. They estimated that families could save about fifty dollars per year by joining the co-op. Working families loved the co-op, but merchants hated it and quickly mobilized to squash it; they refused to sell anything to co-op members and did not respond to the Sovereigns' call for bids. This, plus the fact that many co-op members had difficulty getting cash to pay for goods, made it difficult for the Sovereigns of Industry to survive. In 1875 the organization collapsed.[154]

five

DATA GATHERING AS A MECHANISM FOR UNDERSTANDING THE CITY AND IMPOSING ORDER

Political Arithmetic

In the claims-making process activists have to make the transition from identifying problems and attributing blame to providing evidence to support their claims. The process of ratifying their claims is referred to as *ideational centrality* or *commensurability*.[1] Charity reformers, settlement workers, sanitarians, and other urban activists recognized this and embarked on ambitious data-gathering projects to help them make this step. These statistics and maps were used to help justify claims and amplify framing.

In the early nineteenth century the upper class generally believed that illness and disease were prevalent in the slums as a just requital for the immorality and depravity of the poor. Since the rich were able to create physical, social, and economic distance between themselves and the poor they had the wherewithal to escape the onslaught of these diseases. But when yellow fever, cholera, and tuberculosis struck even the wealthy they realized that they were unable to emerge from all forms of urban epidemics unscathed. Epidemics taught the wealthy that the inhumane conditions of the slums threatened the lives of the rich as well as the poor.

As the century progressed civic leaders recognized that they needed to understand large-scale demographic trends to help them make predictions and prescribe solutions to problems, and so they began to collect a variety of data. The desire to halt the spread of epidemics and find cures for them also provided activists with opportunities to collect data.

Yellow fever and cholera epidemics sparked an interest in the collection of vital statistics about urban populations. Advocates of data gathering felt that it was necessary to have statistical information to help develop more efficient mechanisms for administering municipalities. Thus in 1831 British statisticians began collecting morbidity and mortality rates of diseases to find out how epidemics affected the economic and social fabric of the society. Statisticians framed their arguments in terms of poverty, pauperism, and economic inefficiency. They argued that diseases were a major cause of pauperism and that pauperism led to vice and crime. Illness and premature death increased the number of dependents needing support, which meant that the rich paid higher taxes to support the poor. Statisticians wanted to determine more precisely the cost of ill health; hence they began assigning numerical values to human life and activities. They included factors such as number of days of productive labor lost to illness, the cost of medical care, and funeral expenses. A consequence of the calculations resulting from this kind of "political arithmetic" was the realization that excessive mortality was a preventable loss that was a drain on industry and commerce. Hence one of the frames of the sanitary reform movement was the notion that "health is wealth."[2]

The earliest surveys on health conditions in cities were conducted in Leeds and Manchester in 1831 and 1832, respectively. Edwin Chadwick conducted a survey on the conditions of the working class in England and Wales in 1842 that became a landmark document in the history of the public health movement. Chadwick's report showed a relationship between disease, mortality, and unsanitary conditions such as poor drainage, pollution, contamination of the surroundings with sewage, water contamination, and inadequate garbage collection. He concluded that improved drainage, cleansing of houses, garbage removal, water purification, and proper sewage disposal would improve lives in urban areas. He argued that these reforms should be conducted under the auspices of boards of health working in collaboration with engineers and others with expertise to solve the technical problems presented by these challenges.[3]

In 1854 John Snow developed an elaborate map to help demonstrate that cholera had broken out because people consumed contaminated water from the Broad Street pump. In addition to determining the number of people who died of cholera in London's Soho district, Snow calculated and developed a graphic representation showing the foot traffic to each pump and which residents drew water from which pump. Without the map Snow might not have been able to convince the miasmatists writing health policies that cholera did arise from the consumption of impure water.[4]

In 1841 Lemuel Shattuck, a Boston publisher and the founder of the American Statistical Association, conducted a study of the city's vital statistics for the period 1810–1840. The study showed that there was an increase in mortality and illness in the city over time. Unlike the British and American sanitarians, Shattuck believed in the divine retribution thesis. Consequently he proposed that Boston should focus its reforms on changing the behavior of its citizens to reduce illness rather than improving the city's infrastructure.[5]

Other American reformers took a different approach. John Griscom, a doctor who knew of Chadwick's work, conducted a health study in New York in 1845. One of the founding members of the Society for the Prevention of Pauperism, Griscom may have grown wary of the idea that behavioral modification programs in and of themselves were the most effective way of combating epidemics. He focused instead on environmental reforms as the means of improving public health. He recommended systematic street cleaning and garbage removal and the construction of an underground sewer system designed and supervised by engineers. He also called for an expansion of the Croton River water system to serve the entire city. Following the lead of his British counterparts, in the 1850s Griscom calculated the cost of illness. He estimated that in 1853 alone illness had cost New York City $6,692,640 ($148 million in 2005 dollars). Of this amount, Griscom argued, 60 percent was due to preventable causes related to substandard sanitation and ineffective administration of public health.[6]

Sanitary reformers in the South conducted similar studies. In 1849 Edward Hall Barton, Erasmus Darwin Fenner, and John C. Simonds published a statistical analysis of the costs of the excessive mortality in New Orleans. After the yellow fever epidemic of 1853 Barton headed the New Orleans Sanitary Commission, which conducted an exhaustive sanitary survey of the city.[7]

The framing of excessive illness and premature deaths as preventable losses that created an economic burden on the rich was an effective perspective that motivated wealthy businessmen to action. Because the wealthy bore the tax burden for illness and death they wanted to improve the health and living conditions of the poor; by doing so, they believed, they would reap benefits in the form of more efficient business operations and increased profits.

Cultural Cartography and Social Order

In addition to numerical data gathered from surveys and interviews, civic leaders and reformers sought to collect population, housing, and other data that could be represented spatially on maps. During the nineteenth century such maps played key roles in the social control efforts of civic leaders and their attempts to make cities less chaotic.

For much of human history maps have helped people understand the social, cultural, and physical world. They are a powerful medium through which information about intergroup relations, space, geography, empire, distance, settlement patterns, resources, and transportation routes are transmitted. Maps can organize the world in myriad ways. They depict what appear to be objective snapshots of a given area,[8] yet far from being objective, maps reflect the ideological biases of the maker or commissioner, and they tell stories about the social and political realities of the place and time they depict.

Some of the earliest known maps are city plans and cadastral maps. A wall painting of Catal Huyuk in Anatolia, Turkey, painted around 6200 B.C., depicts a city plan. The Babylonians mapped landholdings and drew city plans on clay tablets as early as 2000 B.C. The Egyptians planned a cadastral survey as early as the thirteenth century B.C. (the survey has not survived). Chinese town maps drawn to scale have been found dating back to the third century A.D. Maps came into popular use in England during the Tudor period, when they were used by landlords and governments to administer landholdings. In America cartographers began mapping Manhattan in the early sixteenth century. The 1527 Maggiolo map depicted the area, as did the Gastaldi map of 1556. A map indicating the location of New England may have appeared as early as 1528. Cadastral maps began appearing more frequently in America in the eighteenth century. They took the form of land plats that documented new settlements and ownership claims.[9]

CONQUEST, COLONIALISM, AND THE IMPOSITION OF FAMILIAR ORDER

Maps are an important tool of warfare, conquest, and control. Invading troops rely on maps to guide them. Often when invading or exploring parties take control of territories, they stamp their imprimatur on the territory by assigning new names to settlements and other landscape features. Hence Columbus and his party assigned names throughout the Americas despite the fact that indigenous tribes had their own names for these places. The ritual of assigning names is one of the rituals of conquest. Thus the demonstration of conceptual dominance cannot be separated from the act of physically dominating people and seizing their land and other resources. This is the case because the renaming of places by conquering groups deprives the conquered group of their own cultural and historic legacy and the meanings the names held for them. Hence maps can be seen as a tool of both physical conquest and conceptual control.[10]

The desire of invaders and explorers to impose names familiar to them on new territories is also driven by a desire to impose familiar order on strange new lands, peoples, and surroundings. Familiar names and objects on maps are used to fill in voids or enclose what might otherwise be blank spaces. On some early maps cartographers filled in unknown spaces guided only by their own fantasies and wishful thinking. This is typical of early maps of America, in which cartographers invented places and landscapes: unexplored mountains and river valleys were filled in with animals, trees, and other features imagined by the mapmakers.[11]

URBAN LAND USE PLANNING

Cartographers have played an important role in defining urban landscapes. In America cartographers represented the ideological and conceptual conquest of indigenous people by replacing the Indian names of trails, lakes, ponds, hillocks, and valleys with Dutch, English, and other European names. Consequently relatively few of the original Indian place names survive today. The maps helped to symbolize and reinforce the conquest and banishment of Native people and their lifestyles.

During the nineteenth century maps took on great significance as tools used by elites trying to maintain order in chaotic cities and administer them more efficiently. As cities grew in size civic leaders wanted to know more about them: how many and what kinds of housing and other building structures were in the city, the number and location of water pumps and stopcocks, which areas of the city were more prone to fires and diseases, and where the open spaces were located. Maps thus helped civic leaders to respond to emergencies, plan growth, and forecast change.

One of the earliest town planning maps of an American city is the 1641 Brockett map of New Haven. The map shows the town laid out around a central Green. The eight squares surrounding the Green were subdivided with the names of the landowners inscribed on the map. The map also identifies two suburban tracts that were being developed. The 1748 town planning map added more details about the town's development. It depicts buildings and the date of their construction, the names of the original thirty-eight land grantees, the names and occupations of the landowners, and a notation about two trees planted on the property of J. Pierpont (a "gent") in 1686.[12] The earliest map of Philadelphia shows evidence of early town planning. Thomas Holme's plat, the *Portraiture of the City of Philadelphia*, was drawn in 1683 and published in London. Another early town planning map depicts the city of Boston in 1722; drawn by Captain John Bonner, it shows the waterfront features in great detail and indicates the location of streets, houses, churches, and public buildings. The legend of the map indicates the dates of major fires (the first occurring in 1653 and the last in 1711) and six small pox epidemics (the first occurring in 1649 and the sixth in 1721).[13]

Massachusetts remained at the forefront of town planning. On June 26, 1794, the General Court of Massachusetts ruled that towns and districts should develop accurate plans of their jurisdictions and turn these in to the Secretary's Office by June 1795. This edict was issued because it was thought that "an accurate map of . . . [the] Commonwealth will tend to facilitate and promote such information and improvements as will be favorable to its growth and prosperity, and will otherwise be highly useful and important . . . for the procurement of the materials necessary for the accomplishment of an object so desirable, and by which the reputation and interest of the Commonwealth will be advanced."[14] The 1811 map of New York heralded the "beginning of the modern city"; it shows not only the layout of the city but the existing and planned open spaces, such as the Battery.[15]

MAPPING DISEASES

As epidemics ravaged cities civic leaders became interested in pinpointing the origins and epicenters of disease outbreaks. The use of maps to chart the course of diseases evolved from merely listing the times that epidemics broke out (as was the case in the Bonner map) to actually pinpointing the location of outbreaks within a given city.[16] One such map appeared in New York, depicting the origin and pattern of the 1832 cholera outbreak. The legend states, "The annexed plan of the city has been sketched for the purpose of enabling strangers to perceive the evidences of non-contagion . . . furnished by the prevalence of epidemic Cholera in New York, in the summer of 1832."[17] The map gives the impression that most of the city was free of the epidemic and that the disease outbreaks were centered in small, concentrated pockets in specific neighborhoods. Prominent black circles indicate the parts of the city where the disease first appeared and where the outbreaks were worst. Mostly major thoroughfares are depicted, except within the cholera circles, where the affected streets are identified in precise detail. The legend also identifies specifically which neighborhoods were affected, in what order the outbreak occurred, and the level of virulence. For instance, the first case was identified at Cherry and Roosevelt Streets, while the last case occurred in the Ninth Ward at Greenwich Village. The map further states that the epidemic "raged with unwonted violence [in the Ninth Ward] after it had nearly subsided in almost every part of the city." The narrative reinforces the notion that disease broke out among filthy people: "Those who are acquainted with the city, need not be informed either of the crowded and filthy state of several of these points, or of the intemperate, dissolute, and abandoned habits of the inhabitants of some of them, or of the proverbial filth of the streets, as well as the houses, in most of these locations. Where these causes were not obviously present, the district is known to be *made ground*; or some local source of the disease existed either in a large filthy stable, as at *Christopher Street*, Greenwich Village, or in the neighbouring marshes."[18]

The street location of the victims is identified and highlighted, while the streets on which the healthy people lived are unmarked. The map goes to great lengths to specify the reasons people contracted the disease: the poor because of their own questionable lifestyle and behavior, and other residents because of environmental factors, that is, factors beyond their control and not associated with any flawed behavior.

In 1865 the Citizens' Association published a map in their report on the sanitary conditions of New York showing the prevalence of typhoid fever in 1864. Unlike the cholera map, the typhus map shows Manhattan from Central Park to the southern tip of the island, with the regular grid pattern of streets. There were sixteen hundred reported cases of typhoid in the city. The map shows the highest concentrations in the Five Points area and along the waterfront. There were fever outbreaks along the West Side almost as far north as Central Park, and a still greater concentration of cases in East Side working-class neighborhoods.[19]

Urban reformers and medical professionals used maps to pinpoint the epicenters of tuberculosis. In 1904 the Maryland Tuberculosis Commission sponsored the first tuberculosis exhibition in Baltimore, featuring a map of deaths from tuberculosis between 1891 and 1900. The map shows that the Biddle Alley neighborhood in Lower Druid Hill, a black working-class neighborhood in northwest Baltimore, had the highest death rate: 950 per 100,000. The rate was 131.9 in the rest of the city. Because tuberculosis infection and mortality was so high in a neighborhood that corresponded with the contours of a poor black community, the disease was effectively framed as a black disease, and the Biddle Alley neighborhood became known as the "Lung Block." The exhibition attracted thousands of people. In 1905 the New York City Tuberculosis Committee opened a similar exhibition. Both the New York exhibition and the findings from the study of Baltimore's Lung Block were featured in traveling exhibits that toured the country and were seen by almost two million people.[20]

Maps were used extensively to track the spread of the bubonic plague and to help develop preventive measures in Honolulu during the 1899–1900 outbreak. The maps were used to outline the boundaries of Chinatown, identify buildings and structures within it, estimate the demographic characteristics of the population, establish the boundaries of the quarantine zone, and help verify compensation claims. Quarantine Island is also identified on some of the maps. While tuberculosis was framed as a black disease, the plague was framed as an Asian disease. When public health officials and the fire department instituted a program of controlled burning of buildings suspected of being contaminated with the plague, maps were drawn up and published in the newspapers pinpointing the location of targeted structures. After the Great Fire administrators used the maps to help them determine the veracity of claims.[21]

MAPS AND SOCIAL REFORM

Social reformers and charities relied heavily on maps to organize their work. They used maps to subdivide cities into small sections, each of which was placed under the control and surveillance of friendly visitors. The maps were also used to identify the poorest neighborhoods and suitable locations to establish charitable institutions and to track the spread of vice and crime. The Progressive Era reformers also made extensive use of maps, but for somewhat different purposes than the earlier charities and tract societies. The settlement workers developed neighborhood maps to show the level of illness in the slums, occupational injuries, overcrowding, poor housing conditions, and poor sanitation. The neighborhood maps were critical in getting the government to improve conditions in cities.

MAPPING FIRES

When fire was the sole source of heat a typical New England family consumed between thirty and forty cords (more than an acre) of wood annually.[22] The widespread use of wood for heating, cooking, and building materials made for a combustible mix. As a result devastating fires swept through cities during the eighteenth century and the nineteenth, creating the impetus for insuring buildings but also identifying structures and areas that were at risk for destruction by fire and hence were too risky to insure.

The earliest known American fire map was done in Charleston, South Carolina, in 1788. However, it was not until the mid-nineteenth century, when fires frequently ravaged cities, that such maps became common.[23] In 1735 there was a failed attempt to sell fire insurance in Charleston. A year later Philadelphia's Union Fire Company offered the first fire insurance scheme in the country, and in 1738 the Philadelphia Contributionship began insuring homeowners against property loss from fire. In each case maps of the city determined insurance risk and rates. Though many homes in Philadelphia were constructed of brick, fire was still a concern because of chimney blazes. Though fines were imposed on householders on whose properties chimney blazes occurred, chimneys remained unswept. The Philadelphia Contributionship believed that the city's trees also constituted a hazard because they obstructed passage, destroyed water courses, and spread fires; hence in 1774 the insurance company began the practice of refusing to insure or renew policies on homes with nearby trees. The insurer reaffirmed that policy in 1781, and the Assembly passed a bill on April 15, 1782, that called for the removal of all trees in streets, alleys, and lanes. In response irate citizens circulated a petition proclaiming that trees enhanced the health of urbanites and were of great public utility. Five months after the tree removal act was passed the Assembly repealed it. Nonetheless the Contributionship stubbornly stuck to its policy. As a result a rival fire insurance company, the Mutual Assurance Company, was formed to insure homes.[24]

Though the number of firefighters increased from six hundred in 1800 to fifteen hundred in 1835, New York City's fire department remained poorly organized and undisciplined. They were often accused of drunkenness, theft, and failure to answer fire alarms. To make matters worse, the firefighting equipment was inadequate and water, already insufficient, frequently froze in cold weather. In 1804 a fire destroyed forty houses, and another in 1808 killed five people; the following year a fire destroyed twenty-five buildings, and in 1816 more than twenty firefighters were injured in a blaze.[25] During the 1820s the fire insurance business grew rapidly and insurers began to demand better fire protection. In 1834 the first fire map of New York City was published. A year later the Great Fire of 1835 broke out on the frigid night of December 16. The fire, which started in Comstock and Adams's store on Merchant Street, burned 674 buildings and 52 acres and leveled 20 blocks at an estimated cost of $20 million ($444 million in 2005 dollars). Intended to help volunteer firefighters learn the fastest way around the city and the location of hydrants, the fire map became a reference guide for insurance companies wanting to identify the buildings they insured (many insurance companies went bankrupt or suffered crippling losses in the 1835 fire). The Great Fire resulted in some changes. Critics of the fire department argued that the fire spread, in part, because of the negligence of the volunteer firefighting companies, and they campaigned for a paid, professionally trained firefighting force.[26]

The Great Fire made it clear that New Yorkers needed a better water supply system; firefighters were helpless during the fire because the water supply was inadequate and cold weather froze the hydrants. At the time the city was heavily reliant on well water; however, well water was contaminated by cesspools and street runoff. Despite the problem with well water, by 1809 the city had 249 pump-operated public wells and scores of private wells.[27]

For a long time New Yorkers had complained about the foul-tasting city water. After the 1832 and 1834 cholera epidemics New Yorkers were desperate for clean water. The 1835 fire was the last in a series of calamities that forced the city to develop an adequate and clean water supply. Construction of the Croton aqueduct began in the spring of 1835 and was completed in 1842. The first water from the system reached the city on June 22. The conduit, 40.5 miles long, carried water to a thirty-one-acre receiving reservoir. By diverting the Croton River in Westchester to reservoirs in Manhattan, New Yorkers were able to drink fresh water. In 1842 the Endicott map depicted all the stopcocks in the city at the time the Croton reservoir was completed.[28]

Sunday, October 8, 1871, turned out to be a fateful and historic night as an enormous conflagration consumed a large part of the city of Chicago. At the height of the inferno one hundred thousand people tried to escape the blaze. The fire burned unabated before it was brought under control on Tuesday. The Great Fire of Chicago is one of the most famous in American history. It started in a working-class neighborhood on the Southwest Side of the city in Patrick and Catherine O'Leary's barn at 137 De Koven Street around nine o'clock and spread quickly. Conditions were perfect for a brisk fire. The summer and fall had been exceptionally dry: only 2.5 inches of rain fell between July 3 and October 9, and a mere 0.11 of an inch fell in the three weeks before the fire. Everything was parched, and the sun-baked wooden structures were like kindling ready to ignite. A strong, gusty wind blowing from the southwest all day propelled the fire toward the city as flames shot a hundred feet into the air. Soon the small blaze turned into the largest fire in the city's history.[29]

At the time of the fire Chicago was still a relatively young town; it was formed in 1833 with a population of two hundred people. The town soon became the fastest growing city in the nation and in 1871 had more than 334,000 residents with 59,500 buildings valued at around $620 million on 23,000 acres. By 1870 more vessels docked in Chicago than in the ports of New York, San Francisco, Baltimore, Philadelphia, Mobile, and Charleston combined. Chicago's rapid growth meant that planning and oversight of building and construction was not always a high priority. As a result, shoddy construction was common, and about two-thirds of the buildings were constructed entirely of wood. Many of those remaining—even those reputed to be fireproof—were constructed with veneers of brick covering essentially wooden structures topped with extremely flammable tar or shingles. Moreover, the decorative elements on most buildings were made of wood painted to look like stone or marble. In 1868 the city's Board of Police and Fire Commissioners noted that the buildings were flimsy and safety was being sacrificed by ornamentation that hid structural weaknesses. Though Chicago had zones in which wooden frame buildings were forbidden, the restrictions were routinely ignored, and the city also ignored the commissioners' pleas to expand the zones. The density of the buildings created an additional fire hazard; the small lots—twenty-five by one hundred feet—were crammed with homes, outhouses, barns, sheds, and other flammable structures. Manufacturing facilities such as paint factories, lumber yards, mills, distilleries, gasworks, warehouses, grain elevators, coal distributors, glue factories, and slaughterhouses were interspersed among the houses. Other fire hazards existed. Because the city was built on the marshy banks of the Chicago River and Lake Michigan which flooded regularly, wooden roads and sidewalks were elevated by as much as seven feet. At the time of the fire, Chicago had more than fifty-five miles of pine-block streets and five hundred and sixty-one miles of wooden sidewalks.[30]

Like other major cities, Chicago was prone to fire. There were one hundred and eighty-six reported fires in the city in 1863, 515 in 1868, and six hundred in 1870. Even more fires were reported in 1871 because of the unusually hot and dry summer. There was an average of around two fires per day in 1870, but by October 1871 an average of six fires broke out daily. By the standards of the time Chicago had one of the best firefighting systems in the country, with seventeen steam-engine companies, six hose carts, four hook-and-ladder companies, and one hose elevator. There were one hundred and ninety-three firemen and twenty-three men working at headquarters in 1871—but even that force could not adequately protect a city that was roughly thirty-six square miles. Cognizant of the fire danger in the city the Board of Police and Fire

Commissioners had earlier lobbied for larger water mains to be laid throughout the city, as well as fire hydrants, more water hoses, limitations on the use of wood as a building material, and a prohibition of the use of tar on roofs, but all recommendations were ignored. As soon as the O'Leary fire was spotted, one neighbor, William Lee, raced over to Goll's drugstore to sound the alarm. However, Bruno Goll, insisting that a fire engine had already been dispatched, refused to sound the alarm. Later, when an official investigation into the fire was launched, Goll insisted that he had sounded two alarms, though no record of either was recorded at the central alarm office at the courthouse. Matthew Schaffer, the watchtower guard at the courthouse, spotted smoke, but he did not follow up because he thought it was just the smoldering embers from a fire that had broken out near the O'Leary home on Saturday night. It took Schaffer a while to realize that the fire he now saw was a new one and he sent out a fire signal at 9:30 p.m. Because the fire was some distance away and it was very dark, he sent the fire trucks to a location about a mile away from the O'Learys. Once Schaffer realized his mistake he relayed a message to his assistant, William J. Brown, to send out a new alarm. However, Brown, believing that a new alarm would confuse firefighters, refused to obey Schaffer's orders. As a result fire engines searched in vain for a fire before heading to De Koven Street, where the fire was now blazing out of control. In addition the fire engines located closest to De Koven Street remained at their station because they did not receive a signal. Two of the local fire engines did go to the blaze that was visible from their station, but they were not well equipped to fight the fire; the newer fire engines that could have helped bring the blaze under control were not sent to De Koven Street.[31]

When it became obvious that the fire engines present could not contain the blaze, a war veteran and former alderman, James Hildreth, obtained the mayor's permission and twenty-five hundred pounds of explosives and began blowing up buildings to create firebreaks. As the fire progressed, as many as twenty blocks and five hundred buildings were ablaze at once. Around 11 a.m. on Tuesday rain began to fall, finally dampening the flames. By then 1,687 acres of the city lay in ashes, and $200 million worth of property had been destroyed ($2.9 billion in 2005 dollars). The fire had burned a swath of the city four miles long and one mile wide. Seventy-three miles of streets and 17,500 buildings were destroyed. The fire left one hundred thousand people homeless, and thirty thousand took refuge in the 230-acre Lincoln Park. Though officials estimated that three hundred people died in the blaze, only one hundred and twenty bodies were recovered.[32]

By Tuesday afternoon rebuilding had begun. Hastily constructed wooden structures replaced those that had just burned down. Crews worked long hours clearing the rubble and dumping the debris into the lake. The rising demand for builders attracted job-seeking carpenters, bricklayers, and farmers to the city. Wages for carpenters and bricklayers rose from five to ten dollars a day. The building boom was so amazing that six thousand wooden shanties, two thousand solid wood-frame buildings, and five hundred brick or stone structures were finished or nearing completion in Chicago by the end of 1871. Though the homes of wealthy Chicagoans who lived downtown were burned, the general perception was that most of the very rich had escaped the conflagration. A day after the fire the *New York Times* noted, "Numerous outlying blocks and many edifices of the better class in the more thinly-occupied [districts] have been spared." Rebuilding efforts were fueled in part by elites who feared the conditions in the hastily erected barracks that many Chicagoans were forced to shelter in after the blaze. Elites were especially worried about the potential for middle-class families who lost their homes to slide into immorality while living in crowded communal quarters with the poor. Images such as these horrified the upper-class reformers. Colbert and Chamberlin described whites so begrimed by smoke that they looked black, and "north side nabobs, herd[ed] promiscuously with the humblest laborer," while Luzerne described "people unclad, half-clad, some wrapped in bed clothing, women dressed in the apparel of the opposite sex, and some protected only by their night-wrappers" wandering the streets. Consequently the Relief and Aid Society set out to rebuild suburban single-family homes as quickly as possible. Despite a harsh winter, the Society built 8,033 homes on the outskirts of the city as the way was cleared for the construction of a new downtown commercial district.[33]

Aid poured into Chicago from all over the country, and the Citizens Committee administered it to any seeking help. However, some reformers worried that the "indiscriminate" disbursement of aid would create a permanent dependent class. Consequently the mayor asked the Chicago Relief and Aid Society, with wealthy merchants such as Marshall Field and George Pullman serving on its board, to take over and administer aid "scientifically." To the Aid Society scientific charity meant applying the model perfected by the Charity Organization Society—dividing the city into districts, collecting detailed information from applications and scrutinizing requests, screening out imposters, and identifying the "deserving" applicants to whom relief was provided. The Aid Society wanted to disperse people residing in what they considered to be the "rude" barracks that the Citizens Committee had constructed. They feared that having too many people in the barracks would create a class of people "dangerous to themselves and the neighborhood in which they might be placed." They wanted to reserve the barracks for "the class who have not hitherto lived in houses of their own, but in rooms in tenement houses." Those left in the barracks would be placed "under the constant supervision of medical and police superintendents" so that "their moral and sanitary condition" would be "unquestionably better than that which has heretofore obtained in that class." This group accounted for about five percent of the displaced. The Aid Society lavished most of its attention on about 40 percent of the displaced who were the skilled laborers and working poor who could have afforded to own homes or who actually had owned homes prior to the fire. The Aid Society described them as the "mechanics and the better class of laboring people, thrifty, domestic, and respectable, whose skill and labor are indispensable in rebuilding the city." For this group the Aid Society built single-family homes as "incentives to industry [and] the conscious pride and independence of still living under their own roof-tree . . . to raise them at once from depression and anxiety, if not despair, to hope, renewed energy, and comparative prosperity." The Aid Society did not help most of the fire victims—those who were deemed either too immoral to deserve assistance or were too well off to need it. It should be noted that though the Aid Society modeled itself after the Charity Organization Society the two differed in important ways: the Aid Society provided significant cash and material benefits to those it did help, and it did not criminalize the behaviors of aid seekers, require people to go into work camps, or place stringent time limits on shelter. Not only did the fire create an opportunity for reformers to exercise more social control over the poor, it created an opportunity to disperse the poor. The Aid Society gave thirty thousand people train tickets to leave town. A new commercial district was carved out downtown, and businessmen eagerly gobbled up real estate at fire-sale prices. New fire-prevention ordinances banned wooden structures from being constructed in the city. Though there was a safety dimension to this, there was an important class implication that should not be ignored: The ban on wooden structures meant that poor people could not afford to build homes in the city. In fact all the cottages the Aid Society constructed for fire victims were wooden and were all placed outside the city limits. Recognizing this, thousands of poor Irish and German immigrants protested the new ordinance on January 15, 1872; the newspapers described the protestors as a "mob," "communists," "north side incendiaries" and the "scum of the community."[34]

Those seeking to impute blame found the Irish an easy target. The newspapers led the way. First, an Irish woman, Mrs. O'Leary, was blamed for causing the fire. Though the O'Learys testified in hearings held after the fire that they had retired for the night when the fire broke out, and several eyewitnesses report that the couple was asleep and had to be woken up when the fire was discovered, the Chicago *Evening Journal* printed an unsubstantiated story on Monday that the fire began when a cow that a woman was milking kicked over a lamp in a stable. The story took on a life of its own as other papers regurgitated it and launched personal attacks on the O'Learys. Patrick O'Leary was described in newspaper stories as ignorant and illiterate. The Chicago *Times* referred to Catherine O'Leary as "an old hag . . . whose appearance indicated great poverty": she "apparently was about seventy years of age and was bent almost double with the weight of many years of toil, and trouble, and privation." The story also imputed the motive of vengeance to Mrs. O'Leary's alleged action, claiming that she was angry because she had been removed from the poor relief rolls. "The old hag swore she would be revenged on a city that would deny her a bit of wood or a pound of bacon." The newspapers reported that Catherine allegedly admitted to being in the barn when the cow kicked over the lamp and that she made weekly applications for aid. The *Workingman's Advocate*, a union newspaper, ran an article in which Catherine rebutted the *Times*'s allegations, she said, "I never had a cint from the parish in all my life and the dirty *Times* had no business to print it." In reality Catherine O'Leary was in her mid-thirties and had never received any poor relief. Both she and her husband earned enough money to have purchased their property in 1864, though they did not have any insurance on it. The couple owned five cows, a calf, and a horse; they lived off the income from Patrick's wages, cash received from selling milk from the cows, and from renting half of their two-story house (which survived the blaze) to another family.[35]

Some newspapers attempted to blame the firefighters, describing them as too inept and inebriated to fight the fire properly. Joseph Edgar Chamberlin of the *Chicago Evening Post* wrote, "The firemen kept at work fighting the flames—stupidly and listlessly, for they had worked hard all of Saturday night and most of Sunday, and had been enervated by the whisky which is always copiously poured on such occasions." Actually the men were exhausted: they had battled the Lull and Homes fire that had consumed four blocks all of Saturday night, and without much food or sleep were called out to fight Sunday night's blaze. Others were blamed for the fire, including the anarchist group Société Internationale, a disgruntled fire extinguisher salesman, and the man who discovered the fire—"Peg Leg" Sullivan—but none of these accusations stuck, and the O'Learys continued to take the brunt of the blame. Cartoonists published caricatures of Catherine, and curious onlookers flocked to De Koven Street to stare at the couple. Feeling trapped and vilified, the O'Learys moved away. The guessing game about who started the fire has continued unabated over the years. Shortly after the fire, Colbert and Chamberlin blamed Mrs. O'Leary, as did Harry A. Musham seventy years later. However, more recent scholarship implicates others. Anthony DeBartolo thinks that a group of boys including Louis Cohn and Mrs. O'Leary's son, James, started the fire accidentally while playing craps in the barn. Richard F. Bales points the finger at "Peg Leg" Sullivan and possibly an accomplice, Dennis Regan.[36]

Boston, with its densely packed wooden buildings, was also particularly vulnerable to fire. Beginning in 1653 several great fires consumed large portions of the city. On October 13, 1860, a daredevil photographer, James W. Black, took the first aerial photograph of an American city, the section of Boston that was consumed twelve years later when the Great Fire broke out on Saturday evening, November 9, 1872, in a dry goods store and hoop skirt factory on Summer Street. The fire consumed the entire business district, with a loss estimated at $75 million ($1.16 billion in 2005 dollars). The fire was catastrophic because the city was short on water, fire horses, and personnel. Firefighters watched helplessly as the fire raged on Saturday night and all day Sunday. Residents climbed onto the roofs of buildings and tried to put out the fire with hoses, but to no avail. In all, sixty-five acres and seven hundred and seventy-six buildings were burned. The militia had to be called in to stem the looting.

Beginning with the Great Fire of 1649 Boston strengthened its fire code after each big conflagration. All families were mandated to equip themselves with leather firefighting buckets and a ladder long enough to reach the ridge pole of the roof of their dwelling; all adult males had to join the bucket brigade as soon as a fire alarm sounded. After the 1711 fire the city purchased three suction engines and organized a public fire department, with specific units assigned to the ten fire wards. Greater fire protection measures were taken after the 1872 fire as well, when wooden structures, frames, roofing materials, and support structures were banned. The city mandated that buildings be constructed with brick barriers between floors and fireproof boiler rooms, stairs, and elevators.[37]

Like other cities, Boston organized fire patrols: citizens walked the streets at night in search of fires. The city also began to mete out harsh punishment for negligent use of fire and for arson. Women and blacks were punished most severely. A farmhand accused of arson in Lynn was sentenced to twenty-one years working for the man whose property he allegedly burned, but a woman accused of arson was burned at the stake when a small boy died accidentally in a fire she allegedly set. A slave, charged with burning the home of Joshua Lamb in 1681, was "burnt to Ashes" on the Boston Common.[38]

Fire maps were developed for a number of other cities. The Aetna Insurance Company commissioned a survey of several towns in Tennessee in 1866. These maps were not published, but a year later the surveyor, D. A. Sanborn, published the *Insurance Map of Boston*. Sanborn formed the Sanborn National Insurance Diagram Bureau in 1868. More fire insurance maps were created for Boston and Philadelphia in the 1870s and 1880s. The maps indicated the location of fire hydrants as well as whether buildings were made of brick, stone, or wood. The Sanborn Map Company became the leading maker of insurance maps.[39]

MAPPING REAL ESTATE

Realtors too developed an intense interest in maps that identified the location of a city's buildings. In 1851 urban maps named Dripps I and II plotted Manhattan's real estate. The maps included information for realtors and insurers about the size, dimensions, and configuration of buildings; lot sizes and locations; and the width of streets. Perris's maps of 1852 to 1854 were also detailed censuses of the city's buildings; insurers used these maps to assess the risks (such as fire) to the properties they owned and the cost of insurance based on those risks.[40]

New England elites who owned country estates developed an interest in large-scale county maps. During the 1840s 28 such maps were published; from 1851 to 1860 327 county maps were produced. These maps grew so popular that Henry Francis Walling alone produced nearly 280 county maps of states from Vermont to Alabama. The maps showed landholdings, settlement patterns, and points of interest within the county.[41]

MAPPING DRAINAGE AND PARKS

Urban cartographers were also interested in mapping the topography of cities. One of the first maps to detail the topography, drainage patterns, and layout of a proposed park appeared in 1855; it was Viele's map of the Central Park site.[42] Viele is responsible for developing two other important topographic and drainage maps of Manhattan, the 1865 *Sanitary and Topographic Map of the City* and the 1874 *Topographic Atlas of the City of New York*.[43]

Viele feared that New York was becoming, in his words, one of "the most disease-ridden cities in history." In 1859 he likened New York to Europe during the plague. He thought that cholera, malaria, and other diseases were caused by disruption of the ecological patterns of the city. He argued that waterways that were blocked or filled in as the city expanded became the breeding grounds of diseases, which were caused by excessive moisture in the ground. A cavalry officer in the Rio Grande area of Texas after the Mexican War, Viele had suffered from yellow fever and malaria. He argued that these diseases were more prevalent in low-lying areas and areas with underground streams. A British statistician, William Farr, also believed there was an association between variations in topography and these diseases. Like Viele, he believed that epidemics were more likely to occur in low-lying areas. Viele thought that the topographical characteristics of Manhattan made it vulnerable to epidemics; he claimed, "The sanitary condition of any city is intimately connected with its proper drainage. . . . Any inquiry into causes or remedies for sanitary evils . . . shall be based upon a thorough knowledge of the topography." Viele thought the problem could be partially resolved by allowing watercourses to have outlets to the tidewater, building wider streets, and planning streets and sewers in accordance with the natural drainage patterns of an area. In 1865 he drew a sanitary and topographic map of Manhattan showing the original watercourses on the island. The map was an accompaniment to a report on hygiene and public health.[44]

MAPPING ENVIRONMENTAL HAZARDS

Nineteenth-century mapmakers also documented environmental hazards in cities, identifying noxious and hazardous facilities such as paint factories.[45] As the story of one mapmaker indicates, urban mapmaking could be hazardous work:

> At one chemical plant I noted that the old map indicated a shed marked "Retort House." I saw at a glance that the dimensions and distances were wrong on the old map, but when I started to measure them for myself and reached the shed I discovered that it was full of chemical retorts. From each retort issued a hazy mist of poisonous gasses. However, the wind was blowing in the right direction, and I got the correct measurements of the shed, counted the retorts and marked them all down, just as the wind changed and a cloud of that suffocating poison gas encompassed me. I threw myself on my face and rolled over the bank, down the towpath of the canal, where I lay until I had cleared my lungs of the suffocating fume. I knew exactly why the other fellow who made the old map did not have the correct distances and dimensions. He had evidently worked at a safe distance.[46]

six

SANITATION AND HOUSING REFORM

Sanitary Reform

Activists grew increasingly concerned with the environmental conditions that caused epidemics, and beginning in the mid-eighteenth century sanitarians such as Thomas Bond and Benjamin Rush urged improved sanitation as a solution.[1] The sanitary reform movement was primarily concerned with health, access to open space, clean water, sewage systems, drainage, waste disposal, and hygiene.

During the 1820s activists campaigned to move urban cemeteries to the outskirts of town after recurring yellow fever outbreaks because doctors and other health reformers feared that decomposing bodies spread noxious gases and diseases. Furthermore, the practice of disinterring bodies, dumping them in trenches to make room for new corpses in overcrowded cemeteries, alarmed hygienists. The first rural cemetery, Grove Street Cemetery, was built on the outskirts of New Haven in 1796; by the 1830s other rural cemeteries such as Mount Auburn (Cambridge, Massachusetts), Green-Wood (Brooklyn), and Laurel Hill (Philadelphia) were built.[2] The 1832 and 1849 cholera epidemics stimulated further sanitation reform. Cities began planning water systems and activists campaigned for urban parks to protect water supplies. In 1849 the newly formed American Medical Association published a sanitary survey of American cities. During the 1840s and 1850s sanitarians intensified their efforts to establish public parks and playgrounds. Advocates believed that parks helped to trap pollution and cleanse the air; even the American Medical Association advocated building public parks. These actions were followed in 1859 by the formation of the New York Sanitary Association and in 1866 by the Citizens' Association of New York.[3]

WATER AND SEWAGE TREATMENT

Though Philadelphia built its waterworks in 1801 and other cities followed suit, as late as 1860 most communities literally drank their own sewage or that of upstream residents. Not surprisingly, typhoid fever, dysentery, and cholera were endemic and urban mortality rates were extremely high. The fear of epidemics drove cities to build sewage systems.[4] The completion of New York's Croton water system in 1842 intensified the debate about whether the city should speed up the development of the sewage system. The installation of indoor plumbing exacerbated the growing inequality between the city's rich and poor; the rich paid for indoor plumbing and water, while the poor relied on public wells and water hydrants. Six years after the Croton system was in operation two-thirds of New Yorkers still had no running water in their homes because the pipes had not been laid in poor neighborhoods and landlords in the tenement districts did not want to pay to install plumbing in their buildings. By 1860, 136 cities had developed water treatment systems, and by 1880, 598 sewer and water systems were in operation.[5]

As most New Yorkers continued to use outhouses and cellar water closets, the city wrestled with the problems arising from improper sewage disposal. New York's sewage, or "night soil," was handled carelessly: the workers emptying the toilets and cesspools at night spilled excrement in the yards and on the streets as they made their way to the docks to dump it into the river. An estimated 750,000 cubic yards of fecal matter was dumped in the river annually; some slips were so filled in that they could not be used for shipping purposes. As the city debated whether to expand its sewer system, the 1849 cholera outbreak killed five thousand people. In response the city intensified its efforts to build sewers, and by 1855 had laid seventy miles of new sewer pipe. Still, in 1859 most poor neighborhoods had no sewers.[6]

MUNICIPAL HOUSEKEEPING, CLEAN STREETS, AND BEAUTIFUL CITIES

The rural life movement inspired Catharine Beecher to write books on efficient suburban home designs with the latest time- and space-saving devices and good sanitation practices. Writing in the 1840s, around the same time Andrew Jackson Downing was advocating improved taste in rural residences, Beecher urged women to keep their homes neat and clean and to plant gardens.[7]

Beginning in the 1850s Ellen Swallow Richards, a Vassar graduate and the first woman to attend the Massachusetts Institute of Technology, applied the concept of *oekologie* to her work.[8] Some of her most significant contributions to our understanding of the environment came in the field of municipal housekeeping and sanitary reform. Using her background in sanitary chemistry and nutrition, Richards focused on the home environment: sanitation, waste, home economics, and food chemistry. She was also concerned with air and water pollution and wrote extensively about the causes of pollution. Her work contributed to the consumer nutrition and home economics movements.[9]

Other women also played key roles in the municipal housekeeping movement. Mildred Chadsey wrote, "Housekeeping is the art of making the home clean, healthy, comfortable and attractive. Municipal housekeeping is the science of making the city clean, healthy and comfortable."[10] The 1880s saw a proliferation of groups such as the Ladies' Health Protective Association of New York pushing the idea of sanitary homes and cities. The municipal housekeeping movement was buttressed by a number of housekeeping and home decoration journals with impressive circulation figures that proliferated in the second half of the nineteenth century. The journals instructed women on cooking and food safety techniques, home design and decoration, and fashion. *McCall's* (founded in 1870), *Ladies' Home Journal* (1883), and *Good Housekeeping* (1885) were widely distributed and very influential. *Ladies' Home Journal* had a monthly circulation of almost one million in 1900.[11]

The condition of city streets became a major focus of the City Beautiful movement. Activists asserted that there was a strong relationship between the physical appearance of a city and its moral state. They believed that civic symbols would go a long way in shaping the behavior and morality of the urban masses. They lamented the fact that, compared to European cities, American cities had a paucity of public monuments and landmarks. They argued that the blandness of cities combined with filthy streets made it difficult for urbanites to have civic pride and take interest in their cities. In the mid-1890s reformers began pushing for the beautification of cities. By 1905 there were at least 2,426 improvement societies operating in cities across the country. This was a loosely coordinated movement in which activists shared ideas about a variety of initiatives, ranging from pollution reduction to removing unsightly billboards, fences, and overhead electrical wires and cleaning up garbage from streets, alleyways, and vacant lots. Activists also planted trees, flowers, and shrubs along sidewalks and sponsored home beautification contests. They agitated for cities to erect statues and fountains, paint murals on buildings, and commission aesthetically pleasing boulevards, bridges, streets, lighting fixtures, and garbage containers.[12]

In New York street cleaning became a quasi-military operation, directed by Colonel George Waring, a war veteran and former Central Park engineer who had worked closely with Frederick Law Olmsted while building the park's drainage system.[13] Waring's "White Wings," clad in white uniforms and hats, removed garbage and snow from the streets. To supplement the work of the garbage corps Waring organized youths into Juvenile Street Cleaning Leagues; more than five thousand children volunteered for the leagues. Under Waring New Yorkers began sorting their trash, recycling ashes and organic materials, removing oils and greases, and placing garbage in sanitary landfills. Waring used his engineering skills and knowledge of sanitation when he built the sewage system of Memphis, where his work helped to rid the city of yellow fever epidemics. City Beautiful initiatives gave way to more comprehensive city planning by the early 1900s.[14]

WASTE DISPOSAL IN CHICAGO'S PACKINGTOWN

Sanitary reformers laid the groundwork for settlement house activists to make Chicago's Packingtown and neighborhoods like it cleaner. Clean water was an elusive dream for Chicagoans. By the early nineteenth century the city's wells were so contaminated that residents got water from wagons hauling clean water from Lake Michigan. In 1836 the Illinois legislature granted permission to the Chicago Hydraulic Company to supply the city with water, but the company did not complete construction on its pumping station until 1842. The water intake pipe extended 150 feet into the lake, but that soon became contaminated with wastes, and fish swam into the intake pipe, clogging faucets. This led residents and visitors to joke about taking a bath in chowder. Chicago was authorized to build its own waterworks in 1851. The city constructed a water intake that extended six hundred feet into the lake, but this too became contaminated.[15]

Though Chicago passed ordinances in 1843, 1849, and 1851 to try to curb the pollution of the Chicago River, the packinghouses ignored these and the laws weren't enforced. Those wanting to circumvent city laws set up their operations just outside the city limits. During the Civil War, as the meat-processing industry boomed, packers began locating their plants on the South Branch of the river, below Thirty-first Street (the city limits). By 1863 the vacant fields in the area were covered with decaying animal matter; countless rats and flies swarmed over the wastes and horrid smells pervaded the community. The city extended its boundaries to Thirty-ninth Street in 1863 and oversaw cleanup efforts. Everything was buried in pits covered with lime and clay. However, within a year the South Branch of the Chicago River was again filled with effluent from the packinghouses. The city responded by flushing industrial and human wastes into Lake Michigan, the source of their drinking water. Conditions were so bad that one irate citizen sent a letter to the *Chicago Tribune* saying, "It is extraordinary that . . . a city of nearly two hundred thousand inhabitants . . . should quietly permit a river running through its entire limits to be converted into a gigantic sewer. . . . Is it possible that . . . laboring men of this city . . . are going to . . . permit a few wealthy men to corrupt the air and the water, which are the common property of all?" The newspaper responded with an editorial saying that Chicago could "better afford to bring its drinking water from Evanston than to drive the packinghouses to St. Louis."[16]

Sanitarians pushed the city to enforce the laws already on the books, but a powerful coalition of nuisance industries—the packinghouses, glue manufacturers, fertilizer makers, and grease manufacturers located along the river—opposed the reforms and the enforcement of the laws. Unlike the nuisance establishments that were zoned out of Lower Manhattan and Boston early on, it took much longer for Chicago to control their industries. By then the meat processors weren't marginal businesses on the periphery of the city; they were among the largest, most prosperous establishments in the city. Many of the owners of the meat-processing plants were among the city's millionaires, so they carried enormous clout. Nonetheless in 1865 the state passed an ordinance requiring the packers to remove animal matter from their facilities and dispose of them in areas that were at least four miles from the city. In addition, in the early 1870s sanitary reformers founded a citizens watchdog group called the Diligent Citizens Association to monitor how well the nuisance industries along the river complied with city ordinances. Despite the increased surveillance, some companies continued to flout the laws. This led activists to campaign to rid the city of its stink factories. In 1877 the city passed an ordinance that allowed it to issue and revoke the licenses of packinghouses and glue and fertilizer factories in the city as well as those within a mile of its borders. The following year the city brought charges against twenty-seven companies under the new "stink ordinance." After grand jury hearings some companies were found guilty of violations and others settled out of court.[17]

The area containing the stockyards and Packingtown earned Chicago the name "Hog Butcher of the World." In 1870 Packingtown had a population of three thousand, but the population swelled to eighty-five thousand by 1889 as workers flocked to the vicinity for jobs. Meat processing was big business in Chicago; in 1892 thirteen million cattle, sheep, and pigs were processed in the stockyards.[18] A City Homes Association report described the surrounding neighborhoods in bleak terms: the houses were built on stilts, and there were no sewers, so sewage from the houses collected in stagnant pools. The foul-smelling stockyards belched heavy clouds of smoke. The alleys in the densely populated black neighborhood of "Hell's Half Acre" (located close to Thirty-sixth Street and Armour Avenue) overflowed with garbage. Rooms were dark, and many people lived in cellars with overflowing toilets.[19]

As Chicago grew, epidemics of small pox, dysentery, and typhoid fever struck regularly. In 1881 568 people died of typhoid; thousands more died throughout the decade. The 1891 death toll in the city reached 1,997, a rate of 178 per 100,000 residents. More people died of typhoid in Chicago than in any other city. Not surprisingly civic leaders fretted as the 1893 World's Columbian Exposition neared. Fearful that people would stay away from the fair because of the city's reputation for typhoid epidemics, organizers piped in fresh spring water from Waukesha, Wisconsin, to the Exposition grounds, a distance of over one hundred miles.[20]

Settlement house workers lived and organized in areas such as these that had severe public health and sanitation problems. Chicago's Twenty-ninth Ward, the Packingtown district, had one of the highest infant mortality rates in the city in the early 1900s: one-third of all babies died before age three. In 1909, one hundred and twenty-seven babies died, nearly half of them within three blocks of the neighborhood dumps. The dump at Forty-seventh and Robey Streets was the largest in the city. Mary McDowell, a former settlement worker at Hull House and head resident at the University of Chicago Settlement (also located at Forty-seventh and Robey), lived in the area and collaborated with activists from nearby settlements, the South Side Settlement League, women from Hyde Park, and doctors to investigate the dumps and develop a comprehensive health survey of the area.[21]

Bubbly Creek, a dead arm of the South Branch of the Chicago River and a virtual open sewer, formed the northern boundary of Packingtown. The garbage dumps, which covered about twelve acres, and meatpacking houses formed the western boundary, and the community was bordered on the east by the "hair fields," an area containing animal hair and other putrefied slaughterhouse waste. The dumps were located adjacent to the homes of Packingtown's large immigrant population. Dumps were established in the big holes that resulted from the round-the-clock excavation of clay to supply a Twenty-ninth Ward alderman with the raw materials for his brick and tile business. Once the clay was exhausted the alderman sold the city the rights to fill the holes with garbage. When the holes were filled the same alderman leveled the ground and sold lots on the property for residential development. Each day in Packingtown a steady stream of malodorous garbage wagons rumbled through the streets on their way to the dumps. Women moved through the trash searching for bottles, iron, rags, and other items to sell, reuse, or consume. Because the dumps were not covered with dirt or disinfectant, flies, mosquitoes, and vermin flourished. On dry days dust clouds from the dumps and excavation operations blanketed the neighborhood.[22]

As Chicago geared up to host the Columbian Exposition activists used the opportunity to pressure city government to clean up the area. The Diligent Citizens Association joined forces with Hull House activists to advocate for sanitary reforms in Packingtown. Just before the Exposition the streets and alleyways east of Halsted were paved and sewers completed, and in May 1893 two scavenger units were dispatched to remove garbage from the area. The city dumped only dry garbage in the pits and covered each load with lime or other disinfectant. Two crematories were opened in the district; a third was on display at the Exposition. Chicago sang the praises of the stockyards, turning them into a tourist attraction that rivaled the Ferris wheel in popularity. Ten thousand people toured the stockyards each day and more than a million altogether visited the area.[23]

Nonetheless when Packingtown residents opposed dumping in their neighborhood, a lawyer argued before the City Commission, "In any city there must be sections reserved for unpleasant things, and the people of such sections [must] become used to the unpleasantness." McDowell was convinced that the dumps and Bubbly Creek were responsible for the prevalence of illness and disease in the neighborhood. She didn't believe that simply closing the Packingtown dumps and placing them elsewhere in the city would solve the problem. Instead she insisted that the city change the way it disposed of garbage. She studied methods of garbage disposal in the United States and Europe and reported to the city on alternative disposal techniques, such as incineration, and methods of garbage reduction. McDowell lectured at mass meetings around the city while women's clubs organized garbage committees to help educate the public about alternative disposal methods.[24]

In addition to undertaking neighborhood epidemiological and demographic studies to establish a link between unsanitary conditions and illness, the Hull House Woman's Club popularized the strategy of environmental monitoring by citizens. When inadequate garbage collection became an unbearable problem in the city, Jane Addams submitted a bid to do the collecting. After the bid was rejected, the Hull House workers organized their own garbage patrol. In the early morning they followed the trucks to the dumps, kept charts and maps, made citizen's arrests of landlords violating disposal ordinances, and complained to contractors. Because of their efforts the city took a serious look at the garbage problem.[25]

In May 1911 the city appointed a commission to study the feasibility of improving street and alley cleaning and garbage removal and disposal. However, not much progress was made until 1913. The reduction process proposed by McDowell was attractive to industries because greases that fetched a high price on the market would be extracted from the wastes before dumping. Buffalo built such a plant in 1886, and by 1911 reduction plants were operating in Cleveland and Columbus. In 1913 a new reduction plant was built in Packingtown and the garbage dumps were closed. The reduced garbage, dry rubbish, and ashes were used for grading and filling along the lakefront.[26]

Though the elimination of the dumps resulted in improved health for Packingtown residents, Bubbly Creek continued to be a health hazard. The creek was filled with sewage from homes, the packing plants, and stock pens; blood and entrails of animals; and refuse from the plants and sewers. The amount of waste dumped into the creek from the packing houses was equivalent to the waste of a million households. Gasses bubbled up through the thick scum that formed on top of the waste. In the beginning the effluent was carried to the lake, where the materials decomposed gradually. But by 1890 Bubbly Creek was nothing more than a stagnant pool of refuse and sewage. The lakefront was also heavily polluted, and the drinking water drawn from it made people ill.[27]

The city reversed the flow of the Chicago River to halt the flow of the garbage into the lake, but the reversal dumped effluent into the Illinois River instead and nearly stopped the flow in the already sluggish Bubbly Creek. The putrefied garbage filled the bed of the river at the rate of six inches per year from 1895 to 1908.

McDowell campaigned tirelessly to improve conditions at Bubbly Creek. During the 1904 stockyard strike she gave reporters tours of the creek, pointing out conditions in the stockyards as well as the residential areas. The efforts of the Hull House and Northwest Settlement House workers and the widespread publicity surrounding the publication of Upton Sinclair's exposé of the meatpacking industry, *The Jungle*, facilitated the passage of the Pure Food and Drug Act of 1906. McDowell also lobbied the War Department in Washington to get them to fill in Bubbly Creek, but because the creek was not navigable it did not fall under federal jurisdiction. She then lobbied the city's health department. In 1910 a consortium bought the portion of the creek west of Ashland Avenue for a railroad right-of-way and filled it in. A few years later the city decided to reduce the waste going into the creek; the packing houses bore 60 percent of the cost and the city paid the remaining 40 percent.

Packingtown is now an industrial park, and the 1879 Stone Gate that once marked the official entrance to the stockyards is a historic landmark.[28] But in the late 1990s Bubbly Creek was still a highly contaminated waterway. At the junction of the South Fork and South Branch is a large expanse of open water originally designed for sailing vessels to turn around. In 1999 the Wetlands Initiative embarked on a project aimed at reclaiming the turning basin to create a wildfowl and fish habitat.[29]

Housing Reform

LIVING CONDITIONS

New York

Despite reformers' complaints living conditions remained appalling in most cities for much of the nineteenth century. In the 1830s landlords in New York began converting warehouses, mansions, churches, and breweries into multi-unit dwellings to accommodate the influx of immigrants. Called "rookeries" these buildings contained small, dark, dank apartments with closet-size unventilated rooms. The economic windfall realized from the rookeries inspired speculators to build larger barracks-style tenement buildings. Beginning in the 1850s four- and five-story tenements, built on lots twenty feet wide and one hundred feet deep, sprouted all over working-class neighborhoods. Like the rookeries, these "railroad apartments" were dark and unventilated and lacked indoor plumbing. Residents shared outdoor privies or cellar vaults. The apartments typically rented for two dollars to three dollars per month. By the 1870s there were twenty-one thousand rookeries and railroad apartments in the city.[30]

Overcrowding was a major problem in the tenements. At the end of the nineteenth century five-story tenements constructed for twenty families often housed as many as one hundred, plus their boarders. The "flophouses" were even more crowded: two hundred flophouses in Five Points contained eighteen thousand bunks. The 1897 census showed that thousands of people lived in rear tenements also. Some of the most destitute lived in stables that rented for about twenty dollars per year. Most families did not have access to indoor plumbing. Only 306 (0.12 percent) of the 255,033 people living on the Lower East Side had a bathroom in their apartment. In 1893 only 2 percent of the

12,434 families in the Centre Street Italian neighborhood had a toilet, even though the average rent was $4.60 per month ($94 in 2005 dollars).[31]

Nonetheless the number of people living in tenements in New York and other cities grew rapidly. New York's superintendent of buildings estimated that from 1862 to 1872 the number of tenement dwellers increased from 380,000 to 600,000.[32] Seven hundred people lived in Gotham Court (Five Points), a building shoehorned into a lot only twenty by two hundred feet. Worse still, one seven-story building on the Lower West Side contained three thousand occupants. In 1880 the Lower East Side had a population density of 200,000 people per square mile; by 1900 the density was 300,000 per square mile. In New York City about one out of every three persons in the Tenth Ward slept in unventilated, windowless rooms. *New York Times* editorials claimed that two thousand people lived in cellars.[33]

Blacks were charged higher rents than whites for even the most decrepit dwellings. As one landlord put it, "We have a tenement on Nineteenth Street, where we get $10 for two rooms [from blacks] which we could not get more than $7.50 for from white tenants previously."[34] Landlords argued that even though their black tenants were neat, clean, and less inclined to destroy the property than were recent immigrants, they charged blacks premium rents because whites would not live in the same house or tenement with blacks, or even one recently occupied by blacks. The landlords, believing that the value of their property was diminished by renting to blacks, charged a premium to recoup those losses. Thus landlords profited handsomely from discriminating against black tenants and charging them for white racism. Jacob Riis's analysis of rents showed that the same tenement that rented to white tenants for $127 per month was rented to black families for $144 per month, about 13 percent more.[35]

Chicago

An investigation of tenements in Chicago between 1883 and 1884 exposed wretched conditions. Thousands of people were concentrated in areas without drainage, plumbing, light, ventilation, or fire safety equipment. The sewage problems were enormous, the outhouses disgusting, and the alleyways and vacant lots contained decaying matter and stagnant pools. In addition, food was prepared in unsanitary conditions. Working-class families paid rents at rates that netted the landowners 25 to 40 percent profit for tiny, substandard flats. There was an average of more than eight persons per dwelling in Chicago; about 17 percent of the dwellings contained three or more families. Nevertheless this housing stock was considered superior to that found in other urban centers.[36]

Living conditions hadn't changed much by 1900. A City Homes Association report described the Chicago tenements as a "wilderness of bad housing and sanitary neglect." The report continued, "The Stock Yards district and portions of South Chicago show outside insanitary conditions . . . [of] indescribable accumulations of filth and rubbish, together with the absence of sewerage, mak[ing] the surroundings of every dilapidated frame cottage abominably insanitary." In the smoky, foul-smelling swamps of South Chicago (the Polish district) the houses were perched on stilts above green, slimy, standing water. Families lived in waterlogged, rat-infested basements. There were no sewers and the streets were unpaved.[37]

Philadelphia

Unlike New York and Chicago, Philadelphia, nicknamed "the City of Homes," had a housing stock that consisted of a large number of single-family homes and owner-occupied dwellings. While New York dealt with its housing shortage by building large tenements, and Boston crammed the poor and new arrivals in old, deteriorating houses in the South End, Philadelphia grew outward, building row houses and cheap, single-family homes. By 1867 forty-five hundred such units were being added to the city's housing stock. This is not to say that Philadelphia did not have slums. Slum neighborhoods could be found in the Southwark and Moyamensing neighborhoods and in the alleyways and courts hidden behind the main streets of the old city. Though the slums were not as concentrated as in other cities or dominated by tenements, the narrow alleyways and dark courtyards of the small three-story houses, known as "Trinity" or "Father, Son, and Holy Ghost" houses, made life miserable for residents. The Trinity houses had one room to a floor and were so crammed together that little light or air got inside. They lacked proper plumbing. Sweatshops occupied the upper floors in Jewish neighborhoods. Chickens were kept in the bedrooms and goats in the cellars in homes in the Italian sections of the slums.[38]

The first American building and loan association opened in Philadelphia in 1831. Between 1849 and 1876 over $50 million in housing investments flowed through such institutions. Philadelphians benefited from having most of the country's building associations (about 450 of 600) located in their city. This allowed would-be homeowners to invest in shares in the associations at a moderate interest rate. While land prices were too high in Manhattan for most working-class families to buy a home, thousands of working-class Philadelphians were able to buy two-story brick row houses for $1,000 to $2,500. In 1891 a Philadelphian earning $25 per week ($513 in 2005 dollars) was considered a potential homeowner. Though few workers earned that much, if several members of the family pooled their wages they could afford to buy a home. In 1882 locomotive workers made $605 per year ($11,500 in 2005 dollars), the average iron or steel worker made $500 ($9,500), and textile workers earned $350 ($6,700) annually. Another factor helped Philadelphians gain access to cheap housing: the custom of ground rent, whereby a homeowner did not have to buy the lot the house was built on but could choose instead to pay a nominal rent for the land.[39]

Despite the fact that it was easier for working people to buy homes in Philadelphia than in other large cities, owner occupancy did not exceed 25 percent. However, Philadelphia was better off than other cities in this respect. Furthermore the density per unit was lower in Philadelphia than in other cities. In 1880 the average occupancy rate per dwelling in Philadelphia was 5.7 people. In comparison, the average occupancy rate was 16.36 people in New York, 8.25 in Boston, and 8.24 in Chicago.[40]

In his study of the black community in Philadelphia W. E. B. Du Bois found that the average annual rent paid by black families rose steadily from 1838 to 1896. In 1838 blacks paid an average of $44 ($717 in 2005 dollars) per family for their annual rent. That number rose to $49.86 ($988) in 1848. In 1896 blacks in the Seventh Ward paid an average of $126.19 ($2,800) per family for their annual rent. Thus between one-fourth and three-fourths of a family's income went to rent. This being the case, many blacks took in lodgers or shared apartments with other families. Du Bois argued that the high cost of rent for black families stemmed from the fact that whites, not wanting to live close to blacks, limited the parts of the city in which blacks could live. Moreover realtors, knowing that blacks had few choices in where they could live, raised the rent for black tenants or refused to rent to blacks altogether. In addition, blacks lived in neighborhoods close to their place of work so that they could commute to work relatively easily.[41]

Of the 2,441 black families Du Bois analyzed only 334 (13.7 percent) had access to bathrooms, and many of these families shared bathroom facilities with other families. There was no plumbing in some bathrooms, and if there was water it was cold. The homes were densely packed together as tenements were built over what used to be small backyard space. Twenty-nine percent of the families either had no yard or shared their yard and outhouse with others.

Eighteen percent of the black families had a private yard smaller than twelve by twelve feet.[42]

According to Du Bois, the back alley Trinity apartments "consist[ed] of three rooms one above the other, small, poorly lighted and poorly ventilated. The inhabitants of the alley are at the mercy of its worst tenants; here policy shops (illegal lottery operations) abound, prostitutes ply their trade, and criminals hide. Most of these houses have to get their water at a hydrant in the alley, and must store their fuel in the house. These tenement abominations of Philadelphia are perhaps better than the vast tenement houses of New York, but they are bad enough, and cry for reform in housing."[43] Du Bois described the Moyamensing neighborhood as forlorn. Most of the black families in the neighborhood were rag pickers. The dark, damp, and dirty cellars were rented for twelve and a half cents per night.[44] He described the small sheds that people lived in:

> At the back of each house are small wooden buildings roughly put together, about six feet square, without windows or fireplaces, a hole about a foot square being left in the front along side of the door to let in fresh air and light, and to let out foul air and smoke. These desolate pens, the roofs of which are generally leaky, and the floors so low that more or less water comes in on them from the yard in rainy weather, would not give comfortable winter accommodations to a cow. . . . They are nearly all inhabited. In one of the first we entered, we found the dead body of a large Negro man who had died suddenly there. The pen was about eight feet deep by six wide. There was no bedding in it. . . . The body of the dead man was on the wet floor beneath an old torn coverlet.[45]

IMPROVING HOUSING CONDITIONS

During the 1990s wrecking balls were busy in cities such as Chicago as crime-ridden, rat-infested housing projects were razed. Activists and city governments were optimistic that the new, mixed-income housing developments that would replace the high-rise projects would provide improved housing for residents. The story behind the demolitions bears a striking resemblance to the story of tenement housing reform a century earlier. Today housing projects are characterized by dark, dank hallways; small, dark, poorly ventilated rooms; overcrowded apartments; infestations of rats, roaches, and other pests; leaking faucets, standing water in the basements, and sewer backups; and broken elevators, garbage chutes, and incinerators.[46] A century ago reformers described similar conditions in the tenements and urged city governments to replace them with better housing. There is one difference worth noting: a century ago housing reformers campaigned to improve housing for white immigrants to the city; today the public housing that many of the Progressive Era reformers pushed for has been transformed into forbidding islands of crime and poverty where poor black and Latino families are isolated, segregated, and warehoused.

Early Reform Efforts

Since the early 1800s housing reformers have been calling attention to squalid housing conditions in cities. One reformer, Samuel Halliday, sounded one of the earliest warnings when he called attention to the housing conditions of the poor in New York.[47] Another reformer, John Griscom, began campaigning for better housing in 1845, when he wrote an exposé on slum conditions in New York. In 1852 Griscom gave a lecture at the Broadway Tabernacle that focused on the importance of proper ventilation. He noted that the city had spent $15 million to secure clean water but had ignored the issue of ventilation. In opposing Central Park, Griscom wrote in 1853 that factories, schools, and apartments were unventilated; thus the poor spent a good deal of time indoor breathing "foul gases." He argued that to improve the health of the poor windows should be cut "where there are none," workshops and schoolrooms ventilated, stagnant water pumped from cellars, drainpipes and toilets connected to sewers, and dwellings painted. Griscom believed that instead of spending money on the park in which the poor could enjoy fresh air, the city could better use the money by ventilating the dwellings in which the poor spent most of their time.[48]

Pressure from activists led the New York State Legislature to pass the 1867 Tenement House Law, the first such law in the country. The law defined a tenement as a multiunit dwelling in which more than three families lived independently of each other. It required landlords to provide at least one privy and a fire ladder for each tenement. The law also specified the maximum portion of the lot that could be covered by buildings, a minimum area for rear courts, and standards for open-air ventilation of interiors; every sleeping room was to have either a window, a ventilator, or a transom. Landlords responded to this law by satisfying only the minimum requirements for ventilation. Furthermore they began connecting the front and rear buildings on a lot; this resulted in buildings that were more than four rooms deep, with interior rooms ventilated only by narrow air shafts or doors fitted with transoms. Landlords also began placing the privies in subterranean vaults; the stench was overwhelming.[49]

Tenement Design Competition and the Birth of the Dumbbell Tenements

In 1878 the periodical *Plumber and Sanitary Engineer* announced a competition for a new design for tenements; 206 designs were submitted by architects from the United States, Canada, and Britain. The winning entry, submitted by James E. Ware, was the "dumbbell" design, so named because the apartments looked like dumbbells when seen from above. Between the front and back rooms the exterior walls were indented, creating airshafts between adjacent buildings. The airshafts extended from ground to roof. The rooms had windows, and each apartment had a toilet with plumbing, vents to guard against sewer gas, and a kitchen. The interior rooms were ventilated by horizontal airshafts.[50] The dumbbell tenements were usually five to seven stories tall, with fourteen rooms on each floor and seven rooms on each side. The four apartments (two four-room apartments in the front and two three-room apartments in the back) on each floor shared two toilets, and the windows in ten of the fourteen rooms opened onto the airshaft.[51]

The dumbbell design did not meet with universal approval. One of the problems identified early on (and one that would plague the tenements built with this design) was with the airshaft. The shafts between buildings were between fifty and sixty feet long but only twenty-eight inches wide. The shaft was enclosed on all sides and was the full height of the building, which was between sixty and seventy feet. The shafts were intended to provide air and light, but no space was provided for air intake at the bottom. Furthermore very little light penetrated to the bottom of the shaft. To make matters worse, residents used the shafts to dispose of garbage and bodily wastes; as a result, the airshafts were foul smelling. They acted as a flue when fires broke out, spreading fires from one floor to another.[52] Tenement residents quickly discovered that the airshafts acted as echo chambers, amplifying the sounds coming from adjoining tenements. Because of the inexorable noise and vile smell of the airshafts, many dumbbell tenement residents rarely opened their windows.[53]

Nonetheless housing reformers incorporated the specifications of the dumbbell design into the 1879 New York Housing Act prohibiting landlords from building on more than 65 percent of the lot. The 1879 act also prohibited the construction of rooms that were not ventilated by a source of outside air. This meant that the Board of Health forced landlords to put forty-six thousand interior windows into buildings citywide. Between 1880 and 1900 nearly twenty thousand dumbbell tenements were constructed in Manhattan and the Bronx. Boston followed New York's lead, but as late as 1900 builders could still construct buildings on 100 percent of the lot in Chicago.[54]

Jacob Riis and New Housing Reforms

The publication in 1890 of Jacob Riis's book *How the Other Half Lives* brought heightened attention to housing issues and catalyzed housing reform. Riis, an immigrant who lived as a homeless vagrant in Five Points soon after his arrival from Denmark, later returned to the neighborhood to write journalistic exposés and a book publicizing the inhumane conditions of the slums. Riis was the first to use photographs and present slide shows of housing conditions.[55]

Riis's exposé was followed by the passage of the New York Tenement Housing Act of 1890, an extensive investigation into housing problems by the State Tenement Housing Commission in 1894, and legislation authorizing the demolition of tenements. As a result of this attention to housing issues, many settlement workers focused on passing legislation to improve housing conditions. For instance, Robert Veiller of University Settlement became a staunch advocate of restrictive legislation (such as housing codes) as a means of housing reform. Veiller wanted to educate the public about how poorly workers were housed, so in 1900 he developed an exhibition of photographs depicting squalid housing conditions. He included examples of well-designed housing as alternatives. Settlement houses in Chicago, Boston, and elsewhere mounted similar exhibits. These elicited great public interest; more than ten thousand people viewed the South End House's exhibit in Boston in a two-week period. The New York exhibit caught the attention of the governor, Theodore Roosevelt, who pushed for housing reform. Veiller drafted a bill that formed the basis of the New York Tenement House Law of 1901, which banned dumbbell tenements, required apartments to be constructed with private bathrooms, and mandated fire protections, more windows, lighting in hallways, and waterproofing of cellar floors. By this time about two-thirds of the city's population (2.3 million people) lived in dumbbell tenements. The 1901 law also required that buildings have side courts at least four feet wide and backyards at least eleven feet deep; this meant that buildings could not cover more than 72 percent of a standard lot. The law also mandated that at least one room in the apartment be 120 square feet or larger, and none could be smaller than 70 square feet. The law established a Tenement House Department to enforce the laws and monitor the upgrading of more eighty-three thousand "old law" structures (tenements built before 1901). At the time Manhattan and the Bronx had about 44,000 tenements, and about two-thirds of the people in these areas lived in tenements. Veiller also urged the Charity Organization Society to create a Tenement House Committee to provide independent oversight. The Committee promoted the regulation of tenements, the enforcement of laws, and good management practices. The New York housing law was copied in New Jersey in 1901 and in numerous cities across the country later on.[56]

The construction of new-law tenements and steam-heated apartments in the outer boroughs in the late 1920s resulted in more than 250,000 people moving out of the Lower East Side. Landlords who still owned old-law tenements in Lower Manhattan had vacancy rates of 15 to 30 percent. Eventually many old tenements were demolished.[57]

Reformers continued to work for improved housing conditions. In New York they helped to pass the 1929 Multiple Dwellings Law, which required landlords to install toilets, fire retardants, and ventilation for inside rooms. Fire retardants had to be installed within six months of construction, and the legal ownership of a building was extended to include mortgage holders. Consequently banks and landlords threatened to abandon about sixty-eight thousand old-law tenements that had survived to the early 1930s. Beginning in 1934 New York Mayor Fiorello La Guardia razed almost ninety-five hundred tenements in a massive slum-clearance project. However, the dumbbell tenements have not completely disappeared; during the 1990s dumbbells still housed tens of thousands of people.[58]

Housing reform was a major issue in Chicago too. Jane Addams and Emmons Blaine, the daughter of Cyrus McCormick, launched the City Homes Association in 1900. Robert Hunter and other settlement workers conducted a comprehensive study of the city's housing conditions that year, which prompted the passage of a tenement house ordinance in 1902. Similarly South End House launched a study of the tenement districts in Boston and worked to have dilapidated buildings torn down. Mary Sales, a Smith College graduate and member of the College Settlement Association, did the first comprehensive housing study of Jersey City in 1902. While conducting the study Sales was threatened by landlords and arrested; however, the published study forced the governor, Franklin Murphy, to appoint a special tenement commission. Other settlement houses, such as Kingsley House in Pittsburgh, College Settlement in Los Angeles, Union Settlement in Providence, Kingsley House in New Orleans, and Northwestern Settlement in Chicago, played similar roles. The publicity generated by the housing studies led the steel executive Henry Phipps to finance Phipps House as a model tenement in Harlem in 1907. With a $10,000 grant from the Russell Sage Foundation, Veiller founded the National Housing Association in 1910.[59]

The National Housing Association, whose thirty-seven board members were drawn from fifteen states, focused on improving urban and suburban housing and reducing congestion. The organization studied the causes of internal migration that resulted in overcrowding in cities and ways of redistributing the population. It also acted as a clearinghouse of information on housing and worked on strengthening local housing associations. The Association's mission was to help enact, enforce, and defend housing laws; improve older buildings; and train workers in housing reform work.[60]

As part of the housing reform movement, activists sought to educate tenement dwellers about their rights, duties, and responsibilities. In an effort to overturn the stereotype of tenement dwellers as slovenly, lazy, irresponsible people who destroyed the dwellings they occupied, reformers developed a pamphlet titled *For You*. One of the overarching themes of the pamphlet was "health is wealth." The pamphlet, which dovetailed with the goals of the municipal housekeeping movement, included instructions on the essentials of sanitation, housekeeping tips, and information on avoiding diseases. Three female inspectors of the Tenement House Department and the Charity Organization Society distributed the pamphlets to tenement dwellers. While on their rounds these inspectors gathered information on violations of tenement laws.[61]

Overcrowding and City Planning

In 1907 Florence Kelley of Hull House helped to found the Committee on Congestion and Population. She mounted an exhibition on the dangers of overcrowding in 1908 at the Museum of Natural History. The exhibit, which traveled to several cities, featured pictures of pale, sickly slum children juxtaposed with healthy-looking country children. While some charities saw outplacement as the solution to overcrowding, Kelley and others called for more parks, playgrounds, schools, improved tenement house laws, better public transportation, and comprehensive city planning that would limit the number of people living in a given area. Overcrowding was still a major problem in cities in 1900. Veiller went as far as advocating height restrictions on buildings, thinking that such restrictions would help to reduce congestion. However, overcrowding in small houses also became a major concern. In one case in Chicago ninety-eight people were found living in four houses; in another case fifteen families lived on a twenty-five-foot lot. The mayor of Chicago and the governor of Illinois appointed commissions on congestion, and the first national city planning conference was held in 1909. The National Association of City Planning was formed at that meeting.[62]

Theoretical Analysis

Though some of the earliest urban evangelical and charity reformers were women, early on men took control of the discourse and defined this arena as a male sphere of activism. Women were asked to step aside from projects they initiated as men took over. Both the evangelical and charity reform movements relied heavily on a rhetoric of rectitude. Immorality and irreligiosity were master frames and the target of action was change in individual behavior. These were alterative movements. A rhetoric of calamity was also prominent in these movements; activists saw epidemics as evidence that God was smiting segments of the population as a form of reprisal. The solution was a change of behavior to avoid future calamity. Activists in both movements collected data, but the charity reformers were more effective in using the data to support their claims and to amplify and extend their framing. Their master and submerged frames evolved as the century progressed. When the claims of the evangelical movement concerning the cause of poverty were challenged, evangelical reformers were not effective in developing alternative counterclaims. Their most effective strategy was to spin off charitable organizations to work within the charity movement.

Both movements arose at a time when city, state, and federal governments had a crisis of legitimacy in this area.[63] Government had not taken a leadership role in providing social services; it was left up to reformers to finance their operations and make and implement policies. Though individual behavior modification was the primary goal of these movements, governments were drawn in to support or buttress the work of the reformers and eventually adopted some of the policies. The political opportunity that those movements exploited was a weakness in government and lack of legitimacy and leadership.

These movements also had to mobilize the resources necessary to accomplish their goals. They did this by convincing wealthy businessmen to lead and finance these movements. They tapped their networks of wealthy compatriots and members of influential religious institutions. It was evident early on that there were overlapping directorates. Founders of evangelical reform organizations also founded charity organizations. Men like John Griscom were involved in evangelical groups and charities; he also had ties to the government. Executives of banks and corporations also played prominent roles in these movements.

Industrialists building company towns believed that they could engineer healthy communities and a more efficient and committed workforce by providing decent, cheap housing and by controlling their workers' lives. They funded their work from their own pockets and dictated how they wanted things to function. When Francis Cabot Lowell sold stock to start his company most of the shareholders were family and friends. Even the utopian industrial communities that were built on an egalitarian model operated with a top-down approach. Workers finally revolted in some of the company towns when owners would not negotiate wages or eliminate harsh work rules. Some of the strikes were quite devastating. The strike at Amoskeag was a major contributing factor in the demise of the factory.

The settlement movement was aimed at making both transformative and incremental changes in government while trying to improve the quality of life in cities. The movement effectively challenged the master frame of the evangelical and charity movements. Though settlement activists still expected the poor to modify their behavior in order to improve their lives, the activists successfully shifted the focus of the discourse. They argued that poverty, illness, and environmental degradation did not stem from immorality and lack of religion but from structural factors in the society. They also argued that problems arose because of lack of government regulation and that the government ought to be more proactive in this arena. The settlement activists relied on a rhetoric of rectitude but attributed blame to sources different from those attacked by the charity reformers and evangelicals. Settlement activists also targeted different actors; in addition to the poor, they targeted the government and corporations. Settlement workers employed a rhetoric of rationality. They collected spatial and statistical data as well as oral histories and used this information in their reports, exhibits, and policy proposals. Activists also relied heavily on the rhetoric of unreason. They identified instances of manipulation and exploitation of the poor and organized campaigns around these. And they framed the issues as urgent. Employing a rhetoric of calamity, they implored all the targets of their action (the poor, government, industry) to act immediately to change conditions to avert impending disaster. These activists were very effective at using the rhetoric of endangerment to frame the workplace as a hazardous place and in pinpointing corporate exploitation of workers, callousness toward them, and wrongdoing.

The settlement activists excelled at framing poor urban residents as people with rights to whom the notion of justice and fair play should be extended. Their use of the rhetoric of entitlement was instrumental in getting the working class to collaborate with them. Settlement activists recognized the agency of slum dwellers and used it in their organizing. They articulated a framing of the relationship between poverty and the environment that differed from the prevailing understanding of the issues. They also helped to foster an atmosphere that facilitated the cognitive liberation of slum dwellers. The settlement house movement was unique in providing a space for female reform efforts and embracing a large number of female activists and leaders. Though a few women, such as Josephine Shaw Lowell, attained leadership roles in the charity movement, the settlement movement provided many more opportunities for female activists to take on leadership roles and gain prominence.

The settlement movement was not as well financed as the evangelical or charity movements. They did not have the deep pockets of the industrialists who financed company towns either. As a result settlements had to raise funds from philanthropists and rely on the government to take over demonstration projects or fund new ones. They had to mobilize people as well as resources. While the evangelicals and charity reformers focused on delivering services to people without organizing collective endeavors with the masses (because they mobilized the wealthy benefactors needed to fund their efforts), the settlement workers had to get the support of slum dwellers. They did this by making extensive use of recent college graduates who helped conduct studies and organize the communities. They tapped into salient identities of the settlement workers: their desire to improve conditions and their sense of obligation to improve civic life.

Settlement activists also exploited openings in the political structure. Where government initiatives and policies were either weak or nonexistent the settlement activists stepped in and developed policies, provided direction for government agents, and forced government action. They also courted government representatives as allies who could help push through policies. There was evidence of interlocking directorates in the settlement movement, as some of the most prominent settlement house activists went on to work on government commissions or in government departments. Unlike the evangelical reformers and charity reformers, who relied heavily on their membership and networks in elite institutions, the settlement activists relied on building movement infrastructure by collaborating with the poor, labor unions, government, and networks of dedicated college graduates committed to progressive social reform.

The charity, evangelical reform, and settlement movements can be considered institutional fields, and there is ample evidence of isomorphism within each. In both the evangelical and charity movements coercive isomorphism was at play as organizations such as the Charity Organization Society tried to coerce other charities to behave a certain way. There is also mimetic isomorphism as evangelicals and settlement groups copied the style and successful strategies of others in the field. There is evidence of mimetic isomorphism in the establishment of company towns as industrialists copied each other's ideas in establishing these. There is evidence of normative isomorphism in the charity, evangelical reform, and settlement movements: activists in all three movements recognized the need for professional staff. Consequently charity and settlement workers sent activists to be trained at designated institutions; they also helped to develop the field of social work.

Part III

URBAN PARKS, ORDER, AND SOCIAL REFORM

In this section I examine the rise of the urban park and recreation movements and cities' attempts to develop open spaces. I analyze how parks fit into the discourse of social control, order, and reform. Chapter 7 focuses on the campaign for urban parks, the link between parks and public health, and the role of cemeteries as models for urban park development.

Chapter 8 examines the development and functions of private parks and squares. I look at the rationale for developing large landscaped parks, how they were framed and their roles publicized, and their funding and siting. I analyze how class, race, and gender dynamics influenced urban park development and explore the relationship between parks and the social and political life of cities. I also analyze the differences between the working class and the elites who built and managed the parks.

In Chapter 9 I consider the rise of the recreation movement. I examine the role of women in developing small parks and playgrounds and the changing meanings of parks and explore working-class leisure styles and the articulation of working-class recreational needs. In the final chapter in this section, chapter 10, I examine the influence of Central Park on modern park development and the evolution of contemporary park building, park functions and financing, and the role of city government in overseeing and maintaining parks.

seven

CONCEPTUALIZING AND FRAMING URBAN PARKS

The Campaign for Urban Parks Begins

THE SIGNIFICANCE OF THE URBAN ENVIRONMENT

Urban elites campaigned for and developed open spaces as a means of improving conditions in the city. With the exception of Theodore Roosevelt and Frederick Law Olmsted (and to a lesser extent Gifford Pinchot), the most well-known American environmentalists were not actively involved in urban environmental affairs. Though urban environmental activism of the nineteenth century and early twentieth has been understudied, some of the earliest forms of environmental organizing and politicking occurred in the cities. Urban environmental politics pioneered important strategies that were adopted by conservationists and preservationists of the time. The open space campaigns and other environmental battles made the city the center of environmental reform during the mid- to late nineteenth century.

It should be noted that though the terms *conservation* and *preservation* are used interchangeably today, at the beginning of the twentieth century they had different meanings. Conservation implied a utilitarian approach to natural resources, a determination to develop and use the resources wisely for the current generation. Preservation implied saving resources for their own sake or intrinsic value as well as the needs of future generations.[1]

While the analytic spotlight has been trained on the environmental activists who shunned the cities and flocked to the mountains for inspiration and camaraderie, some activists stayed in the cities, making urban centers the focus of their activities (see Figure 1 on page 17). Consequently it is important to study

Box 3 Business Environmentalism

Business environmentalism refers to the close relationship between business leaders and environmentalists and to the environmental initiatives spearheaded by business leaders that influenced the development of environmental policies. Many nineteenth-century business leaders were interested in environmental activities that promoted order, improved the quality of life, enhanced their own status, facilitated their recreational pursuits, and furthered their business interests. Some business leaders were also conservationists. Consequently businesses had significant influence on early environmental organizations, the evolution of environmental policies, and the direction of environmental actions. The economic concerns of businessmen were factored into the concerns of the environmental movement because these men had significant influence on the agenda and policy direction of the movement.

urban environmental activism because it provides a lens through which we can better understand the early environmental movement. The urban environment became one of the contested terrains around which a number of ideological battles were fought.

The model of business environmentalism that dominated the early mainstream environmental movement first manifested in the evangelical and charity movements, and later in the development of cemeteries. It was evident during the development of Central Park and other landscaped urban parks. The overlapping directorates (business, government, and environment) evident in late-nineteenth-century and early-twentieth-century conservationism and preservationism also appeared in urban environmental politics in the mid-nineteenth century. Also at that time the dense social networks of environmental power elites were visible in urban environmental politics; conservation and preservation power elites emerged decades later.

Pastoral Transcendentalism, a transitional form of wilderness Transcendentalism, was a design feature of many landscaped urban parks. The ideology of pastoral Transcendentalism was clearly articulated and put in practice by Frederick Law Olmsted and Calvert Vaux years before John Muir popularized wilderness Transcendentalism. Urban elites and intellectuals saw landscaped parks as appropriate places for "manly" exercise several decades before wilderness superseded the parks as the site for the expression of manliness. Preservationists used urban environmental programs to justify their demands. For instance, Muir used the development of and public support for landscaped urban parks to argue that national parks would also be accepted by the public.[2] Although initially there was opposition to urban parks, over time the public came to embrace them, and Muir expected the same would be true for national parks.

Preservationists and conservationists may have learned from urban environmentalists how to effectively target the government to achieve their desired outcome. By the second half of the nineteenth century urban activists figured out that they could mobilize resources more effectively and get more environmental policies developed and enforced if they could get the government involved. Urban environmentalists also developed effective mechanisms to get results in the face of government inaction.

THE CHANGING MEANINGS AND FUNCTIONS OF URBAN PARKS

Frederick Law Olmsted defined parks as spaces used for public or private recreation, more spacious than gardens and with a more scattered arrangement of trees than the woods.[3] This definition was new: the conception, form, and function of parks had evolved over the centuries. Civilizations of the Near East began designing and managing their landscapes before the birth of Christ. In 700 B.C. Assyrian noblemen practiced riding, hunting, and combat skills in reserves earmarked for training purposes. The Persians were influenced by the design of these reserves and used them in the development of royal hunting enclosures. Royal hunting grounds flourished in Asia Minor in 550–350 B.C. The Greeks were the first to landscape public places as plazas or agorae; these tree-shaded plazas, adorned with fountains, were used for public gatherings, rest, and relaxation. The agora can be seen as a precursor to the modern city park.[4]

In medieval Europe open spaces were maintained exclusively for the use of the ruling classes. The term *parc* was then used to describe enclosed tracts of land stocked with animals for the chase. But even this designation grew more complicated. The grounds of the old English manors used to be divided into two parts: the inner grounds, closest to the dwellings, were fenced off and kept as the private pleasure grounds of the owners, while the outer grounds were left open to the public (eventually the public gained right-of-way access to these lands). At one time the term *park* referred specifically to the outer parts of the grounds. Over time the meaning of the term changed to refer to the large private estates and hunting preserves of the aristocrats. Landscape gardening emerged in England in the early sixteenth century. These early parks were characterized by broad stretches of greensward framed by sparsely distributed trees. In Europe many of these parks were created by evicting the poor from the land to create what Williams calls a "rural landscape emptied of labour and labourers . . . from which the facts of production had been banished." Public parks began to emerge in the seventeenth century. In 1652 London's Hyde Park was opened to the public on a fee-paying basis. By the late eighteenth century landscaped parks were being built in cities.[5]

In America gardens, squares, small parks, and commons were beginning to appear in the seventeenth century as forms of urban open space. Governor John Winthrop of Massachusetts negotiated the acquisition of forty-five acres of Reverend Blaxton's (Blackstone) farm and converted it into the Boston Common in 1634. Winthrop wanted "the Commonage" to be preserved for "Common use." This was codified in a 1640 law stating, "There shall be no land granted either for house plot or garden out of ye open ground or Common field." Citizens voted to pass the law and each household was charged six shillings to purchase the land.[6]

During the 1720s Boston took the first step in transforming the Common into a park by planting trees along Tremont Street to form the Mall. From 1816 to 1836 tree-lined malls were developed along Beacon, Charles, Park, and Boylston Streets. By 1860 much of the Common was covered with trees. Public executions were banned in the Common in 1812, and Bostonians ceased using it as a cow pasture in 1830. Elites helped to protect the Common by opposing schemes to bisect it with roads and streetcar lines.[7]

New Haven emulated Boston in setting aside public lands and undertaking park development. However, New Haven didn't just set aside open space on its periphery; its Green was in the center of the city, and the city was laid out around it in 1641. The city began a concerted tree-planting effort in 1759, when 250 buttonwood and elm trees were planted around the Green; this was followed by the planting of trees along Temple Street. In 1790 the city created a committee to develop bylaws for "preserving trees for shade and ornament" in the city. At this time James Hillhouse, a lawyer and real estate developer, began planting trees around the city. In 1798 the city passed an ordinance to protect the Green from unruly geese and Yale students. New Haven was so renowned for its beautification efforts that by 1809 the diarist Samuel Wordsworth described the city as one where "rural charms and city beauties join," where "Art and Nature everywhere combine."[8] In 1821 the Reverend Timothy Dwight noted that "scenery does not often strike the eyes with more pleasure" than in New Haven.[9]

Not only did settlers plant elms because they were hardy and fast growing; some argued that the trees were religious symbols as well, that their graceful form and interlocking limbs were suggestive of Gothic cathedrals. Indeed in his 1907 work, *The American Scene*, Henry James wrote that the verdant columns were so seductive that one may overlook the pervasive poverty in elm cities. He argued that in the winter, when the "leafy shroud" was removed, one couldn't help notice the unpleasantness the trees had concealed.[10]

Though the original design of Philadelphia called for an esplanade along the Delaware and five small squares, housing and commercial enterprises intruded on these planned open spaces. However, during the 1760s Cherry Garden became a designated public open space in Philadelphia.[11]

Bowling Green was laid out in New York during the 1670s. Around 1785 a park advocate writing under the pseudonym "Veritas" suggested that the Battery and the Fields, located on the southern tip of Manhattan, be reclaimed and turned into a public park.[12] At the time the closest things to a park that existed in America were town commons and greens. However, these were still multipurpose spaces that served as pastures; sites of executions, rallies, and protests; and homes for civic institutions.[13] New York's Common (or Fields) in Manhattan also had an almshouse, workhouse, prison, and African American burial ground. The site was converted into City Hall Park, and numerous gatherings, rallies, demonstrations, and riots were held there.[14]

Shortly after "Veritas" called for a park, Pierre L'Enfant suggested building one around the Collect and making the park and pond the focal point around which New York City would grow. The city opted to fill in the pond instead.[15] On April 13, 1807, the commissioners of streets and roads in the City of New York, Gouverneur Morris, Simeon De Witt, and John Rutherford, were given the task of developing a city plan depicting the layout of streets, roads, and public squares. Their plan, completed in 1811, designated few areas for public open space. The commissioners rationalized their decision as follows:

> It may be to many a matter of surprise that so few vacant spaces have been left, and those so small, for the benefit of fresh air and the consequent preservation of health. Certainly if the city of New York was destined to stand on the side of a small stream, such as the Seine or Thames, a great number of ample places might be needful. But those large arms of the sea which embrace Manhattan Island render its situation, in regard to health and pleasure, as well as to the convenience of commerce, peculiarly felicitous. When, therefore, from the same causes, the prices of land are so uncommonly great, it seems proper to admit the principles of economy to greater influence than might, under circumstances of a different kind, have consisted with the dictates of prudence and the sense of duty.[16]

The commissioners did set aside open space to be used for a reservoir. Because they were concerned with social order and defense, they also set aside space for military exercise, "to assemble, in the case of need, the force destined to defend the city."[17]

As these events were occurring activists were beginning to refine the definition of *park* and advocate for open spaces that served a restricted range of functions. As the social construction of the urban park evolved, elites saw parks as mechanisms of social control; they provided moral uplift and "tranquilizing" recreation; improved gentility and civility; socialized the masses into middle-class values and tastes; induced better attitudes toward work and produced more efficient workers; and served as a place for acceptable expressions of manliness. Parks were also seen as repositories for works of art and sources of cultural enlightenment wherein people were exposed to beauty in pastoral settings. Moreover parks were expected to improve health, ease overcrowding by providing breathing space in congested cities, act as urban resorts for people with no access to the countryside, function as the commons or social nerve center of the city, structure the plan and growth of the city, protect the urban water supply, increase property values, mute class conflicts, provide jobs, and function as a mechanism for dispensing political favors.[18]

From the early nineteenth century on, city park advocates trying to acquire public park lands used some or all of these arguments to rationalize and gain support for park projects and allay the fears of their social contemporaries.[19] Andrew Jackson Downing, a renowned landscape architect and horticulturalist, linked park advocacy to the general social reform message of temperance and improved taste, civility, and refinement of the masses: "You may take my word for it, [the parks] will be better preachers of temperance than temperance societies, better refiners of national manners than dancing schools and better promoters of general good feeling than any lectures on the philosophy of happiness."[20]

Despite such claims, the presence of parks did not alleviate the ills resulting from the growing poverty and social inequality in the cities, and riots, labor strikes, and other forms of social conflicts continued.[21] Realizing that the evangelical reforms had failed to restore social order, wealthy New Yorkers searched for other solutions. Some elites, such as Robert Minturn, who funded the tract societies, began giving charity directly to the poor; in the 1830s they flocked to his door for aid. However, in 1843 he succumbed to the dictum of the Society for the Prevention of Pauperism and other charities that argued that individual almsgiving was inappropriate, inefficient, and a "dangerous species of charity." As one of the founders of the Association for Improving the Condition of the Poor he espoused the idea that the "injudicious dispensation of relief" was the main cause of rising poverty; the Association's workers (some of them former volunteers for the tract societies) aided only the deserving poor and sent the undeserving poor, or "incorrigible mendicants," to the almshouses, workhouses, or jails. Businessmen in other cities took a similar course of action.

Not everyone thought the poor should be institutionalized simply for being poor. Some park advocates saw the parks as a gentler way to improve the character of the poor while assimilating them into mainstream society. They emphasized the park's potential for building a more cohesive society with less class tension. Thus Horace Greeley argued that creating "good places for public gathering" would improve workers, creating a more civilized person who would be attracted to "milder and more genial excitements." He attributed the general rowdiness of the working class to a lack of "humanizing and elegant resorts." Expressing a sentiment that Ross echoed years later, Greeley claimed that there was an ideal relationship that should exist between members of a society and that parks were a means of binding the rich and poor in a single orderly and refined republic.[22]

Olmsted too believed in the restorative and calming powers of parks and in their ability to help the classes bond, thereby reducing tensions. Olmsted believed parks would "inspire communal feelings among all urban classes, muting resentments over disparities of wealth and fashion." The park's scenery would "more directly assist the poor and degraded to elevate themselves," calm the "rough element of the society," "divert men from unwholesome, vicious, destructive methods and habits of seeking recreation," and counter "a particularly hard sort of selfishness" and anomie prevalent in the cities. Gregarious recreation, the coming together of thousands of people of various walks of life in the parks, was the remedy for the alienation and hard selfishness of urban life.[23] In 1853 Olmsted urged his longtime friend Charles Loring Brace, "Go ahead with the Children's Aid [Society] and get up parks, gardens, music, dancing, schools, reunions, which will be so attractive as to force into contact the good & bad, the gentlemanly and the rowdy."[24] Other park advocates echoed Olmsted's words. Stephen Duncan Walker, a Baltimore clergyman, saw the public park as "a commonwealth, a kind of democracy, where the poor, the rich, the mechanic, the merchant and the man of letters, mingle on a footing of perfect equality."[25]

Throughout his life Olmsted promoted parks as places where the classes mixed and inequalities were erased. In a talk given to the American Social Science Association he declared, "Consider that the New York Park [Central Park] and the Brooklyn Park [Prospect Park] are the only places in these associated cities where . . . you will find a body of Christians coming together, and with an evident glee in the prospect of coming together, all classes largely represented, with a common purpose, not at all intellectual, competitive with none, disposing to jealousy and spiritual or intellectual pride toward none, each individual adding by his mere presence to the pleasure of all others, all helping to the greater happiness of each."[26]

ELITE RECREATION: PRIVATE GARDENS AND SPORT FACILITIES

Though large urban parks were not a part of the American landscape until the mid-nineteenth century wealthy Americans have had access to private parks, squares, pleasure gardens, and hunting and racing grounds since the seventeenth century. Private and semiprivate open space became more common among elites during the eighteenth century; their interest in open space was fueled in part by their growing interest in gentility and refinement. They educated themselves about history, art, music, and literature. Participation in outdoor recreational activities, especially by men, was one way to demonstrate refinement. The men rode horses and took up hunting while the women did needlework. Gerard Beekman, for instance, wanting to present a refined image, began wearing the best imported silk stockings instead of plain linen and took up the "glorious sport" of hunting. At a time when only a handful of New Yorkers were wealthy enough to own carriages, he took morning rides in his carriage to hunt plovers with a gun imported from London. Although some of New York's wealthy families grew crops commercially on their country estates, many used the estate grounds primarily for aesthetic and recreational purposes. Caught up in the gardening and landscaping fads that preoccupied English estate owners, many American estate owners surrounded their summer homes with expansive lawns, ornate flowerbeds, orchards, greenhouses, conservatories, fishponds, and grottos. Anthony Rutgers laid out his garden with elegant shrubbery in the geometric (gardenesque) style popular with rural gardeners of the day. The garden extended in all directions from the mansion. An orchard was planted on the southern side, and the pasturelands and cultivated fields were on the northern side of the property. The demand for flowers was so great that William Prince established a nursery in Flushing that later became the Linnean Botanic Gardens. In essence the country estates provided their owners with private parks in which they sauntered and played golf, tennis, and cricket. Governor William Cosby of New York declared the 175-acre Governors Island his own private game preserve. The De Lanceys built a race course on their estate.[27]

Elites were so smitten by thoroughbred racing that in 1665 Governor

Richard Nicolls established the New Market Track at Salisbury Plain (now Hempstead, Long Island). A race track, the Church Farm Course, was built in New York City in 1725. Fox hunting also became popular among elites. In the 1760s James De Lancey and Lewis Morris combined their packs of foxhounds soon after De Lancey returned from a trip to England, bringing back horses and hounds. De Lancey's manager, John Evers, organized the Riding Hunt, held weekly in the Bronx, Brooklyn, and Queens. Smaller packs of hounds were used on hunts in Manhattan, but by 1822 all the Manhattan hunts were moved to Long Island or Westchester to take advantage of greater open space.[28]

Some of Lower Manhattan's early country estates did not remain estates very long because they were engulfed by development as the city expanded northward. For instance, the estate of Anthony Rutgers was converted to Ranelagh Garden. Opened in 1765, Ranelagh was a tavern, pleasure garden, and orchard named after a similar establishment in London. The grounds were illuminated at night and used for concerts, fireworks, and other entertainment. Cherry Garden had cherry orchards and facilities for lawn bowling. The Bayard estate, Bayard's Mount, was also converted to a pleasure garden. Ranelagh and Vauxhall Gardens (originally known as Bowling Green Gardens) were frequented by the gentry.[29]

PRIVATE PARKS, PUBLIC SUBSIDIES, MIDDLE-CLASS BENEFITS

As mentioned earlier, wealthy urban residents, like their counterparts in Europe, had access to private parks such as St. John's Park in New York and Louisburg Square in Boston long before public parks were established. However, early in the nineteenth century wealthy urbanites, unwilling to continue building private parks at their own expense, successfully convinced cities to subsidize private gated parks that only they would have access to. At the same time attempts to develop public parks in working-class neighborhoods failed.[30] Rich urbanites moved away from the congested business districts and into exclusive communities, reminiscent of London's Regent Park neighborhood. In the American villa park enclaves, elegant homes surrounded private parks and squares. To accommodate the elite the Common Council of New York modified the city street grid to permit private parks such as Gramercy Park and Union Square to be built. The wealthy were also allowed to enclose parts of blocks with railings and rename the enclosures to distinguish their neighborhoods from other sections of the city; Lafayette, Waverly, Irving, and University Place were created under this private zoning ordinance.[31]

Gramercy Park

Samuel Bulkley Ruggles, a lawyer and public official, was born in Connecticut and grew up in Poughkeepsie, New York; he moved to New York City in 1921 and married Elizabeth Rathbone, the daughter of a rich merchant. In 1831, when he decided to abandon law and become a real estate developer, he used his wife's inheritance to purchase the twenty-two-acre Gramercy Farm from one of Petrus Stuyvesant's descendants. He petitioned the Common Council for permission to create Gramercy Park, arguing that such open spaces would increase property values. The Council responded to Ruggles's proposal by giving him a special tax apportionment to develop Gramercy Park and the surrounding neighborhood. Ruggles divided the property into 108 lots, forty-two of which were set aside for a private park.[32]

Gramercy Farm was formerly Krom Moerasje, a Dutch name meaning "little crooked swamp." The farm, with its hills and swamps, was traversed by a spring-fed stream, the Crommessie Vly, which gouged a forty-foot ravine through the property. Ruggles moved about a million horse cart loads of earth to create the park and surrounding house lots. He spent $180,000 ($3 million in 2005 dollars) developing the site and deeded Gramercy Park collectively to the sixty-six lot owners. He got a tax exemption for the park in 1832 because the Board of Aldermen believed that Gramercy Park would be surrounded by valuable homes that would be taxed. By 1833 the square was fenced in and most of the lots were sold. However, housing construction did not begin at Gramercy until the 1840s.[33]

Ruggles placed restrictive covenants on the properties that "prevent[ed] purchasers . . . from erecting other buildings than private dwelling houses" on their lots. He fenced in the park and maintained it at his own expense. His goal was to "keep it forever unoccupied so as to admit the free circulation of air." As in the case of St. John's Park, only people whose property bordered the park had keys to the gate. The only time nonresidents and their guests were allowed to use the park was during the draft riots of 1863, when Union soldiers were billeted there. (Since the draft riots were in part a class riot in which working-class rioters destroyed the properties of the rich, it was in the interest of well-to-do Gramercy Park residents to have a military presence in their park.) The people who erected the fashionable dwellings around Gramercy Park ensured that their property values remained high by placing additional restrictive clauses in their deeds. The deeds specified the kinds of buildings that could be erected (usually brick and stone) and banned the construction of buildings for "noxious or dangerous trades and business."[34] Two prominent

Box 4 Restrictive Covenants

Restrictive covenants are clauses inserted in property deeds that specify and delimit what property owners can do with their land and buildings. Clauses can specify under what conditions the property can be sold, how and to whom, the height of buildings, how far from the street the buildings may be set, the size of the lot, the percentage of the lot that can be built upon, and the use to which the property can be put.

environmental activists, Gifford Pinchot and Theodore Roosevelt, lived in homes overlooking Gramercy Park. During the 1890s park residents successfully fought off several proposals to build a road through the park. Gramercy remains a private park today.[35]

Union Square

Union Place was a part of Elias Brevoort's farm. It contained a collection of vacant lots, shacks, and a potter's field when Ruggles began developing it and successfully petitioned the Common Council to change the name to Union Square. He also convinced the Board of Aldermen to enclose, regulate, and grade the square (much of the cost of this work was assessed to Ruggles's properties lying between Union Square and Gramercy Park). To enhance access to Gramercy Park and Union Square, Ruggles got the state legislature to build a new north-south artery; he called the northern part Lexington Avenue and the southern portion, which led to Union Square, he named Irving Place. The city provided a fence for Union Square. Ruggles built curbs and sidewalks and sold most of the lots around Union Square in 1839. He then built a four-storey house for himself on one corner of the square. Like Gramercy Park, the homes around Union Square were developed during the 1840s.[36]

Tompkins Square

Though Tompkins Square appears on the 1811 city plan as a public open space, the Stuyvesant family took the lead, albeit reluctantly, in turning the marshland into a park. Not all landowners were willing to underwrite the cost of developing land. Some, such as the Astors, Pells, and Fishes, refused to develop their swamplands, and when prodded by the city to do so threatened to abandon their property. The powerful landowners, especially the Stuyvesants, eventually got the city to allocate $62,000 of public funds to compensate them for improvements to their property. After securing the funds the Stuyvesants set about developing Tompkins Square. They transformed the muddy flats into a park that opened in 1834. The tree-shaded square was surrounded by an ornamental cast-iron fence.[37]

SUBSIDIZED DEVELOPMENTS AND OPEN SPACE

New York City subsidized private developers and allowed for the establishment of private parks because the city was desperate for an infusion of cash. As expenditures for infrastructure building and other budget items soared, city leaders found that their revenues were not keeping pace with expenses. For more than a century the city relied heavily on rental income from its own properties (waterfront lots, wharves, and common lands), licenses, and franchise fees to fund its operations. However, by the 1820s such revenue streams were insufficient to cover the city's operating costs. To make matters worse, property, especially farmland, was taxed lightly during the eighteenth century. That changed in the nineteenth century as land values rose and traditional sources of revenues shrank. Property taxes increased substantially during the 1830s, from just over $200,000 assessed in 1830 to $1.1 million assessed in 1837. With this change aldermen began encouraging and helping entrepreneurs to develop property in the hope that the tax on such property and single-family lots would be greater than that assessed on undeveloped property. Thus the homes built around private parks provided a potential tax windfall for the city. For this reason private parks in planned communities were more attractive to aldermen than parks in tenement districts, where only limited taxes could be generated.[38]

THE CHALLENGES OF DEVELOPING UNIFORM RESIDENTIAL COMMUNITIES

As late as the mid-nineteenth century efforts to develop uniform, elite residential communities still ran into difficulties. Because it was difficult to control land use, owners of expensive homes could find noxious, commercial, or industrial facilities or cheap housing being built next door or close by. Although wealthy citizens had an interest in keeping the residential areas in which they lived exclusive, once a property changed hands residents had no control over what could be done with it. For example, when the Vanderbilts acquired land close to St. John's Park to build industrial facilities, they changed the neighborhood dramatically, eventually resulting in the destruction of the park.[39]

Attempts to develop planned communities with ample open space date back to the seventeenth century in America. Philadelphia was originally planned as a garden city in 1682. The villages of Woodside (Queens) and Williamsburgh

(Brooklyn) were developed as planned communities in 1835 and 1836, respectively. The second half of the nineteenth century saw more concerted efforts to plan communities with parks and other open space. When Olmsted and Vaux designed Riverside, Illinois, in 1868 they included ample open space.[40] So, because of the lack of land use planning and general lack of open space in cities, elites campaigned for the development of large landscaped parks that could be used to anchor more uniform residential neighborhoods and commercial development.

URBAN ELITES, CULTURAL NATIONALISM, AND THE IDEA OF PUBLIC PARKS

During the mid-nineteenth century urban elites campaigned intensely for public parks. Although advocates supported park development for many reasons, one of the major factors was cultural nationalism. American elites, feeling that the country's civility and cultural attainment lagged behind Europe's, sought to enhance America's prestige. Thus they focused on developing and promoting cultural and educational institutions as well as identifying natural wonders that were superior to European attractions. Moreover as gentility spread from the upper to the middle class more Americans collected books and artwork, hired tutors for their children, toured Europe, attended theaters and museums, and joined literary clubs. Communities built ornate churches, libraries, athenaeums, and civic institutions. Americans also began to preserve artifacts, and unique remote landscapes. Thus as the desire for genteel forms of recreation and social interaction spread, the demand for elaborate landscaped parks increased.[41] The famed architect and designer of mansions on Fifth Avenue and in Newport, Richard Morris Hunt, who saw the park as a potential repository of the artistic expression of "great national ideas," declared that a park in Manhattan would encourage the development of "blocks of dwelling houses excelling in beauty and magnificence any we can boast of in the New World."[42] The park would allow "the mind to expand to its utmost possibilities."[43] Others believed that Central Park, along with the other cultural institutions of the city, would create in New York the "national, liberal and cosmopolitan spirit that is generated only by one acknowledged central city of a great country."[44]

Like the early wilderness enthusiasts and wildlife advocates, the upper- and middle-class urban park supporters were strongly influenced by European notions and standards of art, nature, and beauty. These urban environmental advocates were part of an intelligentsia who belonged to exclusive organizations such as the Century Club, Union Defense Committee, Union League

Box 5 Romanticism

Romanticism connotes an enthusiasm for the strange, remote, solitary, and mysterious. Romantics, particularly those who adopted this ideological outlook in the second half of the nineteenth century, showed a preference for wild, untamed places such as the American wilderness, where they could express their freedom. They disdained tamed and manicured landscapes.

Sources: Arthur Lovejoy, *Essays in the History of Ideas* (New York: Braziller, 1955); Roderick Nash, *Wilderness and the American Mind*, 3rd ed. (New Haven: Yale University Press, 1982), 47.

Club, St. Nichols Society, New England Society, New York Club, and the American Social Science Association.[45]

Park advocates were also Romantics. The Romantic influence in urban park conception and design is not surprising since the leading landscape architects of the 1850s through the 1880s were schooled in the European tradition.[46] In addition the landscape architects were influenced by the Hudson River School painters,[47] who were ardent cultural nationalists, and the Romanticism popularized by Emerson, Thoreau, the Transcendental Club, and outdoor enthusiasts. Both Andrew Jackson Downing, who grew up among the Hudson River painters in Newburgh, and Calvert Vaux, who also lived in Newburgh and married the sister of a Hudson River School painter, were influenced by this school of art. Frederick Law Olmsted, another leading landscape architect, met Downing, Vaux, and Luther Tucker, the publisher of the *Cultivator*, on visits to Newburgh. The landscape architects and Hudson River painters had ample opportunities to interact since many of the painters also lived in New York City and joined the same social clubs as other elites in the city.

Olmsted grew up listening to George Perkins Marsh and Emerson lecturing at the Young Men's Institute in Hartford, Connecticut, his hometown. When he was a teenager he read Emerson and Ruskin.[48] While living on Tosomock Farm on the south side of Staten Island, Olmsted was a neighbor and legal client of Judge William Emerson, the essayist's older brother. (Thoreau lived at William Emerson's home and tutored his children.) After the 1832 cholera outbreak and the 1835 fire, wealthy New Yorkers began building homes on Staten Island. The wealthiest residents built their homes on the south side. Thus Olmsted's neighbors also included William Henry Vanderbilt, William Cullen Bryant, and George Putnam, the book publisher, with whom Olmsted would later undertake an ill-fated business venture. Several of the neighbors, including Vanderbilt, consulted Olmsted about landscaping their property.[49]

Later in life Olmsted became a member of the Union League Club and was elected to the Saturday Club in Boston; Emerson, Agassiz, Lowell, Prescott, Hawthorne, and Longfellow were among the luminaries active in the Saturday Club at that time.[50]

The urban park advocates embraced the European Romanticism of Rousseau and Wordsworth. To the earliest generation of American Romantics who remained in the city, "wild" meant the pastoral and picturesque settings of large urban parks. Like Rousseau, they believed that the lives of the urban poor would be improved if they could experience pastoral beauty and rural charms.[51] Emerson, for instance, conceptualized the wilderness and relations between humans and nature in a way that was closer to the European interpretation than to the second wave of American Romantics and Transcendentalists such as John Muir, whose followers became wilderness advocates.[52]

Emerson, Rural Life, and Pastoral Landscapes

Ralph Waldo Emerson, a minister, poet, philosopher, social critic, and Transcendentalist, began writing about nature during the 1830s. A lifelong rambler, he was drawn to nature since his youth. Emerson was twenty years old when his family moved from Boston to the rural Canterbury section of Roxbury, Massachusetts, in 1823. They were still close enough to Boston for him to walk to the city daily. He sometimes carried a hunting rifle as he sauntered through the woods. Emerson described Roxbury as "a picturesque wilderness of savin, barberry bush, catbrier, sumach, and rugged masses of pudding stone."[53]

Emerson came from a modest background but inherited his first wife's estate and earned enough later on from speaking engagements to live very comfortably for much of his adult life. He had an epiphany while visiting the Jardin de Plantes in Paris in 1833 and became interested in plant taxonomy, order in the natural world, and the beauty of gardens and pastoral settings. After his visit to the Jardin de Plantes he began incorporating more ideas about the natural world into his writings. After living in Boston for several years he left the hectic city behind and settled in Concord in 1835 with his second wife. The newlyweds settled on a two-acre property that sloped toward Mill Brook. Emerson considered Concord and Bush, the name of his estate, a refuge from "the compliances and imitations of city society." He liked the "lukewarm milky dog-days of common village life." He loved sauntering around Concord, took daily walks in the woods around Walden Pond, and went on longer hikes to Cape Cod, the Connecticut Valley, and Maine. These experiences provided the context in which he wrote about the natural world and encouraged others to embrace a rural lifestyle.[54]

Box 6 Transcendentalism

Transcendentalism arose in New England in the 1830s. Adherents believed there is a spiritual relationship between humans, nature, and God. The earliest American Transcendentalists were generally the children of Puritans or liberal Unitarian ministers. Transcendentalists believed in the existence of a reality or truths beyond the physical. They believed that divinity was inherent in the human soul and that a person's perceptions and intuition provided knowledge of the divine and that was the basis of truth and moral judgment. Transcendentalists argued that there is a parallel between the higher realm of spiritual truths and the lower one of material objects. Natural objects are important only because they reflect universal spiritual truths. People's place in the universe was divided between object and essence; their physical existence rooted them in the material portion, while their soul gave them the ability to transcend their physical condition. Hence they were aware of a God that was both immanent and transcendent. For Transcendentalists the wilderness was the place where spiritual truths were most pronounced.

While Locke believed that all knowledge and understanding is gained through sensory experiences, Kant argued that certain concepts, such as time and space, are "transcendental," that is, they are innate categories of the mind and are known intuitively. In addition to Kant, Transcendentalists were also influenced by Coleridge and Thomas Carlyle, Emanuel Swedenborg, the French philosopher Victor Cousin, and Hindu writers. Some of the most prominent members of the movement and the Transcendental Club were Emerson, Frederic Henry Hedge, Bronson Alcott, George Ripley, Thoreau, and Margaret Fuller. The Transcendental Club first met in 1836; most of the members lived in and around Boston. The group published *The Dial* from 1840 to 1844. Two other minor Transcendental Club publications were the *Western Messenger* and the *Aesthetic Papers*.

Sources: Ralph Waldo Emerson, "Nature," in *Essays* (Boston: Houghton Mifflin, 1883), 161–88; Henry David Thoreau, *Excursions: The Writings of Henry David Thoreau* (Boston: Houghton Mifflin, 1893); Sherman Paul, *Emerson's Angle of Vision: Man and Nature in the American Experience* (Cambridge, Mass.: Harvard University Press, 1952); Roderick Nash, *Wilderness and the American Mind*, 3rd ed. (New Haven: Yale University Press, 1982), 84–86; Bryan F. Le Beau, "Transcendentalism," and Anne Baker, "Transcendentalist Writers," both in *The Encyclopedia of New England*, edited by Burt Feintuch and David H. Watters (New Haven: Yale University Press, 2005), 806–8, 1032–33.

Emerson's views of nature and wildness were closer to Downing's views than to Muir's. Though Emerson was an established nature writer by the time he read Downing's work, Downing still influenced him. Emerson was an aspiring gentleman farmer when he read Downing's treatise on orchard cultivation and rural taste, *The Fruits and Fruit Trees of America*, and used it to help develop his orchard. Downing focused on domesticating and ameliorating the species he worked with.[55] The urban park advocates borrowed heavily from the Emerson-Downing tradition in their attempts to build tasteful pastoral and picturesque landscapes with rural scenery for genteel recreation within the city.

TO SEE AND BE SEEN: CLASS, CARRIAGE PARADES, AND URBAN PARK ADVOCACY

In the early nineteenth century wealthy New Yorkers reveled in high-fashion parades down Broadway, Fifth Avenue, and the Battery on Sundays and evenings. Wealthy Baltimoreans and Philadelphians also paraded around upscale thoroughfares in their expensive carriages. The carriages became a public stage from which one could see and be seen. The parades also offered a chance to peruse the latest merchandise on sale at the retail stores.

By midcentury wealthy citizens had moved away from the Battery and Lower Broadway and no longer drove their carriages into neighborhoods and along streets filled with working-class people. Consequently the high-fashion promenades declined in popularity as "respectable" citizens ceased to participate in them. This led the New York *Post* to lament that the few remaining female promenaders were "stared" at by hordes of "whiskered and mustachioed chatterers," by which the *Post* meant immigrants. Horse racing and trotting became popular at this time, and though elites lived in residential enclaves anchored by small parks, these open spaces were not large enough for parades or trotting races. Thus the *Post* called for the creation of a new park for more genteel and tasteful public recreation.[56]

Elite park advocates hoped the new park would protect upper-class white women from the masses and that it would be a regulated space for daytime socializing and a respite from the confines of the home. The rich wanted grander public spaces for promenades because fine carriages and country estates had become important symbols of wealth, leisure, and conspicuous consumption. Thus a large park, devoid of the teeming masses, would provide an elegant substitute. As the wealthy also associated parks with an appreciation of landscape art, their association with such a place was another public signifier of elevated and refined status. Furthermore, when the *Post* reported that fashionable citizens of Europe were "not in the habit of walking in the streets," New York elites too wanted to get off the street and into a park to promenade.[57] Ergo the move to build large, centerpiece urban parks was a move to construct a place where the middle and upper classes could recreate comfortably.

The parks were also a de facto white space, as blacks and other minorities were intimidated and harassed if they tried to use the parks. Cultural nationalists, urging cities to build parks and other cultural institutions, promoted the democratizing nature of these institutions. However, they failed to extend democracy to all members of the populace. While upper- and middle-class whites could use such public spaces freely and the white working class was grudgingly admitted, minorities were not welcomed. So, even though working-class whites and blacks lived in similar neighborhoods, they didn't occupy the same social location. Whiteness conferred upon poor whites access to cultural institutions, jobs, social services, and other social and economic opportunities denied blacks. For blacks, parks, museums, transportation systems, and universities were undemocratic public spaces. Blacks eventually challenged their exclusion from these amenities and institutions during the early years of the civil rights movement.[58]

From Cemeteries to Public Parks

CEMETERIES: THE EARLIEST PASTORAL RETREATS

The development of pastoral cemeteries preceded the development of pastoral parks. Consequently park advocates had to recognize the similarities between parks and cemeteries, but at the same time clearly express the need for both kinds of open spaces to serve separate functions. While both types of spaces provided greenery, beautiful vistas, and places for quiet contemplation, the kind of contemplation one did in each space differed. Park advocates also had to articulate functions that parks served that cemeteries could not. Though cemeteries served as models for the design of urban parks, parks had to incorporate new elements into their designs to accommodate the different functions they were intended to serve.

As late as 1857 no American city had a major, completed landscaped park. Thus anyone wanting to experience pastoral landscapes had to go the countryside or the cemetery. As some of the first planned open spaces close to urban centers, cemeteries such as Grove Street in New Haven, Mount Auburn in Cambridge, Laurel Hill in Philadelphia, and Green-Wood in Brooklyn were enormously popular.

Grove Street

Grove Street Cemetery was the first garden cemetery built in the United States. The cemetery was built because of the yellow fever epidemics in New Haven in 1794 and 1795. Since the common burial ground on the Green was too crowded to continue serving as a burial ground, in 1796 a group led by then Senator James Hillhouse acquired six acres of land on the edge of town and built the New Burial Ground, later renamed Grove Street Cemetery. The first burial took place there the following year.[59]

Grove Street introduced new innovations in burials and eased concerns about public health. The cemetery guaranteed burial rights in perpetuity, while users were taxed for the cost of maintaining it. The notion of burial in perpetuity and cemetery landscaping and maintenance were new at the time. The cemetery also introduced the concept of family plots. The cemetery was laid out with avenues running in straight lines and trees were planted within. Its geometric design was intended to mirror the nine squares of the city. Despite these innovations sales of plots slowed considerably after the first three years. As a result not many towns developed this type of cemetery.[60]

Mount Auburn

Mount Auburn in Cambridge popularized the rural garden and pastoral cemetery. In the three decades between the opening of Grove Street and the planning of Mount Auburn, large cities grew much more crowded, and there was greater demand for cemeteries of this sort. The idea of developing Mount Auburn was first proposed by Jacob Bigelow, a prominent doctor, botanist, and professor at Harvard, in 1825, when he called a "few gentlemen" to his house to discuss the building of a new cemetery. Bigelow proposed developing a landscaped cemetery composed of family burial plots, interspersed by trees, shrubs, woods, and flower gardens. The committee first tried to buy land in Brookline, and when these efforts failed, in 1830 they concentrated on a thirty-seven-acre tract of land four miles from Boston on the Cambridge-Watertown border known as "Stone's Woods" or "Sweet Auburn." The group wanted to build a rural cemetery to rival the cemeteries of Europe.[61]

The Massachusetts Horticultural Society, incorporated in 1829, oversaw the operations of Mount Auburn; Bigelow was the corresponding secretary of the Horticultural Society. To fund the construction, lots were sold for $60 each ($1,000 in 2005 dollars). Each lot owner automatically became a life member of the Horticultural Society. There was great demand, and more than a hundred plots were sold by August 1831.[62]

Mount Auburn was built on a hilly grove above the Charles River. The trustees acquired more land over time, and by 1854 the cemetery covered almost 130 acres. Mount Auburn was modeled after the Père Lachaise Cemetery just outside of Paris. It was laid out in the picturesque style, with curvilinear drives, ponds, shrubs planted around the rock outcrops, and hillocks.

Mount Auburn was built at a time when Americans were rethinking how to lay the dead to rest. The term was changed from *burial grounds* to *cemetery*, and instead of crowded churchyards the dead were buried in spacious garden plots. The rural cemeteries were seen as places that memorialized the dead in perpetuity (rather than disinterring them and dumping the bodies in ditches to make way for new burials), thus they were designed as beautiful green spaces where the living could visit their loved one's gravesite. The opening of these new cemeteries was greeted with a mixture of excitement and curiosity. Nearly two thousand people attended the consecration of Mount Auburn in 1831. It became so popular that within months, the trustees banned horses and carriages, except for those attending funerals. Only lot owners and their guests were admitted on Sundays.[63]

Pastoral cemeteries were planned to impose a desired social order. Since many visitors were urbanites seeking open space to recreate and relax in, the cemetery proprietors desired to impose social order on the masses during their visits. The Horticultural Society developed elaborate rules regarding the use of the cemetery and the behavior of visitors within it. Members of the society were upper-class Bostonians whose pursuit of horticulture reflected their refined tastes and breeding. They believed their appreciation of beauty was an antidote to the crass materialism surrounding them. For them the cemetery was a place to express an appreciation for nature without too much intrusion from the outside world. The cemetery was closed at sunset and no refreshments were allowed on the grounds. Anyone making "unseemly noises" or otherwise "conducting themselves unsuitably" was asked to leave the grounds; if such persons refused to leave they were prosecuted. No vehicle could be driven in the cemetery at a pace faster than a walk, and horses could not be left unfastened from designated posts without a keeper. Visitors were prohibited from picking flowers or breaking off the branches of trees or shrubs; anyone found with flowers taken from the cemetery grounds was prosecuted. Visitors were also prohibited from discharging firearms and writing upon or defacing monuments, fences, or structures within the cemetery.[64]

The Rising Popularity of Rural Cemeteries

Laurel Hill Cemetery in Philadelphia was developed on the former estate of the prominent merchant Joseph Sims and designed by John Notman. A decade after its opening in 1836 more than thirty thousand people visited the cemetery annually. The two-hundred-acre Green-Wood Cemetery in Brooklyn, incorporated in 1838, was the largest of the four discussed here. Mount Auburn also inspired the creation of Green Mount in Baltimore; Swan Point in Providence, Rhode Island; and the Valley Street Cemetery in Manchester,

New Hampshire. Green Mount was built on the former country estate of the merchant Robert Oliver in 1838.[65]

Cemetery commemorations were grand affairs. Emerson was asked to give the dedication speech of Sleepy Hollow Cemetery when it opened in Concord in 1855. Sleepy Hollow was a garden cemetery modeled after Mount Auburn. Emerson liked the cemetery because it was an arboretum planted with native trees.[66]

There was a great demand for the rural cemeteries, and their popularity skyrocketed. Famous and wealthy people wanted to be interred in these new cemeteries. As was the case with other types of open spaces, the wealthy controlled and tried to monopolize these cemeteries. They also vied with each other to lay themselves and their families to rest in the newest and most prestigious cemeteries and eschewed the crowded downtown churchyards and crypts. In addition, the rich were in the process of abandoning their old churches and building newer ones on the urban fringes. The rural cemeteries provided a way of escaping the discomforts of the city in life as well as in death. As United States Supreme Court Justice Joseph Story asked in his address at the dedication of Mount Auburn,

> Why should we deposit the remains of our friends in loathsome vaults or beneath the gloomy crypts and cells of our churches, where the human foot is never heard, save when the sickly taper lights some new guest to his appointed apartment . . . ? Why should we measure out a narrow portion of earth for our graveyards in the midst of our cities, and heap the dead upon each other with a cold, calculating parsimony . . . ? Why should we expose our burying-grounds to the broad glare of day, to the unfeeling gaze of the idler, to the noisy press of business, to the discordant shouts of merriment, or the baleful visitations of the dissolute? . . . It is painful to reflect that the Cemeteries in our cities, crowded on all sides by the overhanging habitations of the living, are walled in only to preserve them from violation.[67]

Referring to the rural cemetery as a "field of peace," Story went on to describe Mount Auburn as a place where energy emanated from the abundance of nature. It was an antidote to the old burial grounds:

> There are around us all the varied features of her beauty and grandeur—the forest-crowned height; the abrupt acclivity; the sheltered valley; the deep glen; the grassy glade; and the silent grove. Here are the lofty oak, the beech, that "wreaths its old fantastic roots so high," the rustling

pine, and the drooping willow. . . . Here is the thick shrubbery to protect and conceal the new-made grave; and there is the wild-flower creeping along the narrow path. . . . All around us there breathes a solemn calm, as if we were in the bosom of a wilderness, broken only by the breeze as it murmurs through the tops of the forest, or by the notes of the warbler pouring forth his matin or his evening song. Ascend but a few steps, and what a change of scenery to surprise and delight us. We seem, as it were in an instant, to pass from the confines of death, to the bright and balmy regions of life. Below us flows the winding Charles with its rippling current, like the stream of time hastening to the ocean of eternity. In the distance, the City.[68]

Developers also began building rural cemeteries because in the major cities, burial space had not kept pace with population growth. In New York, where there were ten thousand interments annually, there was a dire need for additional burial grounds. In 1822 there were twenty-two burial grounds south of City Hall in Manhattan, but these small, unkempt grounds were full. Furthermore during the yellow fever outbreaks of the 1820s people suspected that burying the dead in the same ground where the city got its well water might be contaminating the water supply and spreading diseases. Hence in 1823 the city passed a law banning all burials in graves or vaults south of Canal, Sullivan, and Grand Streets. By 1851 all burials south of Eighty-sixth Street, except those in private vaults and cemeteries, were also banned.[69]

Before the advent of the rural cemeteries burials were not permanent. Burial grounds were ripped up and the bodies discarded or reinterred for a variety of reasons. For instance, after the 1811 city plan of New York was approved churchyards were destroyed and paved over to make way for the new streets in the grid plan. As wealthy residents moved away from downtown and their churches followed suit, the old church properties were often sold to developers who frequently dug up and sent unclaimed bodies to potters' fields.[70] In Philadelphia in 1835 burial spaces were so scarce that a crowd of enraged citizens once seized and broke the laborers' tools to prevent them from digging up the bodies buried in the German Reformed graveyard and throwing them into a trench.[71]

For all these reasons the wealthy eagerly bought plots in the rural cemeteries and dug up the bodies of their relatives and reburied them in family plots in rural cemeteries. For instance, Emerson exhumed and reburied the body of his mother and five-year-old son in Sleepy Hollow in Concord.[72] The theft of the retail merchant Alexander T. Stewart's body from St. Mark's Church prompted the rich to build fortress-like mausoleums around their dead, even in the rural cemeteries. Some built their own private cemeteries. For instance, Olmsted was commissioned to design W. H. Vanderbilt's twenty-one-acre Staten Island cemetery.[73]

Green-Wood

Henry E. Pierrepont, inspired by his visit to Mount Auburn and the potential economic windfall of burying New York's dead, proposed to build a "city of the dead" in Brooklyn in 1832. However, high land prices kept him from building Green-Wood for some time. Then the Panic of 1837 enriched some speculators and made it easier for Pierrepont to realize his dreams. As profiteers such as John Jacob Astor snapped up land and houses cheaply and foreclosed on properties when owners fell behind on their payments, Pierrepont acquired the land for Green-Wood from Dutch farmers desperate to sell. Green-Wood was incorporated in 1838 and the first lots went on sale in 1839. Like Mount Auburn, Green-Wood was developed on a hilly knoll as a pastoral retreat with winding paths. The cemetery struggled financially for a few years, but grew in prestige and popularity after the body of De Witt Clinton, the mayor, senator, and governor of New York, was moved there from Albany in 1844. To make it easier for people to get from Manhattan to Green-Wood, Pierrepont began operating a ferry across the Gowanus Creek in 1846. The cemetery was so popular that by the 1850s it was attracting a hundred thousand visitors per year, and by 1860 half a million people were visiting the cemetery annually.[74]

As was the case in Mount Auburn, the behavior of visitors was a major concern for Green-Wood's managers. While tourists complained that the continuous procession of funerals interfered with their enjoyment of the cemetery, Green-Wood's Sunday police watched for violators of cemetery rules and escorted them to the gate. Yet visitors continued to break the rules, lounging on the grass, smoking, picnicking, talking and laughing loudly, picking flowers, and breaking the four-mile-per-hour speed limit in their carriages. The tension between paying homage to the dead and the recreational use of the cemetery was a thorny problem that overshadowed the operation of the rural cemeteries.[75]

CAMPAIGNING FOR URBAN PARKS

Noting the popularity of the rural cemeteries and the general lack of open space in the cities, park advocates such as the poet and newspaper owner William Cullen Bryant and Andrew Jackson Downing stepped up their efforts to get cities to build public parks in the 1840s. At about the same time Americans returning from Europe began campaigning for the construction of public parks. The novelist Catharine Maria Sedgwick wrote that she found it impossible to "enter the London parks without regretting the folly (call it not cupidity) of our people, who, when they had a whole continent at their disposal, have left such narrow spaces for what has so well been called the 'lungs of the city.'"[76] (In 1842 the *New York Mirror* urged the city to build parks because they would become "the lungs of the city.") Bryant's trip to England likewise influenced his desire for a planned park system in New York. He began his campaign in 1844 and intensified it a year later while visiting the public parks of London. (At the time, London had devoted thirteen thousand acres of land to park space.) Bryant was an ardent supporter of Central Park. Influenced by Londoners who saw their parks as the lungs of the city, Bryant, like Sedgwick, used this metaphor to describe the potential role of American parks.[77] He argued that it was a "cause of regret that in laying out New York, no preparation was made, while it was yet practicable, for a range of parks and public gardens along the central part of the island." He urged the city's leaders to create parklands immediately: "The advancing population of the city is sweeping over . . . [the available lands] and covering them from our reach."[78]

Other prominent Americans traveling in Europe were quite taken by the public parks there and urged Americans to develop similar parks. The writer Caroline Kirkland's reaction was similar to Sedgwick's: "Nothing we saw in London made our own dear city of New York seem so poor in comparison as these parks. . . . After seeing these oases in the wilderness of streets, one can never be content with the scanty patches of verdure . . . that [in New York] form the only places of afternoon recreation for the weary, the sad, the invalid, the playful."[79] In 1848 the *Democratic Review* joined the chorus by encouraging the city to plant shrubbery wherever there were street development projects so that New York could become a "city of gardens."[80] Bostonians considered the Boston Commons the lungs of the city and likened it to London's parks, and Philadelphians likened Fairmount Park to European parks.[81] Brooklyn's Walt Whitman campaigned to create a park in the Fort Greene neighborhood, a mixed-race working-class section of the city. In 1847 Brooklyn acquired the Fort Greene site. Washington Park was developed on the site in 1848.[82]

The reformer Charles Loring Brace of the Children's Aid Society advocated the construction of more recreation spaces for the poor. In 1855 Brace urged the city to build parks with fountains, flowers, statuary, and chapels as a "natural" and healthful "way for the laborer to worship."[83] Brace saw the squalid and cramped conditions in which the poor lived and understood the dire need for open space in their communities. Thoreau urged each community to set aside "a park or primitive forest" as a way of keeping "the New World new [and] preserv[ing] all the advantages of living in the country."[84] Yet despite the efforts of these early advocates, no coordinated movement emerged in any American city to implement plans to build landscaped public parks. This is due, in part, to the fact that when the parks campaign began in the early 1840s the country was just recovering from the Panic of 1837; such public works projects lacked political support and were not economically feasible. However, in the early 1850s New York was booming economically and the city's business class grew interested in the park movement. Other cities quickly followed New York's lead.[85]

DOWNING, CEMETERIES, AND URBAN PARKS

Andrew Jackson Downing was born in 1815 in Newburgh, about fifty miles north of Manhattan. He cultivated wealthy friends among New York's elite and in 1838 married Caroline Elizabeth DeWint, the daughter of a wealthy land speculator and investor in rail and ferry lines. The couple built an elegant Gothic revival home on land donated by Downing's parents and with financing from Caroline's father. Downing used his own estate to demonstrate the landscaping techniques he wanted his clients and others to emulate.[86]

Downing became the leading advocate of tasteful and genteel rural living, believing that the rural lifestyle would combat what he saw as the "too great bustle and excitement of our commercial cities." His 1841 publication, *A Treatise in the Theory and Practice of Landscape Gardening*, helped to establish him as the country's premier landscape architect. Though Downing's *Treatise* was heavily influenced by Loudon's *Suburban Gardener and Villa Companion*, the American gentry viewed it as a fresh way to look at and interact with rural American landscapes. Downing was astute enough to speak directly to elites trying to cultivate taste and style at their rural residences. He used the book to demonstrate how to convert a farm into an elegant country seat by installing ample lawns, curving driveways, and trees and bushes strategically placed in the front of the property and orchards in the rear. He suggested that estate owners could make their property appear more elegant by remodeling farmhouses. He recommended accentuating the roof lines and chimneys and adding Gothic elements and porches. Though he eschewed the profession of nurseryman, as his parents had advocated, he developed a nursery and mail-order business to ship plants to aspiring gentleman farmers and gardeners all over the country. Downing also urged property owners to improve their farms to make farm life more attractive. He believed that a more comfortable home life would encourage men to settle down and discourage them from migrating west in search of fortune. Eventually Downing had to sell his nursery; he did not have the financial wherewithal to keep up his Hudson River estate, and the financial wrangling and legal tug-of-war between him and his father-in-law forced him to sell the nursery to pay off his debts.[87]

Downing emerged as the most prominent park advocate in the early 1850s. Blending the arguments of rural lifestyle advocates, the sanitary reform movement, and the emerging park movement, he became one of the first to articulate a comprehensive vision for American urban parks. Like Bryant, he was an ardent admirer of European public parks, and between 1848 and 1851 he published editorials and articles calling for public parks modeled after those in Europe.[88] Soon after park advocates proposed a large landscaped park for New York, Downing wrote a compelling letter of support for the proposal. He argued that New York was "deluded" and had "contented itself with little door-yards of space—mere grass-plots of verdure which form the squares in the city, in the mistaken idea that they are parks." He contended that because of the lack of "any breathing space for pure fresh air, any recreation ground for healthful exercise, any pleasant roads for riding or driving" wealthy Americans sojourned to Europe for years at a time, and others fled the cities at least during the summer.[89]

Downing was an ardent believer in parks as a valuable source of cultural enlightenment. He described the urban park as "republican in its very idea and tendency": "It takes up popular education where the common school and the ballot-box leave it, and raises up the working-man to the same level of enjoyment with the man of leisure and accomplishment."[90] Downing argued, "The higher social and artistic elements of every man's nature lie dormant within him, and every laborer is a possible gentleman, not by the possession of money or fine clothes—but through the refining influence of intellectual and moral culture." He urged the government, "Plant spacious parks in your cities, and unloose their gates as wide as the gates of morning to the whole people . . . [for the] common enjoyments for all classes in the higher realms of art, letters, science, social recreations, and enjoyments."[91]

To reinforce the need for parks Downing pointed to the popularity of cemeteries such as Mount Auburn, Green-Wood, and Laurel Hill and argued that public gardens should also be provided in cities.[92] He designed the Cemetery of the Evergreens on the border of Brooklyn and Queens. Impressed by the way the rural cemeteries were funded, he suggested that cities could fund parks by copying the cemetery model, forming joint stock companies and selling shares in the park. Such parks would be open to shareholders and to nonshareholders who paid a small entrance fee. According to Downing, such a scheme would pay for park construction and provide surplus funds for cities: "Such a project, carefully planned and liberally and judiciously carried out, would not only pay, in money, but largely civilize and refine the national character."[93] It should be noted that Downing's proposal, if followed, would have resulted in parks for the gentry only, as the poor couldn't afford to purchase stock or pay the entrance fee.

Some park advocates did not want to charge entry fees. They contended, "The park is a priceless boon to the weak and invalid of all classes, but particularly to the poor." Free admission was a way of demonstrating that parks were democratic and welcoming to all classes.[94] In this vein, the *San Francisco Bulletin* proclaimed, "The man with a small purse and a large family should be made to feel that he has an equal interest with his richer neighbor in this one spot on earth's surface. This equality can only be assured by demonetizing money at the entrance to the park. The procession of fine turnouts and of fashionably dressed pedestrians does not inspire a sense of inequality so long as appeals are not made for expenditures which the poor man cannot afford."[95]

Downing suggested that cities should acquire parkland by asking rich individuals to donate land; to commemorate their generosity, the city could inscribe tributes to them on statues or marble vases in the parks. Moreover the rich could sponsor one or a group of trees in the parks, while rich ladies could organize tea parties and fairs to fund tree planting or landscaping projects.[96] In this Downing was expressing a sentiment held by the wealthy merchant families whose country estates he designed. Many of those families believed that even if public funds were used to build the parks they should be spaces to exhibit and commemorate the achievements of the rich. With noblesse oblige they pointed out that the parks would be open to all classes and would benefit the poor.[97]

Downing was not alone in calling for the private financing of parks. The San Francisco park commissioners sought private donations of land when they wanted to extend Golden Gate Park to the Presidio on the northern shore of the peninsula. Park contributors included some of the city's most prominent families.[98] Recognizing that such funding schemes could exacerbate inequalities and limit access to urban parks, Olmsted opposed the private financing of parks.[99] He preferred that municipalities fund their parks through public spending rather than relying on the whims of private benefactors,[100] and he championed the idea of raising funds for parks by increasing streetcar fares. Olmsted did not discuss the regressive nature of such taxes or whether the poor had as much access to the park as the wealthy. Despite his opposition to private financing of parks, Olmsted and Vaux did plan Seaside Park in Bridgeport, Connecticut, whose land was donated by P. T. Barnum and Nathaniel Wheeler, a manufacturer and inventor. Olmsted also worked on Beardsley Park in Bridgeport; that park was a gift from James Walker Beardsley, a farmer who donated land to the city for a park.[101]

In 1848 Downing started lobbying in earnest for a public park in New York City. He appealed to the "imaginative and cultivated few," whose "refined minds" could appreciate the "correct taste in art." At the same time, believing that parks were works of art, he argued that a properly planned and managed public park would have a civilizing and refining influence on the lower-class inhabitants of the city. Unfortunately, Downing died shortly after a group of businessmen proposed that New York acquire the Jones Wood tract of land on the East River to build a public park.

During the 1840s Downing's reputation was so great that he was called upon to design the grounds of the Medary, Alverthorpe, and Brookwood, all mansions in Philadelphia. In 1847 his fee was $20 per diem ($411 in 2005 dollars).[102] His most influential works were the Washington Mall, the White House grounds, and the grounds of the Smithsonian.[103] His articles in the *Horticulturalist* and the *Journal of Rural Art and Rural Taste* influenced both Vaux and Olmsted, two men who would go on to become the most influential landscape architects in America. Downing published one of Olmsted's early essays, "The People's Park at Birkenhead, Near Liverpool," in the *Horticulturalist* in 1851 as part of the campaign to develop parks in America. Like Downing, Olmsted believed that parks were an anchor that encouraged the masses to appreciate landscape architecture and art. For Olmsted, parks were also a mechanism to elevate the level of civilization in America.[104]

eight

ELITE IDEOLOGY, ACTIVISM, AND PARK DEVELOPMENT

Business Environmentalism and Park Advocacy

PUBLIC PARKS, MIDDLE-CLASS BENEFITS

In 1851 several events converged to result in a group of park activists organizing to build a large landscaped park in New York. The catalyst was an anonymous "gentleman" recently returned from a two-year sojourn in Europe who expressed his desire to see a great park in New York.[1] The anonymous gentleman was Robert Bowne Minturn, a merchant and codirector of the Bank of Commerce and one of the richest men in New York. Born in New York City in 1805, Minturn left school at age fourteen and entered a countinghouse after his father died. Despite his lack of formal education, he read extensively. In 1825 he entered into a partnership with Charles Green (with whom he clerked) and in 1830 entered the firm of Fish and Grinnell, which later became Grinnell, Minturn and Company. In addition to his work with the Association for Improving the Condition of the Poor Minturn was a founding member of St. Luke's Hospital and the first president of the Union League Club. He was also the commissioner of emigration.[2]

Following the strategy used by Jacob Bigelow in developing Mount Auburn Cemetery, Minturn called a dinner meeting of a select group of wealthy and influential gentlemen at his home to discuss the possibility of a park in Manhattan. Like his upper-class contemporaries who toured Europe, Minturn thought New York City needed a grand park to compare more favorably with those found in European cities. The group, which I will refer to as the

Minturn Circle, picked a "beautiful grove" on the East River that was "covered by trees of about 80 years growth" and "bordered by a high cliff of gneiss rock" known as Jones Wood. They formed a committee to contact the owners about selling their property.[3]

Though the Minturn Circle was a gathering of men, and Robert Minturn is credited for the proposal, some claim that the original proposal for the park came from Anna Mary Minturn, Robert's wife. Anna's grandson declared, "The agitation for establishing Central Park was initiated by her and carried to success by her husband and the friends whose interest in the plan she had aroused and inspired." Other aristocratic women adopted this mode of activism during the late nineteenth century. Forty-five years after Anna worked behind the scenes to mobilize activists to build a park, women adopted a similar strategy with the formation of the Audubon Society. They worked behind the scenes, while men represented the organization in more public roles.[4]

The Minturn Circle adopted a model of environmental activism pioneered by cemetery developers and wealthy sport hunters in the 1830s: gather a group of like-minded, wealthy patricians to achieve the desired environmental outcome. However, the Minturn Circle extended this model to pioneer a form of business environmentalism that was replicated by conservationists and preservationists decades later and lasted well into the twentieth century. Wealthy men without much input from the general public, but in many cases claiming to act on the public's behalf, embarked on park-building projects, bought national parklands, paid rangers to patrol national parks and wildlife refuges, and promulgated environmental laws and policies.

Businessmen became ardent environmental advocates when it facilitated their recreational pursuits, furthered their business interests, and enhanced their status in society. The Minturn Circle was a dense network of aristocrats among whom upper-class endogamy was commonplace; that is, not only were the families intertwined by marriage, but they belonged to the same churches and had overlapping business partnerships. The Minturn Circle included men who were rail and steamship magnates, real estate brokers, bankers, international traders, and large landowners. The Alvords and the Primes owned land close to Jones Wood. Some of the men pushing the park proposal had multiple memberships in exclusive clubs. The University Club, for instance, limited its membership to two thousand; only one thousand men belonged to the Union League and Union Club. Other clubs were even more exclusive, admitting only fifty or fewer members; some had up to five hundred men on their waiting list at any given time.

Several factors motivated the elite to undertake park projects. Urban environmental improvements, such as park building, were seen as part of the larger social reform initiatives being undertaken in cities and was framed as such. Reform-oriented business elites saw park building as a natural extension of their social control efforts. Businessmen were motivated by other factors as well. The wealthy could enrich themselves if they owned land at or close to the proposed park site. Furthermore the park was a way of showcasing a city's culture and cultivation and could refine both the elites and the masses, stabilize and ensure high real estate values, eliminate or retard unwanted commercial development, improve public health, and be used to gain or control political favors and provide jobs.[5] At the time the Jones Wood proposal was crafted the Corporation of New York, which owned nearby Hamilton Square, had begun selling off parts of the square for building lots. Some park advocates feared that if Jones Wood was not converted to a park or cemetery it would suffer the same fate.[6]

The Minturn Circle knew they needed the support of powerful political allies to help sway New Yorkers. Ergo they enlisted the help of the mayor and a senator living in the city. On May 5, 1851, the Jones Wood proposal was put before the Common Council of New York City. The mayor, Ambrose Kingsland (who came from a wealthy mercantile family), argued before the Council, "The public places of New York [are] not in keeping with the character of our city." Thus it was necessary to make "some suitable provisions for the wants of our citizens." Moreover, Kingsland argued, no time would "be more suitable than the present one for the purchase and laying out of a park on a scale which will be worthy of the city." He argued further, "The establishment of such a park would prove a lasting monument to the wisdom, sagacity and forethought of its founders."[7]

At the time of this proposal it was unheard of for public funds to be used to pay for large landscaped parks. New York had small parks and squares of twenty-five acres or less, such as the Battery, City Hall Park, and Washington Square. In addition there were small private parks such as St. John's and Gramercy, located in wealthy neighborhoods. The city's seventeen public squares and private parks constituted about 144 acres, although two-thirds of them were really undeveloped lots in Uptown. The largest were the twenty-four-acre Hamilton Square and twenty-acre Mount Morris Park. The 1811 city plan originally called for 450 acres of open space in Manhattan, but by 1838 the open space acreage had dwindled dramatically. Moreover during the 1840s and 1850s the city sold off some of its public lands; some of that land had to be repurchased later to build Central Park.[8] William Cullen Bryant, a park advocate, called attention to the neglect of the parks and squares in 1850. He argued a year later that although St. John's and Gramercy Parks were "ornaments" in the city, they were really "private gardens" that were "no more public than the houses that surround them."[9]

As soon as the park proposal was made public Bryant's *Post* and Greeley's *Tribune* began promoting the 150-acre Jones Wood Park.[10] James Beekman, a wealthy state senator and neighbor of the Joneses (for whom Jones Wood is named), joined the Minturn Circle and began lobbying for the park. Soon aldermen were urging the city to seek state authorization to acquire the Jones Wood property, located between Sixty-sixth and Seventy-fifth Streets and Third Avenue and the East River.[11]

Though the Minturn Circle had East Side landowners in their inner circle, they were unable to convince the major landowners in the area, the Joneses and the Schermerhorns, to sell. When John Jones died his daughter Sarah inherited 132 acres of the old Louvre Farm. Her husband, Peter Schermerhorn, acquired an additional 20 acres adjacent to Sarah's holdings, and the couple built a splendid country estate with well-manicured lawns. Philip Hone described the musical soirée, complete with visiting opera stars, that the couple held at the estate in 1845 as the best party of the season. Park advocates found the site attractive because, as Bryant put it, "Nothing is wanted [to convert the site into a public park] but to cut winding paths through it, leaving the woods as they are now, and introducing here and there a jet from the Croton aqueduct."[12]

Peter and Abraham Schermerhorn objected to the Jones Wood proposal, which sought to acquire ninety acres of the Jones property and the remaining acreage from the Schermerhorn property. The land in question was estimated to be worth $1.5 million. The Schermerhorns argued that the government had no right to take private property for the purpose of converting it into a place of public amusement: "A park is not sufficient necessity to justify it being taken by the state in opposition to the wishes of the owner and by the violent exercise of eminent domain."[13]

Nonetheless, Beekman, a known nativist, introduced a bill in the state senate to take the Jones Wood properties by eminent domain. It passed on June 18, 1851. In an effort to save their property the Schermerhorns suggested shifting the boundaries of the proposed park north, but James Crumbie, a supporter of the Jones Wood proposal, objected that the new boundaries would encroach on his property. The Joneses were not pleased with the new boundaries either, because although the park would now exclude the Schermerhorns' landholdings, it would still include the Jones mansion.[14] It appears that East

Side park proponents wanted a park near their property, but did not want their property to be turned into a park.

In the meantime the Minturn Circle sought and got the support of more of the city's wealthy merchants and bankers. These men were targeted because the Minturn Circle thought they would "be called upon to contribute most liberally [through taxation] toward the expense of the new park." These later supporters were also business partners, friends, family, and marriage partners of those already in Minturn's inner circle, and they were bound by memberships in the chamber of commerce, charities, and exclusive churches and social clubs. On July 11, 1851, the bill authorizing the city to build a large public park passed the state assembly.[15]

DEMOGRAPHIC CHANGES

At the time of the discussion over the Jones Wood proposal there was no question that the city would benefit from developing public parks. But what was the most expedient and appropriate way to do this, and was public input necessary? The city's population had increased more than ninefold between 1800 and 1860, rising from 66,000 to 814,000. Thousands of immigrants arrived in the city annually. In 1855 only 48 percent of the New York population was native-born; 28 percent were born in Ireland, 15 percent were from Germany, and 5 percent from England, Wales, and Scotland. About 20 percent of the city's residents were entrepreneurs or professionals, 12 percent were artisans, and 68 percent were wage laborers. Those percentages varied dramatically from one ethnic group to another. Of the white residents, native-born New Yorkers were the wealthiest and the Irish-born the poorest. About 34 percent of the native-born were entrepreneurs and professionals, 12 percent were artisans, and 44 percent were wage laborers. In contrast, most of the Irish (85 percent) were wage laborers, 8 percent were artisans, and only 7 percent were entrepreneurs or professionals. The English immigrants were not as well off as the native-born, but were better off than the Germans, who were better off than the Irish.[16] The rapid growth of the city's population made the acquisition of public green space a priority to provide open space for residents and to prevent all the land from being gobbled up by development.

MAKING THE CASE FOR A PUBLIC PARK

Though the aldermen supporting Jones Wood argued that the "necessity" of a park had "long been acknowledged by all classes in the community," the park planning process was really the brainchild of a tightly knit power elite who saw themselves as representing the public's will. The *Journal of Commerce* argued that the park proposal "did not originate with [land] speculators, but with a worthy and excellent citizen who has not other views in the movement, but the public good."[17] Depending on the audience, the Minturn Circle publicized and justified the park on the grounds that it would promote the city's commercial vitality, improve the health and well-being of the citizenry, improve the morality of the working class, create social order in the city, and demonstrate that the city's well-to-do were as cultivated as Europeans.[18]

Park advocates trumpeted these benefits far and wide to counter the perception that they undertook park-building efforts for pecuniary gain and other selfish motives. They also used arguments that British park advocates utilized successfully in their campaigns for increased open space.[19] However, benefits that would accrue first and foremost to the middle class were not widely publicized outside of elite circles. For instance, park advocates saw park development as an early and effective form of zoning, eliminating nuisance industries (such as bone-boiling facilities, slaughterhouses, milk distilleries, pigpens, dumps, and breweries) from residential neighborhoods. Individual landowners tried to exclude such facilities from upscale neighborhoods by putting restrictive covenants on properties they leased or sold, but undesirable industries still permeated residential neighborhoods. A large public park meant the demolition of such facilities, removal of people involved in those industries, and substitution with more desirable businesses, housing, and neighbors. In addition the business elites wanted to develop and live in cohesive, exclusive residential neighborhoods with immediate access to open space. Park advocates welcomed the opportunity to use a large park as the anchor around which they could plan an exclusive neighborhood. Jones Wood provided an opportunity for elite residential development in conjunction with a public park, reminiscent of the villas overlooking Regent's Park in London or private housing in Liverpool's Birkenhead Park. This model was attractive to the wealthy in other cities: mansions overlooked Boston Common and Fairmount Park in Philadelphia.[20]

Not everyone was enthused about developing Jones Wood Park; some questioned the motives of the Minturn Circle and other park supporters. The *Journal of Commerce*, referring to the "construction of a great wilderness in the upper part" of Manhattan, proclaimed that there was "no need of turning half the island into a permanent forest for the accommodation of either gentlemen or loafers."[21] Editorials in the paper claimed that money spent on the park, "this mammoth of the imagination," would result in "enriching the few soil holders" of the area.[22] Others objected to the proposal on the grounds that there was a paucity of "fashionable promenades and drives" in the city on the "scale of magnificence that marks European cities"; they rejected the Jones Wood site because it was too small to have a grand carriage drive. The *New York Times* argued that "Jones Wood was not the best site for a park. Its position is not *central*, and therefore it would not be within the reach of the greatest [number of the city's residents]." The paper also noted that the chosen site "is *on a river*, and therefore such a neighborhood needs no other ventilation. Its river front will soon be needed for commercial purposes." For this reason the riverfront property was "more valuable than other land in the centre of the island." Andrew Jackson Downing supported the idea of a park, but he thought the park should be a minimum of five hundred acres; he objected to the Jones Wood site because it was too small. He suggested an alternative site that included the Croton Reservoir.[23]

BENEFITS ASSESSMENT AND THE PARK PROPOSAL

Property owners also took an interest in the park proposal. Prior to the Jones Wood proposal, large property owners were taxed for land improvements (streets, lights, sewers, development of squares) adjacent to their property. In 1837, as the cost of these benefits assessments soared, large landowners began organizing themselves. They argued that instead of charging landowners, the cost of improvements should be borne by the general public. *Journal of Commerce* editorials argued soon after the Minturn Circle proposal became public, "The expense of public parks, and all other public improvements should be paid for by the public, and not assessed on the lands of a few individuals."[24]

Minturn Circle members wanted to reap the benefits of living adjacent to the park without being taxed heavily for park development. They proposed a general public tax. This was an important break with tradition that would set a new precedent for financing public works projects. Though some elites declined to support the park because of the controversial tax proposal, several large landowners with property adjacent to Jones Wood supported the Minturn Circle's position as the controversy heated up.[25]

THE WORKING-CLASS ALTERNATIVE

Just as the lobbying for Jones Wood neared a crescendo, a proposal for a "central" or "middle" park emerged as a challenge. The battle over the appropriate park site raged for three years. Though the discussions were carried out largely in elite circles, the proposal to build a centrally located park, rather than the "sidelong" Jones Wood Park, opened up the park debate so that a wider range of people got involved in the deliberations. The Minturn Circle continued to lobby for Jones Wood. The downtown merchants held their ground; many wanted no additional park at all and objected to paying for a park through general taxation. *Journal of Commerce* editorials contended that a new park was unnecessary and building a large one could create a housing shortage. Arguing that the number and total acreage of parks already existing in the city were adequate, the paper claimed that the waters surrounding New York provided ample open spaces that were "far better than so much park . . . with its unwholesome dampness in the summer and decaying foliage in autumn."[26]

Middle-class land reformers and working-class artisans questioned the wisdom of a large uptown park and advocated developing smaller neighborhood parks in areas where a larger percentage of the population could have immediate access to them. They also argued that the taxes would be burdensome for the poor and that money could be better spent on affordable public housing. Better housing, they argued, would do far more to improve people's health than the park. Politicians such as Mike Walsh, sensing an opportunity to get working-class support, seized upon the issue. He endorsed the Jones Wood proposal but only on condition that Battery Park was enlarged.[27]

Built on a landfill, the ten-acre Battery Park was located on Manhattan's southern tip at the strategic point where New York Harbor converged with the Hudson River. Originally the site of Fort Clinton (constructed between 1808 and 1811 to protect New York Harbor from invasion) the site ceased to be of military value after the War of 1812. It was converted into a park in 1823, when the land was ceded to the city. After wealthy residents abandoned the area for more exclusive parks and squares the Battery became a working-class park. In addition to being used as a fort, over the years the Battery had been used as a promenade, an entry point for immigrants, a tea garden, a music hall, and a public gathering place. Castle Garden, a popular entertainment spot, was connected to the park by a bridge; P. T. Barnum staged sold-out concerts in Castle Garden's theater.

Shippers and merchants resisted plans to increase the size of the Battery because they thought expansion would consume valuable waterfront real estate. The *Journal of Commerce* wanted to see the land used for wharves and railroad tracks rather than a "deer-park." To pacify Battery supporters the aldermen voted to enlarge the Battery (with fourteen acres of landfill) at the same time they endorsed the Jones Wood proposal. However, the West Side landowners continued to lobby hard for a centrally located park. Activists on each side struck deals, signed petitions, wrote editorials, and resorted to bribes to influence lawmakers and the public.[28]

Labor unions such as the New York Industrial Congress also participated in the park debate. They suggested that, if the city wanted to serve the people's needs, they would be better off buying vacant squares and converting them into parks in the densely populated parts of the city. They argued that this would be cheaper than building a large park. Some labor unionists condemned the Jones Wood plan, questioning the "public spirit" of the real estate speculators promoting the park. Labor leaders also argued that the poor would bear a heavy tax burden for a park they would seldom use because they couldn't afford to visit it.

The working-class position did not receive extensive coverage in the mainstream press. However, ethnic papers such as the German-language *Staats-Zeitung* printed labor's position. *Staats-Zeitung* opposed the Jones Wood park, calling it "a mad project" benefiting "greedy speculators" who would drive through the park in "pompous" carriages. It called instead for many small parks spread throughout the city and questioned whether a single park would improve people's health, as advocates claimed. The paper argued that it was the narrow, dirty streets, not the lack of parks that made New Yorkers unhealthy. The working class supported government expenditures to improve schools and roads; in other words, they wanted to see projects aimed at reducing social inequality.[29]

THE CENTRAL PARK ALTERNATIVE

The Jones Wood coalition began to crumble when the aldermen voted to enlarge Battery Park. The major newspapers opposed the expansion, but the two pro-labor papers, the *Dispatch* and the *Tribune*, supported it. Succumbing to pressure, Mayor Kingsland vetoed the Battery expansion in August 1851. As a result, Battery supporters collaborated with two factions: activists opposing any large park and a second group promoting a central park. In this way they held the Jones Wood proposal hostage to get their wish of seeing the Battery expanded and improved. "No enlargement, no park" was their battle cry. About a month after Kingsland vetoed the Battery expansion bill, the veto was overruled on a technicality.[30]

Plans for Jones Wood continued to hit snags. Wealthy landowners, refusing to sell, went to court to prevent seizure of their property by eminent domain. The Joneses and Schermerhorns, whose primary residences were downtown, were no strangers to opposing public developments near their country estates. They did this to avoid paying benefits assessments. On December 1, 1851, a district court judge ruled that the Jones Wood decree violated due process because an escape clause allowed the city to back out of the deal but the landowners were not given any such options. This ruling opened the way for the creation of Central Park.[31]

In the summer of 1853 a group of leading Upper West Side landowners endorsed and helped to get the Central Park bill passed. However, the question remained: Would the city build one or two large landscaped parks? The West Siders were interested in pursuing the same development strategies pioneered by the East Siders in the Minturn Circle. In 1850 West Side residents had tried unsuccessfully to develop a fifty-acre private park modeled on Regent's Park in London. A similar fate befell the East Side plan for a private park. Thus by 1853 West Siders were convinced that they would get a landscaped park only with government intervention. Although wealthy land speculators stood to benefit from the development of the park, this time the chosen site was not inhabited by wealthy residents, but primarily by sixteen hundred Irish and German immigrants and blacks living in modest homes characterized as shanties. Wealthy New Yorkers who owned holdings on the proposed site tended to lease those properties to poor people, so the landowners stood to gain from the city's purchase of their land. To ensure the support of West Siders one prominent resident warned that if Central Park was not developed, the area would be filled with "a class of population similar to that of Five Points." Hence Central Park was viewed as a "breakwater to the upward tide of population" that would force "persons of limited means" to live elsewhere.[32]

By this time the bankers, shipping magnates, downtown merchants, brokers, and lawyers, some of them former supporters of the Jones Wood proposal, backed the Central Park bill. Central Park would be financed, in part, by benefits assessments on property surrounding the park. Despite setbacks, supporters of the Jones Wood proposal continued to push modified versions of the bill through the legislature. However, the Jones Wood bill was finally defeated. Its demise was partly due to its proposed tax structure; the bill did not levy any benefits assessments on neighboring landowners.[33]

The city began assessing and acquiring land to build on the Central Park site in 1853,[34] and in the following year the state legislature rescinded its authorization to purchase the Jones Wood site. On January 9, 1854, Judge Robert Roosevelt, uncle of Theodore Roosevelt and founder of the New York Sportsmen's Club, made a final ruling on the issue: Central Park would be built; Jones Wood would not. Later that year, when Fernando Wood was elected mayor of New York City, he vetoed an attempt by the Common Council to reduce Central Park to almost half of its proposed size. That same summer the city authorized over $5 million ($111 million in 2005 dollars) to buy the land for the park. This was much more than the early estimates; Downing, for instance, estimated that it would take about $1 million to acquire the land for a park of at least five hundred acres. In addition, the benefits assessed against adjacent landowners was capped at $1.7 million, about one-third of the cost, and was spread over a large area, making it cheaper for adjacent landowners. While landowners on the immediate periphery of the park benefited from the assessment structure, those on the fringes of the assessment area complained that they were being assessed even though they would not receive the immediate benefits of the park.[35]

THE SHIFT AWAY FROM BENEFITS ASSESSMENTS

The move toward funding public works projects through public spending rather than benefits assessment had been gaining momentum for some time, but the Central Park debate became a catalyst that set a new pattern for funding such projects. Prior to 1850 awards paid to landowners whose properties were acquired by eminent domain for parks (such as Washington, Union, Tompkins, and Madison Squares) amounted to $444,106. In addition all but $500 of the cost of building these parks was assessed to landowners whose properties lay adjacent to the new developments. The Croton aqueduct and Central Park projects changed all that. Beginning with the Croton aqueduct, the city began to assume public debt to pay for large public works projects. In 1850 the city's debt was $12.2 million ($271 million in 2005 dollars), most of it to build the Croton aqueduct. As part of the financing for Central Park, the city issued municipal bonds to generate about $5.4 million to buy park land. Between 1850 and 1863 an additional $11 million in bonds was issued for park construction and $6.5 million for street and boulevard construction. Between 1870 and 1871 bonds worth $21 million were issued to acquire lands for Riverside Drive, Morningside Park, the widening of Broadway, and extensions of Madison and Lexington Avenues.[36]

The benefits assessment system began to crumble with the issuance of bonds to construct the Croton aqueduct, built between 1835 and 1842. By the late 1840s landowners were strident in their opposition to benefits assessment. They argued that major projects such as the aqueduct were public goods that should be paid for by the general public. The anti-assessment advocates used the Central Park case to make their point and won their first major victory by influencing the process such that the proposed tax structure of the park did not levy the entire cost onto adjacent landowners.[37]

Several factors accounted for public spending on the Croton aqueduct rather than a complete reliance on benefits assessment. The project began just months before the Great Fire of 1835, which took a heavy toll on businesses.[38] Just as the business community was recovering from that disaster the Panic of 1837 hit. It took until the early 1840s for many businesses to recover. However, financial hardship wasn't the only factor; the nature of the Croton aqueduct project played a role too. Unlike parks or streets, whose development clearly benefited adjacent property owners, the monumental undertaking of diverting the Croton River from Westchester County to Midtown meant that the aqueduct traversed a lot of sparsely populated land before reaching its destination. It was difficult to argue that merely traversing a person's property with an aqueduct constituted a benefit to that property owner. This was particularly true because the aqueduct wasn't supplying water to residents in outlying areas. As an 1842 Endicott map shows, the water pipes and stopcocks were located primarily below Fourteenth Street.[39]

The Croton aqueduct differed from preceding projects in that it would clearly have widespread public benefit. Unlike street paving, lighting, and small parks that benefit primarily a small localized area, the aqueduct would benefit the whole city. Furthermore, it was a project in which the beneficiaries could be taxed directly each time they benefited from the system. The city could pay for the cost of construction through the collection of fees for usage from it in perpetuity, rather than assessing the benefits taxes to a relatively small group of landowners. Once there was a deviation in the way major public works projects were funded, opponents of benefits assessments had the ammunition they needed to argue that public funding of public works projects was more equitable and appropriate than benefits assessment. Hence by the time Central Park funding became a major issue, complete reliance on benefits assessment to fund the project was no longer a politically feasible option.

THE DEBATE OVER THE PROPOSED TAX STRUCTURE

The 778-acre Central Park (expanded to 843 acres in 1863) was almost five times as large as all existing parks and squares in the city combined. Once the decision was made to build Central Park, the *Dispatch* charged that the poor would be burdened by the usage tax. Others, such as the housing reformer Robert Hartley of the Association for Improving the Condition of the Poor, feared that Central Park would consume so much acreage that it would exacerbate the housing shortage, leading to higher rents and more overcrowding, which Hartley argued would create greater public health problems. One estimate claimed the park deprived the city of 13,521 building lots that could house 190,000 people. John Griscom, the city inspector and a health reformer, reiterated the working-class call for neighborhood parks rather than one large park. This approach would be "less aristocratic" and "more democratic, and far more conducive to public health," declared Griscom. Furthermore Griscom questioned whether the park could deliver on the health benefits promised by advocates. He argued that even if Central Park acted as "the lungs of the city" it would provide fresh air for only a limited area. In addition, he argued, it was pointless to expose poor people to the park for an hour or so a day while they dwelled and worked the remainder of the time in windowless, unventilated factories, apartments, and cellars. He argued that to really improve the health of the poor one needed to improve housing, working conditions, and access to open spaces in neighborhood. It took decades for New York and other cities to heed the working-class plea for small neighborhood parks to complement the large, centerpiece parks such as Central Park.[40]

Many wealthy residents loved the large park and the recreational access it provided them. As the *Herald* claimed, "Our citizens will be in possession, for the first time, of a good drive and ride, both of which have been so long needed." The paper predicted, "Every equipage in the city will be daily directed [to Central Park,] soon to become the permanent and habitual resort of pleasure, pretension and fashion."[41]

The First Real Park and the Pastoral Ideal

OLMSTED, VAUX, AND THE CENTRAL PARK DESIGN COMPETITION

Frederick Law Olmsted and Calvert Vaux were the landscape architects chosen to translate the upper-class vision into the reality of a landscaped park that could serve the dual purposes of meeting the needs of the wealthy while lifting up the masses. In the process both of these men elevated landscape architecture to a new level. They saw Central Park as a cultural entity that served important social functions and as a single piece of artwork with a unified theme. Olmsted and Vaux shared Downing's conviction that parks as works of art could play a significant role in helping American society reach a higher level of civility. Like Downing, they strongly believed that parks would bring rural recreation to city residents who had no access to the countryside.

Calvert Vaux, a physician's son born in London in 1820, received his architectural training from Lewis Cottingham, a leader of the Gothic Revival movement. Vaux met Downing in London in 1850 at an exhibition of his work. Downing was impressed with Vaux's sketches and offered him a job. Vaux moved to Newburgh, New York, to work in Downing's firm. He helped Downing design the grounds of the White House and of the Smithsonian Institution and country estates along the Hudson River and in Newport, Rhode Island. Vaux and Olmsted met in 1851 at Downing's house. For four years following Downing's death Vaux stayed with Downing's old architectural firm, but in 1857 he moved to New York City to design a bank building. There he continued Downing's work by pushing for the construction of Central Park and for holding a design competition.[42] He got embroiled in the debate over the merits of Egbert Viele's design for Central Park. Viele was the chief engineer of Central Park until May 1858. He made the first topographical survey and drainage map of the park in 1855 and developed a design for the park in 1856. However, in 1857 Viele's design came under fire from critics who argued that the plan had the mechanistic imprimatur of an engineer; it did not have a naturalistic appeal and lacked taste. Vaux himself criticized Viele's design at a park commissioners' hearing.[43]

In August 1857, a month after Olmsted began working as superintendent at Central Park, a park design competition was announced. Vaux, who was two years Olmsted's junior, convinced him to collaborate on an entry. Olmsted consulted Viele (who was still the chief engineer of the park and who was planning to enter his design into the competition), and when Viele did not object Olmsted began working with Vaux on a new design. Thirty-three designs were submitted, eleven of them from current or former park employees.[44] After nine rounds of voting the Vaux-Olmsted Greensward Plan won.[45] Olmsted and Vaux proposed to build a park that juxtaposed pastoral and picturesque elements to create a sylvan retreat in the middle of Manhattan.

OLMSTED AND THE EVOLUTION OF PASTORAL AND PICTURESQUE LANDSCAPE DESIGN IN AMERICA

In the Olmsted-Vaux partnership Vaux used his architectural background to visualize and design structures such as the famed bridges and archways of Central Park that blend into the landscape, while Olmsted oversaw the construction of the park. Olmsted also brought to the design the ideals of pastoral and picturesque landscapes. He understood the kinds of vegetation to use in particular settings, how to arrange them, and how they would look when they matured. Of the two, Olmsted left a much more extensive record of how he thought about these concepts and translated them into landscape designs.[46]

Olmsted, the firstborn of a successful dry goods merchant in Hartford and whose family was one of the founding families of the city, was steeped in the European park-building tradition. Born in 1822, he got early and frequent exposure to European parks and gardens from the prints his father kept around their Connecticut home, trips to the White Mountains, and other New England tours the family took each summer. At the time Olmsted was growing up many northeastern elites vacationed in the White Mountains and purchased landscape paintings of the region done by artists such as Thomas Cole as a means of cultivating and displaying their refinement and taste. Olmsted was first introduced to landscape gardening through William Gilpin's *Remarks on Forest Scenery* and Uvedale Price's *Essay on the Picturesque*, works by English writers on landscape and agriculture. Olmsted also made lengthy trips to Europe in the 1850s.[47]

From a very early age Olmsted believed that scenery, nature, and parks put one in a contemplative mood, a state of mind he found highly desirable. Throughout his career he sought to foster or induce that state of mind through the designs of his parks.[48] While working in Yosemite he reflected on the natural scenery as a mechanism for counteracting what he considered to be the "severe and excessive exercise of the mind."[49] Echoing Rousseau, Olmsted wrote, "The power of scenery to affect men is in a large way, proportionate to the degree of their civilization and to the degree in which their taste has been cultivated. . . . The severe and excessive exercise of the mind . . . leads to the greatest fatigue and is the most wearing upon the whole constitution. . . . The enjoyment of scenery employs the mind without fatigue and yet exercises it, tranquilizes it and yet enlivens it; and thus, through the influence of the mind over the body, gives the effect of refreshing rest and [is] reinvigorating to the whole system."[50] Olmsted sought to use subtle designs in his parks to create this mood.

Though John Muir, who was much more interested in the grand, sensational scenery of the wilderness than in pastoral urban parks, was somewhat intrigued by the park, he was not as wholly captivated by it as Olmsted was. Muir thought of the landscaped urban parks as "half wild parks and gardens of towns."[51] He was more predisposed to vigorous exercise than tranquilizing urban recreation, and was so out of his element when he stopped in New York briefly in 1868 before sailing to California that he didn't even visit Central Park. Though Central Park had gained a national reputation by then, Muir wrote, "I saw the name Central Park on some of the street-cars and thought I would like to visit it, but fearing that I might not be able to find my way back, I dared not make the adventure. I felt completely lost in the vast throngs of people, the noise of the streets, and the immense size of the buildings. Often I thought I would like to explore the city if, like a lot of wild hills and valleys, it was clear of inhabitants."[52] Muir did take an afternoon drive through Central Park in October 1898, accompanied by Charles Sprague Sargent. He considered the park "a fine wilderness for a town, heartily appreciated by rich and poor." He did not comment much on the scenery or landscape effects.[53] In fact, the thing that interested him most in Central Park was the glacial markings on granite outcrops.[54]

It was not Olmsted's intent to re-create vast and rugged mountainscapes devoid of people. Instead he used the subtlety of domestic arrangements to stimulate the unconscious and elevate people to a higher plane of thought; he wanted park visitors to forget their mundane concerns and explore other thoughts and feelings. As an early Central Park guidebook claimed, Olmsted and Vaux's aim was to capture the "delicate flavor of wildness, so hard to seize and imprison when civilization has once put it to flight."[55]

Olmsted had a clear idea of what an ideal park would look like:

> The landscape character of a park, or of any ground to which that term is applied with strict propriety, is that of an idealized, broad stretch of pasture, offering in its fair, sloping surfaces, dressed with fine, close herbage, its ready alternatives of shade with sunny spaces, and its still waters of easy approach, attractive promises in every direction, and, consequently, invitations to movement on all sides, go through it where one may. Thus the essential qualification of a park is *range*, and to the emphasizing of the idea of range in a park, buildings and all artificial constructions should be subordinated.[56]

Olmsted also believed that pastoral landscapes made the ideal park scenery.[57] Consequently such scenery was prominent in Olmsted-Vaux parks. They describe pastoral scenery as having "comparatively slight variations of surface" and "no protruding ledges of rock, no swamps difficult of drainage . . . no especial bleakness, or danger to trees from violent winds." Pastoral scenery "consists of combinations of trees, standing singly or in groups, and casting shadows over broad stretches of turf, or repeating their beauty by reflection upon the calm surface of pools, and the predominant associations are in the highest degree tranquilizing and grateful, as expressed by [the passage] . . . 'He maketh me to lie down in green pastures; He leadeth me beside still waters.'"[58] This is one of the clearest articulations in Olmsted's and Vaux's writings of the belief that a divine presence exists in pastoral landscapes. Here they express the belief that pastoral scenery not only calms and tranquilizes the mind, but has a divine power of its own. It draws the park user into the scenery without the user's being conscious of what causes him or her to be captivated by the green pastures and still waters. While wilderness Transcendentalists often discussed the spirituality of sublime places, pastoral Transcendentalists did not spend as much time articulating the spiritual significance of the landscapes they created.

To incorporate pastoral scenery in the parks they built, Olmsted and Vaux designed broad stretches of gently rolling greensward edged by irregular borders of trees and shrubs, creating a sense of space and distance. This technique was used to induce a sense of unconscious or indirect recreation, so that the recreationer is absorbed in the park experience without being fully conscious of the process by which it occurs. Olmsted described this process as the "unconscious influence" of "tranquilizing" recreation.[59]

GREENSWARD MAP

Olmsted recognized that the demands of urban life left people with the need to unwind, to stimulate the brain in ways that compensated for work-related stress; however, he did not favor recreation that would lead to overstimulation. The parks, therefore, served as pastoral retreats from everyday pressures. He thought that the "unbending of the faculties" was an important exercise.[60] In their description of pastoral scenery in Prospect Park Olmsted and Vaux expounded on the theme that gentle exercise relieves the brain: "Civilized men, while they are gaining ground against certain acute forms of disease . . . cannot [find the remedy] in medicine or in athletic recreation but only in sunlight and such forms of gentle exercise as are calculated to equalize the circulation and relieve the brain."[61]

Olmsted and Vaux were largely responsible for the spread of picturesque landscapes in America. Rejecting the symmetrical, geometric designs of European gardens such as Versailles that were laid out in the gardenesque style (dominated by floral arrangements and specimen plantings), they opted to contrast wilderness vistas with more subtle arrangements to express their vision of the picturesque.[62] Though picturesque scenery was secondary to pastoral vistas in Central Park, Olmsted used picturesque landscapes to vary the scenery. Moreover picturesque arrangements were much better adapted than pastoral landscapes to the park's rocky terrain.[63] To create the picturesque effect, Olmsted and Vaux planted native foliage to create a complex picture of light and shadow near the eye.[64] For instance, on the rocky hillside south of the Croton Reservoir known as the Ramble they planted shrubs and evergreens profusely, thereby creating a sharp contrast between light and shade. They also used subtle variations in shape, color, and texture to create contrasts and transitions in the landscape. In the *Greensward Plan* they argue that they wanted to obtain "the broadest effects of light and shade which can be obtained upon the ground . . . to produce the impression of great space and freedom." They hoped that people would find "in the broad spaces of green sward" (the pastoral parts of the park) "unrestricted movement."[65] In Prospect Park Olmsted and Vaux also designed pastoral spaces to provide what they saw as "a sense of enlarged freedom."[66]

Drawing on Romantic and Transcendental beliefs that wilderness provided the sharpest contrast with civilization, Olmsted and Vaux sought to incorporate wilderness themes (wilderness scenery was also growing in popularity among elites). Recognizing that pure wilderness would be very difficult to recreate in an urban environment, they opted for a compromise I will call pastoral Transcendentalism. Pastoral Transcendental scenery combined the open, gently sloping, manicured, and domesticated spaces of the pastoral style with the wilder, untamed, more irregular look of the picturesque. Pastoral spaces produced a sense of great openness and freedom and had a quiet, soothing effect, while picturesque spaces hinted at nature's bounty, mystique, and uncertainty.[67] This kind of composition is described in Olmsted and Vaux's discussions of Prospect Park: "We therefore abandon all ideas of contrasting the publicity of the city with the privacy of deep woods, mountains, lakes, and rocky fastnesses, and accept another ideal altogether, that of pastoral rural life, as the most valuable and universally available one, for the purpose we have in view."[68] They explain, "Although we cannot have wild mountain gorges, for instance, on the park, we may have rugged ravines shaded with trees, and made picturesque with shrubs, the forms and arrangement of which remind us of mountain scenery."[69]

Hence as part of the social construction of urban parks, landscape architects adopted a muted form of Transcendentalism, pastoral Transcendentalism, to distinguish it from the more intense form of wilderness Transcendentalism practiced by Muir and his followers. Pastoral Transcendentalism attributed virtues to natural objects such as trees, meadows, and brooks that could be replicated in urban park settings.[70] Rather than developing parks around vast, dramatic natural landscapes that evoked fear, landscape architects designed subtle landscapes for the gentle exercise of the mind. From the pastoral Transcendentalist perspective, the recreation experience was intended to put people into a contemplative mood and tranquilized state of mind. Through exposure to the scenery users were expected to transcend their everyday stress and concerns. Since these were man-made landscapes there was no general expectation that divine forces would manifest in the physical objects in the environment.

THE INFLUENCE OF THE TROPICS IN OLMSTED'S PICTURESQUE DESIGNS

Europe and New England weren't the only places that inspired Olmsted. He drew upon Asia and Central America as well, using the lushness of tropical vegetation as his model for creating picturesque scenery. He was introduced to this kind of scenery on his trip to China, Java, and Malaysia in 1843 and was exposed to subtropical vegetation during his trips to the American South during the 1850s. His experience of crossing the Isthmus of Panama in 1863 (while on his way to San Francisco aboard Cornelius Vanderbilt's dingy, unsafe steamer *Champion*) further intensified his desire to capture the luxuriance of tropical scenery in his landscape designs.[71] Olmsted was not interested in using tropical plants to create lushness (as in the gardenesque style); he intended to utilize native foliage to produce these effects. He wrote to his wife during the crossing that he had anticipated much from the tropical scenery and that it had exceeded his expectations. He noted that the vegetation was "superb and glorious and makes all our model scenery—so far as it depends on beauty of foliage—very tame & quakerish." The scenery produced "a very strong moral impression through an enlarged sense of the bounteousness of Nature." He told his wife that he was "thoroughly enchanted with the trees & vines. . . . They are Gloria in Excelsis with lots of exclamation points, thrown in any where, in grand choral liturgy." In his letter Olmsted wondered whether his attempts to produce the same effect with northern plants in Central Park was a "preposterous" idea. However, he concluded that by carefully assembling and arranging the foliage, one could get the desired effect.[72]

A day later he wrote to Ignaz Pilat, the Central Park gardener, "I have never had a more complete satisfaction and delight of my love of nature than I had yesterday in crossing the isthmus. You will remember that I always had a reaching out for tropical effect, but found the reality far beyond my imagination, resting it as I did upon very inadequate specimens, hastily and imperfectly observed." Olmsted wished he had seen Panama before trying to create his picturesque designs in Central Park: "I wished that we could have seen five years ago what I saw yesterday, and received the same distinct lesson which I did yesterday. . . . The scenery excited a wholly different emotion from that produced by any of our temperate-zone scenery. . . . It excited an emotion of a kind which our scenery sometimes produces as a quiet suggestion to reflection, excited it instantly, instinctively and directly." Olmsted reflected on how the scenery produced this effect: "I think that I was rather blindly and instinctively feeling for it, in my desire to give 'tropical character' to the planting of the island, and luxuriant jungled variety and density and intricate abundance to the planting generally of the lake border and the Ramble and the River Road."[73]

Three years after his travels through Panama Olmsted worked on Prospect Park, where he took the opportunity to create luxuriant, tropical-inspired scenery. In their plan for Prospect Park Olmsted and Vaux wrote, "We may perhaps even secure some slight approach to the mystery, variety and luxuriance of tropical scenery, by an assemblage of certain forms of vegetation, gay with flowers, and intricate and mazy with vines and creepers, ferns, rushes and broad-leaved plants. But all we can do in these directions must be confessedly imperfect, and suggestive rather than satisfying to the imagination."[74]

Olmsted explained in the letters he wrote while traveling through Panama that what impressed him most about the vistas was the powerful emotional response they evoked in him. He explained that his response arose from "a sense of the superabundant creative power, infinite resource and liberality of Nature—the childish playfulness and profuse careless utterance of Nature."[75] Olmsted's letter to Pilat contained detailed instructions on how best to re-create the tropical effect using temperate plants. Here Olmsted expresses his desire to have the scenery in his landscape designs do two things: first, the scenery should be calming and tranquilizing; second, it should be interesting and dramatic enough for the park user to notice it. He believed the scenery should be compelling; it should leave the visitor awestruck and with a sense of wonder. Thus the scenery evokes unconscious (calming and tranquilizing) and conscious (wonderment) responses.

Muir crossed the Isthmus of Panama on the *Nebraska* in 1868, five years after Olmsted. Muir too was impressed with the foliage: "Never shall I forget the glorious flora. . . . The riotous exuberance of great forest trees, glowing in purple, red, and yellow flowers, far surpassed anything I had ever seen, especially of flowering trees. . . . I gazed from the car-platform enchanted. I fairly cried for joy and hoped that sometime I should be able to return and enjoy and study this most glorious of forests to my heart's content."[76]

THE YOSEMITE INFLUENCE

Olmsted spent time in and around Yosemite from 1863 to 1865. He described the pastoral Transcendentalism he experienced there and the images that would influence his later urban landscape designs.[77] Olmsted was more fascinated with the scenic valleys and meadows than the dramatic peaks. He preferred the peaks when the detailed features were obscured by fog because the indistinct and hazy views made the scenery appear infinite and mysterious. He liked the idea of obscuring the details in distant vistas. In accordance with the picturesque theory of landscape design, he avoided calling attention to prominent landscape features or individual trees in his park designs (trees were not labeled; this despite the fact that Downing, who adhered to the gardenesque style, called for labels in order to educate the masses).[78]

REBUFFING THE GARDENESQUE STYLE

During the 1850s horticulturalists working within the gardenesque tradition began to rely heavily on imported flowers and shrubs to make highly decorative displays. Olmsted believed that these showy floral arrangements were being emphasized at the expense of developing coherent passages of scenery. As he saw it, the gardenesque style lacked an important element of effective landscape design: the interest generated from frequent contrasts and surprises. Specimen planting went against the spirit of the place and therefore lacked good taste.[79] Olmsted expounded on these themes in his essay "Park," published in 1861. He discussed the changes made in London's Hyde Park, where park managers were planting gardens in the summer with brilliant displays of flowers, specimens, and subtropical plants. However, the old trees were disappearing and young ones were not being planted soon enough. Olmsted argued that the "streaks of fine gardening here and there offer[ed] no compensation" for the loss of the older version of the park.[80] In his attempts to create picturesque scenery he planted native varieties and avoided the specimen planting characteristic of the gardenesque style.

THE LIMITED APPEAL OF THE CEMETERY MODEL

Olmsted and Vaux also rejected Downing's idea that the parks, like the rural cemeteries, could serve associational functions. While Downing wanted the parks to serve as the repositories for objects d'art and buildings commemorating "great men," Olmsted and Vaux thought such associational features detracted from the rural experience they sought to induce in their parks. Downing was aware that the associational features of the cemetery could interfere with contemplative recreation; in fact, he opined that the "only drawback" of the cemeteries was the "gala-day air of *recreation*" that marred the contemplative intent of such places: "People seem to go there to enjoy themselves, but not to indulge in any serious recollections or regrets." Olmsted shared Downing's sentiment, claiming that the cemetery was a "constant resort of mere pleasure seekers, travelers, promenaders, and loungers." Richard Morris Hunt, another park advocate, believed the park should be more associational than pastoral. Even the rules of the original design competition for Central

Park called for an associational park. However, Olmsted and Vaux boldly strayed from the competition guidelines and held firm to their convictions as they built pastoral and picturesque parks across the country.[81]

Olmsted was so committed to the idea that the atmosphere in the cemetery was incompatible with the ambiance he wanted to create in his pastoral parks that he and Vaux's plan for Prospect Park called for the construction of a building that would block the view of Green-Wood Cemetery, which was located about nine blocks west of the park.[82] However, Olmsted and Vaux did not reject the cemetery model completely. Rural cemeteries such as Green-Wood and Laurel Hill introduced secluded groves, curvilinear paths, and naturalistic ponds. Such landscape features were incorporated into the Greensward design and other later Olmsted-Vaux parks.[83]

Sensing that Central Park, the first major testing ground for his ideas, would be a significant symbol of American parks Olmsted wrote, "It is of great importance as the first real park made in this country—a democratic development of the highest significance & on the success of which, in my opinion, much of the progress of art & esthetic culture in this country is dependent." To keep the park unitary, he argued, "The park throughout is a single work of art, and as such, subject to the primary law of every work of art, namely, that it shall be framed upon a single, noble motive, to which the design of all its parts, in some more or less subtle way, shall be confluent and helpful."[84]

THE CHALLENGE OF BUILDING RURAL PARKS IN THE CITIES

Landscape architects wrestled with the tensions inherent in building pastoral parks in teeming cities. The park for them was a serene, contemplative space intended to improve the lives of the people, but the cities that contained most of these parks were noisy and congested, and some of the potential users of the parks were overworked, uneducated, and highly desirous of active recreation. Olmsted and Vaux's Greensward plan resolved the problem of noise and congestion by "planting out" the city from within the park. This was a common design feature in their parks. They excluded the sights and sounds of the city by planting a barrier of trees and shrubs along the perimeter of the park. This also obscured the long straight sides of the park, thus making the park appear more extensive than it really was.[85] According to Olmsted and Vaux, "No one, looking into the closely-grown wood can be certain that at a short distance back there are not glades or streams, or that a more open disposition of trees does not prevail."[86] Other park advocates viewed the park as the antithesis of the city, not an extension of it. The author and art critic

Clarence Cook claimed, "We want to forget the city utterly while we are in the Park, and we want to get into it as soon as possible. . . . We want to find ourselves, without unnecessary delay, among trees and grass and flowers."[87] In later years Olmsted justified this view in his proposal for Boston parks: "We want a ground to which people may easily go after their day's work is done, and where they may stroll for an hour, seeing, hearing, and feeling nothing of the bustle and jar of the streets, where they shall, in effect, find the city put far away from them."[88]

Another innovative way the landscape architects kept the city from intruding on the park was by submerging all its transverse roads. Such an arrangement did not interfere with the movement of people and the flow of wagons and carriages around the park. This helped to achieve the "range," openness, and suggestion of freedom that Olmsted and Vaux thought was very important to pastoral landscape design.[89] They did not want to obscure the sunlight from any vantage point in the park.[90]

In 1865, after years of collaborating on various projects, Olmsted and Vaux formed an architectural firm. Their company, which lasted until 1872, was responsible for designing scores of the nation's most impressive urban parks. In addition they designed several college campuses, hospital and asylum grounds, and private estates.[91]

Early Zoning and Urban Renewal

Many parks, including Central Park, displaced blacks and unemployed and working-class people who lived in the spaces chosen for the parks. Whether or not these people owned the land on which they lived, these areas were perceived as urban wastelands retarding the development of cities; the people were considered "squatters" and their homes were demolished. For example, Downing's proposal to acquire land for a public park described one potential site as "five hundred acres . . . between thirty-ninth street and the Harlem River, including a varied surface of land, a good deal of which is waste area, so the whole may be purchased at something like a million . . . dollars."[92] According to Bunce, the Central Park site "was a mass of rude rocks, tangled brushwood and ash heaps. It had long been the ground for depositing city-refuse."[93] In both cases, many people lived on the sites before the parks were built.[94]

Egbert Viele, a member of one of New York's oldest Dutch families, the first engineer in chief of Central Park, and the first person to lay out a design for the park, called the land "as uncompromising an acreage as could be found in Manhattan, inhabited by roving animals and about 5,000 squatters, most of them, of foreign birth." He claimed that when he surveyed and mapped the land he armed himself and carried "an ample supply of deodorizers."[95] Olmsted's view was just as pejorative. He described the Central Park site as follows:

> The southern portion of the site was already a part of its straggling suburbs, and a suburb more filthy, squalid and disgusting can hardly be imagined. A considerable number of its inhabitants were engaged in occupations which are nuisances in the eye of the law, and forbidden to be carried on so near the city. They were accordingly followed at night in wretched hovels, half hidden among the rocks, where also heaps of cinders, brick-bats, potsherds, and other rubbish were deposited by those who had occasion to remove them from the city. During the autumn of 1857, three hundred dwellings were removed or demolished by the Commissioners of the Central Park, together with several factories, and numerous "swill-milk" and hog-feeding establishments.[96]

Though Central Park and parts of Manhattan north of it were considered rural in the 1850s, sixty thousand people lived in the area; sixteen hundred people lived on the site of Central Park itself. Because they were immigrants and blacks it was easy for elites to stereotype them. While immigrants constituted 53.6 percent of the city's population, they made up more than 90 percent of the residents in the area proposed for the park. Immigrant whites were disdained by native-born whites for living among and possibly mating with blacks. John Punnett Peters of St. Michael's Episcopal Church in New York described the community in the 1840s as a "wilderness" filled with "the habitations of poor and wretched people of every race and color and nationality." He added that "this waste" had "many families of colored people with whom consorted and in many cases amalgamated, debased and outcast whites." These communities were also disdained by elites because they subsisted on the land. In addition to raising vegetables, crops, and animals, the people fished and collected fruit and nuts. They collected driftwood from the river and branches and twigs from the trees for firewood. Even before the park was completed such subsistence activities were banned.[97]

Though elites described the park site and its inhabitants in disparaging terms, more than two-thirds of the area's inhabitants worked as unskilled laborers or in service jobs; the rest were tailors, carpenters, masons, or other skilled artisans; 10 percent ran small businesses, such as grocery stores and butcher shops. Thus the widespread characterization of the people as criminals, vagabonds, or human waste living "in the lowest depths of filth" bears questioning.[98]

Because of the cost of acquiring parkland, parks were often sited in the cheapest (and often most inconvenient) locations: on the outskirts of town, on garbage dumps, abandoned industrial sites, swamps, hilly or rocky terrain, or land otherwise unsuited for cultivation or other forms of development. Even in New York both Jones Wood and Central Park were on the outskirts of town when the sites were chosen.[99] Park advocates played on city officials' fear when lobbying for parcels of land they wanted converted into parks. They argued that if the land was not used for parks it could "be given up . . . exclusively to shanties, stables, breweries, distilleries, and swine-yards."[100] Thus the parks were used as an early form of zoning and urban renewal, with the goal of preventing additional poor and black people from moving into an area and providing an excuse to remove the ones already living there.[101]

REMOVAL AND COMMUNITY DISRUPTION

When former president Bill Clinton decided to set up his postpresidential office in Harlem in 2001, the move intensified the debate over whether the gentrification of the parts of Harlem close to Central Park would accelerate the displacement of the area's longtime black and Latino residents. The real estate boom in Manhattan, coupled with people's desire to live close to or have a view of the park, drove speculators to acquire property in Harlem. The process that began in the mid-1980s has escalated during the 1990s as young, well-to-do professionals bid up the price of housing and began moving into Harlem. Though many were optimistic that Clinton's presence would be an economic boom to the area, poor residents expressed concern that they would be forced out due to rapidly rising prices. Their concern was grounded in reality. A study by the Corcoran Group found that in the first six months of 2000 the average price of a Harlem brownstone rose 29 percent, to $391,000, from the 1999 average of $303,000.[102] This process is not new. The displacement of blacks and poor whites to make way for Central Park and other urban renewal projects in New York dates back to the early nineteenth century. Even before residents were evicted, they began to run afoul of the new laws banning subsistence activities in the park. In 1856, the newly organized nineteen-member Central Park police force began arresting people for breaking up and selling rocks ("park stones") and for cutting down trees for firewood.

The Irish and German Pre-Park Dwellers

Beginning in 1855 the people living on the land earmarked to become Central Park were evicted. Irish and German immigrants as well as blacks lived in Manhattan on the Central Park site in settlements named Seneca Village, Yorkville, and Pigtown. Despite disparaging stereotypes of the pre-park dwellers, the Irish attended nearby Catholic churches, and many of them grew food, kept farm animals, and sold milk, as they had in Ireland; they also farmed to supplement the meager wages they earned as day laborers building roads and other public projects. After the 1849 cholera epidemic farming activities were curtailed downtown, so families wanting to raise animals moved uptown, where there were fewer restrictions. German immigrants too relied on the land for subsistence; they were also rag pickers or gathered and sold cinders.[103]

Black Pre-Park Dwellers

During the early 1800s blacks in New York faced numerous obstacles to owning land, and state law prevented them from inheriting it. Thus Seneca Village is significant because it was one of the few parts of the city where blacks owned property. In 1825 blacks began buying lots in Seneca Village when the Whiteheads subdivided their farm. Andrew Williams bought three lots and Epiphany Davis bought twelve; soon the African Methodist Episcopalian (AME) Zion church bought six lots. Between 1825 and 1832 the Whiteheads sold fifty parcels, no fewer than twenty-four to black families. Blacks inhabited other parts of Central Park even earlier than the Seneca Village residents. Records show that blacks developed a settlement on York Hill in the second decade of the nineteenth century on a spot located in the middle of the future Central Park. Some blacks living on York Hill experienced earlier community disruptions with the construction of the Croton Reservoir, and some of these families may have moved to Seneca Village.

By the 1840s Seneca Village was known as "Nigger Village." Land ownership was very high among black Seneca Villagers: in 1855 more than half the black households in the village owned their homes. Black Seneca Villagers were thirty-nine times as likely to own their homes as other black New Yorkers. Black Seneca Village residents were also far more likely to own their homes than the Irish residents. Seneca Village also contained important community institutions: two black churches and one racially mixed Episcopal church, a cemetery, and a black school. In 1853, two weeks after the Central Park bill was approved, black Seneca Villagers, unaware of the plans to develop the site for a park, gathered to break ground on the new buildings for the AME Zion Church. As they were erecting the church, government officials were appraising the value of properties in the village. Two and a half years later church representatives protested, unsuccessfully, the condemnation proceedings. By the fall of 1857 the churches, school, and people had vanished from the landscape. Blacks weren't even hired to help build the park; Central Park was built with an all-white, male workforce.[104]

Making Way for the Park and Banning Subsistence Activities

Most of the pre-park dwellers subsisted on the resources of the park. At the time, Manhattan, the Indian word for "island of the hills," was still dotted with fishing holes and numerous streams trickled through the hills. In fact Viele's 1855 map of the park site indicates that the area had numerous hillocks, rock outcrops, surface and underground streams, and standing surface water (in addition to the reservoirs). The pre-park dwellers recycled and reused the garbage by collecting objects that could be used by families or sold on the secondhand market. They fed what was unfit for human consumption to the pigs and goats, and the bones of dead animals fueled the bone-boiling facilities in the park. Though the pre-park dwellers were routinely referred to as "squatters," tax records indicate that at least a fifth of the residents had title to their land and paid taxes on it. Others rented, and some worked out arrangements with landlords; for instance, residents could live on the land without charge in exchange for helping to clear and tend the land.

In 1857 the last of the pre-park dwellers left their homes. Those who owned their land were compensated, but those who rented, leased, or had other arrangements with landowners, or whose improvements to the land could not be assessed, had to start over from scratch. Some moved to rocky and swampy "squatter" settlements west of the park.[105]

FIVE POINTS: A PRECURSOR

The removal of the pre-park dwellers from Central Park can be seen in the context of racial and ethnic relations in New York City during the nineteenth century and earlier "slum" clearances in the city. During the 1830s Five Points and Lispenard Meadows were among some of the most densely populated neighborhoods in the city. When "respectable" tenants refused to move into the neighborhoods, landlords leased the buildings and tenants turned some buildings into brothels, gambling dens, and taverns. Blacks, excluded from the skilled trades and other legitimate businesses, operated businesses in the red light districts. Nonetheless, the districts had a vibrant church and associational life. As new Irish immigrants settled in Five Points (because it was the most affordable and because other whites reviled the Irish) nativists began to express their fear and disapproval of the neighborhood. They began speaking out against the "amalgamation" of whites and blacks, and efforts to promote racial harmony were sometimes met with violence.[106]

Nativists expressed both anti-Catholic and anti-Irish sentiments. The Irish were blamed for much of the degradation and depredations of the city. During the 1850s Irish-born residents constituted about 30 percent of the city's population, yet more than half the people arrested by the police were Irish. They also accounted for 60 percent of the almshouse population and 70 percent of charity recipients. Despite the widespread stereotype of the Irish as vagrants and criminals, in 1855 60 percent of the Irish were employed as domestics, laborers, porters, laundresses, carters, and coachmen. Though many worked hard, their wages were too low to keep them out of extreme poverty. They competed for these jobs with blacks. The Irish and blacks were two of the most maligned groups in the city.[107]

Five Points and other mixed-race neighborhoods like it were considered "nuisances" and symbols of poverty and degradation in the city. The neighborhood was described as "the most notorious precinct of moral leprosy in the city, . . . a perfect hot-bed of physical and moral pestilence."[108] For several years before the antiabolition riots broke out in 1832 and 1834 city officials considered clearing Five Points and removing the people. As early as 1829 activists living outside the area urged the city to seize properties by eminent domain to facilitate clearance of the area. An 1831 report urging clearance contended that all but a few blacks in the area were "proper objects for a Magdalen asylum" for the reformation of "fallen" women. Five Pointers were considered people without rights who should be institutionalized.[109]

Though Five Points wasn't the only poor, racially mixed, deteriorated neighborhood in the city, it was targeted in part because it was one of the most visible. Five Points had a location problem: it was close to the elite homes and cultural centers on Broadway; it was close to City Hall, thus representing a challenge to the city's claim of advancement, civility, progress, and cultural enlightenment; and it occupied prime real estate territory between the city's commercial center and the upscale East Side. To commute to work, social occasions, and political events affluent East Siders had to navigate through or close by the neighborhood.

Sections of Five Points were cleared in 1833 as part of street-widening projects. By the mid-1830s similar projects were being used to justify the clearance of neighborhoods around Chapel and Anthony Streets. These neighborhoods had a high concentration of blacks and business establishments such as taverns and brothels. As Anthony Street was being cleared in 1836, merchants and other businessmen eagerly lined up to redevelop the area. Though the official justification given for clearing these areas was widening of streets to facilitate the flow of traffic, racism, nativism, and the fear of racial mixing played a role also.[110] Thus very early in the process of urban renewal, the homes and businesses of blacks and poor whites were demolished and replaced with businesses and housing for wealthier whites. In the case of Central Park, integrated communities of poor blacks and whites were eliminated to make room for an elite park that would be ringed by upscale housing and businesses. No provisions were made to incorporate low-income housing that could accommodate the displaced poor. This pattern of urban renewal, or gentrification, continues today, whereby poor blacks and other minorities are often displaced to make room for expensive housing that will be inhabited by middle- and upper-class residents.[111]

Parks as Sites of Social and Political Protests

During the great park-building era the condition of the working class was harsh; poverty was ubiquitous and unemployment high. About a fifth of the population had to receive some form of charity. In 1850 New York had the highest mortality rate in the country. At the same time the wealthiest 4 percent of the population controlled more than 80 percent of the wealth. In the winter of 1854–1855, less than a year after funding for Central Park was approved, New York slipped into a recession that lasted until 1857. The severe winter forced thousands of people to seek charitable aid. More than fifteen thousand people lost their jobs, and soup kitchens set up by rich merchants served an estimated twelve thousand people daily. By January 1854 the Association for Improving the Condition of the Poor announced that it had helped fifty thousand people and its funds were almost exhausted. The unemployed began to hold demonstrations at City Hall and City Hall Park almost daily, demanding that the Common Council use the public works contract system, undertake new public works projects to increase employment, institute a guaranteed minimum wage, expand work opportunities in Central Park, curtail grain exports, prohibit landlords from evicting tenants, develop an urban land-grant program that would distribute common lands to workers, and provide half a million dollars to help the poor build "inalienable homes." Demonstrators also wanted the city government to establish a $3 million loan fund to help resettle unemployed people in the West. A Special Committee was established to study the feasibility of contracting out the work to be done on Central Park rather than using day laborers; the committee concluded in 1858 that the contracting system would not work because of the complicated nature of the project.[112]

Not all New Yorkers wanted to expand public works projects. Some questioned the wisdom of spending large sums of money to construct a park during a recession; others suggested reducing the size of the park to save money. However, realizing that the park could provide thousands of jobs, Mayor Fernando Wood, the rabidly racist leader of the New York secessionist movement, vetoed a bill aimed at reducing the size of the park.[113]

The recession worsened in 1857 as 985 businesses failed in New York City, with losses totaling $120 million ($2.5 billion in 2005 dollars). An estimated forty thousand people lost their jobs and most labor organizing ground to a halt. The recession hit female workers harder than males; 50 percent of all female workers lost their jobs, compared to 20 percent of male workers. Food prices soared. To make matters worse, landlords evicted thousands of tenants who couldn't pay their rent. According to an estimate by the Association for Improving the Condition of the Poor, during the three winter months of 1857–1858 forty-one thousand people sought shelter in police stations.[114]

There were daily demonstrations at City Hall in 1857. Mass meetings of the unemployed, some attended by as many twenty thousand people, were held at City Hall Park, Washington Square, and Tompkins Square. Mayor Wood responded to the crisis by suggesting that the city launch public works programs to get people back to work, that is, build and grade more streets, construct engine houses and police stations, repair docks, construct a new reservoir, and move quickly to begin work on Central Park. He suggested that workers on these projects could be paid in flour, cornmeal, and potatoes instead of cash.[115]

As the protests widened, demonstrators turned their attention to Central Park, demanding that the unemployed be given jobs in its construction. In August 1857, when Olmsted took over as superintendent of Central Park, the aldermen authorized $250,000 ($5 million in 2005 dollars) for park construction and gave Olmsted the order to hire a thousand men.[116] Despite the fact that park building was hard work, wages low, conditions severe, and employment short-lived and uncertain (because they were patronage jobs controlled by politicians), Olmsted was inundated with applications. Even before he reported for his first day on the job his flustered servant rushed to inform him that there were twenty men waiting outside his door for jobs and four had already forced their way inside the house. Later that morning Olmsted reported walking past a throng of five thousand men waiting outside his office: "As I worked my way through the crowd, no one recognizing me, I saw & heard a man then a candidate for reelection as a local magistrate addressing it from a wagon. He urged that those before him had a right to live . . . he advised that they should demand employment of me. If I should be backward in yielding it—here he held up a rope and pointed to a tree, and the crowd cheered."[117]

It was impossible for Olmsted to satisfy the demand for employment. Consequently his office was regularly surrounded by demonstrators carrying banners reading "Bread or Blood." The protestors sent him a list of ten thousand men in desperate need of work. Even if they were hired, there was little hope that many of these workers would keep their jobs for long: in an effort to stretch park employment as far as possible, politicians and park commissioners rotated employment. Frustrated with the ever-changing workforce, Olmsted wrote to his father, "We unexpectedly received an order to pay off & discharge all the men this week, & have been doing it. Probably we shall take on even a larger force, a thousand is talked of, next week."[118]

On November 5, 1857, the Common Workers League organized a "work and bread" demonstration to protest the Common Council's failure to act on Mayor Wood's suggestions. Four thousand people gathered in Tompkins Square and marched to City Hall Park. Speakers decried the economic inequality in the city by arguing that wealthy ladies flocked to Broadway to buy silk while poor women and children were starving. The following day five thousand people marched to Wall Street chanting "We want work." As the protests went on, the mayor called out the police and militia to protect City Hall and other government buildings. Despite the heavy show of force, on one occasion demonstrators in Tompkins Square began a bread riot, in which protestors seized bakers' wagons and invaded grocery stores.[119]

Central Park and other parks continued to be the focal points of protest. During the draft riot, protestors held an "indignation meeting" in Central Park. There were demonstrations during the 1873 depression also; fifteen thousand rallied in New York, and twenty thousand marched in Chicago.[120] In 1871 five hundred people held a rally on the New Haven Green during the printers' strike. There were numerous demonstrations and rallies in New York's parks during the 1870s. In 1877 railroad workers tried to regain wages lost during the great strike of 1873; to make matters worse, their wages had been slashed by as much as 10 percent early in the summer before the strikes began. Riots broke out in Baltimore, Philadelphia, Buffalo, and Chicago. Troops fired on protestors in Pittsburgh, killing twenty-five people; in response strikers set fire to the railroad's roundhouse. New York's Workingmen's Party organized a rally for railroad workers in Tompkins Square. The business magnate William Vanderbilt wrote to the mayor asking him to disallow the rally, but Mayor Smith Ely refused to comply. As the rally neared the police and militia were put on high alert and sailors and marines were called in as reinforcements. The Central Park headquarters was surrounded with loaded howitzers. Twenty thousand people gathered in Tompkins Square for the rally on July 25. As the meeting adjourned, police charged for no apparent reason, but the crowd dispersed without further incident. Huge demonstrations were held on the New Haven Green in support of the strike.[121]

As the social unrest increased middle- and upper-class urbanites supported efforts to militarize the cities by funding militias. During the Astor Place riots in New York in 1849, for instance, the Seventh Regiment (known as the "silk stocking regiment") was activated to guard Astor Place and the nearby homes of the rich. The regiment was used to protect the interests of the rich on other occasions too. Shortly after the 1877 strike and riots, when the regiment wanted a new armory, John Jacob Astor (who was known for his unwillingness to support charitable causes), A. T. Stewart, William H. Vanderbilt, the Brown brothers, the Harper brothers, Singer Manufacturing Company, Equitable Life, and Drexel Morgan contributed money to help build the armory at Park Avenue and Sixty-sixth Street. Formed in 1847, the Seventh Regiment had more than a thousand members by the 1860s.[122]

It is not surprising that the parks became a focal point for some of the social unrest of the period. Parks lay at the intersection between work, home, and recreation. They were also the only conveniently located spaces in the cities large enough to hold massive demonstrations; this led park commissioners to limit public gatherings in the parks. Rallies, public meetings, religious services (all deemed to have the potential to excite people and incite rebellious activities) were excluded from park programming. Boston park commissioners did not allow any public meetings in the parks; Philadelphia allowed only religious meetings; the Chicago South Parks forbade public meetings that would lead to speech making and crowds; and Brooklyn permitted gatherings only for parades of Sunday school children. Commissioners wanted the parks to appear apolitical to retain public support. Though public gatherings were banned, parks were built with military parade grounds, and military exercises were conducted in many parks. While public gatherings had the potential to spin out of control, military exercises were seen as a show of law, order, civility, and national unity. Thus the presence of military personnel in the parks could be considered another dimension of social control.[123]

The Laborers Who Built Central Park

As mentioned earlier, Central Park was built with white male labor. The Irish made up the largest part of the workforce, but Italians, Scots, and native-born whites were well represented. During the time the park was being constructed white dockworkers and waiters often walked off their jobs or resorted to violence to protest the hiring of blacks. As the Irish displaced blacks as domestic servants, shipyard workers, seamen, and casual laborers they were derided for doing "nigger work" and for "slaving like a nigger." This increased racial tensions between the Irish and blacks as the Irish tried to assert their whiteness by brutalizing blacks and refusing to work alongside blacks. To forestall racial conflicts the commissioners of Central Park decided not to hire blacks.

Despite the use of an all-white labor force, a dual labor market operated during the construction of Central Park as various types of work were segregated along ethnic lines. Though the Irish managed to secure park employment, once on the job they were the most discriminated against. They were segregated into the most poorly paid construction jobs, though some managed to parlay these jobs into higher paying contractor and foremen jobs.[124]

On the question of slavery Olmsted was a gradualist, which at the time was considered a moderate position. While abolitionists advocated the end of slavery and pro-slavery advocates supported its continuation, gradualists supported a gradual emancipation of blacks. That is, Olmsted thought the new states in the West should be free states in order for them to be admitted to the Union. However, he also believed that slavery had not prepared blacks well for citizenship. He felt that slaves had the right to be free, but their capacity to act as free citizens should be restored before they were emancipated. For Olmsted, freed slaves should be granted liberties after they had shown they deserved those liberties. Given Olmsted's position on slavery and his view of blacks, it is not surprising that he did not advocate for the inclusion of blacks in Central Park's workforce.[125]

Central Park workers labored long hours (keepers worked up to eighteen hours per day) under strict surveillance for small wages. Park workers were fired if they fell ill or were injured on the job.[126] For about a year laborers worked on the park night and day to get construction to a point where opponents could not stop it.[127] In 1867 the workers resisted the degrading work rules and fought for fair wages and an eight-hour workday. That year they won the eight-hour day and their wages were increased to about two dollars per day, from a low of ninety cents per day in 1861. However, in 1872 the Park Department cut its budget and laid off workers or paid them their wages late.

In May of that year unionized city workers wanting to show that they were a critical component of the city's development staged a rally with the battle cry "Park, pipe and boulevard." Workers demanded to be paid on time and asked that their daily wages be increased to $2.25 ($34 in 2005 dollars). Yet as the depression deepened in 1873 more employers laid off workers, cut salaries, and reduced benefits. That year 550 of the Parks Department's workforce of 882 workers were assigned to Central Park. In 1874 the protests turned violent as Irish workers attacked Italian workers for undercutting wages and the eight-hour day by accepting $1.75 ($27 in 2005 dollars) for ten hours of work. Employers continued to put pressure on workers, whom they considered lazy and overpaid. Elites complained that park workers sat around having "picnic parties" while men doing private contract work still earned ninety cents for a ten-hour day. As pressure mounted in 1875, all city departments cut daily wages to $1.60 ($26 in 2005 dollars), but the parks department kept the $2.00 ($32) wage.[128]

During the campaign for the eight-hour day laborers had successfully established the principle that the government had an obligation to meet minimum working standards, including providing a living wage for workers. During the depression private employers complained that the government standards made it more difficult for nongovernment businesses to control their payroll as their workers demanded similar pay. Critics also complained about the bloated city budget. Olmsted, always a taskmaster and not very sympathetic to workers, was horrified at the number of unemployed workers who flocked to the parks for work.[129] Unemployed and homeless people also slept and panhandled in Central Park; others sold dandelion greens collected there. As was the case in 1857, Olmsted cringed at the sight of the "mob of [job] candidates," "the larger part of whom were of the dirty ruffianly and loafish sort." He wrote, "The room with all the windows opened was filled with a sickening odor of them."[130]

MARIPOSA MINING COMPANY: SPLIT LABOR MARKET AND WAGE ROLLBACK

Rolling back wages was a labor-management tactic Olmsted had employed before. Shortly after he took over the mining operations of the forty-four-thousand-acre Mariposa Estate in California he studied the labor costs of other mining companies and discovered that Mariposa's workers earned more than others in the state.[131] Arguing that Mariposa's wages forced other mining companies to pay higher than normal wages he began to hire Chinese workers, whom he paid about half the wages of white workers: $1.75 per day. Within about four months he had hired enough Chinese laborers to reduce costs by 8 percent. He lowered the white miners' wages from $3.50 to $3.15 per day (workers emptying the ore buckets and others in non-mining jobs received $2.40 per day). Olmsted's actions precipitated a three-week strike among the miners. They demanded restitution of the old wage scale, but Olmsted did not capitulate. Instead he paid off almost a hundred workers (about half the workforce) who objected to reduced wages and replaced them with lower wage workers. He also sent sheriff's deputies after four of the strike leaders, whom he described as "discontented and unwilling workers" who hated "regularity, order and discipline." He wanted to expunge them from the workforce because he thought they had a negative influence on the other workers. Olmsted also had a supply of firearms ready to deal with unruly workers.[132]

Hiring Chinese laborers not only reduced labor costs, but it effectively created a split labor market that made it easier for Olmsted to divide and control the workers. Though the workers struck, they did not organize effectively across racial lines. The strike occurred at a time when anti-Chinese sentiments were high in the West. White workers, believing that they were being replaced and their wages undercut by Chinese workers, took out their anger on the Chinese workers. In the Mariposa case angry workers took the wages owed them and drifted off to find work elsewhere rather than organizing and demanding both the reinstatement of wages and wage equity, wherein whites and Chinese workers were paid on the same wage scale.[133]

ATTEMPTS TO ROLL BACK WAGES IN CENTRAL PARK: OLMSTED'S SECOND TIME AROUND

As superintendent of Central Park Olmsted tried to roll back workers' wages and extend the workday as he and some critics of public works employment schemes blamed workers for the spiraling cost of park construction.[134] This time several factors prevented him from hiring black and Chinese laborers as a cost-cutting measure. In the California context Olmsted was operating in a labor market with a severe shortage of workers; hence all white workers who objected to work conditions could find work elsewhere. However, virulent racism and anti-Chinese violence constrained the work options for Chinese workers; this made them amenable to accepting lower wages. In New York unemployment was high, so there was a surplus of white laborers who would work for low wages out of sheer desperation. Thus Olmsted could exploit white ethnic divisions to create split and dual labor markets. In this way the workforce remained white. Given Olmsted's attitudes toward blacks, he was unlikely to hire them, and even if he had wanted to use Chinese workers in

New York there were very few of them living in the city at the time. In 1859 only about 150 Chinese men lived in the city, and as late as 1880 the census recorded only 748 (the press estimated the number to be around 2,000).[135]

As the depression deepened during the 1870s park workers' wages were eventually cut, but during the early 1880s unionized park workers were energized by the labor movement and pressed their claims. In addition to protesting wage cuts, workers were unionizing because they were being pressured to increase output. Beginning in 1880 employers introduced Taylorism, also called scientific management or Fordism, to the factories. As a result assembly lines moved faster, workers lost control over their work, the owners got richer, and the workers saw few material benefits for their increased output. Workers responded by joining unions: from 1897 to 1904 union membership rose from 447,000 to over 2 million. From 1909 to 1918 about 3 million workers joined the Industrial Workers of the World. The United Mine Workers grew from 14,000 in 1897 to 300,000 in 1917; 400,000 garment workers unionized between 1909 and 1913. As with other reform efforts, unionization drives focused primarily on improving the condition of white workers.

The collective response of park workers restored wages for some. In 1882 the wages of skilled workers such as the stonemasons and carpenters were restored; later the plumbers, painters, blacksmiths, and steam engine operators were given wage increases. However, unskilled laborers, who accounted for about 60 percent of the Parks Department workforce, did not recover lost wages. In 1881, after a 10 percent wage increase, they still made only $1.76 per day. Women earned even less; female washroom attendants earned $1.50 for a ten-hour day.[136]

Though Scobey argues that Olmsted was deeply committed to free labor values and references Olmsted's criticism of the southern slave system as evidence of this,[137] the previous discussion indicates that such a conclusion bears questioning. At times Olmsted was quite hostile to labor. He dismissed workers who were injured on the job, fired workers who struck, sent the police after strike organizers, cut back wages, used Chinese workers to undercut the wages of white workers, and excluded blacks and women from his workforce. These practices lead one to conclude that Olmsted was willing to use repressive tactics to bend workers to his will if he felt it was necessary to do so.

Central Park and Social Order

By the 1930s the idea of parks as tools of social control had lost currency. Even the famed park builder and urban renewal proponent Robert Moses claimed,

"We make no absurd claims as to the superior importance and value of the particular service we are called to render, and we realize that budget making is a balancing of comparative needs of competing agencies." Similarly the president of the American Institute of Park Executives said that park managers should no longer view themselves as their brothers' keepers because they could "hardly hope to effect their salvation."[138] But at the time Central Park was being established, Olmsted and Vaux, like their mentor Downing and other nineteenth-century park planning elites, saw parks as important tools of social control.

The use of leisure and public recreation space as instruments of control was not new when it was introduced in American parks in the 1850s. The ancient Romans organized recreation events to control urban populations and guard against social unrest.[139] In modern times, as workers agitated for a reduction in work hours and predictable blocks of time away from the workplace, the middle and upper classes sought to monitor and control what workers did in their free time. They did so in part by controlling how, when, and how many parks were provided; the distribution of parks; and their size, layout, and management. Olmsted, who once described himself as "a moderate drinker, semi profane swearer [and] a Sabbath cracker" who would "gamble if . . . [he] had brains enough," sought to develop the model for and set the precedence of parks functioning as effective agents of social control.[140]

In a talk before the American Social Science Association Olmsted declared, "The difficulty of preventing ruffianism and disorder in a park to be frequented indiscriminately by such a population as that of New York, was from the first regarded as the greatest . . . [challenge] . . . the [park] commission had to meet, and the means of overcoming . . . [disorder] . . . cost more study than all other things."[141] Consequently Central Park was designed to maximize desired behavior and limit or eliminate undesirable behavior. The park was heavily monitored and supervised, and bad behavior was punished with arrests and fines. Although the park was promoted as serving all the classes, Central Park and others like it were built to accommodate the interest and desires of the upper and middle classes to a greater extent than those of the lower class.[142] The working class and the poor were forced to abide by middle-class mores in order to use these parks. Members of the middle class, however, were not expected to adjust their behavior and expectations.[143]

ENFORCING SOCIAL ORDER

From the outset park advocates, designers, and managers combated working-class styles, values, and needs. City parks were some of the few spaces in many cities where all the classes met and mingled (even in a limited way) outside of the workplace. Despite pronouncements of park elites to the contrary, these encounters were awkward and sometimes hostile. The gentry responded by establishing rules of behavior, thus setting the stage for greater confrontations and rebellion. Upper- and middle-class interference with working-class park use fueled discontent among the poor and helped stimulate working-class environmentalism and political resistance.[144] These conflicts arose because many of the park building projects were conceptualized and managed by the elites and built with working-class labor.[145]

Olmsted and Vaux used a variety of means to enforce social order in their parks. First, they used indirect techniques to include and exclude facilities and activities. In the *Second Annual Report* to the commissioners of Central Park, "the motive of the park" was described in the following terms:

> The primary purpose of the Park is to provide the best practicable means of healthful recreation for the inhabitants of the city, of all classes. It should present an aspect of spaciousness and tranquility with variety and intimacy of arrangement, thereby affording the most agreeable contrast to the confinement, bustle, and monotonous street-division of the city . . . always remembering, however, that facilities and inducements for recreation and exercise are to be provided for a concourse of people. . . . No kind of sport can be permitted which would be inconsistent with the general method of amusement, and no species of exercise which must be enjoyed only by a single class in the community to the diminution of the enjoyment of others. Sports, games and parades, in which comparatively few can take part, will only be admissible in cases where they may be supposed to contribute indirectly to the pleasure of a majority of those visiting the Park.[146]

Olmsted argued further, "The park is not simply a pleasure-ground . . . but a ground to which people may resort for recreation . . . *which will be conducive to their better health*."[147] He wanted the parks to be taken seriously both as works of art and as public spaces where people followed prescribed behavior. Consequently elaborate bridges and entrance gates were built in Central Park (and other parks also), and details such as the ornamental tile floors in the esplanade were added to give the park an aura of elegance and importance.[148] In addition, Central Park's gates were guarded by uniformed keepers who tracked the flow of visitors.[149]

Second, Olmsted devised more explicit ways of maintaining social order. The rules and regulations of Central Park provide one example. To ensure that the park was used in the "proper way," regulations prohibited visitors from walking on the grass; picking flowers, leaves, twigs, fruits, and nuts; defacing structures; throwing stones; annoying birds; using foul language; selling merchandise; behaving in a lewd manner; entering or exiting unauthorized places; climbing the walls; or being in the park after closing. Carrying firearms, playing musical instruments, and displaying flags, placards, or banners were also prohibited. Park visitors could not play games of chance, make speeches, or engage in any indecent act. Violation of these rules could result in fines or imprisonment.[150] Regulations prohibiting walking on the grass were also found in Chicago's parks, but "Keep off the Grass" signs were banned in San Francisco's parks.[151] An outright ban was not placed on these signs in Central Park until the late 1890s.[152]

Third, liquor sales were strictly controlled. Because critics had predicted that the park would be a haven of wild drinking if it was ringed by liquor stores, liquor was sold at concessions in the park's interior, where visitors could drink in a controlled environment. This discouraged the establishment of liquor stores around the perimeter of the park. Olmsted was careful to point out to critics that the park was not surrounded by grog shops and that the socially acceptable park liquor concessions were not encouraging churchgoers to forsake church in order to indulge in alcohol.[153]

Fourth, Olmsted relied on the criminal justice system to punish deviant behavior and keep social order. He lobbied for a park police force modeled after the force that patrolled London's parks and American cemeteries. He argued that New Yorkers, unused to a large park, had "to be trained in the proper use of it, to be restrained in the abuse of it."[154] In 1858 Olmsted hired twenty-two keepers to patrol the park; this number had increased to fifty-five by 1860.[155] The keepers recorded 228 arrests in that year; almost half were for violations of the ordinances, such as walking on the grass, picking flowers, and using foul language; another third of the arrests were for drunkenness and disorderly conduct; assaults, battery, and petty theft accounted for the rest. While on duty the Central Park keepers behaved more like a military unit than the park liaisons and educators they were touted to be. Among other things, keepers could not leave their post or beat until they were relieved by another keeper, and then only to capture an offender. They were on duty up to eighteen hours per day. Keepers could not speak casually with each other or consume alcohol in the park, and when off duty they could not visit bars or get drunk. They were ordered to keep an eye out for drunkenness and apprehend and report anyone violating the park rules. They were to appear at all times as a model of "studied respect and vigilance." They were required to keep their uniforms clean and neat and carry themselves "erect according to instructions received at drill, and march at a quick step from one part of [their] beat to another, except when it is necessary to move slowly or to halt entirely for the observation required in [their] duty, or for the detection or apprehension of offenders." Keepers had to salute their superiors in military style. They were not paid for absences due to illness or injury, even those occurring on the job.[156] Keepers found violating any of the rules were fired at the end of the day. According to Olmsted, the keepers exercised "a distinctly humanizing and refining influence over the most unfortunate and lawless classes of the city."[157]

Brooklyn's Prospect Park was patrolled by fourteen keepers; in addition, forty-eight gardeners assumed park keepers' duties during peak visiting hours.[158] In 1897 the Central Park keepers were replaced by city policemen who patrolled the park. Today, about seventy-five full-time and forty seasonal police officers are based at the Central Park precinct on Eighty-sixth Street.[159]

RATIONALIZING INCREASED SURVEILLANCE OF PARK USERS

Olmsted used the popularity of the park and perceived incidences of deviant behavior by park users to increase police supervision of the park. In a letter to the Board of Commissioners of the Central Park he argued that the existing police force was inadequate to counteract deviant, depreciative, and criminal behavior as the park keepers were overworked and underpaid. Olmsted analyzed the distribution of park visitors as part of his request for more keepers. He explained that the park was so heavily used and that users needed to be watched to keep them from destroying the park. He noted that it received the largest number of visitors in the late afternoons and on Saturdays and Sundays. The largest police force was deployed during these peak visiting hours. Olmsted estimated that in 1860 the park attracted 2,000 carriages and 10,000 people during the afternoon and 100,000 on special occasions. The popularity of the park continued to increase. By 1863 an estimated 4.3 million people visited the park annually, up to 90,000 entering on foot during a single day. The average number of carriages per day was about 1,000 (the largest number counted was 9,460, and the largest number of saddle horses on a single day was 1,640). In all, about 7.6 million people visited the park in 1865; that number increased to almost 11 million in 1872. It is estimated that 87.9 million people visited the park from 1863 to 1873.[160] There was no question that the park was heavily used; the question plaguing park managers was how best to serve the public while protecting park resources.

The crowds were not evenly distributed throughout the park. Some activities, such as ice skating, were extremely popular. On two occasions in December 1860 an estimated ten to fifteen thousand people showed up to skate in the park in one day. The commissioners and the tram and buggy companies hung out white flags with red dots when the ice was skatable. Even as Central Park's managers increased surveillance of park users' activities and sought to restrict what they could do in the park, they created a demand for more recreational activities that the private market filled. The park's skating rink was so popular that it spawned a cottage industry of private owners advertising their ponds as skating rinks. During the 1860s the trustees of St. John's Park consented to flooding the park to create an ice-skating rink. The public was charged ten cents to skate; sleighs lined the park to watch. Soon ice skating spread to Chicago's parks, where small lakes were flooded to make ice rinks.[161]

In May 1870 the Tweed ring, a corrupt group of politicians who had been in control of Tammany Hall and the municipal government since 1866, abolished the Board of Commissioners of Central Park (the state-run body that had overseen operations of the park since its inception) and replaced it with the Department of Public Parks. The new department was filled with Tammany Hall politicians. Though Olmsted and Vaux were named chief landscape architects and advisors to the Board, they had almost no contact with the park or board for more than a year after the new department was formed. During this time the administrators trimmed trees, cleared understory plants, smoothed rocky areas, planted flower gardens with annuals, and built restaurants and other structures in the park that Olmsted and Vaux thought ruined the atmosphere. The collapse of the Tweed ring in 1871 provided an opening for Olmsted to play a larger role in Central Park once again, and in January 1872 he and Vaux were reinstated as general superintendents of the park. Though the title was shared, Olmsted acted as the superintendent while Vaux assumed a minor role.[162]

Olmsted immediately set about reversing the changes the Parks Department had made. He tried to reorganize the park keepers (now filled with Tammany Hall appointees); because he was concerned with rising crime and the defacement of park structures, which he thought occurred during peak visiting hours, he scheduled constant patrols during those times.[163] He also wanted to extend police supervision of the park to nighttime hours, arguing that this was a common practice in Europe. Despite Olmsted's extensive use of the park police, people were still violating park rules. Olmsted argued that there was a strong connection between the erosion of behavioral standards and ineffective supervision. He went on to suggest that the change in the public's behavior was due to an indifferent Parks Department and an improperly trained police force. He suggested retraining the park police force.[164]

To this end Olmsted instituted a strict system whereby keepers walked at a brisk pace to complete the seven-mile circuit of the park. Each keeper did three rounds of the park. This new regime drew sharp criticism. The *New-York Daily Tribune* attacked what it saw as the "Olmsted Chain-Gang System" and argued that Olmsted had turned the keepers into "human velocipedes" who could hardly stop to suppress crime because they were preoccupied with completing their rounds. The paper claimed that while keepers struggled to complete their rounds in time, young "ruffians" stole flowers and stoned visitors and libertines roamed the park insulting unescorted women. Olmsted responded that his system had restored order to the park and that Civil War reports showed that soldiers and surgeons improved their health by marching twenty miles per day. However, the controversy cost Olmsted dearly; in 1873 he was stripped of much of his authority by the newly appointed park commissioners, who replaced Olmsted's system of rounds with beat patrols and authorized keepers to carry clubs. Olmsted tendered his resignation in September 1873 but later withdrew it. Though the relations between Olmsted and the commissioners ranged from frosty to fractious, he continued working at Central Park until 1878. The park was finally completed in 1876.[165]

The system of social control established by Olmsted had the quality of a panopticon.[166] Park visitors were placed under intense surveillance, and they knew they were being watched either by guards stationed at the gates of the park, by those on patrol, or by Olmsted himself. As Foucault would argue, the awareness of being watched was a powerful controlling influence over the park user. Some visitors responded by modifying their behavior such that they did not violate park rules or get caught in any transgressions.

Replicating Central Park

Central Park was a model for the development of municipal parks across the country. Its organization, policing policies, regulations, funding structure, and design were copied.[167] Once Central Park was established, Olmsted embarked on a crusade to influence other American cities to build parks. Evoking cultural nationalistic themes, he broadened the social construction of the urban park as an amenity that was good for national pride and crucial to the development of cities. In 1861 he wrote an article in which he laid out the rationale for expanding the public parks system in America. He argued that almost every large town in the "civilized world" had public parks but that in the

United States there was "scarcely a finished park or promenade ground deserving mention." He noted that "in the few small fields of rank hay grasses and spindle-trunked trees, to which the name is sometimes applied, the custom of promenade has never been established." True promenades existed only at the Capitol, the White House, and the still unfinished Central Park.[168]

In an editorial urging San Francisco to establish a public park he made favorable comparisons of Central Park to European parks and used that as a selling point for establishing a park in the city.[169] He also argued that Europeans, having visited Central Park, were moving to New York and becoming naturalized citizens because of the park. He credited Central Park with improving the lives of both poor and rich. A network of elites supported Olmsted's claims. Prominent business leaders with establishments adjacent to the park claimed that their property values had risen substantially because of the park.[170] The Sanitary Commission reported that women's and children's lives had improved because they were spending parts of the day in the park. Olmsted wrote, "There is no doubt that the park has added years to the lives of many of the most valued citizens and many have remarked that it has much increased their working capacity."[171] Olmsted's appeal to cities to build parks did not go unheard. Even before Central Park was completed residents in Brooklyn, San Francisco, Baltimore, Boston, Buffalo, Philadelphia, and Chicago began lobbying for and building parks of their own.[172]

The Olmsted-Vaux parks had ample plantings, promenades, and places for strolling and carriages. Except for cricket grounds the designers resisted placing facilities for active games and sports in the parks. However, there was room for the upper-class activities of horseback riding, carriage rides, sleighing, and ice skating. Still, Olmsted and Vaux's quest for rural and pastoral vistas continued to dominate their thinking, and they continued to urge that parks have a rural character and be used for a prescribed range of activities.[173] While they provided space in Washington Park for public gatherings, talks, and military exercises, parts of Prospect and Fairmount Parks were designed primarily for rural recreation.[174]

The Rising Demand for Active Recreation Spaces

As the public demand for active recreational spaces grew, Olmsted, Vaux, and other landscape architects began designing areas in parks for active recreation. Early on in the process Olmsted was in favor of allowing public school children to play ball and croquet in Central Park if they presented a certificate showing that they were in "good standing and regular attendance."[175] In

1865, when Olmsted toured the College of California at Berkeley's campus with a view to developing a park there, he suggested that a grassy space be reserved for "athletic games and agreeable exercises."[176] Two years later Olmsted and Vaux's plan for Fort Greene Park designated a section for ball fields. The following year Olmsted designed the Buffalo Park system; one park had a baseball lot. Recognizing the need for open space in poor communities, he wanted some of the parks in Buffalo's park system to be close to the East Side working-class neighborhoods.[177] Olmsted urged the construction of parkways to connect the parks in the city: "At no great distance from any point of the town, a pleasure ground will have been provided for, suitable for a short stroll, for a playground for children, and an airing ground for invalids, and a route of access to the large common park of the whole city."[178]

In 1870 Olmsted and Vaux accepted the futility of trying to keep people off the grass. In New York, for instance, newspapers decried the "iron rule" that made the grass "meadows only in name" or "panoramic beauties to be gazed upon but not enjoyed."[179] Even the *New York Times*, a staunch supporter of the park rules in the 1860s, began complaining about their rigidity. An 1871 article noted, "The exceeding care used to keep the grounds in order has sometimes had the effect of curtailing the enjoyment of visitors." The article continued, "A park . . . where one is not permitted to fling oneself at length on the soft green grass that offers itself so invitingly to the tired and heated wayfarer is somewhat disappointing." The article compared the jealously guarded lawns of Central Park to elegantly appointed New England parlors that are dusted and admired but sparingly used.[180] Another 1875 article lamented, "The rules about not walking upon the grass are now enforced so rigidly that sending children to the Park is rather a punishment for them than a treat."[181] Though Central Park administrators tried to prevent people from using the grass as late as 1901, people ignored them en masse, and on Sundays large numbers of people spilled over onto the grass.[182]

Long before then Olmsted and Vaux began experimenting with ways to relax restrictions on people's movements in parks but still maintain the turf in good condition for a reasonable price. In designing Walnut Hill Park in New Britain, Connecticut, they argued, "There is nothing which people desire more in a park than to walk upon the turf; there is no regulation so offensive or so difficult to enforce as one requiring them to keep off from it. When it is attempted even with an expensive police force, unless the walks are absolutely fenced in, encroachments upon the turf near them are seen to be made." Thus they proposed to "lay the larger part of [the] ground completely open to the public."[183] In planning Chicago's South Park in 1871 they noted that public grounds were being used for athletic sports such as baseball, cricket, football, and running games. They also noted that where large groups of people congregated it was impossible to prevent shrubs and plants from damage, thus they recommended leaving large expanses of turf open for people to use.[184]

Olmsted designed the Boston park system, called the "emerald necklace," from 1878 through the 1880s. After building the Back Bay fens, the Country Park, and Playstead sections of Franklin Park with no accommodations for active recreation, he included some of these elements in Charlesbank, a recreational ground in the West End tenement district. He also included active recreational sites for the remaining sections of Franklin Park. When he designed South Park in Buffalo in 1888 he argued that there was great demand for active recreational space in the park, so it was prudent to build facilities for such a purpose. Olmsted now argued that workers needed more than tranquilizing scenery; they needed access to active recreational opportunities. This shows the extent to which his thinking evolved in three decades of park building.[185] In general, by 1895 more active and unstructured recreation was common in parks. Cranz found that racing, jumping, polo, bicycling, tobogganing, ice skating, rowing, circuses, merry-go-rounds, shooting matches, tennis, croquet, baseball, and lacrosse were allowed in parks across the country.[186]

With their design of Central Park and several other major parks and public grounds in over twenty different cities across the country, Olmsted and Vaux successfully propagated the park as a work of art, a tool of social control, a mechanism to improve health and well-being, and an apparatus to produce better workers. They laid the groundwork for the later park and playground movement and set the tone for the aggressive control of park users' behavior by using police, laws, supervision, limited operating hours, layout and design, the inclusion or exclusion of activities, and the tracking of visitors. As later chapters show, the levels of confrontation escalated as the working class asserted their will regarding access to open space and desired park behavior. Still, Olmsted and Vaux had a profound impact on the social construction of parks and the way people came to understand them. They used their social location as elite white males entrusted with enormous power and discretion to implement their moral, cultural, and social agenda. They were a part of the cadre of ethical elites Ross believed emerged to guide cities to higher levels of civility.[187]

Olmsted and Vaux dominated park building from the 1850s to the 1880s, designing many large, expensive public and private projects. Though the design competition for Central Park called for it to be built for $1.5 million, by 1863 $8 million had been spent; Central Park eventually cost $14 million ($235 million in 2005 dollars).[188]

nine

SOCIAL CLASS, ACTIVISM, AND PARK USE

The Rise of the Recreation Movement

There was a tremendous need for parks as a source of employment and recreation, and the continuing influx of immigrants intensified that need. Between 1860 and 1920 28.5 million immigrants came to the United States.[1] Most cities had difficulty funding the elaborate Olmsted-Vaux parks; for example, in 1867 Newark balked at paying more than a million dollars to build a city park.[2] It was also becoming clear that in addition to grand, landscaped parks there was a need for smaller neighborhood parks designed for active recreational use, which the working class had advocated in the 1850s. Consequently a new breed of park advocates and designers emerged in the 1880s to respond to this need. While the landscape architects and early park advocates made general claims about the therapeutic nature of parks, they did not make any direct interventions to change the actual health status of urban residents; once the parks were built the health improvements heralded by the builders were left to chance. But as living conditions in the cities remained grim, middle-class reformers sought to more deliberately link the goals of the sanitary reform movement with those of the parks and recreation movement. They believed that, in addition to visiting parks, other steps had to be taken to improve the health of park users.

These reformers changed the social construction of urban parks to reflect a new focus on building small neighborhood parks and playgrounds. This entailed a shift in control of the park discourse and agenda from wealthy businessmen and elite landscape architects to women and working-class people. Instead of building parks that primarily catered to the needs of native-born, upper- and middle-class adults, the reformers focused on poor, immigrant children. They believed children should be instructed as a part of the park or playground experience. Drawing on human development theories, reformers argued, "The boy without a playground [will be] the father to the man without a job."[3] These smaller parks were not intended for upper- and middle-class use; instead, they were part of a package of charitable aid directed at working-class clients. The socialization role of the parks was not left to chance either. These newer parks were not built with an eye toward the picturesque or for the purpose of tranquilizing recreation, but for active recreation. Progressive Era reform-oriented activists formed the core of the recreation movement. Their charitable work also represented a shift from earlier efforts that focused primarily on fixing the character flaws of the working class; along with changing people's character, they believed that changing environmental conditions was critical to improving people's lives. Hence they actively engaged in restructuring the urban environment.

Recreation movement activists augmented the work of the landscape architects by encouraging cities and towns to build both large and small parks and playgrounds. They were also responding to epidemics, the lack of public open space, the sight of children being arrested for playing in the streets, massive immigration, and severe overcrowding in working-class neighborhoods. The quest for recreation space became intimately connected to campaigns to improve public health, environmental conditions, and the assimilation of new immigrants.[4]

The reformers who got involved in the playground and parks movement were primarily Protestant civic and religious leaders intent on restoring social order. Many of them were club women and settlement house activists who were concerned with what children and adults did in their spare time. A 1905 survey of settlement workers found that 88 percent were active church members and nearly all said that religion played a dominant role in their lives. More than half of the settlement workers were Congregationalists, Presbyterians, Episcopalians, and Unitarians. The study found that 71 percent of other social reformers were also active church members.[5] Though reformers were religious and believed in improving morality and order among the poor, they differed from the early-nineteenth-century evangelical reformers in that they did not believe that religious conversion in and of itself could cure the social ills prevalent in society. They believed that government and environmental interventions were essential, so they concentrated on those arenas of social change.

These reformers not only marked a change in the ideology of building urban parks, but they represented a significant gender shift. While the landscape architects were males working on grand designs which were supported by large outlays of public funds, the Progressive Era reformers were dominated by women undertaking smaller, underfunded projects. Their activities sought to link the home and work environments with civic institutions and the playgrounds.

Some of the Progressive Era reformers viewed many of the leisure activities of the working class as unhealthy and immoral. As Mann argues, "The history of social reform is the history of the perception of social evil and the will and the way to remove it." Reformers wanted "healthy recreation" to replace activities that reduced the efficiency and readiness of factory workers. They subscribed to the Downing-Olmsted-Vaux doctrine of the virtues of parks as social control agents with a health-giving character. They focused on the idea that healthy workers made better workers and that parks were the key to moral and physical health and cleanliness. Joseph Lee, a leading reformer, clearly articulated this perspective: "The battle with the slum is primarily a battle against the obvious evils of drink, overcrowding, immorality, and bad sanitary arrangements."[6] In effect, these social reformers engineered what Cranz calls the reform park and the reform park movement.[7]

Many of the businessmen and newspapers that supported larger parks opposed smaller parks geared toward immigrants and working-class use.[8] Most cities struggling to finance one centerpiece park did not see the value in small neighborhood parks and consequently did not fund them. Most cities hadn't gotten to the point of developing park systems yet. The social location of the female reformers played an important role in their social construction of urban parks and the way they developed their programs. Being middle- and upper-class women who managed their homes and social clubs, they brought the concerns of the family, youth education, and mainstream socialization to the programs they developed. Because they did not have access to the funds the landscape architects had to develop huge parks, the female reformers relied on their local power base to help establish their projects: community organizations, churches, and schools.

As was true for the Children's Aid Society, children were the favorite targets of the playground reformers because they were deemed more controllable and because reformers believed that if children were converted in their youth, there would be no need to reform them in adulthood. The Massachusetts Emergency and Hygiene Association (MEHA), one group involved early on in developing playgrounds, is typical of organizations these activists formed. Spearheaded by Dr. Marie Zakrzewska, who had seen a "sand garden" in Berlin,

MEHA placed the first pile of sand in the yard of the Parmenter Street Chapel, a mission in Boston's North End, in 1885. The nation's first playground was located in a tumbledown neighborhood frequented by gamblers, pimps, prostitutes, sailors, paupers, and the homeless. The converted single-family homes were inhabited by several families sharing bathroom facilities. The alleyways were strewn with garbage, disease ran rampant, and infant mortality was high. The MEHA volunteers, all of whom belonged to the New England Women's Club, used the structured environment of the sand garden to teach children morality, manners, hygiene, and social skills. They also provided instruction to adults to help them stay healthy.[9] In this line of social work many groups did not seek input and advice from or attempt to collaborate with the people they were helping. Instead they approached the establishment and management of sand gardens as a paternalistic act of charity for which the recipients were or ought to be grateful.[10] This led Thorstein Veblen to argue that, despite their best intentions, reformers functioned "to enhance the industrial efficiency of the poor" and to teach them, "by precept and example," upper-class manners and customs.[11] The following excerpt from a MEHA annual report illustrates this attitude:

> [The sand garden was] opened three mornings in the week for three months to the children of the neighborhood between four and ten years of age. About twenty-five little ones averaging an attendance of fourteen shared the privilege thus offered, under the care of a lady to exercise over them wholesome authority while at their play, and to teach them some useful lessons and morals and manners. . . . No doubt much permanent good is done to the little ones thus gathered in from the street and placed for a few hours each alternative day of the week under influences calculated to develop and strengthen the better qualities and forces resident within them.[12]

The following year Bostonians set aside a portion of the Charles River embankment as a children's play area. The city also dumped piles of sand at various sites in working-class neighborhoods.[13] According to MEHA records, the children responded to the playgrounds with a mixture of delight and cynicism. For example, on the first day the Eliot School yard opened, "it required Mrs. Tobey, the superintendent, Miss Morley the matron, three janitors, and a policeman to entertain and subdue the excitement of the one hundred children who rushed in wild with delight at the novelty offered them."[14] However, some children resisted the notion that they should be grateful to MEHA. A MEHA document reports the following exchange at the Baldwin School playground: "The children . . . were of the most untamable material, and refused to be in the least appreciative when the matron spoke of the kindness of the women who had arranged the playgrounds. 'Pooh! They are paid for it,' one boy remarked. 'Oh no, they do it to give you a good time.'—'Well, they are fools, then,' was his comment."[15]

As was common for many social service organizations, MEHA lacked the resources to fully implement projects; consequently they sought government sponsorship. In 1888 they abandoned their practice of placing sand gardens and rudimentary playgrounds in missions and convinced Boston city officials to place them in public school yards. A similar pattern occurred in New York, where by 1910 playgrounds were being built and operated by the city.

The children and adolescents resisted the reformers' preaching and morality lectures. Moreover the reformers were not successful in breaking up street gangs and affinity groups. In fact by providing sand gardens, playgrounds, missions, and settlement houses, they inadvertently provided the meeting places where youths could consolidate their identities, making it easier for them to mark off their gang turf and stake claims to certain territories. The children quickly figured out that the social workers needed them as much as or even more than they needed the social workers. They deliberately broke equipment and furniture and watched in amusement as flustered social workers threatened to throw them out of the settlement houses, only to turn around and provide new equipment in short order. Without children and adolescents in need of charity and moral uplift the mission of these social workers would be nonexistent.[16] Table 6 contains a brief chronology of the playground and parks movement.

Working-Class Park Use and Competing Definitions of Leisure Behavior

Though the working class labored long hours for little pay to build public parks, they and their families were not free to use these parks as they pleased. Tensions between the rich and poor grew so high in Central Park that German immigrant children threw stones at wealthy families as they paraded around the park in their carriages. Despite these acts of protest, aristocrats such as Caroline Astor continued to take daily carriage rides in the park as late as 1906.[17]

By the 1880s the working class was primed and ready to be more assertive about their recreational needs. Henry George, the author of *Progress and Poverty*, helped to galvanize working-class support for increased leisure time. George wrote, "Give all the classes leisure, and comfort, and independence, the decencies of life, [and] the opportunities of mental and moral development." He ran for mayor of New York in 1886 against a Democrat, Abram S. Hewitt, and the Republican candidate, Theodore Roosevelt. Campaigning in working-class neighborhoods, he argued, "The children of the rich can go up to Central Park, or out of the country in the summer time; but the children of the poor, for them there is no playground in the city but the streets."[18]

The poor also lacked access to parks in Worcester, Massachusetts, in the late 1800s. The majority of the city's working class consisted of poor immigrant families, while civic and business institutions were controlled by the native-born upper class. The city's population was growing rapidly; in 1790 it was only 2,095, but by 1890 the population had grown to 85,000 and by 1900 it was 120,000. Civic leaders tried to keep pace with the provision of social services; they began setting aside open space early on. The first open space, the twenty-acre Worcester Common (or City Hall Common), was designated in June 1669 as "common open space." However, the Common served few of the functions usually associated with contemporary parks. The first structure built on it was the meeting house, which served religious and civic purposes. The Common was also the site of the city's main cemetery from 1730 to 1795; it was ploughed over in 1854. An animal pound, gun house, hearse house, firehouse, and cattle sheds were built on the Common, and it also served as a military training ground. In 1840 railroad tracks (removed in 1877) were built across it.[19] Worcester used public funds to acquire land for its second open space in 1854. The twenty-eight-acre Elm Park was used as the city dump prior to being purchased for parkland.

Though a Commission of Public Grounds was established in 1863, both open spaces remained derelict. The Commission requested money twice to drain the low-lying areas of Elm Park, but was ignored by the city. The Commission was forced to raise funds by selling hay and apples collected from the land, renting it out to farmers and the agricultural society, and holding circuses on it. Things began to change when Edward Winslow Lincoln, whose father was one of the parties who sold the land for Elm Park, became the city's park commissioner.[20]

In 1870, the year Lincoln assumed the post, the Common, whittled away over the years, was about eight acres in size. Lincoln neglected it because he focused on developing Elm Park, located on the West Side, close to the homes of the city's wealthy residents. Lincoln himself had extensive landholdings close to the park. Like wealthy New Yorkers before them, Worcester's elites wanted to use an elegant landscaped park to transform their neighborhood into an exclusive residential enclave. At the time Lincoln was aware of Olmsted's writings and work on Central Park, and though he was strongly influenced by

Table 6 Chronology of the Playground and Parks Movement

Year	Event
1811	Union Park, Manhattan, is planned
1812	Philadelphia acquires the first section of Fairmount Park
1815	Improvements begin on Washington Square Park, Philadelphia
1820–40	Schools and colleges start providing outdoor gyms and sports areas
1823	Battery Park, Manhattan, is expanded
1835–50	Tompkins Square, Manhattan, is landscaped
1844–1868	Philadelphia acquires Lemon Hill; Sedgely purchased in 1857 and donated to city as parkland; land on the west bank of the Schuylkill acquired in 1868 for park
1847	Fort Greene (Washington) Park, Brooklyn, is designated
1853	Central Park authorized
	Work begins on the Hartford, Connecticut, park (renamed for Howard Bushnell in 1876)
1854	Worcester, Massachusetts, acquires land to build new town commons (Elm Park)
1857	Frederick Law Olmsted begins working on Central Park
1858	Olmsted and Calvert Vaux's design for Central Park approved
1859	Boston secures land for Public Garden
	Missouri Botanical Gardens and Tower Grove Park is opened to the public
1860	Baltimore acquires Druid Hill and opens it as a public park
1861	Chicago's first park is laid out in the southern section of Lincoln Park
1863	Park just outside New Haven, Connecticut, is planned
1865	Prospect Park, Brooklyn, designed and built by Olmsted and Vaux
1871	Brookline, Massachusetts, authorizes acquisition of land for city parks
1882	Survey of play opportunities for children is conducted in Boston
1885	First sand garden opens at Parmenter Street Chapel, Boston
1887	Golden Gate Park is completed
	New York City appropriates funds to build small parks
1889	Land is acquired for Edgewood Park, New Haven, Connecticut
1890	Society for Parks and Playgrounds is formed in New York
	Kansas City hires George Kessler to plan the park system
1892	A model playground opens at Hull House in Chicago
	Seward Park, Manhattan, opens
	Greater Metropolitan Park Commission builds the Greater Boston Park System
1898	Outdoor Recreation League is formed in New York
1900	Chicago passes a resolution that all new schools be built with playgrounds
1905	Ten South Park centers open in Chicago
1906	Playground Association of America is organized

Table 6 Continued

Year	Event
1908	Massachusetts playground act is passed
1924	Conference on Outdoor Recreation, is called by President Calvin Coolidge
1930	National Recreation Association is formed
1932	First National Recreation Congress is held
1933	Expansion of recreation facilities and services through national work programs begins

Sources: "Public Parks of Leading Cities," part 2, *New York Times*, September 1, 1895, 28; John Kelly, *Leisure* (Englewood Cliffs, N.J.: Prentice Hall, 1996), 154–60; Charles E. Beveridge and David Schuyler, *The Papers of Frederick Law Olmstead*, Vol. 3, *Creating Central Park, 1857–1961* (Baltimore: Johns Hopkins University Press, 1983), 66, 362–63; Victoria Post Ranney, Gerard J. Rauluk, and Carolyn F. Hoffman, *The Papers of Frederick Law Olmsted*, Vol. 5: *The California Frontier: 1863–1865* (Baltimore: Johns Hopkins University Press, 1990), 431–32; David Schuyler and Jane Turner Censer, *The Papers of Frederick Law Olmsted*, Vol. 6: *The Years of Olmsted, Vaux & Company* (Baltimore: Johns Hopkins University Press, 1992), 24–25, 242; Allen F. Davis, *Spearheads for Reform: The Social Settlements and the Progressive Movement: 1890–1914* (New York: Oxford University Press, 1967), 63; Roy Rosenzweig, *Eight Hours for What We Will: Workers and Leisure in an Industrial City, 1870–1920* (Cambridge: Cambridge University Press, 1983), 144–45.

Olmsted he deviated from some of his main tenets. Lincoln developed Elm Park with both pastoral and gardenesque elements; the pastoral elements were consistent with Olmsted's designs, the gardenesque elements were not. Once the rubble was cleared from the land, three ponds with placid water were constructed with decorative bridges spanning them. Consistent with the theme of blending native and exotic elements, the ponds were stocked with Peking ducks, Toulese geese, great blue herons, swans, and other waterfowl. An amalgam of plantings characterized the flora of the park, and a tree nursery was established. Elm Park was adorned with specimens from Charles S. Sargeant of the Arnold Arboretum and other horticulturalists. Lincoln, himself a horticulturalist, propagated plants and bulbs in his nursery that were transplanted in the park. The formal gardens were planted with peonies, azaleas, tulips, and more. Boats were sometimes available for residents to paddle across the ponds. Lincoln installed jets to provide a smooth ice-skating surface in the winter; he even shoveled the snow himself.[21] Active and commercialized leisure pursuits, such as circuses and carnivals, were banned in Elm Park.

In 1876 Lincoln petitioned the police to patrol the park more diligently, arguing, "This [Park] Commission will exact and enforce . . . decent behavior from all who frequent the Public Grounds, which is not only seemly . . . but is rightfully expected by the community." "Keep off the Grass" signs were posted, and though baseball was allowed its days were numbered; park advocates wanted to move baseball to the working-class and immigrant neighborhoods of Bell Hill, Greendale, and Quinsigamond Village. As Lincoln saw it, "The game of baseball as now played is perilous at best, scarcely supplying the redeeming merit of a dreary amusement of the spectators. It is believed that the city might purchase an acre or two in different sections [of the city] . . . for the express purpose [of providing ballfields]."[22] Although Olmsted and Vaux, the leading proponents of passive park recreation, were already experimenting with active recreational designs and greater use of grassy areas at the time Worcester was developing its parks, the park administrators in Worcester continued to prohibit use of the grass and discourage active recreation in its centerpiece park.[23]

The tensions around leisure space in Worcester brought to the fore two distinct and conflicting definitions and perceptions of the park. While the upper class emphasized passive and refined leisure pursuits and cultural improvement, the working class sought active recreation, fun, and games.[24] Despite the fact that park planners in Worcester did not fully understand or sympathize with the reasons the working class related differently to parks than the middle class did, working-class park behavior was not difficult to understand. According to the compensatory theory of leisure, work is the dominant force in a person's life and leisure compensates for the rigors, monotony, and brutality of the job. Excessive drinking, exuberant park play, demonstrations of power, and loud, rowdy behavior are preferred because they represent the opposite of the routinized danger and boredom of the job.[25] The spillover theory of leisure argues that the alienation the worker experiences in the workplace extends to all aspects of his or her life. The boredom, mental stupor, and fatigue that characterizes work also characterizes leisure. Therefore workers engage in activities that numb the senses and blur their judgment.[26] The spillover theory may explain the motivation of park users who lie prostrate on the grass in a drunken stupor or semiconscious daze.

These differing perspectives on alcohol use, public drunkenness, and rowdy behavior influenced class perceptions of appropriate park behavior. When the working class used parks in middle-class communities, the middle class aggressively sought to teach the working class the "proper" rules of decorum, grace, and charm. Correct park behavior was defined as quiet, orderly, and inoffensive; inappropriate behavior was immoral, rude, loud, disorderly, and obscene. Activities and behaviors such as drinking that the middle and upper classes indulged in privately or in exclusive social clubs were disdained or banned in public. If the wealthy drank in the park, they did so at legal concessions such as the ones established in Central Park. Those concession prices were too high for the working-class people.[27] The middle class was not willing to make allowances for the working class by recognizing the fact that the latter did not have the space or facilities to privatize certain kinds of leisure activities. Consequently park superintendents such as Olmsted, despite his disdain for the working class, felt compelled to defend keeping the parks open to poor and rich alike. Upper- and middle-class people who opposed Olmsted wrote scathing editorials condemning the practice. One such editorial argued that it would be difficult for Central Park to be like the great parks of England or France because of the blurring of class lines and the fact that middle-class values were in danger of being overwhelmed in the park:

> Here [in America], we order things differently. Here, we have no "lower orders," nobody has any "superiors," we know no "nobility and gentry": nothing but a public which is all and everything, and in which Sam the Five Pointer is as good a man as William B. Astor or Edward Everett. Further, whatever is done by or for the public aforesaid, is done by or for Sam as much as any one else, and he will have his full share of it. Therefore, when we open a public park, Sam will air himself in it. He will take his friends, whether from Church street or elsewhere. He will enjoy himself there, whether by having a muss, or a drink at the corner groggery opposite the great gate. He will run races with his new horse in the carriage way. He will knock any better dressed man down who remonstrates with him. He will talk and sing, and fill his share of the bench, and flirt with the nursery girls in his own coarse way. Now we ask what chance have William B. Astor and Edward Everett against this fellow-citizen of theirs? Can they and he enjoy the same place? Is it not obvious that he will turn them out, and that the great Central Park, which has cost so much money and is to cost so much more, will be nothing but a huge beer garden for the lowest denizens of the city—of which we shall yet pray litanies to be delivered?[28]

Olmsted responded by attacking what he saw as the "fallacy of cowardly conservatism," the belief that any cultural, educational, or recreational site that was open to the general public would become uncomfortable for the upper and middle classes because of the uncouth behavior of the lower classes. Olmsted wrote, "There has been much careless puffing of the park and so much ignorant and mistaken fault-finding. . . . If you determine upon it, I shall of course be glad to furnish the fullest information, if any should be wanted, both with regard to the design, & work, as well as the *working* of the park with the people—the phenomena of which already should explode much, somewhat popular, fallacy of cowardly conservatism."[29]

Having no counterpart to the backyard, the mansion, the country estate, or the social club in which to privatize their fun, the working class went to the park, where they were deemed disorderly and out of control. This is not to say that working-class people did not want to use the parks for strolling, lounging, and resting; indeed they did. Peiss found that strolling and relaxing in the parks were two of the most popular leisure activities among New York's working class.[30] The same was true in Worcester, where relaxation was coupled with other kinds of active, organized, and spontaneous sports and games. Working-class park use was characterized by mixed-age, mixed-gender, single and married groups of people recreating together. The following excerpt about Worcester's parks gives an indication of the complexity of the recreational experience:

> Before the 12:05 whistle blows, the crowd begins to arrive from Washburn and Moen's, the envelope shops, electric light station, and many other establishments north of Lincoln Square. After eating, a good romp is indulged in by the girls, running and racing about, with now and then a scream of laughter when some mishap, a fall perhaps, occurs to one of their number. Some of them wander about in pairs or groups, exchanging girlish confidences, or indulging in good-natured banter with their masculine shop-mates. Occasionally a boat is secured by some gallant youth, who rows a load of laughing maidens about the pond, the envied of their less fortunate friends. The younger men try a game of base ball or a little general sport, jumping, running, etc., while their elders sit about in the more shaded spots, smoking their pipes. But when the whistles blow previous to 1 o'clock there is a general stampede to the shops and in a few minutes all of those remaining can be counted on one's fingers.[31]

As this excerpt demonstrates, the parks served a large range of social functions for the working class: exercise, games and sports, social gatherings, courting, and resting.

Working-Class Attempts to Gain Access to Parks

FREE SEATS VERSUS HIRED CHAIRS

As more working-class people began to use Central Park in the 1870s elites complained that well-heeled park users felt uncomfortable using the seats the poor used. Influenced by a European tradition of charging a fee to use park seats, *Appleton's Journal* called for the removal of the free seats from Central Park and the substitution of chairs that could be rented for a small fee:

> The [free] seats draw to the parks nearly every idle and dirty vagabond of the town. Blear-eyed and bloated topers, ragged and vicious tramps, soiled and untouchable wretches of all kinds . . . stretch themselves upon the ever-ready seats. . . . A slightly-better class—that is, a class just above begging and vagabondage—go there to smoke their rank pipes [and] to eject their filthy tobacco-juice right and left over the promenade, and to . . . render the places noisome and offensive. . . . Now the remedy for this evil is to remove the free seats, and substitute therefore chairs at a small charge. . . . If free seats mean lounging places for all the worthless wretches of the city, the parks have lost one essential democratic feature—they have ceased to be places of resort for the whole people, inasmuch as the reputable class are practically excluded therefrom. . . . The parks are designed for [the] respectable mass of people . . . and not for vagabonds—a class who have no rights that anybody is called upon to consider or respect. . . . It would be right . . . to order the exclusion of every man who comes in rags or dirt, who makes a pool of tobacco-juice upon the pavement, who salutes the nostrils of unoffending citizens with the horrible aroma of a filthy pipe, or any other way makes himself an object of abhorrence to decent folk. . . .
>
> [In European parks] the seats are usually chairs, which are furnished . . . at a nominal price—a penny in England. . . . This price, as it is, serves to exclude vagabonds, and acts as a sort of natural selection in the class of people it brings to the parks.[32]

When elites felt they were being squeezed out of the park they harkened back to language framed in democratic ideology to point out their perceived exclusion and to advocate the maintenance of parks where elites felt welcome and where their tastes and feelings were paramount. Unlike Olmsted, who used democratic arguments to justify opening the parks to the poor as well as the rich, some nineteenth-century elites believed public spaces should not be open to the poorest and most downtrodden citizens of the city.

Chair rental became a reality in New York's parks on April 1, 1901, when George C. Clausen, president of the Park Commission, granted a license to Oscar Spate to rent out green, cane-bottomed rocking chairs for a nickel ($1.11 in 2005 dollars); plain chairs were rented for three cents. Anyone renting a chair was issued a ticket that was good all day. There was no public notification or discussion of the bidding process. Spate, of the Comfort Chair Company, agreed to pay $300 ($6,600 in 2005 dollars) annually for the five-year contract. The chairs began appearing in city parks in late June. Men in gray uniforms patrolled the areas where the chairs were placed and demanded that anyone sitting in them pay or get up. For years New Yorkers had been requesting more seats in the parks, but most of the public did not envision that they would have to pay for them. The Park Commission president noted that 4,612 free seats were already in the city's parks, so the addition of rental chairs was not problematic. Thirty-seven rental chairs were placed in Madison Square, 220 in Central Park, and several hundred more were back-ordered.[33]

Not all city officials supported the plan. Randolph Guggenheimer, a real estate lawyer, school commissioner, and council member, called it "ridiculous." He argued, "Parks belong to the people, and should be free to all. If there are not sufficient seats the city should provide more." He continued, "There is no propriety in providing elegant seats for those who can pay for them, and allow those who cannot pay to put up with poorer seats or no seats at all."[34] Spate defended the plan by arguing that filthy, lazy park users who had monopolized the free seats now had all the room they needed to lounge on the park benches. Park users who found the benches uncomfortable could rent chairs and sit wherever they pleased.[35]

New Yorkers found the idea of renting chairs contemptible and vented their anger on the guards and chair vendors. The guards, who were paid one dollar a day ($22 in 2005 dollars), quickly grew tired of the situation. Though the guards were very reluctant to chase old women and children from the chairs, they asked others to leave if they did not pay the rental fee.[36] On July 1, not long after the chairs were placed in the parks, eighteen-year-old Abraham Cohen was arrested in Madison Square for refusing to pay to sit on a chair. The free benches had been moved from the shadiest spot and placed in the sun, and the rental chairs had been placed in the shade instead. Cohen, a student at City College, not knowing that there was a charge for the new chairs, sat in the shade in his usual spot. When he refused to pay, a policeman was summoned. Cohen argued with the officer. "I've got a right to sit here," he said. The policeman replied, "You've got a right to sit on a bench." A crowd gathered to watch the commotion. Martin Helm, one of the onlookers, began to laugh loudly at the police officer; Helm too was arrested. Both were set free the following day.[37]

A survey of the placement of the rented chairs in the parks and along Riverside and East Drives found that in most instances the public benches had been moved to sunny spots and the rented chairs were placed in the shade. The issue of the placement of the chairs was brought to the attention of the police commissioner Michael Murphy, and on July 3 he ordered the police not to arrest people refusing to pay for the chairs unless they were being disorderly. On the following day people attending the Fourth of July celebrations sat on the chairs without paying.[38]

Not everyone objected to the chairs. While some abhorred the drawing of "social lines" with the chairs, a few park visitors paid the rental fee. As one Central Park user said, "I hope to see this thing [chair rental] a success. It is largely in the nature of an experiment. . . . It is an accommodation to that part of the park-loving public that wants a private chair that will be cared for by an attendant."[39] Another park user wrote, "The class that desires more comfort and is amply willing and able to pay for it had heretofore been debarred from the privileges of the parks by reason of the lack of facilities that would meet their views."[40] A letter to the editor declared:

> I cannot understand why people have become so hysterical because a few chairs have been placed in our parks. Are the parks intended for the poor and the idlers only? And should not those who are willing to pay for a seat be allowed the privilege? . . . It should be a great boon to many who wish to spend an hour or two in the park to be able to procure for 5 cents a clean and comfortable chair which will be placed where they will not be compelled to inhale the smoke of ill-smelling pipes, nor nauseated by the expectoration of the "gentlemen" who monopolize the benches. New York is the only place I know of where a decent woman cannot sit down in any of the parks to rest or enjoy the fresh air without running the risk of being shortly joined by one or more rough and dirty loafers.[41]

J. Arthur Holly, a park user, also expressed his support for renting chairs:

> I have used the armchairs ever since they have been placed in the parks, and do not believe a single bench has been moved [to make way for them]. . . . The parks have by this means [chair rental] been made available to hundreds of self-respecting and by no means aristocratic people, who were unwilling to subject themselves to the ordeal of sitting next to unclean, often drunken and generally foul-mouthed loafers. While these unfortunates are in great need and have their rights, have the self-respecting portions of the community none? Certainly 5 cents is not a prohibitive charge, but it is enough to keep the "bums" away and that is all that it is intended to do, and so far as it serves the wants of hundreds

of people . . . I cannot see how any argument can be made against this innocent means of recreation.[42]

An editorial noted, "Our parks are primarily intended for all self-respecting citizens and strangers, poor and rich alike, who may wish to rest awhile amid the trees and flowers, but they are not primarily or exclusively intended for the wholly objectionable characters—loafers and young hoodlums—who infest them at present."[43]

Others were more ambivalent. A commentary in the *New York Times* agreed that the chairs should be removed because they were displacing the benches from the best spots in the park, people were not paying to use them, and those arrested for refusing to pay were being discharged by the magistrate anyway. However, the paper pointed out, New Yorkers should be aware that chairs were being rented in parks in Europe without inciting public hostility.[44] Tecumseh Swift, a reader of the *New York Times*, expressed similar ambivalence. He argued that there should be chairs for hire in parks such as City Hall Park where there weren't enough seats available for women and children to sit. But then he added that because the rental chairs were relegating the benches to the unpleasant parts of the parks, the rental chairs should be removed.[45]

Despite the police commissioner's order not to arrest park users, on July 8 in Madison Square Park there was a confrontation between the chair guards and a small boy to whom a man had given his chair ticket (the tickets were nontransferable). After chasing the boy around the chairs, an exasperated guard allegedly struck the boy in the face. As a crowd gathered in the park the police were called in.[46] There was heavy police presence in the park the following day, as well as more confrontations. Boys overturned chairs and tossed wads of wet paper at the uniformed chair guards. In the afternoon one of the chair guards punched a man who sat on a chair and refused to pay. A riot broke out and the police made arrests, but the crowd swelled to more than twelve hundred screaming, jeering people who followed the police to the station. As the confrontations between the chair guards and the public escalated, Spate implored the police to arrest park users who refused to pay to sit in the rental chairs.[47]

Around the time of the disturbances in Madison Square, the park commissioner George Clausen proposed to put five thousand additional free chairs, estimated to cost $30,000 ($664,000 in 2005 dollars) in the parks. Half of the chairs were slated for Central Park, one thousand would be placed in Riverside, and the rest were to be distributed to other parks. The chairs were expected early in 1902.[48]

The chair-rental experiment was short lived. On July 10 Commissioner Clausen ordered the removal of the rental chairs from Manhattan's parks. Ten thousand people gathered to celebrate the announcement. Rally organizers had applied for a permit to hold the celebration in Madison Square but were denied. When they appealed to Police Commissioner Murphy, he said he had no authority to grant permits in Madison Square, but he did have jurisdiction over the streets around the park. Organizers quickly applied for and got a permit to hold the celebration in the nearby streets. The *New York Morning Journal* provided a band and fireworks and helped book speakers for the festivities. Oscar Spate objected to Clausen's decision; he argued that he had already spent about $21,000 ($465,000 in 2005 dollars) on the chair-rental venture and stood to lose that money.[49]

WORKING-CLASS CAMPAIGNS FOR NEIGHBORHOOD PARKS

By the 1880s the working class began to take a more proactive role in agitating for neighborhood parks. They also became more vocal about the design and use of that space. Working-class advocates in New York pointed out the importance of having parks where people could have access to them. One unnamed activist wrote of these concerns to the park commissioners in 1871:

> To that large class of our population whom necessity does not permit to leave the City, even for a day, the verdant lands and cool, dense shrubberies of the Central Park offer a most grateful relief. There the humbler citizen may catch for a brief moment at morning or evening something of the delight and refreshment which others go to seek in the woods and mountains, or at fashionable baths or seaside watering-places. Indeed, with the help of a lively fancy and glass of mineral water . . . one may enjoy a miniature Saratoga at the cost of a few cents. So, too, a stroll by the borders of the pretty lake, or a row on its placid waters, may convey to a contented and imaginative spirit some notion of the pleasures of Lake Mahopac and kindred resorts.[50]

The case of Worcester is instructive in exploring the tensions that arose over the use of open space. In the 1880s in Worcester the two contrasting visions of park design and usage clashed. While the industrialists urged the city to acquire and develop parks on the West Side for the purposes of fire protection, health, civic pride, and social control and to stabilize real estate values, working-class residents demanded leisure space for more active, play-oriented activities. The need for recreation space was so great in the burgeoning working-class communities that children played in the streets. By 1880, five years before MEHA started the first sand garden and ten years after the wealthy West Siders got their landscaped Elm Park, East Siders began to campaign for new parks and better maintenance of the Common. An 1882 petition drive netted 140 signatures (74 percent from blue-collar workers) requesting a "few acres of land" for "the less favored children." Petitioners declared, "There is no public ground in that vicinity [the East Side] where children or young men can resort, either for health or amusement." That same year children were arrested for playing in the streets and on vacant lots.[51]

Editorials in the *Worcester Daily Times* accused the city of siding with the West Siders and neglecting the East Siders. The editorials noted that open sewers ran through the neighborhoods and that cholera and diphtheria were common in the cramped East Side. As the campaign progressed the East Siders started demanding "every inch of space" the city "could afford them."[52] The lack of access to outdoor space spawned a movement of Irish working-class activists who demanded public playgrounds. They embarked on a letter-writing campaign in the *Worcester Evening Star*, where they complained about Elm Park (pejoratively renamed "Lincoln's Patch"). They charged that they had to stand when they visited the Common because the "people's seats" had been removed to Elm Park, which they described as a "desolate spot where nobody will use them excepting the crows." Letters satirized Lincoln as "the Earle of the frog ponds" and the "grandiloquent Earle of model pools."[53]

Exasperated by the stalling tactics of the city government, in 1884 the East Side representatives on the City Council used a tactic pioneered by Battery Park supporters in New York. Working-class representatives withheld their votes for park improvements in upper-class neighborhoods until parks were approved in working-class communities. Thus East Side representatives opposed the acquisition of Newton Hill, a sixty-acre parcel adjacent to Elm Park.[54] Letters to the *Worcester Sunday Telegram* summed up people's frustrations. Simmering resentment over environmental inequalities and the way the city's actions exacerbated conditions led to letters like the following, which linked park access to environmental health and social and economic inequality:

> Our wealthy citizens live in elegant homes on all the hills of Worcester, they have unrestricted fresh air and perfect sewage [disposal], their streets are well cleaned and lighted, the sidewalks are everywhere, and Elm Park, that little dream of beauty, is conveniently near. The toilers live on the lowlands, their houses are close together, the hills restrict the fresh air, huge chimneys pour out volumes of smoke, the marshy places

give out offensiveness and poison the air, the canal remains uncovered, the streets are different, the little ones are many. While the families of the rich can go to the mountains or to the sea during the hot months of the summer, the families of the workers must remain at home.[55]

When the author of the letter referred to wealthy residents living on the hills of Worcester, he might have been thinking of aristocrats such as Andrew Haskell Green, who moved to New York in 1857 to help oversee the development of Central Park. Green inherited 287 acres of Green Hill in 1848 from his father. The Greens had been buying up portions of the hill since 1754, when they made an original purchase of 180 acres. Andrew Green continued the family tradition and quickly increased the acreage of the property to over 300 acres by the 1850s. Green was a gentleman farmer. He turned the property into a country estate with beautifully landscaped grounds. In 1850 he converted the original family homestead into a forty-two-room mansion overlooking the undulating terrain that sloped down toward the water's edge. In 1872 Andrew's brother, Martin, began managing the estate. Martin, an engineer, dammed Bear Brook in 1878 to create Green Hill Pond. When Andrew Green died in 1903 the nieces and nephews to whom he deeded the 549-acre property sold it to the city.[56]

The economic divide separating the rich and poor was palpable. At the same time the city's wealthy lived in splendor, the poor lived in hovels close to the fetid, open Blackstone Canal that once linked Worcester to Providence. The canal, which operated from 1828 to 1848, became an eyesore and an open sewer soon after it closed. Not only did the outhouses lining the canal's banks drain directly into it, so did sewage from other parts of the city. It took several decades for the city to complete the sewer system and bury all of it.[57]

The threat to block the purchase of Newton Hill worked. In 1884 a Park Act was passed and a comprehensive park plan developed two years later. The plan sanctioned a scheme of dual-tract parks. Heavily influenced by Olmsted and Vaux's notion of parks as instruments of social control, health, and moral uplift, the planners argued that parks would be settings of "healthful recreation" designed to elevate and refine the workers and improve their intelligence, thereby making them better workers.[58]

Similar language can be found in park plans of other cities. For example, the 1910 New Haven civic improvement plan (coauthored by Olmsted's son) argued

If the people of the city, if in particular the women and children, are to have the benefit of a place where they may habitually get a little healthful

> recreation out of doors under agreeable and refreshing surroundings, as a part of the ordinary routine of life; if the children are able to make such use of a playground; if their elders are to get with tolerable frequency even a little walk in a park or square for air and for refreshment from the dulling routine of life in factory, store, office, and cramped dwelling house or flat; if the mothers are to get out occasionally to a pleasant park bench with their sewing or what not, while the children play about them [New Haven should build a park].[59]

Newton Hill was finally purchased in 1888. Though Lincoln and other park advocates had pushed for its acquisition to build a reservoir, as soon as the property was purchased Lincoln built a carriage path to the top of the hill so that the gentry could enjoy new forms of genteel recreation close to home. The area was planted with trees carefully placed so as not to impair the view from the summit.[60]

THE EMERGENCE OF MULTIPLE-USE PARKS

While Worcester's working-class residents were defining the kinds of playground they wanted and calling for multiple-use recreation space, less than fifty miles away, in Boston, MEHA activists were laying out sterile sand gardens in fenced yards. One can see the sharp distinctions between the working-class notion of a playground and the middle-class perception of what kinds of playgrounds should be constructed for the working class.

Even in Worcester, where the working class was quite active in park decisions, middle- and working-class neighborhoods got different kinds of parks. Worcester park plans called for the development of scenic parks on the West Side and two playgrounds on the East Side. When most of the park budget was spent on the West Side, East Side residents declared that Worcester had created a separate and unequal system of "class parks."[61] An 1887 letter to the *Worcester Daily Times* called attention to the unequal access to parks: "Our suburban retreats are dotted all over with notices to 'Keep off under Penalty of Law.' . . . Only the rich can afford excursions to the . . . seashore and mountain. Where then are the masses of people to seek . . . rest and recreation, sunshine and the refreshing breezes of summertime[?]"[62] That same year the city purchased 12.7 acres on the East Side and built Crompton Park; the open field was converted to a playground, but this did little to ease the demand for open space. In fact people wanting to use Crompton Park's baseball diamond camped on the field overnight to secure it.[63]

Because of the limited dialogue between the middle and working classes, the middle class misunderstood the concept of the multiple-use park that the working class wanted. They also failed to understand the demographics of their clientele, that mixed-age groups and both genders wanted to recreate in the same space at the same time. Such groupings desire a variety of recreational opportunities in a given park space. So the middle class went from designing passive, contemplative pastoral retreats to sterile playgrounds oriented to active, male-dominated sports. In addition to space dedicated to playing sports and games, there was a great need for children's play areas, paths to stroll and promenade, trees for shade and benches to sit on, flowers and shrubbery to admire, and contemplative pastoral spaces to relax in.

By the time the second wave of ethnic groups started to demand parks they were rewarded with stripped-down, minimally landscaped ball fields. It wasn't long before other ethnic groups in Worcester copied the Irish East Siders in their quest for park space. In the late 1890s the Swedish wire workers of Quinsigamond Village on the city's South Side campaigned for a playground in the center of their neighborhood. The city finally acquiesced and purchased Greenwood Park in 1905; it was proclaimed a "park for sport" in which little attention was paid to landscaping. After additional acquisitions the park was expanded to 14.9 acres in 1917. Other ethnic working-class neighborhoods in Worcester launched campaigns for park acquisition. In 1901 the English carpet weavers petitioned for a park on College Hill.[64]

Ironically this discriminatory approach to park development, whereby the working class had access to fewer, poorly equipped and maintained parks, increased the level of discontent and ensured that the poor continued to seek out and use parks in middle-class neighborhoods. Consequently when Worcester's working class continued to use parks outside of their neighborhoods the battle over proper park behavior also continued. In poor neighborhoods working-class park behavior was understood, condoned, or ignored. The cramped living conditions forced people to conduct many activities in public that are privatized if one has ample living space. In addition, high unemployment rates contributed to the number of people hanging around the parks. During the 1893 recession more than four hundred jobless men were counted on the Common in one afternoon.[65] The middle class sought to criminalize or label these activities (and those partaking in them) as immoral and inappropriate. Thus conflicts were frequent and tensions were high.

URBAN PARKS AND CAMPAIGNS ELSEWHERE

In the greater New York area merchants continued to oppose what they saw as unwarranted park development, while wealthy residents sought parks to anchor upscale neighborhoods. In the 1880s, when rich Bronx park advocates campaigned for a park by claiming that the city could not have too many pleasure grounds and "breathing spaces," downtown merchants objected by saying that the costs would "cause serious injury to the business interests of the city." Bronx park advocates prevailed by stressing that the park would improve property values. However, by this time politicians were beginning to realize that they had to recognize working-class demands for parks. Thus Mayor Abram S. Hewitt argued that it was the "city's duty to provide at least as many facilities for the poor as it does for the rich." Cognizant of the fact that Henry George, whom he defeated in the 1886 mayoral election, got working-class support by campaigning for parks in poor neighborhoods, Hewitt endorsed an 1887 law authorizing a million-dollar annual appropriation to build small parks in poor neighborhoods in an attempt to capture working-class votes.[66]

William S. Devery, the former Bronx chief of police campaigning in 1902 for the Ninth District, made parks a central issue also. At one campaign stop he said, "Well, what has this district got? We have no public bathhouse, no small park, or nothing else. Either Goodwin or Sheehan [former councilmen] could have got them if they'd gone to the front, but they didn't. I'll show them what plain Bill Devery can do. . . . If I'm elected leader, the people of this district will have a public bath, a recreation pier, and a park, and I'll get 'em."[67]

The Worcester and New York campaigns influenced other cities to develop park systems. In 1889 Donald Grant Mitchell purchased the 360-acre Edgewood farm in New Haven and turned it into a public park. New Haven's elites and visitors from all over the country came to see the park. New Haven also developed East Rock, West Rock, Fort Hale, Waterside, and Bayview Parks. Instead of perceiving the city as the antithesis of the bucolic countryside, New Haven's civic leaders envisioned the city as a place where rural charms and urban life could coexist symbiotically. Mitchell articulated this ideal in an essay published in 1884: "[The ideal city] should have its little nucleus of business quarters upon a bay, or a river. . . . An outlying circle of green, jealously guarded, would project its rays, or avenues of traffic athwart this circle; and those avenues of traffic, by their accretions of lesser and lighter business, would demand zebra-like cross-bars of space and greenness and foliage, to be flanked with files of houses such that a man could not go to work without sight of trees, or a chance to put his foot to the live earth; while all schools and courts and hospitals should have their setting of green."[68] Other social observers echoed these sentiments. Samuel Wordsworth saw New Haven as

Table 7 Total Acreage of Ten Urban Park Systems in 1895

Cities	Total Park Acreage	Cities	Total Park Acreage
New York City	4,711	San Francisco	>1,040
Philadelphia	3,505.93	Buffalo	718.5
St. Louis	2,126	St. Paul	>400
Chicago	1,900	Milwaukee	400 (approx.)
Minneapolis	1,552 (approx.)	Cincinnati	300
Baltimore	1,267 (approx.)		

Sources: "Public Parks of Leading Cities," part 1, *New York Times*, August 4, 1895, 20; "Public Parks of Leading Cities," part 2, *New York Times*, September 1, 1895, 28.

a place where rural charms and city beauties join.[69] After a visit to the city in 1842 Charles Dickens wrote that the trees seemed to "bring about a kind of compromise between town and country."[70]

In 1890 Kansas City hired George Kessler to develop a complete park system. In 1892 the Massachusetts legislature created the Metropolitan Park Commission to develop the Greater Boston park system. By 1897 Philadelphia had constructed thirty-one playgrounds in the city.[71] An 1895 report on urban parks found that New York had 4,711 acres of parkland, St. Louis 2,126 acres, and Chicago 1,900 acres.[72] See Table 7 for the total acreages of several urban park systems and Table 8 for a list of the largest urban parks.

By 1895 Philadelphia had more than fifty parks ranging in size from a quarter of an acre to nearly 3,000 acres, for a total of 3,505.93 acres. The city's parks department also employed about a thousand people. There was at least one park in all thirty-seven wards. Efforts were made to place parks in the most densely populated parts of the city. Across the nation park advocates continued to demolish housing in poor neighborhoods and build parks in their place. Consequently, Philadelphia's park advocates urged the city to tear down dilapidated housing in low-income areas and build parks.[73]

St. Louis's Forest Park, developed at a cost of $2.5 million, had 1,372 acres. In 1895 as many as forty thousand people used the park on Sundays. Forest Park had five picnic grounds, and park users were allowed to collect the wild flowers and freshly cut grass. Buffalos, elk, prairie dogs, and several other kinds of animals were kept in the park. Numerous sports were allowed. While park activists were able to prevent race tracks from being built in Central Park, the

Table 8 Size of the Largest Urban Parks in 1895

Parks	Cities	Park Acreage
Fairmount Park	Philadelphia	2,791
Forest Park	St. Louis	1,372
Golden Gate Park	San Francisco	1,040
Central Park	New York City	843
Druid Hill	Baltimore	671
Jackson Park	Chicago	586
Franklin Park	Boston	561
Prospect Park	Brooklyn	515

Source: "Public Parks of Leading Cities," part 2, *New York Times*, September 1, 1895, 28.

Gentleman's Trotting Club held weekly trotting races on tracks built in Forest Park. Forest Park was patrolled by a head park keeper and twenty-two assistant keepers.[74]

St. Louis also had the 267-acre Tower Grove Park, as well as fifteen smaller parks, most of which were located in the densely populated central portion of the city. The parks were fenced and locked at midnight. Though "Keep off the Grass" rules existed they were not strictly enforced.[75]

Though Chicago devoted nineteen hundred acres to parkland, as late as 1900 the city still had an inadequate park infrastructure because almost all of the developed parkland was along the lakefront. A report by the City Homes Association indicated that there were no parks in the tenement districts and very few in the city proper overall. A few vacant lots existed in working-class neighborhoods, but they were low-lying, swampy, rubbish-strewn fields. Parts of the city had population densities of up to five hundred people per acre with no open space, not even backyards or alleyways. If people in the tenement districts wanted to visit a park they had to travel miles to do so. Partly to compensate for this, in 1900 Chicago passed a resolution that all future schools had to be built with a playground.[76]

The city had six large parks, including Jackson Park in the southern and Lincoln Park in the northern portions of the city. The Southside Commission allowed people to use the grass freely in the 524-acre Jackson Park and 371-acre Washington Park. The Westside Commission set aside "the Children's Common," which children and women could use freely and from which men were excluded. In Lincoln Park visitors could use grassy areas marked "common" without fear of violating park rules; a zoo was also built in the park. Throngs of fancy carriages transported wealthy park users through the parks while poor residents arrived on foot.[77]

The city of Minneapolis had 1,552 acres of parkland, including the popular 125-acre Minnehaba Park, which received thousands of visitors daily. Lake Harriet drew up to twenty-five thousand visitors daily during the summer months.[78]

Though St. Paul began acquiring parklands in the 1860s nothing was done to develop parks until 1887. This was due in part to the fact that people protested benefits assessment taxes for park development, and there wasn't a management structure in place to oversee development. That year the Board of Park Commissioners was created to acquire additional parkland and develop the existing lands. The city's largest park, Como Park, was 396 acres. The city also acquired the 586-acre Phalen Park.[79]

Baltimore funded its 1,200-acre park system through a public transportation ticket tax. The city's largest park was the 671-acre Druid Hill Park. The Baltimore park system had few restrictions regarding the use of the grass.[80]

Golden Gate Park, San Francisco's largest, contained 1,040 acres. It was estimated that about one-fourth of the population went to the park on Sundays, which were designated "park day." There were no restrictions on the use of the grass, and there was a children's playground.[81]

The crown jewel of Buffalo's 718.5-acre park system was the 350-acre Delaware Park, located in the northern portion of the city. All the parks were within twenty minutes of the city center. North Park contained a zoo with buffalos, deer, and other wild animals. Around twenty thousand people visited North Park at peak periods. Park users were allowed to use the grass freely.[82]

Before 1889 Milwaukee had only a few scattered vest-pocket parks, but by 1895 the city had acquired almost four hundred acres to build seven parks. Though the park system was still in its infancy, thousands of people flocked to the completed parks. Park users were allowed to use the grassy areas of the parks.[83]

Cincinnati lagged behind other major cities in the acquisition and development of a park system. The city began acquiring park land late, and the cash-starved park management board was barely operational. Consequently in 1895 the city had only about three hundred acres of parkland. The oldest park in the city, Lincoln Park, was only ten acres in size. Park users were allowed on the grass in some parks during the "on-the-grass" season each year. There were about 1,025 city residents for each acre of parkland. Brooklynites fared worse, with 1,061 residents for each acre of parkland. However, cities similar in size to Cincinnati had more parkland per person. The ratio was about 900 persons per acre in Indianapolis, 337 per acre in Pittsburgh, 190 per acre in St. Louis, and about 150 per acre in Louisville.[84] Park advocates clamored for more parks. One argued, "A tired city man in search of recreation should have, first of all, a change of scene. Just such a change we should be able to give him in our parks. It is possible for the city to obtain from 1,000 to 1,200 acres at a small cost which would serve admirably for a park of the kind most desirable for Cincinnati."[85] Another wrote

> Our first great need . . . is truly rural parks. Such parks must be easily reached by street railway, and the cost of reaching them must be reduced to a minimum. . . . For residents of the hilltop suburbs, Cincinnati's lack of parks does not mean so much, for they have plenty of elbow room in their own neighborhood, but for the thousands who live in crowded quarters on the lower levels, it is a very serious matter. Below the hills the population is denser than in almost any other city in the union. And for these occupants of crowded quarters we have only the park provisions offered by the few acres of Lincoln and Washington Parks and the two or three other tiny ones that are in that part of the city.[86]

Middle-Class Reformers, Working-Class Advocacy

THE SHIFT TO COMMERCIAL LEISURE PURSUITS

In 1877 a working-class Worcester local, describing himself as a "Liberty Loving Citizen," questioned the prerogative of the middle class to prescribe working-class behavior and monopolize the lakes and other recreational resources. He argued that the rich were being unreasonable in expecting the working-class to spend Sundays in church rather than relaxing in the parks and playgrounds. Linking class inequality with recreational inequality, he wrote

> I question very much the right of one class of people to dictate to another class the manner in which they shall worship God or spend the Sabbath. It may indeed be very well for people of ample means, who have plenty of leisure on week days to ride around and enjoy the country air and scenery, to attend church and worship God there on Sunday, but these gentlemen must bear in mind that all the good people of Worcester are not thus favored. A very large majority of all who visit the Lake on

Sunday are people who are confined to shops ten hours in a day, six days in the week, myself among that number. Now, if the gentlemen were to have their way, when should we ever see our beautiful Lake Jewel, to bathe in its limpid waters, to glide over its glassy surface, or ramble in its groves or upon its verdant shores? Do these gentlemen think that working people have no love of the beautiful, have no admiration for God made manifest in the flowers, grass, or trees? Must my appetite be satiated by church ceremonies when it craves the broad open fields, the waving trees, the fragrant flowers and God's pure air and azure sky, filled with its myriad warblers who know no Sunday, but praise Him as loudly at one time as at another?[87]

The writer expressed the frustration and resentment that drove the working class to seek out alternative forms of leisure. Though commercial recreational venues cost money and the parks were free, a number of factors converged to make the working class flock to commercial venues: increased wages, shorter work hours, overcrowding, inadequate outdoor spaces in working-class neighborhoods, and oppressive and intrusive park rules.

It is true that conditions remained miserable for most of the working class; as late as 1929 more than one-fifth of American families lived in poverty. Nonetheless average real wages for nonfarm employees increased by more than half between 1870 and 1900. Although laborers did not fare as well as other, upwardly mobile workers, their incomes increased along with the rest of the labor force. Nationwide manufacturing wages increased another 25 percent over the next two decades. Furthermore the mean number of hours worked declined from sixty-six per week in 1850 to forty-eight in 1920. The shorter work week and increased income resulted in more free time and disposable income for the working class to spend on recreation. By embracing commercial recreational opportunities the working class simultaneously revolted against the imposition of upper-class cultural norms (elite park rules and behavior) and expanded their recreational repertoire. The new recreational pursuits were unconstrained. They involved thrills, escapism, and fantasy, the opposite of the monotony and drudgery of work and the stricture of genteel park behavior. To tantalize potential customers fairs introducing a host of mechanized amusements were held in Philadelphia (1876), Cincinnati (1883), New Orleans (1885), Chicago (1893), Atlanta (1895), Omaha (1898), Buffalo (1901), and St. Louis (1904).[88]

Amusement parks became very popular. Vanity Fair in Providence was a lavish theme park and a precursor of the Disney-style parks. Wonderland was an elaborate ocean-side theme park on Revere Beach in Massachusetts. Similar amusement parks all along the Atlantic Coast became well-known seaside resorts and pleasure beaches.[89]

Circuses, vaudeville shows, and similar amusements grew increasingly popular among the working class. P. T. Barnum, one of the world's greatest purveyors of entertainment, knew how to identify, package, and market amusements for mass audiences. His shows featured circuses, bizarre and grotesque human and animal curiosities, and melodramas. The traveling show exposed millions of rural customers to commercial recreational opportunities. In working-class urban areas pleasure gardens, dance halls, and entertainment spots such as the Bowery in Five Points competed with parks as recreational attractions.[90]

The mechanized attractions unveiled at the fairs found their way to amusement parks and other commercial recreational venues. Family picnics at commercial lakefront parks became very popular. As early as 1875 a crowd of seven thousand working-class Worcesterites went to the lake for Fourth of July picnics, clambakes, steamboat rides, billiards, bowling, and beer drinking. Other commercial recreational areas surrounded Worcester. A similar phenomenon was evident in New York, where picnic grounds, beach resorts, and amusement parks ringed Manhattan, despite the fact that Central Park administrators declared the Fourth of July "the People's Day in the Park."[91] Similar activities were popular in New Haven; in 1891 twelve hundred union members held a Labor Day march on the Green, after which as many as fifteen thousand people spent the day at Savin Rock's waterfront, dance pavilions, and athletic competitions.[92]

Worcester's Lake Park was an instant success as an urban water park. One hundred and ten acres of land on the shores of Lake Quinsigamond was donated to the city in the summer of 1884 (shortly before the vote on the city's Park Act) by Horace H. Bigelow and Edward L. Davis, wealthy manufacturers who were quite involved in city politics. Park Commissioner Lincoln was ecstatic about the gift because he thought such a park could be used to "promote popular enjoyment; to develop a taste for the beauties of nature; and to refine and soften, by cultivating, humanity itself." Bigelow stood to benefit from his donation: he owned several lakeside amusement establishments and the Worcester and Shrewsbury Railroad, the only transportation to the lake. A developer, he also had extensive lakeside property. Between 1870 and 1890 weekend use of the lake increased from one hundred to twenty thousand people per day. Within two years of the gift of the land, the value of parcels on the eastern shore of the lake increased from $35 to $500 per acre. Once the city accepted the gift plans were made to build a rural carriage drive and picnic groves. Lincoln was reluctant to build changing facilities at the lake because he associated public bathing "in breech-cloth or tights" with the poor and immodest. He claimed that the parks were intended to "put a polish upon the face of the earth," not "to scrub the scruff from its inhabitants."[93]

Heavy drinking characterized Worcester's lakeside outings, despite regular police patrols and intermittent bans on liquor sales at the lake. Not surprisingly drunken and rowdy working-class behavior at commercial lakeside resorts provoked middle-class complaints. However, upper-class commercial resort operators, who cared more about making money than abiding by middle-class conventions, also angered their social contemporaries. In 1896 several boat clubs petitioned for better police protection at the lake, complaining, "There are a certain class attracted [to the lake] . . . who care nothing for law and order and who conduct themselves in ways and manners as to terrorize, annoy and disgust the law abiding citizens and destroy and depreciate our property." In 1907 an article in the *Worcester Evening Gazette* claimed, "The atmosphere reeks with rottenness and pollution, which is fast driving away respectable people and giving this beauty to the wicked, lawless, and ignorant."[94]

Because of continuing class conflicts, the workers' fight for parks, and the obvious need for more open space in the city, by the end of the nineteenth century Worcester's parks commissioners began to endorse the petitions for more public recreation space throughout the city. In 1907 the newly elected mayor of Worcester declared in his inaugural address, "Modern industry and commerce should bear its share of the cost in providing a suitable place, conveniently located near the home of the workman, where, after the day's toil is ended, he can with his wife and children breathe a little of God's pure air." In 1908 voters helped to pass the Massachusetts Playground Act, which mandated that cities provide at least one playground for every twenty thousand residents.[95]

JONES WOOD REDUX: A WORKING-CLASS PLEASURE GROUND

Though Central Park was a popular recreation site, some working-class people couldn't afford to or did not have enough free time to go there. Moreover, by the late 1860s some New Yorkers, especially the immigrants, preferred to spend their time and money in recreational spaces that were more tolerant of working-class modes of expression. Ironically Jones Wood became a popular working-class hangout. Not long after the Joneses and Schermerhorns refused to sell their property to develop Jones Wood Park, development engulfed the site. Road construction increased access to the area and merchants established businesses on the periphery of the neighborhood. As the neighborhood changed the families leased parts of their estates for commercial picnic grounds and hotels. In the late 1850s the Jones mansion was converted into a restaurant and hotel encircled by picnic grounds with swings, hobbyhorses, a dance pavilion, quoits grounds, bowling alleys, a billiard saloon, and a shooting range. The grounds were often rented out to ethnic groups. Germans, for instance, often used Jones Wood for gymnastics exhibitions, parties, and festivals; some events drew more than fifty-five thousand German immigrants. The Irish and Scots also recreated at Jones Wood. Newspapers characterized the working-class Jones Wood crowds as "thousands of thieves, pickpockets, murderers," and "pimps." Even the *Tribune* made disparaging comments about the crowds.

Like Central Park and City Hall Park, Jones Wood was the site of political activism. In May 1865 the Workingmen's Union assembled fifty thousand people there for a picnic to launch the campaign for the eight-hour day.[96]

THE WORKING CLASS REDEFINE PARK USE

In less than thirty years the tide had shifted in favor of the working class. No longer did they have to fight to get a scrap of land for recreation. Most cities across the country, believing in the health-giving, restorative nature of parks and in their capacity for social control, acquired parkland in many locations and sought to build parks for active and passive uses. Take, for example, the rationale given for proceeding with comprehensive park development in New Haven in 1910. The plan echoed familiar Olmsted themes as well as calling for active recreation, something Olmsted and Vaux had began experimenting with even though they weren't fully committed to it.[97]

> Facilities for . . . recreation must be provided within easy walking distance of every home in the city. Any plan that deliberately stops short of such provision and leaves any considerable neighborhoods permanently without the benefit of accessible parks and playground for local use, while providing other districts with such facilities at the general expense, is in so far illogical, unjust, undemocratic and unwise. We may say, then, that in a reasonably well-planned city there should be, within easy reach of every family of citizens, local or neighborhood grounds; places for active games and exercises, for public concerts and other similar passive pleasures, and for the enjoyment of spaciousness, of refreshing beauty, of the freshness of verdure and especially of airiness, in so far as such enjoyment may be attainable under the controlling conditions. . . . The

> city of Chicago is now proceeding upon the principle that no dwelling should be more than half a mile from grounds adapted to serve the local purposes. . . . Indeed, since mothers and babies in need of recreation grounds and young boys and girls in want of playgrounds cannot be induced to walk more than a very short distance, such neighborhood grounds ought if possible to be brought within a quarter of a mile of every dwelling.[98]

The continuing concern over working-class leisure styles, combined with a general realization that active sports did not necessarily mean idleness, immorality, and depravity, slowly changed the way the middle class viewed playgrounds. In fact the growing sentiment at the beginning of the twentieth century was that exercise and activity were healthy.[99] Despite their relentless drive to produce better workers and more responsible citizens through structured passive recreation, reformers realized that they no longer controlled or defined working-class recreation. However, far from surrendering, the reformers turned their attention to setting the standards for and choosing and laying out the equipment in working-class playgrounds. They focused on designing parks wherein the activities and the participants could be easily regulated and monitored. One such group, the American Institute of Park Executives, founded in 1898, took a more interventionist approach. Social service organizations, such as the Boy's Clubs, started to offer supervised recreation, and companies began to sponsor sports teams. Reformers created the Playground Association of America in 1906 to concentrate on children and playgrounds. By working with children they hoped to accomplish two ends: reduce juvenile delinquency and socialize children into their roles as workers and citizens.[100] They redesigned playgrounds, laid out equipment to maximize control over the type and range of activities that could occur there, and specified how many playgrounds a city should have. Playgrounds were fenced and supervised to further constrain spontaneity and increase surveillance. As one young user noted, "On the vacant lot we can do as we please, when we have a fenced playground it becomes an institution."[101]

The working class rejected the middle-class model of park behavior and recreational activities and defined a vibrant recreational alternative for themselves. Conceding that the trend was irreversible, reformers adopted a strategy similar to the one they used when they realized that they were losing control over working-class women's leisure: they abandoned attempts to confront and influence groups directly and instead sought to exercise control through the policy arena, by setting standards, controlling the physical environment in which play occurred, and increasing the level of direct supervision. Even with this more oblique approach, however, working-class children still rebelled against the intrusion into their lives.

The new social control efforts did not go unnoticed. In 1912 the *Worcester Telegram* ran a series of articles criticizing the excessive scrutiny and supervision of the poor. The articles contained excerpts of interviews with twenty children, all of whom expressed disdain for the supervision and attempts to control their activities on the playgrounds. Many said they ignored the supervisors or had stopped going to the playgrounds. One eleven year old stated, "I can't go to the playgrounds now, they get on me nerves with so many men and women around telling [you] what to do." A fourteen-year-old boy said, "I can't see any fun playing as school ma'ams say we must play." One group of boys formed their own baseball team outside the auspices of the playground league. When asked what they thought about the storytelling events at playgrounds children claimed that they found the stories silly but applauded because they were told to do so.[102]

Social control efforts were more successful in Cleveland. In 1914 the mayor used the police force to monitor dance halls, make arrests, and shut them down, but this approach was ineffective. As soon as one establishment was padlocked, another opened elsewhere. The police were not making any headway in curbing the drunkenness and illicit sexual behavior associated with the dance halls, so the mayor tried a new approach: he opened several attractive, well-lit, municipal dance pavilions in city parks. Though the pavilions were chaperoned and alcohol-free, they were wildly successful. Crowds flocked to them, leaving the older dance halls practically empty.[103]

Gender, Class, and the Contested Terrain of Women's Leisure

Working women, many of whom were young and single, presented a significant challenge to the reformers. Between 1870 and 1920 the number of female factory workers increased from 34,000 to more than 2 million. By 1920 there were more than 8.6 million women working outside the home. Single working women, particularly those in large cities, had newfound independence and were challenging long-standing beliefs about women's leisure and womanhood. Riis estimated that in 1890 about 150,000 women and girls earned their own living in New York. The estimate would have been higher if he had included the number of women who worked to supplement the family income but did not earn enough to be completely self-sufficient.[104]

For much of the nineteenth century urban public spaces, particularly those associated with the middle and upper classes, were extremely gendered. The worlds of politics and business were male domains. Even civic parades and similar events were largely male affairs. Middle- and upper-class women spent a lot of time at tea parties and in sewing circles; when they ventured out in public they tended to cloister themselves in a network of safe and "respectable" spaces. New York offered the Ladies' Oyster Shop, the Ladies' Reading Room, and the Ladies' Bowling Alley; even the banks and post offices had special ladies' windows. Taylor's — seen as a "women's restaurant" — served an average of three thousand women per day. As mentioned earlier, the desire to create a safe and respectable space in which middle- and upper-class women could recreate was one of the factors motivating activists to advocate for building Central Park. Keenly aware of the need to separate genteel male and female recreational spaces and the need for greater outdoor recreational opportunities for women, Olmsted designed a "ladies' pond" at the skating rink in Central Park.[105]

Aristocratic women founded elite social clubs similar to those of refined gentlemen. Sorosis, founded in 1868, was a women's club that inspired members to engage in social reform activities. In 1890 Sorosis members convened a meeting of sixty-three delegates from clubs in seventeen states, and out of that meeting emerged the General Federation of Women's Clubs. Florence "Daisy" Jaffray Harriman and Anne Morgan, daughter of J. Pierpont Morgan, were concerned with the lack of recreational spaces for genteel women. Noting that many elite clubs barred women, the ladies founded the Colony Club in 1903. They raised half a million dollars and quickly signed up a thousand upper-class female members. They purchased a lot near the Morgan mansion and hired the architect Stanford White to build a clubhouse. Elsie de Wolfe decorated the club; it had a swimming pool, rooftop garden, gymnasium, squash courts, library, card room, cocktail bar, smoking room, and dining facilities. The reception area had a musician's gallery. It also had ten bedrooms. The club was such a success that by 1913 they needed more space.[106] Suffragists founded the Women's City Club of New York in 1915; members included Mary Garrett Hay, Katharine Bement Davis, Mary E. Dreier, Bell Moskowitz, Frances Perkins, Eleanor Roosevelt, Genevieve Earle, and Dorothy Kenyon. These suffragists were wealthy women — the wives and daughters of business moguls or politicians. Membership reached several thousand around 1917. Their clubhouse was a mansion on Park Avenue designed by Stanford White.[107]

Working women did not adhere to the middle- and upper-class norms prescribed for females. Their labor, clothing, relationship to men, courtship rituals, and recreation were changing rapidly. Working-class women were among the "New Women" questioning the division of women's and men's lives into separate realms of work and social activities. As Peiss writes, "The image of the flashily-dressed working woman, joking and flirting with men, spieling late into the night, enjoying a new-found sense of social freedom, resonated uncomfortably within the middle-class public." Women were an active part of the emerging mass culture; they flocked to dance halls and vaudeville performances. Shop girls and shoppers constituted between one-third and one-half of all vaudeville audiences in 1910.[108]

The West Brighton section of Coney Island was a popular working-class hangout where women lost some of their inhibitions. In addition to the rollercoasters, eateries, and dance halls, men and women flocked to the beaches. Wives rented blue flannel bathing suits and straw hats and could be seen cuddling with their husbands, while single women cavorted in tight, skimpy swim wear. At aristocratic resorts each sex spent time on the beach according to a fixed schedule; during the summer season at Newport, a red banner was raised on Easton's Beach at noon to indicate male-only, nude swimming time. While it was appropriate for aristocratic men to swim together nude during the 1880s, elites frowned on working-class men and women, clad in bathing suits, enjoying the ocean and beaches together.[109]

Cultural conflicts ensued as the working-class definition of womanhood ran afoul of the middle-class definition. Leisure was one terrain over which these differing perspectives collided. Reformers referring to West Brighton as the "Sodom by the Sea" were appalled by the behavior in the dance halls and brothels (which covered a ten-square-block area called the Gut), the romancing on the beach, and the frolicking in the waves. Coney Island was particularly irksome because, like other parts of New York and other major cities in America, wealth and poverty rubbed elbows there. On Coney Island the titillating working-class world of West Brighton gave way to the middle-class retreats of Brighton Beach, where lawyers, clerks, and other white-collar workers enjoyed the ocean. Not far from Brighton Beach the upper-class arrived by private trains to the gated enclaves of the Manhattan Beach Hotel and the Oriental Hotel.[110]

The reformers had limited success trying to redefine working-class women's social behavior. Their efforts often met with indifference and hostility; at best, there was limited enthusiasm for specific features of the programs the reformers designed for working-class women. The owners of commercial recreational establishments also resisted attempts to reform working women's recreation. The owners of amusement parks and pleasure gardens understood that they would lose business without female patrons. Instead of lamenting the sad state of working women's behavior, they encouraged risqué behavior at their establishments.[111]

The reform programs that successfully reached working women did so because they incorporated the ideas of the target group. Reformers eventually reframed their morality-laden message into a broader public health message. They founded fresh air organizations, and vacation societies sent working women to a network of country houses for one- or two-week holidays. Feeling compelled to safeguard single young women from the perceived dangers of the city and the streets, reformers also created working-class recreational clubs that mimicked the upper-class women's clubs. This strategy contributed to the redirection of working women's recreation reform at the beginning of the twentieth century. While clubs, friendly societies, YWCAs, and countryside vacation cottages remained active, new reform agencies emerged to tackle the popular recreation of working women. The Education Alliance, the People's Institute, settlement houses, and other organizations located in the tenement districts launched reform programs too.[112]

As time went on reformers realized that working-class women could not be dissuaded from pursuing certain leisure activities, so they focused less on preventing working women from participating in questionable activities and more on supervising the conditions under which women participated. In this way reformers tried to ensure that working-class women's leisure occurred under strict scrutiny and in morally healthy environments.

Demands for Increased Access to Recreation after 1950

Having won the right to the forty-hour work week and paid holidays with the passage of the Fair Labor Standards Act in 1938, workers had large blocks of predictable free time and could indulge in a wide range of recreational pastimes. Consequently they spent less time in bars and more time outdoors. Middle- and working-class families took long cross-country family excursions, and with private space at their disposal could relax in their backyards. They also spent more time fishing and hunting, playing sports, and visiting parks. In Gary, Indiana, steelworkers joined organizations such as the Izaak Walton League and the Lake County Fish and Game Association.[113] In Johnstown, Pennsylvania, after steelworkers won their "eight hours for work, eight hours for sleep, eight hours for what you will," they used their free time to fish, hunt, garden, do home repairs, and play sports.[114]

Working-class families became more concerned about pollution and environmental hazards as they spent more time outdoors. As was the case in Worcester, both middle-class and working-class Gary residents campaigned for better parks; the result was a system of separate and unequal "class parks" to fulfill different needs. During the 1950s some middle-class whites moved out of polluted neighborhoods in Gary to the eastern suburb of Miller. Once there they organized environmental groups to keep the nearby Indiana Dunes unencumbered by development. Their framing of the issue did nothing to reconcile or bridge the "jobs versus environment" dilemma that most workers were concerned with. In fact, by opposing development so strongly, middle-class environmentalists were perceived by workers and the unions as a threat to job security. Furthermore, middle-class activism focused on actions that brought immediate benefits to middle-class suburbanites while ignoring the occupational health and safety concerns of workers. The Steelworkers Union, which supported environmental measures selectively, gave lukewarm support to some of the efforts of Miller's middle-class citizens to preserve the Indiana Dunes. The union did not want to see the dunes destroyed; neither did they want to see all commercial activities blocked. Though Miller's beaches and the dunes could have been a recreation destination for Gary's working-class families, the union did not capitalize on that fact because only a few workers lived in Miller, and most working-class families used Wolf Lake just west of Gary. Instead union leaders focused on the job creation possibilities that involved commercial development of the lakeshore. Several local carpenters, glazers, and ironworkers unions, all much smaller than the Steelworkers Union, sensed an opportunity for job growth and lobbied openly to prohibit expansion of the Indiana Dunes or a wholesale prohibition of development. Indiana Dunes was designated a National Lakeshore in 1966 under the protection of the National Park Service[115]

Instead, the union championed the cause of the "working man's" park at Wolf Lake. Although white working-class groups had been campaigning for access to parks for decades, the inequalities between Gary's middle-class parks and those of the working class were quite striking. Wolf Lake was not as picturesque as the Indiana Dunes, and working-class families had to drive several miles west of Gary to reach it.[116] Neither the middle-class residents of Miller nor working-class activists supported the other group's case for improved recreational space. Even when middle- and working-class families organized campaigns around fishing, the social context of the issue varied, and that manifested itself in different framing, demands, and strategies and a failure to forge cross-class coalitions. Differential practices and access to resources divided the priorities of each class. Whereas middle-class anglers usually fished from private boats, the working class fished from jetties along the shore. Though access to jetties was a major working-class concern, the middle class did not worry about it. Thus by the late 1960s working-class environmental groups started agitating for access to fishing piers along Lake Michigan in Gary and to its west.[117]

Local working-class labor and environmental groups, such as the Workers for Democracy, Calumet Community Congress, Calumet Action League, and Calumet Environmental and Occupational Health Committee, focused on ending pollution and reducing workplace hazards. At the same time industrial encroachments on fishing grounds stimulated Gary's working-class anglers to form environmental groups to protect access to fishing. Indiana's 1964 lake-stocking program had resulted in a sharp increase in the number of anglers on the lake. For the first time in decades the lake teemed with game fish, and the working class took every opportunity to fish. However, anglers were apprehensive about industrial pollution from steel mills and power plants.[118] The anglers joined groups such as the Lake County Fish and Game Association and the Izaak Walton League, transforming them from benign social clubs into political action groups campaigning against development of the waterfront, strengthening water pollution regulations, and conducting their own studies of water quality.[119]

In early 1970 the desire of the working class to fish resulted in a clash with U.S. Steel over public access to fishing piers along Lake Michigan. Working-class anglers in Western Gary and Hammond became increasingly frustrated that they couldn't fish closer to their home and that industrial facilities constructed along the city's waterfront curtailed their lack of access to Lake Michigan. A group of citizens tried to get access to Buffington Pier, a mile-long jetty that extended into Lake Michigan from U.S. Steel's cement plant. The Buffington Pier Community Coalition implored U.S. Steel to either lease or donate the jetty and contiguous harbor to the public. The group argued that the company rarely used the property and that the facilities would function better as a recreation complex with a fishing pier, bathing beach, and boat basin. The group collected over three thousand signatures and solicited the help of the Gary Parks Board. The Calumet Community Congress sent representatives to the Gary Parks Board meetings and wrote letters to local officials. U.S. Steel did not agree to the plan, and the Buffington Pier group was dissolved.[120]

However, the issue was not put to rest. In 1971 a loose coalition of fishermen from northwestern Indiana revisited the issue of access to fishing grounds by demanding use of a pier at the industrial port of Indiana Harbor. Borrowing some of the tactics used in the fishing rights and civil rights struggles of Native Americans and African Americans, over three hundred anglers, equipped with fishing gear, threatened to take over the property located several miles east of Gary and stage a massive "fish-in." Indiana's governor capitulated and declared the harbor open to the public. In 1973 city council representatives from Western Gary tried once again to open Buffington Pier to the public. The measure passed, but U.S. Steel objected and the matter went no further. Finally, in 1975, the Lake County Fish and Game Association managed to convince the Northern Indiana Public Service Company to allow fishing on its lakefront property.[121]

Dying to Get to the Beach: Blacks and Access to Recreation

Though working-class whites had to fight for access to parks and beaches, the battle was even more intense when blacks and other minorities tried to enjoy such amenities. In many instances blacks came face to face with white working-class and middle-class opponents who refused to share public resources with them. The opposition was so great that whenever blacks tried to take advantage of outdoor recreation opportunities, the beaches, parks, and woods often became the sites of deadly conflict. For example, as Chicago's black population grew, there were increasing attacks on blacks in schools, in parks, on beaches, and on public transportation. Washington Park (nicknamed "Booker T. Washington Park" by condescending whites) was the site of many clashes. The *Defender* defiantly asserted that the beaches were free for all to enjoy. However, when blacks tried to use public facilities lifeguards and stone-throwing white mobs prevented them from entering the water. As early as 1913 investigators for the Juvenile Protective Association reported that "even the waters of Lake Michigan [were] not . . . available for colored children." The report pointed out that black children were not welcomed by white children at the beaches. It added that late in the summer of 1912 a little black boy attempting to bathe at the Thirty-ninth Street beach was beaten so badly that the police had to send in a riot call.[122]

In the sweltering summer of 1919, even as the *Defender* and the *Whip*, two of the leading black newspapers, encouraged the city's burgeoning black population to visit the beaches and the parks, the city was about to be engulfed in a series of deadly race riots. It began when four teenage boys walked past the designated black beach at Twenty-fifth Street to go to their favorite spot, dubbed "Hot and Cold," at Twenty-sixth Street. Located behind the Keeley Brewery and Consumers Ice, the spot got its name from the potent discharges from the brewery and the ice company. The toxic discharges temporarily bleached black swimmers. Women stayed away from the spot. On the afternoon of July 27, the boys got onto a raft they had built and decided to pilot it toward a marker several hundred feet offshore. At about the same time several black men and a woman strolled onto the white beach at Twenty-ninth Street determined to take a swim. They were quickly chased away by rocks hurled at them and other threatening gestures. Within minutes a large number of blacks appeared at the Twenty-ninth Street beach, this time hurling rocks themselves. Whites fled and blacks temporarily took possession of the beach. Soon a large crowd of whites returned; they too threw rocks at the blacks on the beach.

Unaware of the melee on the beach, the boys guided their raft to the marker. A white man at the end of the breakwater was throwing rocks at them, but they thought it was a game, and as long as they could see him and duck in time, they remained unharmed. After several minutes one rock hit one of the boys, Eugene Williams, on the side of the head and he slid into the water. The man ran onto the Twenty-ninth Street beach. Half an hour later lifeguards recovered the body of the dead boy from the lake. Accompanied by a black policeman, blacks marched to the Twenty-ninth Street beach and pointed out the man they thought was the perpetrator. The white police officer on duty refused to arrest the man and prevented the black officer from arresting him. As the officers argued, some blacks ran back to Twenty-fifth Street to tell the others what was happening, and a large crowd of blacks ran to the Twenty-ninth Street beach. The white police officer arrested a black man whom a white swimmer had complained about. In the meantime rumors spread around the surrounding white neighborhoods that a rock-throwing black man had struck and killed a white swimmer, while in the "Black Belt" rumors spread that the white officer had caused Eugene's death by preventing expert swimmers from rescuing him and by holding blacks at gunpoint while white swimmers stoned them. Hundreds of angry blacks and whites swarmed the beach. By the time police reinforcements arrived the whites and blacks were stoning each other. A black man fired his revolver at the police officers, striking one of them. The black man was killed on the spot, and the race war began in earnest as a gun battle erupted on the beach. The rioting lasted five days, spilling over into the stockyards, where angry mobs of whites attacked black workers. Before it was all over, an additional sixteen blacks and fifteen whites were killed and over five hundred people were injured. There was $250,000 in property damage, and one thousand people were left homeless. The athletic clubs, whose members were mainly the sons of Irish stockyard workers, had played a significant role in inciting the violence.[123]

After their investigation into the cause of the riot the Chicago Commission on Race Relations, a nonpartisan, interracial group, recommended that the police, state militia, state's attorney, and courts correct the "gross inequalities of protection" at beaches and playgrounds. The Commission also charged the courts with careless handling of cases involving blacks and the police for discriminatory arrests of blacks. The Commission recommended the abolition of the athletic clubs and that the city get rid of vice in the Black Belt (many shady establishments moved to the black neighborhoods and stayed there because this section of the city escaped the Great Fire). The Commission asked the City Council to condemn and raze all buildings unfit for human habitation (most of these were in the Black Belt) and to improve garbage removal and street repair in the shamefully neglected black neighborhoods. The Commission recommended the construction of more recreational facilities in the black community, but at the same time they asserted that blacks should have access to all recreational facilities in the city and should be protected while there.[124]

For decades after the 1919 race riot whites continued to object when blacks used public recreational facilities, particularly when they tried to use facilities outside of black neighborhoods. In 1929 black civic leaders and politicians organized to halt the threats to blacks using the public beaches. Still, in 1931 black newspapers carried stories of blacks being harassed along the lakeshore in Hyde Park. Several cases were reported of blacks being threatened, insulted, and molested by whites on the beach. There were also charges that police officers stationed at South Park ignored black bathers' complaints, signaling to whites that their behavior was appropriate. Fences were erected on some beaches to separate black and white swimmers. The commissioners claimed that the fence separated the private beach from the public beach and that anyone could enter the private beach by paying a fee. In reality white beachgoers stayed on one side of the fence and blacks on the other. In the summer of 1935 a group of white University of Chicago students decided to challenge the validity of the fence. They were called communists and accused of trying to incite blacks.[125]

Even children joined the conflict. White youngsters chased minority youngsters from beaches and organized gangs to fight for what they perceived to be their exclusive space on the beaches and in the parks. The Chicago NAACP reported that at a children's party sponsored by the City Park District at San-ford Park, a park official tried to separate the black and white children from each other. The mayor and other city officials tried to assure the NAACP that no racial discrimination was intended by the park official's actions. Several other incidents with racial overtones occurred at recreational facilities between 1935 and 1940. In one case a black woman and four girls were denied a locker at a park pool and white children ripped up their clothes. In another case thirty black and Latino children and their recreation director were driven away from a lakeside beach by white youths. The attending police refused to make any arrests, stating that he had orders to "put all colored off the beach." The park superintendent denied that such orders were given, but blamed "Communist agitators" for the presence of blacks on a beach that they had not previously frequented.[126]

The director of a park in an Irish neighborhood adjacent to the Black Belt said in a 1937 interview:

> In the last few months I've noticed they [the blacks] don't come in here so often. Well, the only reason I can give is the [Irish] community just won't stand for it. In the summer there were quite a few, but the younger fellows just rebelled about them. They came to me about it and I told them that I couldn't do anything about it. One of the leaders said that they would see about it. I cannot say that we have had much trouble with the different races. The Negroes just stopped coming here. The recent outbreaks didn't happen in the park, from what I understand; they happened out on the street. If I find that any one of the people working here had anything to do with keeping them away from this park, I would do all in my power to get him fired. I don't believe in any prejudice whatsoever.[127]

Most blacks and Mexican Americans in Chicago did not want to risk being assaulted or killed by mobs thinking the public parks and beaches were their private domain, so they swam at black beaches and in the Jim Crow sections of mixed beaches and recreated at Black Belt parks. However, several liberal organizations—black, white, and communist—protested these forms of recreational discrimination. In addition the NAACP filed suits on behalf of people wanting to challenge the system.[128]

Gary's black neighborhood, Midtown, had a severe shortage of recreational facilities, and the two existing neighborhood parks were very poorly maintained. Blacks were barred from using other city parks. Marquette Park's public beach was guarded to ensure that whites could have exclusive use of the facilities. When whites used intimidation and vigilante tactics to deny blacks use of recreational facilities, the police did not protect the rights of blacks. As early as the summer of 1949 a multiracial group of about a hundred residents and labor activists calling themselves the Young Citizens for Beachhead Democracy rallied at City Hall, then drove to Marquette Beach to take control of the beach. As the caravan neared the park the protesters encountered an angry mob of whites wielding bats, clubs, iron pipes, and rocks. Still the caravan continued to the beach. The protesters spread out their blankets, hung banners, and planted an American flag in the sand. However, their beach takeover was short-lived. Police arrived, claimed that the beach was closed for the day, and ordered the picnickers to leave. The group left but printed flyers about their excursion and distributed them at the gates of the steel mills.[129]

The beach takeover was controversial; mainstream civil rights groups such as the NAACP criticized the action as unnecessarily provocative and militant. The Anselm Forum rejected invitations to participate because they thought that revolutionaries had masterminded the plan. The Midtown Youth Council charged that the rally was orchestrated by "pinkos and radicals" solely for the purpose of creating dissension. With the black community in disarray and Latinos silent on the issue of the beach protest, city officials ignored the problem. Within a few years, however, the campaign to integrate Marquette Park had the full backing of civil rights groups. This occurred in part because by the mid-1950s labor activists from the Steelworkers Union had joined the NAACP and formed a large voting block in the organization. Black ministers had become convinced of the need for action, and the Urban League had decided to broaden its agenda to include fighting discrimination in the community. In 1953 all these organizations focused on integrating Marquette Park.[130]

In July of that year two carloads of young black women from the NAACP visited Marquette Park. While they were at the beach their cars were vandalized, and when they returned to the parking lot a gang of white youths awaited them. The youths assaulted and threatened the women. As they were about to overturn the cars, the police arrived and dispersed the crowd, but made no arrests. Civil rights leaders used the incident to demand that the city protect black beachgoers. The mayor promised protection, and a few weeks later, when black representatives from the Urban League and the Interdenominational Ministerial Alliance visited the park, there was no trouble. Although some blacks continued to experience hostility, there was no full-scale violence.

Over time, as the police presence at the park receded, whites escalated their attacks, and again blacks did not feel safe visiting the park. In 1961 another incident forced civil rights leaders to revisit the issue of Marquette Park: on

Memorial Day a black man was severely beaten on the beach by whites as the police looked on. Five hundred blacks jammed City Council chambers demanding that the mayor investigate the actions of the police, integrate the police force, and issue a public statement deploring the actions of the whites and the inaction of the police. The mayor refused and urged blacks to be patient. Even during the 1960s and 1970s, when Richard Hatcher, the first African American mayor of the city, was in office, blacks were still wary about using Marquette Park as sporadic racial violence still occurred there. Adopting an avoidance strategy, blacks stayed away from the beach when whites were there and used it late in the evening and at night. They also used less popular beaches to the west of the city and congregated in one section of the beach regardless of which beach they were using. Rather than enjoying the environmental amenities blacks were concerned with amassing numbers to stave off violent attacks.[131]

ten

CONTEMPORARY EFFORTS TO FINANCE URBAN PARKS

Elite park advocates tried to exert social control and impose order in cities in two ways that are relevant to this discussion. They did this through park design, rules, and surveillance. However, park advocates also achieved social control by taking charge of the financing and management of the parks. Though park advocates and managers became less concerned with using the parks as instruments to "civilize" and "uplift" the masses over time, elites still play an important role in determining the financial status and functioning of the park. Today, issues related to private financing and undue influence of financiers in park planning and management, access to parks for poor urbanites, and programming for the masses versus the upper and middle classes still dominate the conversation. As budget-conscious cities search for ways to fund their parks, the questions about who should pay for the parks and how to guarantee equity in public input and access have once again taken center stage. Increasingly parks are turning to private financing models that are leaving cities with systems of "class" parks reminiscent of the nineteenth-century divide between well-endowed parks in wealthy neighborhoods and their derelict counterparts in poor neighborhoods.

Today Central Park is again at the forefront of the debate over the financing of urban parks. Given its place in American urban park history, it is not surprising that Central Park continues to be a model for urban park development, financing, and management.

A Room with a View

In February 2001 many Americans were intrigued as former president Bill Clinton resolutely pursued his goal of getting an office with an unobstructed view of Central Park. After the House Appropriations Subcommittee, which oversees former presidents' budgets, balked at paying $800,000 a year for his preferred office space on West Fifty-seventh Street, Clinton took his search to Harlem. Determined to have an office with a panoramic view of the park, Clinton set his sights on a space that rented for $210,000 per year. However, before he could get used to the idea of setting up his office in the rapidly gentrifying but decidedly less glamorous neighborhood of West 125th Street, Mayor Rudolph Giuliani (Hillary Clinton's former political opponent for a Senate seat) laid claim to the office space in the name of the Administration of Children's Service, a child welfare agency. In true Clintonian fashion, the quest for an office with a view of Central Park played out as a public drama that was widely reported in the media. Finally, after high-level negotiations between Clinton, Giuliani, Governor George Pataki of New York, Congressman Charles Rangel (who persuaded Clinton to put his office in Harlem), and possibly President George Bush, the Administration for Children's Service was provided space on the sixth floor of the building and Clinton got his office on the fourteenth floor.[1]

Clinton settled for a Harlem office with a view of the park rather than a Manhattan office without because he knows that the view of Central Park, even if it is from Harlem, is more valuable than the actual use of the park itself. Donald Trump would agree with Clinton. In 1988, when Trump bought and renovated the Barbizon-Plaza Hotel on Central Park South into fashionable condominiums selling for $4 million each, he replaced the small windows with giant picture windows. As he put it, "Those openings are immensely valuable . . . because a great view is worth a small fortune."[2] Elizabeth F. Stribling, president of Stribling and Associates, a prominent real estate development company, agrees with Trump. Speaking of Central Park, she says, "To have access to beauty and solace, and be able to view the changing seasons from your window, creates an enduring real estate premium."[3]

Central Park and Modern-Day Power Elites

The Clinton-Giuliani tug-of-war demonstrates that 150 years after it was conceived Central Park is still a potent symbol of wealth, power, and prestige. It was exactly that prestige that Elizabeth Barlow Rogers sought to capitalize on when she became the administrator of Central Park in 1979. At the time the city's commissioner of parks, Gordon Davis, described the park as a "disaster zone." Senator Daniel Patrick Moynihan described it as a "national disgrace" and threatened a federal takeover. Some wanted to make Central Park into a state park, while Moynihan wanted it to be taken over by the National Park Service. Rogers, an advocate of public-private funding and management of parks, created the Central Park Conservancy in 1980 to help enhance the image of the park and transform it into a prestigious institution that wealthy people wanted to associate with and donate to.[4] During the early 1970s, when others despaired about the condition of the park, Rogers got a taste of how one could successfully raise funds for Central Park. As a volunteer youth program leader for the Parks Council she wrote an article titled "33 Ways Your Time and Money Can Help Save Central Park." Within a week of publication $25,000 in contributions flooded into the Parks Council office.[5] As head of the Central Park Conservancy Rogers capitalized on people's desire to contribute to Central Park.

During the 1850s park advocates envisioned Central Park as an elite, landscaped park ringed by the dwellings and businesses of the wealthy. That idea is still so firmly entrenched that a century and a half later some of the richest, most powerful people in America live on Central Park's eastern, western, and southern borders. Though the wealthy continued to live adjacent to the park during the twentieth century, they weren't as involved with the park as their nineteenth-century counterparts. Consequently by the 1970s the park had deteriorated, and the cash-starved city was unable or unwilling to provide funds for repairs and maintenance. Nonetheless the park remained popular. Today more than 25 million people visit annually.[6]

From her earlier work raising funds for the Central Park Task Force Rogers knew that the rich had an affinity for the park that could be tapped to help alleviate its cash-flow problems.[7] A longtime Olmsted and Central Park scholar, she was fully aware of the role that the city's business elite had played in developing the country's first landscaped park. She set out to create an organization whose members and board would form what I will call the Barlow Circle, a re-creation of the nineteenth-century park lobbyists on a much larger and more deliberate scale. The Barlow Circle is primarily a network of corporate executives and wealthy New Yorkers living close to Central Park who donated time and money to the park and made it their cause célèbre.

The Central Park Task Force was created to raise funds for the park's operation. The Task Force and the Central Park Community Fund, both founded in 1975, were replaced by the Central Park Conservancy. The Conservancy has raised and spent more than $300 million restoring the park. Rogers harkened back to the 1850s and classic Downing-Olmsted-Vaux ideology as she set out to rehabilitate the park "as the original creators saw it—a scenic retreat, a peaceful space that would be an antidote to urban stress." She claimed that "the park is foremost a work of landscape art," and she was "not prepared to accept that it's anything else." Rogers and the Conservancy presented the park as an elite cultural institution in which "the trees and the monuments and the gardens are [the art] collection."[8] Framing the park in nineteenth-century Romantic ideology has been so successful that New York's mayor, Michael R. Bloomberg, sees the park as a "treasured oasis and respite from bustling city life . . . where people's spirit . . . [can] be restored by the beauty and power of nature."[9] In reference to the bombing of the World Trade Center the Central Park Conservancy wrote, "When crisis strikes . . . many thousands [of] New Yorkers . . . flock to another sanctuary for comfort and renewal—Central Park. It is nothing less than their secular cathedral."[10]

By pursuing and landing large individual and corporate donations the Conservancy has transformed Central Park into a cultural institution that is one of the major targets of charitable giving. Philanthropists put it on par with the Museum of Modern Art and Carnegie Hall.[11] Understanding how and why the wealthy give was critical to the Central Park Conservancy's success. First, the Conservancy understood that within elite circles there is tremendous social pressure to give, to make donations to each other's pet causes, and to give to elite charities. There is also a mentality that once a charity has attracted prominent donors, others are drawn to it for fear of being excluded from the most prestigious networks. Moreover, organizations that are on sound financial footing are more likely to leverage additional funds than cash-starved institutions.[12]

Richard Gilder, a stockbroker who grew up playing in Central Park and now lives beside it on Fifth Avenue, gave $17 million to the park. He is typical of the kind of donor the Conservancy courts: rich individuals who give to what they are familiar with and to what gives them personal satisfaction. Because so many of Central Park's wealthy neighbors grew up in the fashionable dwellings around the park, visited the park as children, can view the park from the homes and offices, or jog or stroll in the park, the Conservancy believed they would have tremendous interest and a personal stake in helping to restore it to its former glory. It is not surprising that 80 percent of donations to the Conservancy come from about 20 percent of the donors, and that hundreds of the largest donors live within a block of the park.[13] The Conservancy also has an estate planning program that allows supporters to remember the park in their will. Like Gilder, Janet Kramer lived across the street from Central Park and regarded it as her backyard. When she passed away she left $2.9 million to help with the restoration of the park's Historic Playground Landscape.[14]

Douglas Blonsky, the new president of the Central Park Conservancy, described the affinity supporters have for the park as he reflected on his personal experiences there: "Every morning at 6:30, I walk through Central Park on my way to work and savor the view from the Great Lawn. It's a beautiful sight, and it's different every day." Blonsky recalled his revulsion at the state of the park when he visited it decades earlier as a university student: "It was another park and another era when I was a university student and our horticulture class made a field trip to Central Park. It was in such disrepair—landscapes were reduced to bare ground, historic buildings and structures were dilapidated and covered with graffiti, garbage was strewn everywhere. . . . I could never have imagined that a quarter of a century later Central Park would be restored to its former glory."[15]

As one wealthy park neighbor commented, "Central Park is our only estate where I didn't have to hire a hundred gardeners." According to Mayor Bloomberg, a Central Park Conservancy board member and major donor to the park and the founder of Bloomberg Financial Markets, "The park is visible, you can touch it, you can partake of it, you can brag about it, you can see the difference your money makes." For Gilder giving to the park "is a natural for people who live near the park. . . . There are so few things you can give money to that aren't in desperate shape." Charitable giving can also help to protect and enhance the value of investments. As one park donor noted, "In the real estate field, it is just smart business for us to try and improve the quality of life in the city where we have our major holdings." Research supports the pattern of charitable giving to the Central Park Conservancy thus far. Ostrower found that the rich give to universities, symphony orchestras, art museums, and other elite charities that they use and that enhance their social status and business networks and provide opportunities for them to move in the exclusive circles of charitable benefits and social events.[16]

Understanding that the rich, in true Veblenian fashion, emulate each other and apply tremendous social pressure on each other, more than an outsider can apply, the Conservancy sought to involve rich and powerful people in running the organization. William S. Beinecke, formerly chief executive officer of Sperry and Hutchinson, and Henry R. Kravis, the billionaire leveraged-buyout specialist, are men whose phone calls and requests for aid cannot be ignored. According to Beinecke, who was "shocked" at the park's appearance, he had no trouble recruiting the top executives at Morgan Stanley, American Express, and other corporations to sit on the board of the Conservancy (they later donated money). Several donors attest to the fact that when Kravis, a trustee of the Conservancy, requested money, they could not ignore him. "If he helps your charity, then you must help his," said Howard J. Rubenstein, the multimillionaire head of a public relations firm. "In the upper echelons of millionaire's row, it [charitable giving] is almost a trade." Other philanthropists and longtime park supporters such as Iphigene Ochs Sulzberger, the wife of a former publisher of the *New York Times*, and organizations such as the DeWitt Wallace–Reader's Digest Fund, the Arthur Ross Foundation, Time Warner, Tropicana Products, and Trump Organization have given large donations to the park.[17]

The rich are not interested in merely giving money to Central Park and leaving park matters alone; they are quite active in advising park administrators. For instance the Central Park Conservancy announced a $50 million capital campaign in 1989, spearheaded by Kravis. By 1990 the campaign had reached the $42.5 million mark. In 1991 the capital campaign and the Campaign for the Central Park Conservancy (begun in 1987) concluded, having secured more than $50 million in contributions. In 1993, after Richard Gilder gave the Conservancy a $17 million challenge grant, the City of New York pledged an additional $17 million in matching funds. Since Gilder's gift was contingent on the Conservancy's raising at least $17 million in private funds within a three-year period, the Central Park Conservancy, under the guidance of Daniel P. Tully, kicked off another capital campaign. The Wonder of New York Campaign, as the fundraising drive was called, concluded in 1996 after raising $77.2 million.[18] Hence Gilder's original $17 million was used to leverage millions more for the park.

Tea Parties, Luncheons, Benches, and Trees

When Downing suggested in 1848 that the wives of wealthy men host tea parties to raise funds for public parks, he might not have dreamed of the Central Park Conservancy's Women's Committee raising over $30 million for the park and hosting the annual Frederic Law Olmsted Luncheon, attended by more than a thousand guests. It brought in over $2 million in 2002. The first Olmsted spring luncheon, organized in 1983, raised $172,000 for the Conservancy.[19]

According to Phyllis Cerf Wagner, the widow of a cofounder of Random House, the luncheon is a highly anticipated event: "It is in May. . . . Everyone has new spring clothes. They have new hats." In 2001 the Sunday edition of the *New York Times* carried eighteen pictures of the elaborate hats and fashions on display at the Frederick Olmsted Awards Luncheon, attended by 1,150 guests. The *Times* described the affair as an "exquisite fashion spectacle" with an "emphasis on pastel-colored clothes and extravagant Ascot-style hats." Surprise fashion statements included "blouses with jabots" and "caviar-bead flower broaches." The article continued, "Though the fanciful millinery is an anachronism, the event remains an enchanting scene."[20]

The Conservancy's Women's Committee has evoked and re-created the nineteenth-century Romantic history of the park, when wealthy men and women repaired to the park in their latest fashions to socialize. Wagner and the Women's Committee also realized another of Downing's visions: allowing supporters to endow park benches, trees, and structures. In 1991 the Women's Committee established the Adopt-a-Bench program and the Central Park Tree Trust. More than twelve hundred of Central Park's nine thousand benches have been endowed already; for $7,500 one can endow a standard bench and for $25,000 a rustic one. More than four hundred of the park's twenty-six thousand trees have been endowed, at the rate of $5,000 for mature trees and $25,000 to $100,000 for historic trees or groves.[21] Because naming opportunities are lucrative fundraising ventures, sprinkled throughout Central Park are spots such as the Lila Wallace–Reader's Digest Terrace, Doris C. Freedman Plaza, Wagner Cove, Kerbs Boathouse, Robert Bendheim Playground, Rudin Family Playground, Bernard Family Playground, Abraham and Joseph Spector Playground, and the Charles A. Dana Discovery Center.[22] It is ironic that a century and a half after Olmsted argued that the parks should not commemorate the generosity of the wealthy in the parks with monuments, as such features distract from a contemplative recreation experience, Central Park and many others around the country are adorned with statues, fixtures, groves of trees, and other objects indicating corporate sponsorship or hefty donations from business moguls, their heirs, or other wealthy individuals.

The Spread of Public-Private Partnerships for Parks

The Central Park Conservancy and its precursors, the Central Park Task Force and the Central Park Community Fund, are among the oldest groups of their kind. They have also been very influential in propagating the public-private partnership model and the spread of commercialization in the parks. Brooklyn's Prospect Park has a tree program which costs contributors $750 to plant an understory tree, $1,000 for an overstory tree, and $10,000 for a grove of trees. Contributors can also adopt an existing tree. For $10,000 contributors can adopt one of the horses on Prospect Park's historic carousel. The Prospect Park Alliance launched an Adopt-a-Bench program in 2001 at $5,000 per bench.[23] Battery Park and Riverside Park have also created fundraising organizations to help with park operations.[24]

In San Diego Forever Park built a $2 million endowment fund for Balboa Park by focusing on planned giving opportunities.[25] The Golden Gate Parks Conservancy has raised more than $100 million to help with the upkeep and maintenance of Golden Gate National Park and the Presidio; it has assembled a board with numerous corporate executives and established planned giving opportunities. For $1,000 one can join the William Kent Society, while it takes a minimum of $7,500 to dedicate a bench.[26] The National AIDS Memorial Grove, founded in San Francisco in 1991, charges $1,000 to add a name to the Circle of Friends; crescents range in price from $5,000 to $25,000, and one can adopt a bench for $15,000.[27] The San Francisco Parks Trust has raised over $25 million through a commemorative bench program, planned giving, membership opportunities, and corporate partnerships.[28]

Friends of the Public Garden have focused on three Boston green spaces: the Boston Common, the Public Garden, and the Commonwealth Avenue Mall. The organization has created an endowment fund for the parks. Membership costs range from $25 to $2,500.[29] The Emerald Necklace Conservancy, founded in 1996, operates in neighboring Brookline with a mission to focus on the parks that form the "necklace."[30] One can become a member for a donation of $25; corporations can gain membership status for donations beginning at the $1,000 level.[31] Smaller, less well-known parks are also using the public-private financing model. The Friends of Buttonwood Park in New Bedford, Massachusetts, allows subscribers to adopt a tree for $250 or become a regular member for $5 to $500.[32]

Some parks have gotten quite creative with their funding opportunities. The South Suburban Park Foundation in Littleton, Colorado, has a "Trail for Sale" program in which donors can sponsor one foot of trail for $1,000.[33] The Parks Foundation in Cincinnati has raised over $7 million to fund projects in the city's parks.[34] Forest Park Forever has raised more than $94 million for Forest Park in St. Louis; membership levels range from $45 to $10,000.[35] Chicago's Garfield Park Conservatory Alliance charges up to $500 for a membership, $1,000 for a membership in the Jens Jensen Club, and $2,500 to $5,000 to dedicate a bench.[36] In Indianapolis the Friends of Garfield Park have developed programs with corporate and community sponsorship opportunities that range from $1,500 to $10,000.[37]

The Fairmount Park Conservancy in Philadelphia has programs to adopt trees and historic buildings; there is also a robust corporate investor support program.[38] In Richmond, Virginia, the Maymont Foundation has giving opportunities that include the "adopt-a-living-thing" program, tree and bench memorials, carved granite pavers, and planned giving.[39]

Clearly, since the 1970s there has been growing interest in and use of public-private partnerships as a way of funding and managing parks. Though the Central Park Conservancy is the most successful park management organization, several models of public-private ventures are being tried all over the country. These partnerships, in which nonprofit and private groups partner with cities and counties to help run public parks, fall into five general categories.

ASSISTANCE PROVIDERS. These are groups such as Friends of the Public Garden and Golden Gate National Parks Conservancy that typically help a city's parks department with public education, programming, and volunteer work.

CATALYSTS. Some nonprofits, such as the National AIDS Memorial, working with public agencies, act as catalysts in envisioning and creating public park spaces by helping with fundraising, design, and construction.

CO-MANAGERS. In some instances nonprofit groups work with parks departments to help develop capital projects and manage the parks. Co-managers such as the Central Park Conservancy and Prospect Park Alliance may be responsible for staffing, maintenance, programming, fundraising, renovations, and restoration.

SOLE MANAGERS. Under this arrangement a city turns over the management, planning, fundraising, design, construction, maintenance, and renovation of a public park to a private group. As is the case with the Maymont Foundation, the city retains marginal control and oversight of the park.

CITYWIDE PARTNERS. In this arrangement a nonprofit group or a coalition, such as the San Francisco Parks Trust, advocating for a city's entire park system or several parks in the system enters into an agreement with the city to help raise funds for and maintain the parks.[40]

The first wave of partnerships, those that emerged in the 1970s, were by and large assistance providers. The 1980s ushered in the era of conservancies; the Central Park Conservancy led the way. After decades of neglect and declining budgets, by the 1970s many urban parks were in a state of utter disrepair.

Some cities had all but abdicated their role as stewards of the public parks. At this juncture, wealthy citizens, corporations, and nonprofit organizations emerged to partner with cities to help manage and maintain the parks. The public-private partnerships were attractive for a number of reasons. While some city parks departments were mired in red tape, bureaucracy, and inaction, the nonprofit organizations could make decisions fast, raise funds, and save money; they could also be strong park advocates and pressure groups. They marketed parks effectively and were more effective fundraisers than city governments. As was the case in Central Park, wealthy donors reluctant to give money to local governments were much more willing to give money to nonprofit organizations where their money went to the upkeep of a specific park or for specific projects, not to general funds.[41] As was the case in the mid-1800s, the wealthy give to the parks to be charitably inclined, but they also give because they benefit handsomely from their largesse. In the case of the parks receiving the greatest attention and large gifts, the donors either live or work close to the parks (in which case they can use the parks regularly or see them from their home or work), realize an increase in their property values, or have greater stability in cities (as a result of reduced crime or enhanced aesthetic appeal) and more predictable business and living environments. In other words, they can count on more order.

The Central Park Conservancy: A New Model for Managing Public Parks?

Because of the success of the Central Park Conservancy, Central Park is held out as a model of private-public park management, but can and should this model be replicated across the country the way the Central Park design was propagated in the nineteenth century? To assess whether the Central Park Conservancy can be a viable model for urban parks to copy, an examination of the budget will be helpful. The Conservancy, working in concert with the parks commissioner and City Hall, oversees the restoration and maintenance of the park, construction of new facilities, and programming. In 2002 the Conservancy had 205 employees and a payroll of $12.22 million. The Conservancy's payroll grew to $14.54 million in 2005. That year Central Park had a staff of 240, all but 4 of whom were paid for by the Central Park Conservancy.[42]

As Table 9 shows, the Central Park Conservancy had almost $26 million in revenues in 2000. The budget nose-dived in 2001, when the stock market crashed after the World Trade Center bombing and the Conservancy lost almost $24 million in investments. Despite the heavy losses the Conservancy

Table 9 Central Park Conservancy's Operational Revenues and Expenses (in Thousands of Dollars), 2000–2005

Revenues and Expenses	2005	2004	2003	2002	2001	2000
Net assets at the beginning of the year	$124,708	$109,541	$98,940	$99,861	$118,362	
Net revenues	45,033	39,682	32,804	21,186	5,933	$25,817
Total expenses	29,172	24,514	22,203	22,107	24,434	24,876
Total net assets at the end of the year	*140,569*	*124,708*	*109,941*	*98,940*	*99,861*	
Revenue Stream:						
Contributions	23,883	16,464	16,067	18,659	13,149	13,055
Revenue from the City of New York	3,670	3,650	3,353	2,807	3,702	2,926
Special events	2,965	4,744	3,458	2,336	2,392	2,515
Interest and dividends	757	977	3,343	2,785	3,959	
Net loss/gain on sale of investments	1,007	11,207	-5,325	-9,452	5,654	
Change in unrealized loss gain on investments	4,141	494	10,828	3,085	-23,960	
Change in value of split interest agreements	-13	-28	-26	-4	-9	
Other revenue	8,623	2,174	1,538	970	1,046	2,068
Net assets released from restrictions						2,547
Endowment fund income						1,924
Administrative cost recovery						782
Annual revenues before adjusting for losses	45,033	39,682	38,587	30,639	29,902	25,817
Total Investment Losses	*-13*	*-28*	*-5,321*	*-9,456*	*-23,969*	

Source: Central Park Conservancy, *Annual Reports*, 2001, 12; 2002, 14–15; 2003, 37–39, 49; 2004, 15–20; 2005.

Table 10 Percentage of Distribution of Revenues in the Central Park Conservancy's Budget, 2000–2005

Percentage of Distribution of Revenues Before Adjusting for Losses	2005	2004	2003	2002	2001	2000
Contributions	53.03	41.49	41.43	60.89	43.97	50.57
Revenue from the City of New York	8.15	9.20	8.69	9.16	12.38	11.33
Special events	6.58	11.96	8.96	7.62	8.00	9.74
Interest and dividends	1.68	2.46	8.66	9.09	13.24	
Net loss/gain on sale of investments	2.24	28.24			18.91	
Change in unrealized loss/gain on investments	9.20	1.24	28.06	10.07		
Other revenue	19.15	5.45		3.17	3.50	8.01
Net assets released from restrictions			3.99			9.87
Endowment fund income						7.45
Administrative cost recovery						3.03
Total Revenues	*100.00*	*100.00*	*100.00*	*100.00*	*100.00*	*100.00*

Source: Central Park Conservancy, *Annual Reports*, 2001, 12; 2002, 14–15; 2003, 39; 2004, 15–20; 2005.

still managed to have almost $6 million in net revenue and ended the year with almost $100 million in its endowment.

By 2005 the Conservancy's revenues and assets were robust, with more than $45 million in revenue and $140.6 million in total net assets. Contributions constitute the largest revenue stream (53 percent in 2005) in the Central Park budget (see Table 10). Despite getting $3.67 million from the city in 2005, this accounted for only 8.2 percent of the Conservancy's budget. Under its management contract with the Department of Parks the Conservancy can get a maximum of $4 million annually from the city to operate the park.

In 2001 77 percent of the donations to the park came from individuals, 13 percent from corporations, and 10 percent from foundations. More than one hundred corporations contribute to the park. In addition the Perimeter Association comprises 115 residential buildings, hotels, and clubs that surround the park. The Perimeter Association, in collaboration with the Women's Committee, helps to maintain the park's six-mile perimeter.[43]

Though all agree that the Conservancy has done a magnificent job of restoring Central Park and insulating it from budget cuts, critics are troubled by some aspects of the Conservancy's role in the management of the park. They are doubtful that the model can be applied to a wide variety of parks. The Conservancy has been brilliant in securing funds for Central Park, but Central Park is unique, and that is a big reason it can be a fundraising success. Many parks will have a difficult time emulating the Conservancy's strategies because not all parks are surrounded by wealthy people or have the cultural and artistic stature of Central Park. For instance, Central Park's paid staff is assisted by more than three thousand volunteers who contribute more than thirty-five thousand hours of work annually. The volunteers are quite dedicated to the park; in 2005 thirty-eight of them had volunteered in the park for ten or more years.[44]

The Central Park Conservancy also raises funds successfully because of the board of trustees it has assembled. Of the sixty-two members, thirty-eight were executives at major corporations such as Marsh & McLennan, Lehman Brothers, Deloitte & Touche, AOL Time Warner, Ernst & Young, JP Morgan Chase, Goldman Sachs, Estée Lauder, Morgan Stanley, Bank One, Credit Suisse First Boston, Rubenstein Associates, Bloomberg, and American Securities.[45] The Central Park Conservancy combines donations from individuals with corporate donations and government funding to fulfill its budgetary needs and build its endowment. In 2002 twenty-five thousand individuals, foundations, and corporations made donations to Central Park. While some contribute as little as $35 to become members, others contribute millions. By 2003 the number of financial supporters had grown to thirty thousand.[46]

Other New York parks have tried to copy Central Park with much less success. The Prospect Park Alliance, founded in 1987, had a budget of $8.5 million in 2005 (see Table 11) and an endowment of $5 million. The 526-acre Prospect Park lags far behind Central Park, despite the fact that it is an Olmsted-Vaux park and a historic park, because it lacks the neighbors, location, and social cachet of Central Park. In 2002 Prospect Park had about seven million visitors. It had a staff of 101, 23 of whom were paid for by the City of New York and the Department of Parks and Recreation. Eight hundred new members joined the Prospect Park Alliance in 2002.[47]

Table 11 Prospect Park Alliance Budget: 2002, 2003, 2005

Revenues and Expenditures	2005		2003		2002	
	Amount	%	Amount	%	Amount	%
Total assets	$8,516,255				$6,818,920	
Total revenues	7,221,450		5,414,522		4,937,022	
Total expenses	8,208,224		5,414,522		4,937,022	
Net assets at the start of the year	5,959,492					
Net assets at the end of the year	4,972,718					
Revenue Stream						
Foundations	1,873,299	25.9	1,087,531	20.1	1,215,093	24.6
Individuals	833,498	11.5	1,610,141	29.7	1,108,150	22.5
Concessions and revenue	.		1,226,678	22.7	997,705	20.2
Design and construction	320,362	4.4	595,877	11	887,748	18.0
Corporations	417,311	5.8	660,763	12.2	425,253	8.6
Funds reserved from prior year					137,067	2.8
Government	276,859	3.8	214,272	4	130,205	2.6
Interest, dividends, and gains	109,864	1.5	19,256	0.4	35,801	0.7
Investment income	77,434	1.1				
Fundraising events	415,402	5.8				
Donated goods and services	624,114	8.6				
Visitor services and events	2,223,307	30.8				
Total revenues	*7,221,450*	*99.2*	*5,414,522*	*100.1*	*4,937,022*	*100.0*

Source: Prospect Park Alliance, *Annual Reports*, 2002, 19; 2003, 18; 2005, 15.

Prospect Park also has several corporate executives on its board of directors, and the Alliance gets donations from more than one hundred corporations, foundations, and associations. Since 2001 twenty-six donors have given $125,000 or more, yet these figures lag behind Central Park's, whose donors give in the millions. Whereas 3,519 individuals made contributions to Prospect Park in 2002, only 3 of those gave $50,000 or more. In contrast thousands of individuals and foundations made donations at that level to Central Park in 2002. As Table 11 shows, about 66 percent of the park's revenues were generated from corporate, foundation, and individual donations in 2003. As was the case in Central Park, Prospect Park managers are placing emphasis on increasing the percentage of contributions each year.[48]

The Central Park Conservancy's multi-year management contract allows it up to $2 million from concession revenues each year; other parks do not have this arrangement. For other parks revenues generated from concessions disappear into a city's general funds, so it is very difficult for a parks department, much less a single park, to see revenues from concessions. In 1995 Mayor Giuliani signed an agreement with the New York City's parks department to allow all the city's parks to keep a portion of the concession revenues (which did not apply to Central Park because of its special multi-year deal). The agreement was in place for only one year, but it increased budgets dramatically. Though the Prospect Park Alliance spent more than it took in in 2005, there was still about $5 million in assets at the end of the year.[49]

Other New York parks do not rely solely on the citywide partner, City Parks Foundation, to generate additional revenues for their administration and maintenance. The Riverside Park fund works on behalf of Riverside Park, and the Battery Conservancy focuses on Battery Park. However, not all New York parks receive this kind of special attention. Marcus Garvey Park, located in Harlem, and others languish in the shadows of fundraising behemoths such as Central Park. Though Marcus Garvey Park evokes its Olmsted roots and illustrious history on its website, neighborhood park advocates struggle to raise funds for the park. Volunteers who live close to the park try to raise money from local businesses and have recruited volunteers to clean up and maintain the park. Marcus Garvey Park's October 1999 fundraising dinner brought in $7,000; Central Park's Halloween Ball raised $805,000 that same month. A fundraising event at Prospect Park brought in $15,000 that year.[50]

In an effort to raise funds for parks other than Central Park, the City Parks Foundation was created, but the organization hasn't even come close to matching the success of the Central Park Conservancy. In 2005 the City Parks Foundation spent about $7 million and had an endowment of about

$13.3 million (see Table 12). In comparison the Central Park Conservancy had an endowment of $140.57 million in 2005 and expenses of $29.2 million.[51] Other parks host fundraising events similar to those held by the Central Park Conservancy. In September 2001 the Young Friends of the Public Garden's annual Swing into Fall Gala attracted 250 people and raised $26,500 for the parks they support.[52]

Despite the skepticism about and criticism of the Central Park Conservancy's model, budget cuts are forcing many urban parks to close or explore private management and revenue-generating options. Detroit made headlines with its "park repositioning" proposal that called for the city to sell off 92 small "vest-pocket" city parks—some in the poorest section of the city. Detroit has roughly 6,000 acres of parkland in its 367 parks. The parks being sold total about 124 acres. The city's decision rests not on the health and well-being of community residents, but on the revenues it can save or gain. It hopes to reap $8.1 million in revenues and save about $540,000 in maintenance bills for the parks being considered for sale. Detroit also expected to generate about $5.4 million in tax revenues from new development on the land. To counter opposition and criticism city officials point out that they renovated 11 parks and spent about $16 million to rehabilitate a recreation center in 2007.[53]

While some of the neighborhoods in which the small parks are being closed are among the poorest in the city and have the least access to open space, Detroit has spent millions sprucing up its waterfront parks and open spaces (where its major businesses are headquartered and where most tourists flock). So, a century after activists convinced cities to build small neighborhood parks so that everyone could have easy access to open space, cities have allowed the vest-pocket parks to fall into disrepair and are outlining plans to sell the land and close them. Once again, the poorest residents are vulnerable to being left without access to public open space. With the shift in philosophy that parks should be funded privately, it will be increasingly difficult for poor neighborhoods to keep their parks. In the case of Detroit the closures were announced before community members of the affected parks had an opportunity to come up with alternatives.

In their quests to make their park systems more solvent, cities look to New York and Chicago because these are the two largest fee-generating urban park systems in the country. New York City parks have an annual budget of $167 million, 22 percent of which is generated from private revenue sources. Thirty percent of Chicago's $307 million parks budget is generated from private sources. New York and Chicago also outsource much of their park operations; in fact, these two cities have contracted out more park operations

Table 12 Comparison of the Revenues, Expenses, and Assets of Private Park Partners

Name	Year Founded	City Located in	Park(s) Supported
Central Park Conservancy	1975/1980	New York	Central Park
Golden Gate National Parks Conservancy	1981	San Francisco	Golden Gate, Presidio
Maymont Foundation	1975	Richmond	Maymont Estate
Houston Parks Board	1975	Houston	City parks
Forest Park Forever	1986	St. Louis	Forest Park
City Parks Foundation	1989	New York	City parks
Friends of the Public Garden	1970	Boston	Boston Common, the Public Garden, Commonwealth Avenue
San Francisco Parks Trust	1971	San Francisco	City parks
Prospect Park Alliance	1987	New York	Prospect Park
Greater Newark Conservancy	1987	Newark	City parks
Piedmont Park Conservancy	1989	Atlanta	Piedmont Park
Hermann Park Conservancy	1990	Houston	Hermann Park
Cincinnati Parks Foundation		Cincinnati	City parks
Riverside Park Fund	1986	New York	Riverside Park
ParkWorks	1977	Cleveland	City parks
The Battery Conservancy	1994	New York	Battery Park
Garfield Park Conservatory Alliance	1994	Chicago	Garfield Park
The Greening of Detroit	1989	Detroit	City parks

Source: Charity Navigator, *Charity Navigator Rating* (Mahwah, N.J.: Charity Navigator, 2006).

(running parking garages, stadiums, marinas, skating rinks, swimming pools, golf courses, etc.) to private companies and concessions than any other urban park system in the country. Outsourcing can be profitable for city parks departments; for example, the New York Department of Parks generated $21.5 million in concessions and fees in 1979, and by 1997 that amount had grown to $36 million.[54]

Other city parks departments have explored fees and private revenue sources. For instance, the Baltimore Parks and Recreation Department has a

Table 12 Continued

Total Revenues ($)	Total Expenses ($)	Excess or Deficit ($)	Net Assets ($)	Fiscal Year
40,954,120	29,220,720	11,733,400	140,569,894	2005
17,386,163	14,592,813	2,793,350	22,910,820	2005
3,223,927	4,010,106	-786,179	21,842,029	2004
2,261,747	2,832,865	-571,118	15,555,976	2005
7,839,250	3,193,495	4,654,755	14,839,094	2004
7,575,528	7,013,836	561,692	13,303,286	2005
1,936,466	465,882	1,470,584	6,842,686	2004
6,754,459	6,171,243	583,216	6,247,090	2005
6,022,128	6,485,641	463,513	5,959,492	2004
1,351,188	1,060,621	290,567	5,380,222	2005
3,384,817	3,376,604	8,213	5,276,481	2004
1,689,642	1,295,598	394,044	3,828,421	2005
827,632	596,358	231,274	2,090,627	2004
1,896,553	1,135,508	761,045	1,582,701	2004
1,466,235	1,436,782	29,453	692,467	2004
1,946,797	1,289,284	4,189,757	657,513	2004
1,194,146	1,115,985	78,161	503,140	2005
992,120	1,062,221	-70,101	306,012	2004

$15 million budget but generates only about $135,000 (less than 1 percent) of that in fees. Similarly the Providence Parks Department generates only about 3 percent, or $250,000, of its $7.6 million budget in fees. At the other extreme is Wheeling, West Virginia. The Wheeling Parks Commission is one of the most fully self-supporting city parks departments in the country. Wheeling gets only about $190,000 (less than 10 percent) of its $20 million budget from the city. In fact, G. Randolf Worls, chief executive officer of the Oglebay Foundation in Wheeling, predicts that private funds will become the dominant form of financing for urban parks. The Oglebay Foundation raises funds to help administer and maintain Wheeling's Oglebay Park and Wheeling Park. Worls's goal is to increase the parks' endowment to $75 million, at which point he believes the parks can operate without any outlay of public funds.[55]

Increasingly, urban parks are relying on private partnerships for financing as city parks departments broaden their funding bases.[56] In fact a 2000 study of sixteen public-private partnerships nationwide showed that most raised relatively small sums of money and had small operating budgets. Only six had an operating budget of over a million dollars.[57] However, an analysis of the 2004 and 2005 budgets of seventeen such partnerships shows that all but one have an operational budget of at least a million dollars (see Table 12, above). Moreover all but four also had endowments over a million dollars. Table 12 also shows the huge gap between the Central Park Conservancy's operational budget and endowment and those of other park partners.[58]

Critics of public-private partnerships argue that they create a two-tiered park system of well-endowed parks and parks with little or no money. In New York, for instance, the Central Park Conservancy has raised vastly more money than other city parks. In other places nonprofits select and promote the best parks while the others languish. In Massachusetts, for instance, a deliberate attempt was made to trade on Olmsted's name in deciding which parks to rehabilitate. A study of the 280 public spaces in the state designed by Olmsted's firm was conducted, and 10 parks were chosen to represent Olmsted and his firm's work. Hence Friends of Buttonwood Park formed a partnership with the City of New Bedford to rehabilitate that Olmsted park. Similarly the Louisville Olmsted Parks Conservancy highlights the Olmsted-designed parks in their work.[59]

The Central Park Conservancy's presence buffer's Central Park against budget shocks, a luxury not afforded other parks. In February 1998 the Conservancy signed a renewable eight-year management agreement with the City of New York and the Department of Parks and Recreation. Under the contract the Conservancy must raise and spend $5 million annually for maintenance and repairs, public programs, landscaping and rehabilitation, or repair of existing facilities. If the Conservancy meets these requirements the Department of Parks and Recreation will pay the Conservancy up to $2 million, depending on how much the Conservancy exceeds its $5 million threshold in any given year. In addition the Conservancy gets 50 percent of net concession revenues earned in excess of $6 million. The amount the Conservancy can receive from concession revenues is capped at $2 million per year.[60]

Some argue that organizations like the Conservancy allow cities to pare away park budgets and to step away from their commitment to provide public recreation to all citizens. Critics also argue that the Central Park Conservancy undertakes projects its trustees are interested in, and that it can raise money for projects that are not necessarily in the interest of the long-term sustainability of the park or that would benefit the majority of users.[61]

However, as Elizabeth Barlow Rogers sees it, we are moving into an era of contract management for parks, and given the projected budget shortfall for urban parks, many cities may not have much choice but to look at public-private management arrangements.[62] Some city parks departments are not even at the level of funding they were before the financial crises of the 1970s. For instance, though New York's Park and Recreation Department has been innovative in getting funding, it is not back to the level of funding it had in 1978. In the current economic and political climate market-based funding for recreation is gaining support around the country.[63] However, decisions such as Detroit's to sell off its parks raise concerns among many about the survival of the urban park system.

Urban Parks: Developer Exactions, Landfill Conversions, and Public-Private Management Arrangements

Not only is it exceedingly difficult for cities to generate funds to maintain their parks, but it is even more difficult to build new parks. Consequently one of the most common ways in which cities increase their open space is through developer exactions, whereby developers of new housing or commercial enterprises donate or develop parks or other open space as part of their contract to build. However, most of the parks being developed through exactions are being built in the suburbs or counties on the urban fringes.[64]

Despite the growing popularity of developer exactions as a tool to offset the impact of new development and acquire and develop urban parks, a study done by the Trust for Public Land found that even when cities have exaction ordinances, such initiatives do not always result in land acquisition. The study found that the formula to exact fees or land from developers varied widely around the country. For instance, Chicago requires about 1.7 acres of land from developers for every one thousand residents a development houses; Austin mandates 5 acres per one thousand residents, while San Diego seeks about 20 acres per thousand. In six of the twelve cities studied (Austin, Forth Worth, San Antonio, Long Beach, San Jose, and Portland, Oregon) exactions resulted in the creation of 1,572 acres of new parkland. However, according to the exaction ordinances in these cities 2,594 acres of new parkland should have been generated. That means that only about 61 percent of the parkland the exactions stipulated was actually generated. Portland generated about 2.33 times more parkland than its exaction ordinance stipulated because it issued bonds to acquire the land with the hope that the bonds would be paid off by future exaction taxes. Portland expected to generate 39 acres of parkland through exaction but actually acquired 90.86 acres by issuing bonds for land acquisition. Other cities generating a relatively high percentage of parkland in relation to the stipulations of their ordinances were Austin (77 percent) and Fort Worth (63 percent). In contrast, San Antonio acquired no parkland, although 227 acres should have been generated as per the ordinance. Six of the cities in the study (Albuquerque, Atlanta, Chicago, Los Angeles, San Diego, and Miami) had very poor tracking of whether the ordinances had generated parkland and how much land had been acquired.[65]

Even in cities that did track and generate new parkland from exactions, several factors hindered the generation of new parkland. Ordinances require that the park be close to the subdivision that generated the tax for it, but it is sometimes impossible to find nearby land that can be converted to a park. Cost is another factor; some cities underestimate the cost of acquiring land, especially in places such as California, where the price of land is very high. Sometimes new parkland is not generated because cities use the funds obtained from exaction fees to restore existing parks rather than acquire new ones. Some cities provide exaction exemptions to developers who construct other kinds of facilities, such as dependent care complexes, shelters, or group homes, that are deemed beneficial to the community.[66]

Parks that will be built through developer exactions include The Great Park at El Toro, which is being planned for Orange County, California; Stapleton's parks in Denver; and Laurel Hill in Fairfax County, Virginia.[67] The Fresh Kills Park being planned for Staten Island will result from landfill conversion.[68] Though these locales are developing gigantic urban parks that will dwarf historic parks like Central Park, Prospect Park, and Franklin Park, these new parks will not be among the largest urban parks. There are currently twelve city parks that range in size from the 5,554-acre Umstead State Park in Raleigh, North Carolina, to the 495,996-acre Chugach State Park in Anchorage, Alaska. In addition, there are sixty-nine urban parks that are between 1,000 and 5,000 acres.[69]

Some of the new parks are made possible because of the closure of military bases (El Toro), airports (Stapleton), prisons (Laurel Hill), and landfills (Fresh Kills). These newer parks, though influenced by Central Park's design and funding structure, are coming up with new variations of public-private co-management partnership arrangements. The parks feature mixed-use development and residential communities tucked away amid green space. In addition to recycling industrial land and structures, they are at the cutting edge of adopting green technologies into their design and operation.[70]

These parks are breaking with tradition in other ways: their cities are completely outsourcing the work of designing, building, managing, and maintaining these parks; corporations and corporate leaders have significant influence over the development of these parks; and the financing of these parks rests heavily on property taxes (residential and commercial) and user fees (for activities such as golf). In these park developments, park advocates focus on minimizing city funding or raising funds from individual donations. However, concerns have been raised about these new parks, such as lack of public oversight and accountability and lavish perks enjoyed by committee members.[71] Of even greater concern are the ways in which these new parks are sometimes treated as private or semi-private by the developments they anchor. Not only are the poor priced out of the subdivisions that ring these parks, the provisions for recreational activities such as golf courses, horseback riding, and marinas often exclude the poor as well. This is reminiscent of the use of parks to anchor upscale developments in cities in the first half of the nineteenth century. If access to urban parks is increasingly driven by one's ability to pay to use park amenities, to purchase a home overlooking a park, or to donate to a park, it rekindles the debate New Yorkers had over paying to use chairs in the park. The questions Americans will face in the coming years are: To what extent should cities be held responsible for providing their residents with parks? Or will access to such amenities be driven by purely market forces?

Theoretical Discussion

NINETEENTH-CENTURY POWER ELITES AND PARK ADVOCACY

The story of the origins of Central Park demonstrates how a small group of park advocates formed a very effective power elite who propagated several influential ideas: first, American cities needed large landscaped public parks; second, the parks should be funded by general taxation; and third, if landowners refused to sell, then land should be taken by eminent domain. Elites also popularized the notion that parks were effective agents of social control, had health-giving capabilities, and improved the culture and civility of the city and the nation. Though activists advocated the development of large landscaped parks and floated potential funding options before the emergence of the Minturn Circle, it was the Minturn Circle that articulated the immediacy of the problem and put forward an effective solution. The Minturn Circle effectively changed the debate from *whether* a large landscaped park should be built to *where* such a park should be built. The discourse shifted so rapidly that opponents were caught off guard. Though the site and tax structure chosen by the Minturn Circle did not eventually prevail and the use of eminent domain was successfully challenged, it was the Minturn Circle that inoculated the parks debate with these ideas. In fact the Central Park proposal was successful because the site was larger and more centrally located, and, after hearing objections to the Jones Wood tax proposal, Central Park advocates modified the Minturn Circle's proposed tax structure to gain approval for their chosen site.

The Minturn Circle had an impact because its members were a power elite composed of the city's most influential businessmen, politicians, landowners, and wealthy families. This power elite had in its fold some of the most powerful individuals in the most critical constituencies needed to develop massive public works projects in a city such as New York at the time Central Park came to fruition. While the Great Fire drove civic leaders to complete the Croton Reservoir, there was no such catalytic event to trigger mobilization around the park idea. Park advocates had to mobilize resources and supporters with skillful framing and the force of their arguments. These men knew what they wanted and had the power and resources to leverage it. There was also evidence of overlapping directorates in the dense network of family, business partners, neighbors, and friends who were the most ardent park advocates. Kingsland, for instance, was mayor of the city, an influential merchant, and a member of the Minturn Circle. He was able to argue the case for the park in front of the city's governing body. Similarly, Beekman was a landowner in the vicinity of the chosen park site, a senator, and a member of the Minturn Circle at the same time. He was able to introduce and push forward bills supporting the Jones Wood proposal in the state assembly. With the key park advocates having large and overlapping influence in the mayor's office, the state assembly, the Chamber of Commerce, churches, banks and other financial institutions, and the major newspapers, the group was able to propagate their ideas and make the political and policy changes necessary to bring them to fruition.[72]

MOBILIZATION OF RESOURCES OR MOBILIZATION OF PEOPLE?

Schwartz and Paul distinguish between the mobilization of resources and the mobilization of people.[73] The Minturn Circle concentrated its efforts on mobilizing resources. Though the original Minturn Circle broadened its base of supporters to include powerful business and political allies, the early urban park movement was not a mass movement. Park advocates did not court grassroots support; indeed, they ignored suggestions from grassroots advocates to build small neighborhood parks. The fact that the Minturn Circle did not invest time or resources in developing broad public support for its proposal played a role in the defeat of the Jones Wood proposal. Though the Jones Wood opponents were weak and disorganized for much of the campaign, the Central Park supporters effectively collaborated with Battery Park advocates to get their demands met. Eventually the group that mobilized both resources and people was more effective in getting its park proposal approved.

The Central Park supporters were also more adept at framing the issue, publicizing their position, and developing a more acceptable funding model. Even though Central Park was in a more remote, hilly, and landlocked location than the Jones Wood waterfront site, Central Park supporters framed the park as a "central park" and Jones Wood as a "sidelong" park that was more difficult to reach. Their master frame of "central versus sidelong" captured the essence of the conflict between the two sets of park advocates, and the Minturn Circle did not develop an effective counterframe.

Central Park supporters also capitalized on the master frame of size. Knowing that size mattered to park advocates wanting space for carriage rides and trotting races, supporters publicized the size of the Central Park—close to 800 acres in its original conception compared to the 150 acres of Jones Wood—and used this to their advantage. The Minturn Circle could have counterargued that it would cost much more in money and time to build Central Park, but it didn't; nor did it compare the tax assessment structures of each proposal.

WORKING-CLASS ACTIVISM AND THE INJUSTICE FRAME

Working-class activists were effective at getting city governments to develop urban parks. Over the course of their campaign for increased access to parks the working class grew bolder. They became more sophisticated in the way they used the media and framed their message. They supported these demands with surveys, scathing newspaper editorials, political campaigns, and the strategic use of their city council votes. They appropriated the Olmsted-Vaux arguments (as well as the arguments of their park administrators) and used them to frame the issues. The working-class activists argued that they too needed the health-giving benefits of the parks. Working-class Worcesterites used injustice framing to make the issues salient to their supporters.[74] They pointed to the inequitable distribution and maintenance of parks in the city, unequal delivery of services, and the increased health risks for the working class. Activists further argued that confinement to the city made working-class demands for relief more urgent.

Working-class activists focused less on mobilizing resources in order to strategically mobilize powerful political allies. They also targeted the government for greater oversight, policy development, and greater deployment of resources. Having no money to build their own parks, this was one of the few options open to them.

As the twentieth century progressed blacks began to challenge their exclusion from parks, beaches, and other recreational amenities. To be effective they developed cross-race, cross-class coalitions with progressive whites. They used black-owned media and black institutions to publicize the cause and mobilize their constituents and focused on getting the government to act on their behalf to ameliorate conditions.

The Central Park Conservancy is the parks partnership group that most embodies a modern-day power elite. This group was able to get a contract from the city to take over the operations of the park and to have a portion of the lucrative park concession revenues returned to them; though other park partners have tried, they have not succeeded in getting concession revenues. Even with this unusual feature in its contract, the Central Park Conservancy has been able to renew its management contract with the city. The Conservancy has also been able to launch the most ambitious capital campaigns and raise the largest sums of money. They have done this far more effectively than any other park partner. Despite mimicking the Central Park Conservancy, none of the other park partners have emerged as a true power elite in the way the Central Park Conservancy has. The Golden Gate Parks Conservancy and Forest Park Forever have the potential to exert their will as power elites.

Though wealthy property owners in the mid-nineteenth century rebelled against funding parks solely through benefits assessment taxes, modern park advocates and administrators are promoting park financing schemes that rely heavily on benefits assessments levied against landowners adjacent to parks. The developer exactions fees that will result in the development of new parks such as The Great Park at El Toro, Stapleton, and Laurel Hill are contemporary forms of benefits assessments levied against the property of primarily middle- and upper-income homeowners. The difference is that the benefits assessment is capped at a very low level; thus homeowners, who are aware of these taxes before deciding to purchase in a development, actually choose whether or not to pay for these parks. In a way, modern methods of levying assessments against an entire development is similar to the nineteenth-century

Central Park supporters' strategy of levying the assessments against a larger area than earlier practices dictated. Spreading the benefits assessments taxes to a wider group reduces the tax burden on each.

Modern park partners have taken sides in another debate that has been around since the mid-nineteenth century. While Olmsted felt that commemorative items, name plates, and the like detracted from the contemplative mood he wanted to evoke in the park, Downing felt these were lucrative financing opportunities that parks should capitalize on. Ironically Central Park, whose administrators and supporters most wanted to recapture the nineteenth-century pastoral, contemplative park the original designers intended, led the way in creating naming and giving opportunities. A host of other urban parks have adopted this model.

Park partnership groups have focused on mobilizing resources as well as people. Though these organizations encourage the public to use the parks they promote public participation through financial contributions and donated time. The websites of the parks discussed in this chapter indicate that they place a premium on cultivating financial relationships with potential park supporters. Increasingly parks rely on user fees generated from services provided. Even participation in volunteer groups is not always free; many of the volunteer park support groups have a membership fee. Still, park advocates and administrators are careful to court this financial relationship while at the same time framing the parks as democratic spaces for all people. To this end they develop programming to help the public develop an affinity for the park.

Since urban parks are still evolving, it is likely that new models of financing will emerge in the future. New kinds of parks are also emerging; these are multi-use spaces serving a variety of recreational, business, educational, and residential functions. Given the current state of city park budgets it seems likely that new park developments will involve some level of public-private partnerships in the future.

Part IV

THE RISE OF COMPREHENSIVE ZONING

Part IV examines the evolution of planning and zoning in cities. It examines the relationship between aesthetics, class, race, and space. Chapter 11 shows how concerns about class, exclusivity, and aesthetics led property owners to rely on restrictive covenants as a means of developing uniform neighborhoods. Chapter 12 focuses on the rise of comprehensive zoning in America. The discussion also shows how zoning became a tool to implement racial and class segregation in the cities in the early twentieth century. This chapter also analyzes the zoning movement in the context of how it fits into the discourse on order and social control in the city. The chapter examines early attempts at zoning, restrictions on land use, and the emergence of groups devoted to urban planning. Part IV considers the ways in which the ideas of planning and zoning advocates were challenged and how planners responded to those challenges.

CLASS, RACE, SPACE, AND ZONING IN AMERICA

Early Attempts at Zoning

ELITE RESIDENTIAL CLUSTERS

Since the seventeenth century elites have been concerned with managing land use and open space. Planned cities such as New Haven began to emerge in 1641. Boston began segregating land use and regulating the location of nuisance industries in the 1650s, and the mayor of New York expelled the stink factories from Lower Manhattan in the 1670s. When William Penn laid out a plan for Philadelphia in 1682 he tried unsuccessfully to develop a green city.[1] Wealthy urbanites took a step in this direction during the middle of the eighteenth century when they began acquiring land on the fringes of cities and building country estates surrounded by gardens, parks, orchards, farms, stables, and hunting grounds.[2] This might have been the closest urbanites came to approximating Penn's idea of green country towns for gentlemen farmers.

Their country seats notwithstanding, the wealthy still wanted to see more order in the city. Hence wealthy urbanites and town planners made more concerted efforts to establish zones in cities for high- and low-income housing as well as commercial and industrial land uses during the nineteenth century. Yet despite their best efforts it wasn't until the early twentieth century that comprehensive city zoning took effect.

As discussed in chapter 4, starting in the early 1800s wealthy urbanites attempted to develop exclusive residential communities anchored by parks, where they could live with others of like race, ethnicity, and socioeconomic status. But they had little or no control over who moved next door, to whom properties were sold, or what would be developed in the neighborhood.[3] Ergo rich urbanites conceived of ways to plan residential communities more deliberately. In this vein Hezekiah Pierrepont acquired sixty acres in Brooklyn Heights overlooking the East River and subdivided it into lots of twenty-five by one hundred feet. The plan was to build masonry row houses, each home having a backyard; some homes were built on double lots.[4]

American civic leaders looked to Europe for urban living arrangements that incorporated residential living and open space; they found the "villa parks" attractive. These were residential enclaves with clusters of single-family homes built on large, unfenced private lots. The grounds were landscaped in a unified theme, and families living in these enclaves often shared a common recreational open space. Some of the villa parks included semi-detached and row houses. Clapham, outside of London, was developed as a villa park in the late eighteenth century by families that shared similar religious beliefs. In 1823 John Nash developed the Regent's Canal Village. Nash also collaborated with James Morgan to develop the Newbold Comyn Estate at Leamington in 1827. Joseph Paxton and James Pennethorne designed Prince's Park in Liverpool in 1842. A year later Paxton began working on Birkenhead Park outside of Liverpool.[5]

In America, Hudson Square in New York, conceived as a villa park, became the model for developing exclusive residential communities around open space and controlling land use on a neighborhood scale. In 1827 Hudson Square lots were sold around the semiprivate St. John's Park.[6] Though this was a planned residential development there were still no restrictions on the kind of house that could be built. However, within a year developers began taking steps to ensure that buildings looked more uniform in planned communities in terms of height, setback, design, and building materials. They also began placing restrictive covenants on those buildings to preserve their uniformity.

In 1828 the developer Isaac Pearson built and sold a row of houses on Bleecker Street. Each house sold for $12,000 ($128,000 in 2005 dollars). He placed a restrictive covenant on each property stipulating that the ten feet between the front wall of the property and the street should remain open space, unoccupied by any buildings, walls, fences, or railings. In so doing Pearson created a parklike setting on the street. He also convinced the Common Council to rename the block Leroy Place.[7]

PLANNED COMMUNITIES

Developments similar to Hudson Square and Leroy Place soon appeared in other cities as developers experimented with creating entire communities from scratch. In 1835 John R. Pitkin, a wealthy Connecticut merchant, began buying up large tracts of land east of Brooklyn along the shores of Jamaica Bay. Pitkin wanted to build a city, to be called East New York, to rival Manhattan. He laid out streets and lots and prepared a city plan with housing, industrial buildings, schools, and parks. Pitkin also sought to develop a model village in Woodside (later renamed Woodhaven), Queens. In 1836 a group of men bought an estate and a farm in Williamsburgh (now a part of Brooklyn) and erected fourteen upscale model homes on the site. That year John Haviland acquired land and developed New Brighton on Staten Island. He built rows of villas on a hillside overlooking the water and the Manhattan skyline.[8]

Efforts to develop planned communities were sporadic until the 1850s. Andrew Jackson Downing generally advocated building country mansions surrounded by expansive, landscaped grounds, but he conceded that smaller rural villas could be built if grouped around a park. A group of wealthy New Yorkers decided to experiment with Downing's idea of a Romantic suburb in the mid-1850s. Alexander Jackson Davis helped Llewellyn Haskell, a wealthy drug merchant, to develop Llewellyn Park in New Jersey. Davis had authored the 1837 book *Rural Residence*, and some of his designs were included in Downing's *Treatise*. Llewellyn Park was a collection of villas set on curvilinear drives ringing the Ramble, a fifty-acre park. Located in the foothills of the Orange Mountains the elaborately landscaped development was a success; it offered beautiful vistas of brooks, waterfalls, meadows, and wooded glens. The retail baron A. T. Stewart founded Garden City, Long Island, in 1869 along similar lines.[9]

At a time when landscape architects, developers, and planners focused on the external features of the house and lot, Catharine Beecher did pioneering work advocating improvements and increased efficiency inside the home. She included many innovations in her home designs to improve the lives of the occupants. Beecher urged women to consider the home as the woman's sphere and to take control of it. She advocated using the home as the venue for instituting moral, religious, sanitary, and social order in the family.[10]

Olmsted's Planned Suburban Communities

Frederick Law Olmsted was involved in planning subdivisions and suburbs in the second half of the nineteenth century. In his 1866 plan for the College of California at Berkeley he argued that a suburb should have park-like spaces to serve as the "social rendezvous of the neighborhood" and should also provide "domestic seclusion." The gardens surrounding such homes should not be considered mere "ornamental appendages of a house" or "marks of the social ambitions of the owner" but "essentials of health and comfort." Olmsted argued that if there were undesirable features close by (dirty roads, ugly buildings, or the haunts of drunken or undesirable people) trees and other greenery could keep such features out of sight and hearing.[11] He used the planting-out method to block the sights and sounds of the city from intruding on recreational experiences in Central Park.[12]

Olmsted's plan for a subdivision around Berkeley was never adopted, and his proposal of a similar plan for Long Branch, New Jersey, was not built either.[13] He finally got his chance to build a suburban community in Riverside, Illinois, in 1868. Olmsted believed that the "essential qualification of a suburb is domesticity." He argued that property values were higher near parks and that the density of urban dwellings was a "prolific source of morbid feebleness or irritability and various functional derangements" among urban dwellers. In planning Riverside, located nine miles from Chicago, Olmsted used curvilinear roads and oblique-angled intersections to "imply leisure, contemplativeness and happy tranquility."[14] On undesirable land use that marred the aesthetic appeal of residential communities he wrote

> Line a highway . . . with coal yards, breweries, forges, warehouses, soapworks, shambles, and shanties, and there certainly would be nothing charming about it. Line it with ill-proportioned, vilely-colored, shabby-genteel dwelling-houses, pushing their gables or eaveboards impertinently over the sidewalk . . . and it would be anything but attractive to people of taste and refinement. Line it again with high dead-walls, as of a series of private mad houses, as is done in some English suburbs, and it will be more repulsive to many than the window-lighted walls of the town blocks. Nothing of this kind is wanted in a suburb or rural village. Nothing of this kind must be permitted if we would have it wholly satisfactory. On the contrary, we must secure something very different.[15]

Recognizing the difficulty of controlling what people built on their lots, Olmsted continued

> We cannot judiciously attempt to control the form of the houses which men shall build, we can only, at most, take care that if they build very ugly and inappropriate houses, they shall not be allowed to force them disagreeably upon our attention. . . . We can require that no house shall be built within a certain number of feet of the highway, and we can insist that each house-holder shall maintain one or two living trees between his house and the highway-line.[16]

Olmsted believed that communities should be designed with village greens, commons, and playgrounds rather than an enclosed park. At regular intervals on each road, he wanted openings large enough for a natural group of trees, croquet or ball grounds, sheltered seats, and drinking fountains.[17] He developed plans for such a community at Tarrytown Heights, New York, but the company underwriting the development went bankrupt in the Panic of 1873, and Olmsted's plan was never realized.[18]

CEMETERIES AND ZONING

Ironically the first examples of successful zoning occurred in rural cemeteries such as Brooklyn's Green-Wood in the 1830s. As was the case in Mount Auburn in Massachusetts, the trustees of Green-Wood outlined rules governing use of the cemetery by lot holders and visitors. However, Green-Wood's rules were more elaborate. The trustees wanted the cemetery to have open space and high-quality, durable monuments; hence they barred the use of veneers, specified the maximum curb around plots, and banned pet burials. Those who owned plots had to receive permission to plant shrubs and trees. They were told to build the foundations of their monuments solid and deep and to avoid "tame and uninteresting monuments." They could not use posts and chains because these rusted and children swung on them. Furthermore the trustees reserved the right to remove "any injurious or offensive structure or monument" placed in the cemetery.[19]

Restrictive Covenants

EARLY USE OF RESTRICTIVE COVENANTS

During the early eighteenth century large landowners routinely inserted clauses in their leases requiring lessees to remove all structures they built on the property within ten days of the expiration of the lease. However, later in the century landowners wanting to develop their land without expending large sums of money began encouraging their lessees to build permanent, high-quality buildings. They placed covenants in the leases requiring lessees to build a house to specifications that were spelled out in the contract. A typical covenant required the lessee had to build a "substantial, workmanlike and well built" two-storey brick or brick-front house within a year of taking occupancy; at the end of the contract lessees could purchase the structure or renew the lease. By the nineteenth century covenants specified that only one house, usually of a mandated size, could be on the lot; this prevented lessees from building houses on the rear of the lot and renting them out. Though a second house would have helped the lessee financially, landowners claimed that it attracted poorer tenants and lowered their property values. The goal of the building covenants was to coerce lessees to invest a substantial amount on construction and to limit the long-term leases to those who could afford to build and maintain the property. This meant that lower-middle-class and working-class families could not afford to lease such properties.[20] For instance, covenants prevented the artisans and small businessmen who leased lots from Trinity Church from building more than one house, as was common in the 1770s and 1780s.[21]

LAND ACQUISITION AND RESTRICTIVE COVENANTS

Though the cemeteries were able to develop effective zoning beginning in the 1830s, it was much more difficult to develop effective zoning in urban and suburban communities. Central Park and the wealthy residents who flanked it eventually played a critical role in the development of zoning laws and the spread of such laws in America. Central Park represented an important step in the process of controlling land use in the area. To build the park more than eight hundred acres were cleared of houses, small-scale industries, animals, and garbage dumps, and about sixteen hundred poor blacks and whites were evicted.[22] These actions helped to create some order even before the park was completed in a part of Manhattan that was relatively undeveloped. Once this large swath of land was cleared and the development of a large landscaped park got under way, wealthy residents moved to the surrounding blocks and began taking steps to organize the land use in their neighborhoods.

The section of Fifth Avenue north of Fifty-ninth Street that ran along Central Park became one of the last escapes for New York's wealthy. But the street had both elite residences that changed hands regularly and upscale commercial establishments. As businesses and millionaires vied to get a Fifth Avenue address, land values skyrocketed and developers responded by trying to squeeze more buildings into the shrinking space. Vacant land north of Fifty-ninth Street appreciated by 200 percent between 1868 and 1875; some lots quadrupled in value even before they were developed. Just as the wave of mansion building peaked in the 1890s developers began demolishing the homes to build luxury apartments. Residents responded by seeking greater control of development changes in the area; when private efforts to control space failed the residents organized and requested government intervention.[23]

As chapter 3 indicates, wealthy Manhattan residents began migrating uptown in the mid-eighteenth century. Commenting on the procession of elites hurdling each other to get uptown, George Templeton Strong, a Gramercy Park lawyer and diarist, wrote, "When I was a boy, the aristocracy lived around the Battery . . . [and] on the Bowling Green. . . . Greenwich Street, now a hissing and a desolation, a place of lager beer saloons, emigrant boarding houses, and vilest dens, was what Madison Avenue is now. . . . We [the elites] are a nomadic people, and our finest brownstone houses are merely tents of new pattern and material."[24]

Even before Central Park was finished Mary Mason Jones (Edith Wharton's aunt) built her mansion at Fifth Avenue and Fifty-seventh Street. At the time the area was still undeveloped, and shantytowns, slaughterhouses, and asylums were more common there than grand homes. However, it was not long before "goatville" gave way to "Millionaire's Mile." The mansions that sprouted in the area were three- and four-storey Italianate buildings topped with a mansard; they ranged in price from $75,000 to $150,000. The Astors moved from Washington Square to Thirty-fourth Street to Fifth Avenue and then to Sixty-fifth Street. Mrs. Astor's move to upper Fifth Avenue set off a virtual stampede of emulators seeking to acquire the most fashionable address in New York. Andrew Carnegie built his mansion at Fifth Avenue and Ninetieth Street, just south of Harlem. By the mid-1880s the millionaires Harry Payne Whitney, Charles Harkness, Jay Gould, Collis P. Huntington, Benjamin Altman, Robert Goelet, Solomon Guggenheim, Russell Sage, and William Rockefeller lived on upper Fifth Avenue. The most intense period of mansion building on the upper avenue was from 1880 to the 1920s.[25]

By this time elites had begun to realize the limits of their exit strategy: if they wanted to live in Manhattan, and most did, they were running out of places to flee to. While they relinquished Battery Park, Bowling Green, and Hudson Square, these "door-yards of space," "grass-plots," or "scanty patches of verdure,"[26] to working-class housing and industrial development, they weren't about to abandon their view of and access to Central Park. Nonetheless the pressure from commercial encroachment intensified with each passing year. In 1869 developers applied for a permit to convert a Fifth Avenue mansion into a boardinghouse, and during the 1870s commercial enterprises—clothiers, piano salesmen, jewelers, dry goods merchants—engulfed Broadway as far north as Twenty-third Street.[27] At this point the wealthy began to look for more effective means of stabilizing their neighborhoods, maintaining exclusivity, and protecting their investments.

One approach was to refuse to sell. This strategy was not particularly successful, though, because few families opted for it; consequently the holdouts were often left with homes surrounded by commercial enterprises, and they too eventually succumbed to the pressure to sell or the threat of eminent domain.[28] Although for a while they gained a moral victory for holding out, the character of the community they sought to preserve changed rapidly around them.

RESTRICTIVE COVENANTS AND THE COURTS

Because holding out was ineffective, elites continued to use restrictive covenants to control land use. Restrictive clauses stipulated limits on building size, type and quality of buildings that could be constructed, and setbacks. Because the covenants stayed with the land they bound all future owners. By the early twentieth century most of upper Fifth Avenue was covered by restrictive covenants. As time went on the wealthy combined large land purchases with the use of restrictive covenants to control larger spaces. For instance, when the Catholic Orphan Asylum was sold in 1902 and construction began on an eighteen-storey hotel, William Vanderbilt bought the land and built the Marble Twins on it instead. He also purchased the Langham Hotel when it came on the market, tore it down, and placed covenants on the land stipulating that the property had to remain residential for at least twenty-five years from the time of purchase.[29]

Restrictive covenants grew in popularity from the 1830s onward, and many were drafted with the intent of creating and maintaining openness, green space, and residential neighborhoods. In 1836 Anna Prince, Susan Lawrence, and Margaretta Willoughby, heirs of Johannes Debevoise, a prominent landholder and member of the Dutch Reformed Church, subdivided their large tract of land in Brooklyn and placed a restrictive covenant in the deeds. It read

> No dwelling house, store house or other building or structure of any kind or description whatsoever (excepting fences) shall at any time or times hereafter be erected on any lot of ground fronting on or otherwise adjoining Debevoise place, Bond street, northwardly of Schermerhorn street or Hanover place within the several distances hereinafter specified from the lines or sides of said places and the street . . . with regard to Debevoise place within fifteen feet . . . the intent and meaning in this respect . . . being on the one hand to insure an open space . . . between the lines of the buildings fronting on or adjoining the said streets and spaces, respectively, and the observance of uniformity in the location of such lines.[30]

By the 1840s covenants specifying setbacks were common in Manhattan. Their intent was to maintain open space in the front of buildings and uniform building lines. In 1846 owners of lots north of Thirteenth Street between Broadway and Fifth Avenue placed restrictive clauses on properties specifying an eight-foot setback.[31] Property owners also wanted to exclude certain types of buildings and land use from their neighborhoods. John Wendel and his wife went a step further in trying to create a sense of openness and prevent noxious facilities from being built on a property they sold to Henry Hurlbut in 1859. Their restrictive covenant forbade the buyer, his heirs, or assigns from "erect[ing] any buildings within forty feet of the front of said lots, except of brick or stone, with roofs of slate or metal, and will not erect or permit upon any part of the said lots any slaughter house, smith shop, forge furnace, steam engine, brass foundry, nail or other iron factory, or any manufactory of gun powder, glue, varnish, vitriol, ink or turpentine, or for the tanning, dressing or preparing of skins, hides or leather, or any brewery, distillery or any other noxious or dangerous trade or business."[32]

However, restrictive covenants did not control development in the manner intended; neither did they result in cohesive neighborhoods. When an entire area was not covered by similar covenants a hodge-podge of building types and land uses arose. Moreover owners of tracts of land covered by covenants began violating the covenants or challenging their legality in the courts.

In a landmark case, *Trustees of Columbia College v. Thacher*, the courts examined whether it was appropriate to uphold the 1859 covenant on a property located between Fifth and Sixth Avenues at Fiftieth and Fifty-first Streets. This case laid the groundwork for how the courts would balance the issues related to the enforcement of restrictive covenants. The plaintiff, Joseph D. Beers, sued his neighbor, Thomas Thacher, for violating a covenant. The covenant stated that property owners should not erect, establish, or carry on "in any manner, or any part of the said lands, any stable, school-house, engine-house, tenement or community house, or any kind of manufactory, trade or business whatsoever, or erect or build, or commence to erect or build, any building or edifice with intent to use the same, or any part thereof, for any of the purposes aforesaid." Thacher erected a four-storey dwelling. One room in the basement was used as a real-estate office and another was used for deliveries; both offices posted signs on the building. Beers sued Thacher for violating the covenant by permitting businesses barred by the covenant to operate from his dwelling. While the case was in court an elevated railway was built along Sixth Avenue and a train station was built across the street from the premises. The court ruled in favor of the defendant, arguing that the railway station affected the premises injuriously, making it unsuitable for a dwelling house but profitable for business operations. Judge J. Danforth's opinion stated, "The premises may still be used for dwellings, but occupants are not likely to be those whose convenience and wishes were to be promoted by the covenant, persons of less pecuniary ability, and willing to sacrifice some degree of comfort for economy, transient tenants of still another class, whose presence would be more offensive to quiet and orderly people who might reside in the neighborhood. . . . The land in question furnishes an ill seat for dwelling-houses, and it cannot be supposed that the parties to the covenant would now select it for a residence."[33] The court noted that the character of the neighborhood had changed so much, as numerous businesses already existed in the vicinity of the disputed property, that the goal of the covenant was already defeated. The court also believed that it would be unfair to deprive the defendant of the opportunity to develop his property in accordance with others in the neighborhood.

The covenant on the Debevoise property played an important role in legal decisions about restrictive covenants. Suit was brought in 1896 to halt the construction of a building in the courtyard strip. When the case, *Zipp v. Barker*, was brought to trial in 1899 the defendants claimed that they had a right to build in the easement because the character of the neighborhood had changed so dramatically since the time the covenant was signed that the property was no longer useful for residential purposes but was far more valuable for business purposes. They argued that if the covenant was enforced it would result in a loss to all property owners in the neighborhood, some of whom had already violated the covenant by constructing buildings on the courtyard strip. The court found that the covenant on the Debevoise tract was not a covenant against a particular class of buildings suitable to a neighborhood, but was aimed at maintaining a condition that was just as valuable for a business as for a residence. As long as the plaintiff's right to light, air, and an unobstructed view, as per the covenant, was not denied, buildings could be constructed on the courtyard strip. Rosa Zipp lost her case.[34]

Roth v. Jung involved the violation of an 1858 covenant covering properties on Sumner and Willoughby Avenues in Brooklyn. Roth sought to restrain Jung from building a four-storey, eight-family tenement on his property on the building lines of both streets. (Jung bought the property from Roth in 1902.) The covenant stated

> No store or grocery shall be erected or kept on said premises, nor any workshop, manufactory or stable, nor any erection or building that is usually deemed a nuisance or that shall be offensive in a neighborhood

occupied for residences, and that only dwelling houses shall be built upon said premises (except that neat greenhouses or graperies may be built thereon), and further, that no dwelling houses shall be erected thereon that cost less than twenty-five hundred dollars, and that any house or erection that shall be placed upon said premises shall be set or placed back at least twenty feet from the line of the street on which the same shall be placed, and so as to leave a yard of at least twenty feet between any such house or erection.[35]

The court ruled in favor of the defendant, refusing to uphold the covenant. The court found that the neighborhood had changed dramatically since the covenant was written: there was an elevated trolley system, an asylum, high brick fences, a brewery, and several tenements in the immediate vicinity of the defendant's property. While the signers of the 1858 covenant may have envisioned a neighborhood of villas, the neighborhood had developed in a dramatically different fashion. To deprive the defendant of the opportunity to build the tenement would deprive him of the opportunity to improve his property in a fashion similar to adjacent properties; such action by the court would be inequitable. The court also noted that the case did not show any detriment to the plaintiff.[36]

In 1904 John McClure sued Robert Leaycraft to prevent him from constructing a six-storey, forty-two-unit apartment building at 145th and St. Nicholas Avenue in New York. McClure's property was subject to an 1886 covenant forbidding the erection at any time of buildings except brick or stone dwelling houses, and specifically forbade the construction of "any tenement, apartment or community house" on the property. But as in the earlier cases the neighborhood had changed dramatically since the covenant was signed. There were large apartment buildings with commercial enterprises on the first floor on the three corners directly opposite the defendant's property; there were also many flats and tenements in the vicinity. The court found that constructing an apartment building would increase the value of both the plaintiff's and the defendant's property and would not make the neighborhood less desirable or decrease the property value of neighboring properties. The court also found that the changes that the neighborhood had undergone since 1886 made it unsuitable for the erection of a private dwelling house and that enforcing the covenant could not restore the neighborhood to its former character.[37]

The court made a similar ruling in *Schefer v. Ball.* Finding that the neighborhood around Thirty-seventh Street between Fifth and Sixth Avenues was no longer residential, the court refused to enforce a covenant that would have prevented Thomas Ball from building a ten-storey brick building, part of which was constructed in the seven-foot easement specified in an 1846 covenant. The case was appealed and the judgment affirmed in 1908.[38]

Another hotly contested restrictive covenant case went before the court in 1908 and was finally decided in 1914. *Batchelor v. Hinkle* involved a violation of an 1849 restrictive covenant on a property located between Broadway and Sixth Avenue and Twenty-fifth and Twenty-sixth Streets. The covenant stipulated that a five-foot easement should be maintained on the lot and that no buildings should be constructed in the courtyard space. Shortly after acquiring the property in 1908, the defendants, Eugene and Terry Hinkle, began constructing a twelve-storey loft building that extended out to the street line. At the time they began construction all the buildings on the south side of the street observed the five-foot setback. The court ruled in favor of the defendants in 1908 by refusing to grant a preliminary injunction, and the defendants completed construction on the building. The decision was appealed, and in 1909 the court ruled in favor of the plaintiff, finding that the restrictive covenant created an easement that the plaintiff's and defendant's properties were subject to and that the plaintiff was entitled to enforce that easement.[39] That decision was also appealed, and in 1910 the court again ruled in favor of the plaintiff. The judges found that building on the easement amounted to an appropriation of the property. An easement appurtenant to real property cannot be taken against the will of the owner except by eminent domain; as the encroachment was not for public use, the defendant was not entitled to acquire the easement by eminent domain.[40] The case was brought to trial again in 1912, and the judges affirmed the 1910 decision.[41]

The final decision in *Batchelor v. Hinkle* came in 1914, when the judges refused to enforce the covenant, reversing the 1910 and 1912 decisions and finding that the outcome the signers of the 1849 covenant sought had been defeated; the area was no longer residential, hence preserving the courtyard space was not economical for business establishments that wanted to maximize their use of space. The court also found that the plaintiff (whose property was located 115 feet from the defendants' and had also violated the covenant) had not suffered any substantial damage from the erection of the Hinkles' building. Indeed the court noted that the rental value of all the properties on the block, including the plaintiff's (whose building was used for business purposes), had increased. It was also noted that the easement had been built on when many of the buildings in the vicinity were remodeled. Hence enforcing the covenant would not benefit anyone or restore the character of the neighborhood but would inflict damage on the defendants. All the property owners on the block, except one, signed a document expressing their desire to dispense with the 1849 agreement.[42]

Restrictive covenants became such a contentious issue that buyers began avoiding properties that had them. For this reason, despite the area's exclusivity, vacant lots dotted Fifth Avenue. The courts became increasingly reluctant to enforce restrictive covenants stipulating setbacks and building type, size, and height in neighborhoods that had lost the residential character the covenants sought to preserve. Consequently property owners removed the restrictions and began selling to commercial enterprises. Some sellers were more deceptive, passing on property to unsuspecting buyers without disclosing that the property was bound by a restrictive covenant. Ironically the Vanderbilts began undermining their own policy by selling and leasing their lots to commercial concerns and by removing restrictions from the land they sold. For instance, in 1916 Morton Plant, one of the residents to whom the Vanderbilts had sold property, sold the mansion he had built back to the Vanderbilts; they in turn leased it to the Cartier jewelry company. Once the Vanderbilts seemed to abandon their use of restrictive covenants there was a domino effect as other Fifth Avenue families began tearing down mansions and selling and developing their properties. Even some of the most ardent advocates of the mansion neighborhood began converting their properties into luxury apartment buildings.[43]

twelve

LAND USE AND ZONING IN AMERICAN CITIES

Comprehensive Zoning in Cities

Though civic improvement groups gained prominence in New York they had existed elsewhere decades earlier. New Haven organized attempts to beautify the city through systematic tree-planting efforts. In 1833 prominent citizens formed the Society for Rural and Architectural Improvement, dedicated to the "improvement of architecture and scenery, in the structure of buildings and laying out of grounds, the cultivation of refined taste, and the permanent embellishment of our city." This group was the precursor to groups such as the Fifth Avenue Association that emerged almost a century later to help develop comprehensive citywide zoning policies. The Society for Rural and Architectural Improvement helped renovate the Grove Street Cemetery and oversee architectural developments in the city.[1]

When it became obvious that individual efforts were not effective in halting unwanted development on Fifth Avenue organized groups began to advocate planning and zoning improvements. One such group, the West Side Association, organized by William Martin in 1866, a few years after the draft riots, focused on real estate issues north and west of Central Park. The group was composed of real estate professionals such as Martin, large landowners, developers, merchants, financiers, and politicians. The West Side Association agitated for uptown improvements, tax relief, a renaming of the avenues west of Central Park, and rapid transit service in Upper Manhattan.[2]

THE FIFTH AVENUE ASSOCIATION

Restricting Businesses and Traffic

As the area began to change rapidly, the merchants on upper Fifth Avenue organized a movement to control development. Like the mansion owners the merchants wanted to preserve the unique character of the avenue, but instead of seeing the area as Millionaire's Mile, the merchants saw it as America's most magnificent shopping thoroughfare. Thirty-seven property owners, residents, and merchants formed the Fifth Avenue Association in 1907 with the stated goal of "conserv[ing] at all times the highest and best interest of the Fifth Avenue section." By 1910 membership had reached 276; in 1916, when the Association helped to pass the zoning resolution, it had 700 members. Though many members owned Fifth Avenue property, not all did. The goals of the Association were as much about enhancing the allure of the street as an upscale residential, shopping, and retail district as about increasing property values.[3]

Members of the Association had a grandiose vision for the avenue. In 1910 they drew up plans for lighting, safety, the expulsion of noisy, smoky automobiles from the thoroughfare, and street cleaning to make Fifth Avenue "the most distinctive commercial thoroughfare, the most delightful promenade, the most charming boulevard in the world."[4] This vision of the avenue endured for some time. When three hundred members gathered at Delmonico's for the first Autumn Luncheon in 1916 speakers proclaimed Fifth Avenue to be "the greatest street in the greatest city of the greatest country of the world." Richard H. Waldo of the *Tribune* said that Fifth Avenue should become a "trademark street of commercial success, architectural beauty, and material idealism which all America should be desirous to emulate." Recalling Fifth Avenue's less glamorous past as a "muddy, dirty road beside a bog," Waldo claimed that the avenue had gone from "sublimity to squalor."[5]

Ultimately the Fifth Avenue Association wanted to create an exclusive residential and retail zone devoid of beggars, immigrants, and cheap amusement spots (such as those engulfing Broadway). Though formed during the Progressive Era, it wanted to maintain the racial and class exclusivity that drove the millionaires to move to the area in the first place. Activists pushing for the conservation of the avenue argued that Fifth Avenue was still unique among commercial strips in New York. Though it represented the zenith of commercial culture in the city it remained remarkably "uncommercial" in its appearance. It represented an idea of commerce that differed from other commercial districts around the country. Association members did not want to see Fifth

Avenue turn into Broadway, with its oversized, gaudy signs and mélange of shoppers. While Broadway and other shopping districts focused on selling a large volume of cheap merchandise to the masses, Fifth Avenue retailers catered to a small loyal clientele wanting high-quality merchandise and willing to pay high prices for it. Hence the Association wanted to keep Fifth Avenue serene and unhurried, unlike the cluttered and frenetic atmosphere that characterized Broadway and other shopping districts.[6] As a 1912 document of the Fifth Avenue Commission, a quasi-government group also concerned with the preservation of the avenue, read, "If, however, our indifference to the appearance continues, we may expect that Fifth Avenue will cease to retain even its present commercial prominence but will become another . . . cheaper Broadway, with a garish electric sign display and other undesirable accompaniments."[7]

The Fifth Avenue Association was far more successful in preventing an influx of manufacturing plants and low-end commercial establishments onto the avenue from Forty-second to Ninetieth Streets than the rich and powerful businessmen who had tried for years to stabilize the neighborhood. By holding the line on unregulated development and growth, Fifth Avenue became one of the most valuable residential and commercial districts in the country. In addition the Association helped to author some of the most far-reaching land use policies of the early twentieth century and helped to set in motion zoning ordinances for distinct types of economic activities and sweeping land use controls that helped to shape the development of the city. The Association helped to pass the 1916 zoning resolution, the first citywide ordinance in the country.[8] But members were not about to rest easy. Waldo urged members at the luncheon not to feel that they had accomplished everything because of their success to date. He reminded them that the avenue was not yet safe from all threats, but that builders were beginning to recognize that the beauty and architectural charm of the thoroughfare were assets.[9]

The Fifth Avenue Association was successful because it worked with property owners and the city government to develop zoning laws and on street widening projects and policing. Members paid close attention to the demolition and construction of buildings; they wanted to retain mansions and low-rise apartment buildings overlooking Central Park, and they were happy to see old, uninteresting brownstones replaced with luxury apartment buildings. From the very first meeting the Association considered traffic congestion, garbage disposal, public nuisances, and street widening to be critical issues to focus on to enhance the neighborhood's appeal. The Association was also heavily involved in overseeing the landscaping along the avenue. Thus for many years the Fifth Avenue Association acted as a self-appointed police and traffic department, public art commissioner, and city-planning commissioner for the area, from Fifth to Madison Avenue and all cross-streets between. The Association provided information to members, helped to pass a number of ordinances, restricted traffic on the avenue, forcibly removed beggars and peddlers, removed tasteless signs, and had input on the architectural design of new buildings.[10]

The Fifth Avenue Association thought the "growing inconvenience caused by the great numbers of peddlers and beggars now infesting Fifth Avenue at all hours of the day and night" required that "further active steps be taken to rid the Avenue of this growing nuisance."[11] Members argued that the "indiscriminate use of Fifth Avenue for street parades" resulted in "a very serious loss of business to merchants along the thoroughfare, without any real compensating gain to the public."[12] In a letter to Mayor William Gaynor they wrote, "We object to all parades except patriotic and civic parades and parades in general on holidays or at night and the disastrous affect on business at other times makes it imperative that this protest be made. The loss to merchants during the year is enormous, accounting to millions of dollars, and we respectfully request that all other than patriotic and civic parades except on holidays or at night be directed to other thoroughfares."[13] The Association's stance prompted an angry letter to the editor from a reader in 1913:

> Noting that the Fifth Avenue Association proposes to destroy the ancient repute of that thoroughfare as the city's principal parade ground by forbidding all parades in its midst, so to say, I am moved to inquire where in thunder else can we have our parades? And we have got to have parades or the people will blow up in some other manner, perhaps more violently and viciously. . . . I suggest at no expense that all parades, after Fifth Avenue becomes restricted to business to the exclusion of pleasure, be ordered to march around Central Park. This will give six miles of display on wide streets not disturbed by the presence of business houses and afford considerable stoop room—especially on the east side—to spectators, not to mention six miles of nice comfortable park wall seats.[14]

At the Association's fourth annual dinner, held at the Waldorf Astoria in 1913, a letter from Mayor Gaynor, who had recently died, was read to the four hundred in attendance: "I am thoroughly sympathetic with the business men of your avenue in desiring to have the number of parades there decreased. . . . I do not see why Riverside Drive should not serve this purpose admirably instead."[15]

The Fifth Avenue Association also restricted the number of taxis and delivery trucks driving through the area as well as delivery times, tried to prohibit empty vehicles from using the avenue, and pushed hard to restrict the number of buses. Members sought the formation of a traffic commission, participated in planning for street extension and widening projects and improved traffic flow, and conducted traffic counts to force the city to reduce congestion in the area. In October 1916 they counted almost twenty-five thousand vehicles going through the intersection of Fifth Avenue and Forty-second Street in a single day. The Association also studied how effectively the Traffic Squad directed traffic through busy intersections; as traffic congestion got worse, the Association placed traffic towers at key intersections in the area in 1922.[16] That same year the Association hired three men from a detective agency for its own private police force to patrol the avenue and planned to hire 130 more "special policemen" and establish their own headquarters.[17]

After Fifth Avenue residents complained about traffic noise in the early hours of the morning, the Fifth Avenue Association sought to widen the transverse roads in Central Park to make it easier to get from one side of Manhattan to the next, thereby lessening the amount of traffic on Fifth Avenue. As one prominent banker lamented, "I paid $200,000 for the ground on which my house stands; the house cost $160,000 . . . and we can't sleep after 4 o'clock in the morning." Another resident complained, "I have a home on Fifth Avenue and one on Long Island but I have to keep the family at the Long Island home, because we can't keep the windows open on account of the noise from the early morning hours on." Yet another said, "If they [the trucks] keep on they will make it impossible to maintain Fifth Avenue as a residence district of the character planned, and ultimately would cause a depreciation of the property."[18] Ironically in 1912 it was the Fifth Avenue Association that pushed for deliveries to be made to the four hundred or so area merchants in the morning hours as a means of relieving daytime congestion on the avenue.[19]

Enhancing the Aesthetic Appeal of the Neighborhood

The Fifth Avenue Association was determined to regulate the visual appearance of the avenue. Along with the Municipal Art Society, an organization dedicated to the beautification of the city, the Association concentrated on monitoring the design of new buildings along the avenue. Though the Association did not have the legal authority to enforce building restrictions, its architectural committee evaluated plans for new construction and renovations.[20] Minutes of a 1911 Association meeting declared that the overarching goal for Fifth Avenue was "the beauty of the Avenue as a whole, rather than the beauty of each particular building, important though the latter be. The development of Fifth Avenue along the lines of beauty is largely a matter of the willingness of architect and owner to sacrifice their own interest for the benefit of the whole—in other words to erect buildings which will contribute to the beauty of the Avenue in its ensemble, and not with the purpose solely of making conspicuous their own establishment. . . . There must be a certain amount of self sacrifice to bring about a generally satisfactory effect." The Association did not want to see a "jumble of buildings of greatly varying height and greatly varying color, without any consideration of neighboring construction" on upper Fifth Avenue.[21] It also formed a vigilance committee charged with eradicating objectionable signs from the avenue.[22]

The Fifth Avenue Commission worried that the avenue was deteriorating: "A Noble approach to our finest Park and a real parkway has been permitted so to degenerate that we must abandon for the time being at least, all thought of making it the counterpart of any of the splendid avenues of Paris or other great cities abroad. Yet while the time has gone by for such a hope, we may nevertheless . . . still make of Fifth Avenue a dignified street. . . . It need not be without impressive features."[23] The organization believed that controlling the signs—their number, size, style, and orientation—was a critical part of maintaining an aesthetically pleasing residential district. Consequently it monitored the area strictly, opposing "for sale," gas and electric illuminated signs, and those attached to roofs or on the sides of buildings. It did not object to certain shop window displays it deemed appropriate. After more than a decade of lobbying to control the lighting and signage on Fifth Avenue, in December 1921 the Fifth Avenue Association convinced the mayor to sign an ordinance that eliminated almost all illuminated signs along Fifth Avenue. Not all merchants were happy with the ordinance, and some challenged it in court, but the law prevailed.[24]

THE PASSAGE OF THE ZONING RESOLUTION

The passage of the 1916 zoning resolution, which the Fifth Avenue Association helped to secure, had effects around the country, as it opened the door for government to dictate and regulate a range of functions related to land use, the character of neighborhoods and cities, and the way cities grew and developed. Association activists campaigned tirelessly for height restrictions on buildings and a reduction in the construction of loft buildings on the avenue;

they claimed that very tall buildings and too many lofts threatened to ruin the character of the avenue. The Fifth Avenue Commission wrote a height-limitation ordinance in 1913, but it didn't pass. However, the Heights of Buildings Commission, formed in 1913, took up the issues addressed by the Fifth Avenue Commission. Another commission, the sixteen-member Commission on Building Districts and Restrictions, was formed in 1914 to divide the city into zones and outline the boundaries of each. That year the Commission on Building Districts and Restrictions wrote a report urging comprehensive building height and mass limitations in the city. The zoning resolution went into effect on July 25, 1916. It produced specific height limitations on buildings and the zoning of different neighborhoods into residential, business, and unrestricted sections. Almost all of Manhattan below Central Park was divided into zones in which buildings could rise no more than two times the width of the street. However, Fifth Avenue, which was given special consideration throughout the process, had lower height specifications; there buildings could rise no more than one and one-quarter times the width of the street. The zoning resolution also focused on segregating industrial and residential areas.[25]

Within months of the passage of the zoning resolution a lawsuit challenging its constitutionality was filed. In the case of *Anderson v. Steinway* Estelle Anderson signed a contract with Steinway & Sons on July 13, 1916, to sell the company property on West Fifty-eight Street between Sixth and Seventh Avenues. Steinway made a down payment on the property, which had no restrictions on it; the company intended to build a six-storey warehouse and loft to manufacture and sell pianos. However, between the date the contract was signed and August 1, the date that the sale would be finalized, the zoning resolution was passed and the property in question fell in a residential zone. This meant that Steinway could not conduct its business from this location. As a result Steinway refused to finalize the sale and demanded that the plaintiff refund the down payment of $3,000. The court ruled in favor of Steinway, claiming that it would be unjust to force the company to buy property it couldn't use for the purposes intended. The court made this ruling in spite of the fact that neither buyer nor seller knew that the property would be in a residential zone at the time the contract was made.[26]

During the hearing Anderson argued that the zoning resolution was unconstitutional since it restricted the uses to which she could put her property and prevented her from maximizing the value of the property. The value of real estate, she argued, lies in putting it to the right use. Anderson contended that she was being deprived of her property without due process or compensation. The court responded by stating that the decisions of prior court cases raising similar questions indicated that the courts were not sympathetic to the plaintiff's arguments.[27]

BUILDING HEIGHT RESTRICTIONS

The Fifth Avenue Association wasn't the first group to push for restrictions on the heights of buildings in New York. In 1885 a state law was passed mandating that the "height of all dwelling-houses and of all houses used, or intended to be used as dwellings for more than one family, thereafter to be erected in the city of New York, shall not exceed eighty feet upon all streets and avenues exceeding sixty feet in width."[28] The law was challenged in court three years after its passage, when the city's superintendent of buildings refused to grant a permit to the Buckingham Hotel to erect an addition that would have exceeded the height limit. The court ruled that the legislature had the right to exercise its police powers to pass the act; the question was whether the law applied to hotels. The judge agreed with the defendant, Albert D'Oench, that the term *dwelling house* did not include hotels and that the law was intended to regulate the height of apartment buildings. Since there were no height restrictions on stores, factories, warehouses, or office buildings and since most hotels in the city were already taller than eighty feet, it was unfair to deny the defendant a permit.[29]

Boston courts recognized the legality of police powers in imposing restrictions on use of signs and heights of buildings, but helped to frame the conditions under which property owners would be restricted in the use of their property. In a case that went before the court in 1906 Francis Welch challenged the constitutionality of the 1904 and 1905 laws that restricted the height of buildings in Boston.[30] There were three main questions before the court in *Welch v. Swasey*: In exercising its police power, can the legislature limit the height of buildings in cities so that none can be erected above a specified number of feet? Can the legislature classify parts of the city so that different heights are prescribed in different sections? Is it legal for a commission to be established to determine the boundaries of different height zones in the city? The court ruled that in the exercise of its police power the legislature may regulate and limit personal rights and the rights of property in the interest of public health, public morals, and public safety. The court noted that the construction of very tall buildings, especially when built on narrow streets, can exclude sunshine, light, and air, and in so doing affect public health. Very tall buildings may also increase the risk to people and property in the event of a fire. For these reasons the legislature has the power to regulate the height of buildings.[31] The court used the precedent set in the New York case *People v. D'Oench* to support its conclusion.

The Supreme Judicial Court of Massachusetts ruled that it was within the power of the legislature to designate different heights of buildings for different neighborhoods. The law specified a height limit of 100 feet for buildings in residential areas and 125 feet for buildings in business districts. The court found that because land was more expensive and in greater demand in business districts it was reasonable to allow taller buildings in those areas.[32]

However, the courts made it clear that restrictions on the use of private property would not be upheld if these were done solely for aesthetic purposes. In the case of the *Commonwealth v. Boston Advertising Company* the defendant was charged with violating a 1903 regulation of the Metropolitan Park Commission by placing a sign at Revere Beach so large that it would be seen from the parkway. The park commissioners argued that the public parks and parkways were created and maintained to enhance the health and pleasure of the people. The judges concluded that the parks' primary function was to provide pleasure to users. Insofar as they incidentally enhanced health by providing fresh air, sunlight, and a means of exercise, the health of park users was not affected by the presence or absence of signs and billboards. In effect park rules promulgated solely for aesthetic reasons would not be upheld if they infringed on the use of property. Such rules amounted to the taking of property for public use without compensating the property owner.[33]

The judges in *Welch v. Swasey* argued that property owners could not be forced to give up their property rights or pay taxes for purely aesthetic goals, but that considerations of beauty and taste may enter as secondary concerns. They concluded that dividing the city into residential and business districts, each having different building height limitations, was appropriate and constitutional and was not done purely for aesthetic reasons. The court also ruled that it was legal to establish a commission to divide the city into business and residential districts and that the actions of the commission were constitutional.[34]

Other cities also restricted building heights. Buildings could not exceed 175 feet in Baltimore or twelve stories in Denver except for towers, belfries, spires, smokestacks, and special fireproof buildings. Los Angeles limited building heights to 150 feet, except for public buildings, monuments, and other exempt structures. Portland, Oregon, limited buildings to 100 feet in height. Rochester, New York, stipulated that a building should not be taller than four times its width. San Francisco and Providence limited the height of the building according to the character of its construction.[35]

Several lawsuits challenged the height restrictions on buildings brought about by the zoning resolution. On November 25, 1921, the zoning board changed the building height restrictions along Fifth Avenue from East Sixtieth Street to East Ninety-sixth Street, lowering it to 75 percent of the width of the street. *Palmer v. Mann* was brought before the court to challenge the new height restrictions. The question before the court was whether the frontage on the west side of Fifth Avenue that runs alongside Central Park was covered by the new legislation altering the heights of the buildings. The plaintiff, Laura Palmer, brought suit against Tenement House Commissioner Fran Mann because Palmer wanted approval to construct an apartment building that was 150 feet tall. The commissioner refused to approve the plan or issue a permit on the grounds that the building would violate the new height limitations, estimating Fifth Avenue to be a hundred feet wide.[36]

The zoning board had the power to change the building height limitations of a given zone. However, if at least 20 percent of the property owners (owning property on, opposite to, or behind the frontage earmarked for alteration) submitted a signed protest, the board could approve changes only with a unanimous vote of that body. The plaintiff in *Palmer v. Mann* claimed that 20 percent of the owners had filed petitions opposing the new height restrictions, but the court found they hadn't. Leaving the question of petitions aside, the defendant argued that it was legal to build a 150-foot building because the 1916 zoning resolution stated, "The width of the street is the mean of the distance between the sides thereof within a block. Where a street borders a public place, public park or navigable body of water the width of the street is the mean width of such street plus the width, measured at right angles to the street line of such public place, public park or body of water." The defendant claimed that the width of Central Park, estimated at five hundred feet, should be added to the width of Fifth Avenue. If that was done, the proposed building would not violate the new zoning guidelines. The court found that Fifth Avenue was a hundred feet wide and though part of it faces Central Park, no addition would be made to its width.[37] The court reached a similar conclusion in *Thorofare Developing Corporation v. Deegan*, finding that prior attempts to amend the zoning law to permit the construction of very tall buildings had failed. The judge added that the governor's veto in 1926 of amendments to the zoning law "stressed the importance of preventing the erection of such towering structures upon streets bordering on parks and open places in cities."[38]

Ridding Fifth Avenue of Loft Buildings and Factory Workers

Beginning in 1910 the Fifth Avenue Association began to complain that the garment workers were clogging the sidewalks and determined that limiting the height of buildings would stop the spread of loft factories on Fifth Avenue. Lofts—multistory buildings used for manufacturing, storage, and showrooms—were already plentiful on lower Fifth Avenue and around Washington Square. One of the most famous of these buildings, the Triangle Waist Company, played a role in the attempt by the Fifth Avenue Association to eliminate lofts from the upper avenue. Located in the Asch Building almost on the site where Henry James grew up, the Triangle Waist Company burned in March 1911, and the blaze killed more than a hundred workers. The Fifth Avenue Association used the fire to push for limits on the number of people working in the factories, greater fire prevention measures, and the elimination of lofts altogether.[39] The Association's president, Robert Grier Cooke, explained, "When the Asch fire came we saw our opportunity. We said: 'Let us bring about proper fire regulations. By so doing we may not only serve a great public purpose, through the help that we may give to workers, but also our own purposes of keeping manufacturers of garments—"sweat shops"—off Fifth Avenue.'"[40]

Shortly before the Triangle fire William Kendall, a member of the Association's architectural committee overseeing the development of Fifth Avenue, summarized the problems facing the avenue: dirty, dusty roads, the poor condition of the pavement, and the throngs of workers. Kendall elaborated: "Perhaps the greatest evil, and one which the association has been combating for a long time, is the loitering of great crowds of employees; especially at the lunch hour. It goes without saying that Fifth Avenue will cease to be a great retail shopping street if this nuisance continues. I know of nothing that has more important bearing upon the future development of the avenue along lines of beauty than this last consideration."[41]

In a statement to the Fifth Avenue Commission in 1913 the Association argued that the loft buildings "have practically ruined that part of the Avenue" on which they had been built. They "have utterly changed its former high-class character, and have had a derogatory effect upon the entire neighborhood." A 1914 study by the Association found that there were about nine hundred factories between Twenty-third and Fifty-ninth Streets between Sixth and Madison Avenues. The Association objected to the lofts and the people using them.[42] In an appeal to the Fifth Avenue Commission, Bruce Falconer, the Association's lawyer, wrote, "The buildings are crowded with their hundreds and thousands of garment workers and operators who swarm down upon the Avenue for lunch hour between twelve and one o'clock. They stand upon or move slowly along the sidewalks and choke them up. Pedestrians thread their way through the crowds as best they may." Falconer added that the influx of immigrants into the district had frightened away upper-class female shoppers, lowered property values, and prompted an exodus of "high-class shops and stores." The Association's response was to try to preserve the neighborhood and protect it from the loft buildings and their employees, who were viewed as a nuisance and a menace.[43]

Robert Grier Cooke wrote an opinion on the subject for the *New York Times*:

> Once these factories became firmly established the character of Fifth Avenue changed. The noonday crowds, numbering tens of thousands, filled the sidewalks, overflowing into the street and practically blockading the avenue from Fifteenth to Twenty-third Street for an hour on every working day. This made shopping difficult. Entrances to stores and offices could be reached only by vigorous pushing methods and naturally men and women preferred to do business in neighborhoods where they did not have to fight their way through dense masses.
>
> The Fifth Avenue Association sounded its warning years ago. No efforts were spared to combat this evil. High-class retail stores were forced to move, office buildings became deserted and property prices as well as rental values began to drop. It looked as if Fifth Avenue, especially below Twenty-third Street, was doomed.[44]

Cooke contended that the most serious problem facing Fifth Avenue was the garment manufacturers in the district: "Crowds of loiterers issue to block the sidewalks at the luncheon hour, put[ting] a blight on the great thoroughfare. . . . The high-class retail trade was suffering from such conditions."[45]

In a debate printed in the *New York Times* Ernest Flagg and Robert Grier Cooke discussed the pros and cons of building skyscrapers in New York. Flagg, also a member of the Fifth Avenue Association, wrote in favor of the skyscrapers. He contended that critics opposed them because of the increased fire risk, the increased traffic they generated, their ugliness, and because they cut off light from streets and surrounding buildings. But Flagg argued that it was senseless to limit the height of buildings in New York because the city already had many tall buildings and it was too late to stop them from proliferating. He pointed out that limiting height would not ensure light to buildings and the streets; to attain this goal buildings would have to be limited to about four stories. Limiting height would not reduce the risk of fire either; buildings were at risk for fire because they were constructed with cheap, flimsy, flammable materials. Building materials had to change to reduce fire risk. Flagg also argued that height limitations did not reduce congestion; he pointed out that Paris is very congested even though buildings are limited to six stories in height.[46] Cooke responded by arguing that restricting building height would result in a more harmonious architectural style and would limit the number of garment manufactures, loiterers, and workers on Fifth Avenue.[47]

Although the Fifth Avenue Association had forcibly removed beggars from the avenue, the organization was still concerned with the type of people sauntering along Fifth Avenue. Poor, immigrant workers were not the class of people they wanted walking along the thoroughfare, so at the instigation of the Association the police began arresting garment workers on their lunch break. Beginning in 1910 the Association began holding meetings with the Cloak and Suit Workers Union and with the Cloak and Suit Manufacturers Association to discuss various methods of crowd control along the avenue. Proposals were made to cordon off sections of side streets for workers or to shorten the workers' lunch hour to forty-five minutes. However, union representatives opposed any reduction in the lunch hour, claiming that workers had fought long and hard for the one-hour lunch break and would not give it up. The mayor also objected strenuously to attempts to arrest workers; hence the Association resorted to posting placards (written in several languages) around the neighborhood discouraging loitering and tobacco spitting.[48]

A 1910 article claimed that though it was costly to establish a garment factory on Fifth Avenue, manufacturers wanted the prestige that came from engraving a Fifth Avenue address on their clothing labels. The article went on to say, "The factories should be placed in the inexpensive quarters of the East Side, where the [factory] hands might have time and opportunity to do their shopping, and where spectacles of wealth and fashion would not impress them with impossible standards of living."[49]

Not many people spoke on behalf of the workers' rights to walk along Fifth Avenue's sidewalks. Mayor Gaynor was one public figure who tried to counter the attempts to rid the avenue of poor workers. Gaynor wrote in a June 1913 letter to the Fifth Avenue Association, "The recent migration of businesses northward has created difficult problems for you [Fifth Avenue merchants] in the overflowing of great armies of workers on your sidewalks from the side streets at the noon hour. It is difficult to decide how to alter that condition for although it is not pleasant for you business men to have your business effectively blocked for a busy hour each day, the workers too, have their rights of fresh air and sunshine at the lunch hour. I should be very thankful for any helpful suggestions from you."[50]

Another politician, Borough of Manhattan president-elect Marcus Marks, also asked the Fifth Avenue Association to be more considerate of the workers. Speaking at the Association's annual dinner Marks said

> Associations such as yours may be very helpful in ameliorating conditions, provided they do not consider merely the interests of the owner and tenant, but also give full weight to the human element which enters into the problem. Crowds on the street may be unpleasant and uncomfortable, but we cannot dismiss the unpleasantness or the discomfort without remembering that these crowds are made up of men and women who not only need but are entitled to the advantages of fresh air, exercise and recreation during the noon hour. Before you can correct the present situation you must provide other means of supplying these needs of the working people, which incidentally include social intercourse and the discussion of problems of mutual interest.[51]

Challenging Zoning Laws

Not all merchants and Fifth Avenue residents agreed with or wanted to abide by the restrictive covenants and height restrictions that dictated where, how, and what type of buildings could be constructed in cities. Within a few years of the passage of the zoning resolution several challenges were filed in the courts. Challengers were emboldened by legal decisions that showed the courts were hesitant to enforce covenants when neighborhoods in and around the areas covered by covenants had changed dramatically, rendering the covenant void. However, the courts were also reluctant to undermine city ordinances. They refused to allow residents to violate building height restrictions or place illuminated signs in forbidden places.

NEIGHBORHOOD CHANGES

In the case *Bouvier v. Segardi et al.* suit was brought against Cornelia Segardi and others for erecting a large show window on West Forty-sixth Street in 1906. The flimsy, temporary-looking window encroached on a five-foot easement stipulated in an 1852 restrictive covenant and impinged on the view of the adjoining building. Unlike prior cases, such as *Batchelor v. Hinkle*, in which the courts ruled that because the neighborhoods had changed the restrictive covenant was unenforceable, the court in *Bouvier v. Segardi et al.* ruled that the neighborhood of West Forty-sixth Street had not changed so much as to render the restrictive covenant obsolete. Though the street was not exclusively residential, the wide appearance caused by setting back the buildings provided a distinct advantage to the upscale businesses on the street. Hence the observance of the setback benefited adjacent properties. The court also noted that local property owners had formed an association to maintain the character of the street. Though it would cost the defendants about $2,500 to remove the window, the window reduced the plaintiff's property value by about $20,000, plus about $3,000 annually in diminished rental value. The court enforced the covenant by ruling in favor of the plaintiff, Michel Bouvier.[52]

In the case of *Wallack Construction Company v. Smalwich Realty Corporation* Wallack asked the court to stop Smalwich from building in an eight-foot setback, as required by an 1846 covenant. The lower court granted a temporary injunction restraining Smalwich from building; however, the Supreme Court of New York overturned the lower court ruling, finding that since the restrictive covenant was written the neighborhood had changed from a residential to a business district.[53]

Even as late as the 1920s residents in the vicinity of Fifth Avenue filed law suits in an effort to get the courts to enforce restrictive covenants barring the construction of buildings other than dwelling houses. In *Bowers v. Fifth Avenue and Seventy-seventh Street Corporation* Spotswood Bowers tried to prevent construction of a fourteen-storey apartment building between Seventy-sixty and Seventy-seventh Streets. Bowers argued that the property was bound by an 1871 restrictive covenant barring the construction of any building other than an "ordinary first class dwelling." The court ruled that the term *dwelling house* in the covenant was broad enough to include apartment buildings. The court disagreed with the plaintiff's assessment that the neighborhood hadn't changed much since the covenant was signed and comprised exclusively private residences, observing that many apartment buildings were on the same block and near the plaintiff's property. Consequently the court refused to enforce the covenant.[54]

ILLUMINATED SIGNS

When the permits allowing merchants to display illuminated signs on their businesses on Thirty-fourth Street between Fourth and Seventh Avenues expired on December 23, 1921, the city clerk refused to renew the permits. A new sign ordinance prohibited illuminated signs in this area. When merchants refused to remove the signs from their buildings the city threatened to forcibly remove them. The Oppenheim Apparel Corporation sued, complaining that area merchants were being discriminated against because the theaters and other places of amusement were allowed to keep their illuminated signs while the merchants were being forced to remove theirs. The court found that though the use of illuminated signs may be good for business, "the multiplication of those outstanding signs in this very busy section of the city easily can become an eyesore, a nuisance and an improper use of the air space over the thoroughfare." The court held that the plaintiff was not being discriminated against because theaters and other places of amusement were allowed to use illuminated signs. The complaints were dismissed.[55]

The Racialization of Zoning

CHALLENGING THE RESIDENTIAL SEGREGATION ORDINANCES OF CITIES

In the second decade of the twentieth century the courts signaled that they would uphold municipal ordinances that partitioned cities into zones or limited building heights. Covenants aimed at enhancing the character and aesthetic appeal of neighborhoods and those fostering an upscale business climate were upheld if they continued to serve the purpose for which they were originally intended. However, the courts refused to uphold covenants if the character of the neighborhoods had changed so dramatically that affirming the covenants would not maintain or restore the neighborhoods.

Around the time municipalities were developing comprehensive zoning laws city councils began passing ordinances that zoned cities along racial lines. In addition, private parties began signing restrictive covenants aimed at creating and maintaining racially exclusive neighborhoods. On May 15, 1911, the City Council of Baltimore passed Ordinance No. 692, which designated certain city blocks for whites and others for people of color. The ordinance stipulated that whites could not live in any house, building, or structure on blocks designated for people of color, and minorities could not live on blocks designated for whites. The ordinance was passed "for the preservation of peace, the prevention of conflict and ill-feeling, between the white and colored persons in Baltimore City, and for promoting the general welfare of the City."[56]

John Gurry, a person of color, was indicted for violating the ordinance. The Maryland Court of Appeals ruled that the ordinance was valid because the different racial groups had similar restrictions. Furthermore it was the duty of the city to protect the general welfare of city residents; since animosity might arise and violence erupt between whites and blacks living beside each other, cities had the power to take action to forestall conflicts and violence. Consequently it was appropriate to use an ordinance to separate the races.[57]

However, the court did rule that one aspect of the ordinance was unconstitutional: it ignored the property rights of some residents. For instance, if a black person bought property in a block that was later declared a white block, he or she would not be allowed to take possession of the property. It would be unlikely that a black person would be able to sell the property to a white person, and the property could not be sold to another person of color. In effect, the black person would be left with property he or she could not use or sell. Though a white person with property in a minority neighborhood could not take possession of it and could not sell or rent to other whites, he or she might be able to sell or rent to people of color. In addition if members of either racial group inherited property in blocks in which they were prohibited from living they could not take possession of their property. Under such conditions the ordinance was declared unconstitutional because it was tantamount to the taking of property.[58]

A similar case went to the Supreme Court of North Carolina. On July 5, 1912, the board of aldermen of Winston passed an ordinance that made it illegal for whites to live on blocks where the majority of the residents were people of color and for blacks to live on blocks where a majority of the residents were white. In 1913 William Darnell, a person of color, moved into a house on a street where the majority of the occupants of the other houses were white. Darnell was charged and fined for violating the ordinance. The court dismissed the case against Darnell, arguing, "An act of this broad scope, so entirely without precedent in the public policy of the State and so revolutionary in its nature, cannot be deemed to have been within the purview of the legislature." The court did not believe that the General Assembly intended to confer such "broad and arbitrary a power" on the aldermen. The judges found that though the aldermen claimed to act for the general welfare of the city, the ordinance was arbitrary. It opened the door to making laws prohibiting people of different political parties or religious groups from living on the same street. The court also found that the ordinance prevented property owners from selling or renting their property to whomever they pleased and that such an ordinance could result in an exodus of the most enterprising blacks from the city, leaving it with "the unthrifty and less desirable element."[59]

Likewise, the Supreme Court of Georgia, overturning a lower court decision, ruled in *Carey v. Atlanta* that a similar ordinance adopted by the city on November 3, 1913, denied "the inherent right of a person to acquire, enjoy, and dispose of property" and violated the due process clause.[60] However, the Supreme Court of Virginia came to opposite decisions on racial segregation ordinances passed in 1911. Mary Hopkins and John Coleman, both African American, were separately charged with moving into white blocks after ordinances were passed designating separate blocks in Richmond and Ashland, respectively. The Virginia Supreme Court upheld two lower court decisions that questioned the validity of both ordinances. The judges in *Hopkins v. City of Richmond* (heard at the same time as *Coleman v. Town of Ashland*) ruled that the ordinances were valid and exemplified a reasonable use of the police powers of the municipalities in question.[61]

A similar segregation case went to the U.S. Supreme Court. *Buchanan v. Warley* challenged the segregation ordinance passed by the city of Louisville in 1914 that established separate blocks for whites and people of color. In this case a white person sold property to a person of color on a white block and the buyer was prevented from occupying the house. In reversing the judgment of the Kentucky Court of Appeals the U.S. Supreme Court justices found that denying persons of color the right to occupy property they own because it was in a white neighborhood amounted to a taking of property and was not a legitimate use of the police powers of the state. Such an act violated the Fourteenth Amendment: "The constitutional guaranty of equal protection, without discrimination on account of color, race, religion, etc., includes 'the right to acquire and possess property of every kind.'" In effect, the Louisville ordinance denied rights which were vested before the ordinance took effect without compensating residents for it.[62]

Despite the *Buchanan v. Warley* decision cities continued to pass racial segregation ordinances similar to those the courts had struck down. New Orleans passed two ordinances in 1912 and 1924 that segregated the residences of whites and people of color. In addition the ordinances empowered the city engineer to deny building permits to whites wanting to build in black neighborhoods and vice versa. The ordinances were challenged in the case of *Tyler v. Harmon*. In 1925 Joseph Tyler sought to prevent Benjamin Harmon from modifying his cottage to create an additional apartment that Tyler believed Harmon intended to rent to blacks. The ordinances prohibited blacks from living in that section of town. The Supreme Court of Louisiana ruled that the racially restrictive ordinance was valid. The court also issued an injunction against Harmon, barring him from renting his cottage to blacks. However, in 1927 the U.S. Supreme Court ruled, in concert with *Buchanan v. Warley*, that the New Orleans ordinances alienated vested property rights and were unconstitutional on those grounds.[63]

The same ordinances were challenged by the Land Development Company of Louisiana in 1926. The district court dismissed the case, but it was heard on appeal in the circuit court in 1927. Relying on the *Buchanan v. Warley* decision, the court of appeals reversed the lower court decision and struck down the ordinances.[64] The U.S. Supreme Court also relied on *Buchanan v. Warley* and *Harmon v. Tyler* in reaching a similar decision in *Richmond v. Deans*.[65]

CHALLENGING PRIVATE RACIALLY RESTRICTIVE COVENANTS

Despite the many rulings against racial segregation ordinances, these rulings did not affect the private agreements that homeowners entered into among themselves to create and preserve racially exclusive neighborhoods. In deciding racial covenants cases there were three fundamental questions before the courts: Were racially restrictive covenants valid? Could municipalities and states use their judicial and other police powers to enforce racial covenants between private parties? Were racially restrictive covenants enforceable when the neighborhoods they were intended to protect had undergone substantial demographic changes?

The courts upheld racially restrictive covenants between private parties for decades. In 1918 the Supreme Court of Missouri found that a covenant covering property on Wirtman Place in Kansas City was valid and that Elizabeth and August Koehler had violated the covenant by renting their property to blacks.[66] The following year the Supreme Court of California ruled in the case of the *Los Angeles Investment Company v. Gary*. Alfred Gary and his wife, both African Americans, had acquired a property covered by a deed restriction stating that "no person or persons other than of the Caucasian race shall be permitted to occupy the property." The court declared that the covenant was valid and enforceable.[67]

The Michigan Supreme Court also found racially restrictive covenants between private homeowners valid and enforceable. When Anna Morris and her husband, both of whom were black, bought property in the Ferry Farm subdivision of Pontiac, residents of the neighborhood filed suit against them. In the case of *Parmalee v. Morris* the court ruled that the covenant which read "said lot shall not be occupied by a colored person" was valid and enforceable.[68] Similarly the Court of Appeals of the District of Columbia and the U.S. Supreme Court found that a covenant signed by twenty-nine homeowners in 1921 that barred blacks from occupying property in their neighborhood was valid.[69]

A racially restrictive covenant was also upheld in the case of *Grady v. Garland et al.* in 1937 in Washington, D.C. Grady brought suit on behalf of himself and five other lot owners seeking to get the restrictive covenants binding their lots to be declared "clouds upon the titles of the owners thereof, impeding the free use and enjoyment of their properties." The covenants were placed on the lots between 1901 and 1905, when they were built by the real estate firm Middaugh & Shannon. Grady wanted the covenant to be "cancelled, removed, and held for naught"; he pointed out that minority families were already living to the west of the properties bound by the covenant. In upholding the covenant the U.S. Court of Appeals for the District of Columbia argued that the covenant was intended to prevent people of color from living in the square with the eight lots: "[The covenant] furnishes a complete barrier against the eastward movement of [the] colored population into the restricted area—a dividing line." Given the vague arguments made by the plaintiff, the court refused to strike down the lower court's ruling because such action would serve to destroy the value of the defendants' property.[70]

On September 30, 1927, five hundred homeowners in the Washington Park subdivision of Chicago signed a restrictive covenant that barred property owners from leasing, selling, or having any part of their premises occupied by people of color except janitors, chauffeurs, or servants. The neighborhood consisted of 583 parcels covering twenty-seven city blocks. In 1934 Olive Burke filed suit against Isaac Kleiman for leasing an apartment to James Hall, an African American. At the time of the suit three other properties were in violation of the covenant (which ran with the properties until January 1, 1948). Three buildings housed ninety-six minority tenants within 650 feet of Burke's property. Kleiman argued that the character of the neighborhood had changed so dramatically that the covenant should no longer be enforced. The Court of Appeals of Illinois disagreed, ruling that since only 4 of 583 parcels had violated the covenant, that did not constitute a change big enough to warrant voiding the covenant. In addition there was action pending against the other violators to stop them from violating the covenant.[71]

Residents of the Washington Park subdivision were back in court in 1937. By then they had formed the Woodlawn Property Owners' Association to help enforce the restrictive covenant. Olive's husband, James Burke, president and executive secretary of the Association, had changed his mind about the use of covenants and resigned after a quarrel with other members. Having announced that he would get even with the Association by moving people of color into the neighborhood, Burke embarked on a scheme to do so. Carl Hansberry, an African American, acquired a property covered by the covenant through a third party white individual who bought the property with the intention of passing it on to a holding institution or reselling directly to a person of color. Hansberry, who was already leasing an apartment in the restricted area, was ordered by the court to vacate that premise. He moved into a Rhodes Avenue property, at which point Anna Lee filed suit against him. Hansberry was prevented from collecting rent from the two white tenants in his building (the tenants were ordered to deposit their rent with the clerk of court). In the event that the white tenants moved Hansberry was also prohibited from renting to black tenants. Hansberry was given ninety days to vacate that property also.[72]

Hansberry argued in court that the covenant was not valid because it required that 95 percent of the property owners in the restricted area must sign the agreement for it to take effect and that 95 percent had not signed. Relying on the findings in *Burke v. Kleiman* that asserted that 95 percent of the property owners had signed the covenant, the Appellate Court of Illinois did not probe Hansberry's claim. Moreover the court found that Hansberry had entered into a conspiracy with Burke to take possession of the property even though he knew the property was covered by a restrictive covenant. The judges concluded that Hansberry acquired the property illegally.[73]

Hansberry appealed the ruling to the Supreme Court of Illinois and again argued that the restrictive covenant was not valid. Investigation into Hansberry's claim found him to be correct; only about 54 percent of the property owners had signed the covenant. Yet despite the error of the appellate court, the Illinois Supreme Court upheld the lower court ruling.[74] Hansberry's case then went to the U.S. Supreme Court, which in 1940 ruled that the Illinois Supreme Court decision denied Hansberry due process guaranteed by the Fourteenth Amendment. The condition that 95 percent of the property owners sign the restrictive covenant for it to go into effect was not met. Consequently the decision of the Illinois Supreme Court was reversed.[75]

THE USE OF POLICE POWERS TO ENFORCE COVENANTS

Another series of cases going all the way to the U.S. Supreme Court challenged the use of police powers to enforce private covenants. In 1948 four such cases were argued before the Court at the same time: *Shelley et ux. v. Kraemer et ux.*, *McGhee et ux. v. Sipes et al.*, *Hurd et ux. v. Hodge et al.*, and *Urciolo et al. v. Hodge et al.*

The case of *Shelley v. Kraemer* originated in St. Louis. In 1911 thirty of thirty-nine property owners on Labadie Avenue signed a restrictive covenant stating that no non-Caucasian could use or occupy the properties for fifty years. The covenant specifically mentioned that "people of the negro or Mongolian Race" should not occupy the properties covered by the deed. There were fifty-seven parcels in the subdivision, and the thirty property owners who signed the covenant held titles to forty-seven parcels. At the time the agreement was signed five parcels were owned by blacks; one parcel had been occupied by blacks since 1882. On August 11, 1945, J. D. Shelley and his wife, Ethel, who were African Americans and who had no knowledge of the restrictive covenant, bought property in the restricted area. In October of that year Louis and Fern Kraemer sued to prevent the Shelleys from moving in. The Shelleys questioned whether all the signatures needed to make the covenant effective were obtained. The trial court dismissed the petition because all the property owners had not signed the covenant. The case was appealed to the Supreme Court of Missouri, which reversed the trial court ruling, finding instead that the covenant was valid and that its enforcement by the court did not violate the Fourteenth Amendment.[76]

The case of *Sipes v. McGhee* originated in Detroit. In September 1935 homeowners in the Seebaldt subdivision signed an agreement stating that the lots should not be occupied by people of color. The agreement also specified that 80 percent of the property owners must sign for it to become effective. Orsel and Minnie McGhee, African Americans, purchased a house in the restricted area in 1944. Benjamin Sipes and other property owners brought suit against the McGhees to have them vacate their property. The defendants argued that the covenant was not signed by 80 percent of the property owners, and therefore it was not valid, but the Supreme Court of Michigan found the covenant to be properly executed and enforceable.[77]

The U.S. Supreme Court reversed both these rulings in 1948. First the Court found that "the restrictive agreements standing alone cannot be regarded as a violation of any rights guaranteed to petitioners by the Fourteenth Amendment. So long as the purposes of those agreements are effectuated by voluntary adherence to their terms, it would appear clear that there has been no action by the State and the provisions of the Amendment have not been violated." However, the Court then ruled that once an action went beyond the realm of voluntary compliance and the judicial system was used to enforce the covenants, such an action was unconstitutional.[78]

Two similar cases arose in the District of Columbia. In 1906 twenty of thirty lots on Bryant Street were sold with deeds prohibiting blacks from occupying the properties. James and Mary Hurd, an African American couple, bought a house in the restricted area in 1945, and Raphael and Florence Urciolo, white realtors, sold property to three African Americans in the restricted area. At the time of these sales eleven lots—not covered by the deed restrictions—were already owned by blacks. The district court ruled that the deeds of the African American petitioners were null and void and ordered them to vacate their property within sixty days of the decision. The Hurds appealed the case, but the U.S. Court of Appeals for the District of Columbia upheld the district court ruling. The case then went to the U.S. Supreme Court, where the lower court decisions were reversed.[79]

In the *Hurd v. Hodge* case plaintiffs had tried to argue that the neighborhood had undergone a racial transformation great enough to void the restrictive covenant. However, the court of appeals was not convinced by this argument and ruled to enforce the covenant.

DEMOGRAPHIC TRANSFORMATION OF THE NEIGHBORHOOD

Like the covenants restricting businesses, private-party restrictive covenants were also challenged under the aegis of neighborhood change. In cases where dramatic neighborhood changes had occurred the courts were willing to nullify the covenants, but where such changes were not apparent the covenants were enforced. In the case of *Hundley v. Gorewitz* six homes on the west side of Thirteenth Street in the District of Columbia held a restrictive covenant dating back to 1910 that prohibited property owners from renting, selling, or leasing their property to blacks or other people of color. Frederick and Mary Hundley bought one of these properties in 1941, and their neighbors Gorewitz and Bogikes filed suit against them. The district court found the Hundleys in violation of the covenant and cancelled their deed. The case went to the U.S. Court of Appeals for the District of Columbia, which reversed the district court ruling on the grounds that the neighborhood had changed so much that enforcement of the covenant was not beneficial: "When it is shown that the neighborhood in question has so changed in its character and environment and in the uses to which the property therein may be put that the purpose of the covenant cannot be carried out, or that its enforcement would substantially lessen the value of the property, or, in short, that injunctive relief would not give a benefit but rather impose a hardship, the rule will not be enforced."[80]

Similarly, in the case of *Mays v. Burgess*, the court of appeals ruled to enforce the restrictive covenant because the neighborhood had not undergone significant changes.[81]

Theoretical Discussion

During the late nineteenth century elites realized the limits of individual attempts to develop uniform, aesthetically pleasing neighborhoods and commercial thoroughfares. Though they relied heavily on restrictive covenants and used the courts to enforce these, they could not attain their goal. Recognizing this, businessmen and urban planning activists collaborated on developing a comprehensive vision of city planning. One group, the Fifth Avenue Association, grew enormously influential in zoning and planning politics in New York City. They had a vision of how cities could be planned to be more efficient, orderly, and beautiful and were able to impose that vision on New York because the city's politicians and administrators had no alternative to offer. Activists in other cities, also finding a vacuum in leadership in this arena, developed zoning and planning laws.

The Fifth Avenue Association was so effective because it was composed of wealthy people who wanted to live, worship, run their business establishments, and shop together without being intruded upon by the poor or minorities. They struck a critical compromise: instead of fighting for an exclusively residential neighborhood they conceded that upscale and tastefully designed retail and commercial enterprises could coexist with mansions and lavish apartments without reducing property values. Once they resolved the residential versus commercial debate, which could have been quite divisive, they collaborated to garner the resources necessary to meet their goals.

They framed the issues clearly and effectively. Fifth Avenue was framed as the grandest, fairest avenue in the world, with eye-catching vistas and a feeling of spaciousness; crowded, noisy streets detracted from the aura the Fifth Avenue Association wanted to convey. Opponents did not come up with effective counterframes to challenge the Association's. When it decided that lofts and poor immigrant workers were an anathema to Fifth Avenue, many of the loft owners capitulated, either moving away or making the changes the Association demanded. The business owners who challenged the Association did so through the courts. While the Fifth Avenue Association fought in the courts it also developed important political alliances and got allies appointed to positions in city government who could help push through their policies and legislation. The Association called for commissions to study various issues and made sure its members played key roles on those commissions. The Association also collected data to justify its claims and sent the reports to governing bodies. These data were persuasive because the city often had no data to counter the findings.

The Fifth Avenue Association courted the support of crucial people, mobilized both resources and people quite quickly, and got wealthy residents to participate. The group consisted of about seven hundred people; they did not solicit supporters from outside the neighborhood. The group was quite effective in mobilizing who it perceived to be the affected constituents. It was also able to mobilize vast monetary resources to launch multiple campaigns. It targeted the government to become more proactive in zoning and planning issues; its members to keep the neighborhood exclusive, upscale, tasteful, and uniform; nonconforming businesses to relocate; and poor people to work elsewhere.

The Fifth Avenue Association was also effective because its members were a tight network of activists who were able to influence each other's business practices. For instance, the bankers in the Association refused to finance undesirable business operations in the neighborhood, and merchants refused to buy or sell the products of boycotted businesses or lease space to them.

Cities quickly recognized the power of zoning and adopted a variety of zoning ordinances. Co-opting the frame of neighborhood character from early zoning activists, racist city government operatives extended the framing of zoning laws to argue that racially restrictive covenants were needed to separate the races, reduce animus, and safeguard public health. The master frame of conflict reduction, safeguarding public health, and property values was attractive to those wanting racially exclusive neighborhoods. For them, class exclusivity wasn't the only way to protect property values; racially explicit covenants were also tools used to enforce homogeneous residential communities. Eventually blacks denied housing and whites angered by their inability to sell to the highest bidder challenged the racially restrictive covenants. Cases such as *Spies v. McGhee* and *Shelley v. Kramer* went to the U.S. Supreme Court as blacks expressed their outrage at the racially motivated city ordinances that were passed in several cities and at the private agreements that restricted the occupancy of dwellings. Using the rhetoric of justice and fairness blacks challenged these ordinances through the courts with the help of the NAACP and other institutions and by articulating quite clearly that they expected their civil rights to be respected and upheld. Opponents of private restrictive covenants attacked their Achilles' heel: the covenants relied on individuals or groups to comply with the restrictions. Such covenants were undermined when individuals began to break them.

Part V

REFORMING THE WORKPLACE AND REDUCING COMMUNITY HAZARDS

Part V examines the hazards urban workers faced on the job as well as efforts to reform the workplace. From the reformers' perspective the workplace was another aspect of urban life that needed to be controlled and changed in order to make cities more livable. In chapter 13 I examine the garment industry, industrial accidents, and the way working- and middle-class reformers organized to improve working conditions. I also explore the relationship between working-class labor activists and middle- and upper-class reformers and the role of gender in workforce participation and activism.

Chapter 14 focuses on pollution. I examine factories' attempts to pass air pollution ordinances and analyze air and water pollution and other hazards in the workplace, particularly in the steel mills. I explore how ethnicity, race, class, and gender influenced exposure to pollution as well as the kind of activism engaged in and the challenges involved in regulating industrial giants. Using the asbestos industry as a case study I discuss the relationship between workplace hazards and the rise of industrial medicine and analyze how corporations responded to illnesses arising from exposure to hazardous conditions in the workplace. I close the chapter with a discussion of contemporary working-class environmentalism.

thirteen

WORKPLACE AND COMMUNITY HAZARDS

The Triangle Fire: Flaming Skirts Billowing in the Wind

On February 17, 2001, Rose Freedman, the last survivor of the Triangle Waist Company fire, died in her apartment in Beverly Hills. She was 107 years old. On March 25, 1911, Freedman, then two days shy of her eighteenth birthday, escaped death by following the factory owners and other workers to the roof of the blazing building, where she was rescued.[1] There were 14,002 fires in New York City in 1911. Large and deadly fires were not surprising since there were 13,603 buildings listed as dangerous by the Fire Department, but only 2,051 had been inspected by the department's forty-seven inspectors in 1911. The Triangle fire stands out in history because it shocked the nation and forced the government to make the workplace safer.[2]

The workplace was a major venue for reform during the second half of the nineteenth century. Not only were vast numbers of workers killed and injured on the job, but environmental hazards from industrial facilities spilled over into surrounding communities. Around the time of the Triangle Waist Company fire approximately a hundred people were dying on the job each day. In 1907 fifteen hundred people died and another six thousand were injured in fires. Yet desperate immigrants continued to pour into cities such as New York at the rate of eighteen thousand per month. Indeed, on November 26, 1910, an eerily similar disaster struck a four-story shirtwaist factory in Newark that employed between two hundred and three hundred women. That fire killed twenty-five young women who were either incinerated or killed when they jumped out the windows to escape the blaze. Right after the Newark blaze more than twelve hundred New York factories were inspected; significant fire hazards existed in virtually all of them. Ironically Triangle was found to be one of the safer factories among those inspected.[3]

The Triangle fire that broke out in the Asch Building at the corner of Washington and Greene Streets gained notoriety because it symbolized the callousness with which corporate executives treated workers and the inhumane work conditions that prevailed. In addition to influencing urban planning policies the fire exposed gender exploitation and highlighted the connection between working- and upper-class feminism of the early twentieth century, labor organizing, and occupational safety. The case of the Triangle fire also illustrates the complex nature of ethnic and class relations in urban centers. It is a story about inter- and intraethnic discrimination, exploitation, and upward mobility of immigrants. It also illuminates some of the inter- and intraclass dynamics of the time.

THE ORIGINS OF THE TRIANGLE WAIST COMPANY

By the time the fateful fire broke out the owners of the Triangle Factory, Max Blanck and Isaac Harris, were rich men living in adjacent luxury townhouses near the Hudson River and traveling in chauffeur-driven cars. Harris's family of four had four servants; Blanck hired five servants for his family of seven. Shortly before the fire Blanck moved to his brand-new home near Prospect Park. Despite taking on the trappings of an upper-class lifestyle Harris and Blanck didn't belong to the class of patrician entrepreneurs who dominated the economic and political life of the city; they were members of a new breed of once poor, immigrant moguls who struck it rich in their entrepreneurial forays into the garment industry. They were immigrants who got their start as exploited workers in the sweatshops of the Lower East Side and who lived in crowded hovels when they first arrived in New York. Both were born in Russia, Harris in 1865 and Blanck three or four years later. In the early 1890s both immigrated to New York.[4]

Economic hardships and persecution resulted in large numbers of Russian Jews immigrating to New York City in the early 1880s. Later waves of Jewish immigrants from Russia and Poland were more skilled than their predecessors. Under the czars Russian Jews were restricted in their occupational choices; one occupation open to them was altering and mending secondhand clothing. This meant that a large proportion of the later waves of Eastern European Jewish immigrants, such as Blanck and Harris, were skilled garment workers.[5]

Blanck got his start as a garment manufacturer when he opened a small sweatshop on the Lower East Side while living in a tenement with his wife, Bertha, on East Eighth Street. Details on Isaac Harris are sketchy, but he married Bertha's cousin, Bella, and joined forces with Max Blanck. In 1900 Harris lived in a small apartment on East Tenth Street. Together Blanck and Harris had all the skills necessary to enter the booming shirtwaist business and do well. They were about to capitalize on one of the fashion revolutions of the 1890s: the shirtwaist, a woman's blouse. The blouse-and-skirt combination was a fashion sensation that revolutionized women's clothing. Garment manufacturers could barely keep pace with the demand for the simple, practical outfits that could be worn to work, social events, and other occasions. With these outfits women could jettison the corsets, hoops, bustles, and other cumbersome feminine accoutrements. In 1910, when the U.S. workforce hovered around 90 million, 5 million workers were women and they needed clothing they could move around in comfortably. About a third of all factory workers in New York at the time were women, and most of them wore shirtwaists and skirts to work.[6]

Blanck and Harris opened the Triangle Waist Company in a small shop on Wooster Street, a few blocks from their apartments, in 1900. In 1902 they moved their operation to Washington Place, half a block east of Washington Square, and into a ten-story loft building designed by Joseph Asch. They leased the ninth floor, nine thousand square feet, a huge amount of space dedicated to garment making. At the time the typical garment manufacturing shop was about thirty times smaller. The space had large windows and twelve-foot ceilings. In this space Blanck and Harris helped to revolutionize the garment industry.[7]

With the advent of loft buildings manufacturers could fit their entire manufacturing process under one roof. The Asch Building alone had ninety thousand square feet of space, and hundreds of loft buildings dotted Manhattan. Between 1901 and 1911 approximately six new loft buildings were completed each month, and nearly eight hundred new skyscrapers were built in that decade. Though originally built for storage, manufacturers quickly realized the potential of these buildings and eagerly leased space in them. The open floor plan meant manufacturers could partition the space to suit their needs, and the high ceilings meant they could squeeze more workers into the space while still meeting the minimum requirement of 250 cubic feet of air per person. The open space allowed garment manufacturers to attach long rows of sewing machines to a single electrical motor by using a drive shaft and flywheels. This resulted in marked efficiency over the small sweatshops using nonmotorized sewing machines. Triangle Waist quickly expanded its operations, and in 1909 the company leased the entire eighth and tenth floors as well as the ninth.

The Triangle Waist Company came into being at a time when the demand for ready-made clothing was skyrocketing. In the first decade of the twentieth century the amount spent on these garments tripled, to roughly $1.3 billion. The rise of modern manufacturing facilities such as Triangle made it possible to meet this demand.[8]

At the time of the Triangle fire Blanck and Harris owned other shirtwaist factories in New York, New Jersey, and Pennsylvania, including the Imperial Waist Company, the Diamond Waist Company, and the International Waist Company. Triangle was the largest blouse-making facility in New York; there more than five hundred workers shipped two thousand garments per day, about $1 million in goods annually. The Triangle factory was nicknamed the "prison" because of the harsh conditions under which workers toiled.[9]

As horrid as conditions in loft garment factories were, they were an improvement over the cramped, squalid sweatshops. Before the loft factories, not only were workers crowded in small basement apartments making garments, but no single shop made the entire garment. Preteen boys worked for low wages as runners; weighed down by enormous bundles, they toted fabric from shop to shop. One sweatshop cut the fabric, another basted, another stitched, yet another set in the lining, and another added decorations. The lofts introduced revolutionary changes in the garment-manufacturing process. Not only were large numbers of workers now assembled in one work space, but the entire garment was manufactured on site. While these changes made manufacturing garments more efficient and netted owners greater profits, it also made them more vulnerable to union organizing. Although it was almost impossible to organize small sweatshops with only a few workers, organizers could reach hundreds of workers at each loft. The same lofts that allowed manufacturers to increase their production exponentially also greatly increased their capital investment in machinery, materials, leases, and insurance. They were under great pressure to keep the production lines operating, and they felt the need to settle or break strikes quickly.[10]

Triangle occupied the top three floors of the building. Harris and Blanck, whose offices were on the tenth floor, patrolled the factory frequently. Primarily Eastern European Jewish immigrant women and Italian women worked for them, six days per week up to fourteen hours per day with a half-hour lunch break. On average the women earned about four dollars weekly; they were issued day-old apple turnovers for overtime pay. The Triangle Company used the "bundle system," a variation of the piecework system, in which workers were issued a small paper ticket when they completed enough shirtwaists to form a bundle. Because workers frequently misplaced their tickets, they were not always paid for the work they completed. Furthermore, workers were required to rent their materials, chairs, lockers, and equipment from the company, and these rental fees reduced their earnings. Management whittled away at workers' earnings even more by fining them for tardiness, spending too much time in the restroom, talking during work, or not sewing fast enough.[11] A split labor market existed because women earned less than men when they did the same jobs. There was also a dual labor market as males were hired into the most skilled position of fabric cutters. Women were not allowed to cut fabric and were frequently classified as "learners" for as long as four years, regardless of their level of skill.[12]

While workers in the lofts toiled high above the din of the streets and dank basements, those on the seventh floor or higher labored above the reach of any fire ladder in the city. This cost many workers their lives.[13]

THE STRIKING SUMMER OF 1909

New York has always been an important center of labor activism. One of the earliest strikes occurred in 1677, when twelve carters struck; they were held in contempt of court.[14] During the early twentieth century a wave of strikes hit the garment and other major industries in the city and elsewhere around the country. Between 1880 and 1920 the needle trades were the third most strike-prone industry behind mining and construction. New York's garment district was hobbled by strikes during the summer and fall of 1909. That year more than a quarter-million laborers worked in city's garment industry, producing about two-thirds of the clothing worn in the United States. At the time Manhattan's blouse manufacturing industry earned $50 million a year. The garment industry as a whole doubled in size in the first decade of the twentieth century. Despite the boom, factory owners kept wages low; in 1909 training wages were around three dollars weekly, barely enough to provide room and board for a worker. Even the most highly skilled workers in the industry earned only $20 weekly (and the wages of many had declined during the 1907 depression). Because garment making was a seasonal business, many workers were laid off for months at a time with no wages.[15]

Workers in the garment factories were routinely followed when they went to the bathroom and told to hurry back to work; they were cheated on their wages and mocked when they complained. Factory owners used many tricks to shorten the lunch hour and fix the time clocks to lengthen the workday.

Female workers had to endure leering foremen. Workers were searched daily. Women suspected of stealing materials were strip-searched while male employees looked on; in 1909 two women brought suit against the Triangle owners after enduring one such ordeal. These and other horrid working conditions drove working-class women such as Clara Lemlich to begin organizing for higher wages, shorter working hours, and better working conditions. Lemlich, a draper at Louis Leiserson's shirtwaist factory, led a strike against the company. For this she was brutally beaten one evening by a gang of men hired to assault and intimidate strikers as they left the picket line.[16]

Jewish garment workers began organizing in the 1880s, forming the United Hebrew Trades. In 1900 seven unions came together to form the International Ladies' Garment Workers Union (ILGWU). Despite the name, this was a predominantly male union composed of skilled garment workers (cutters, makers of custom wear, and cloak makers); in contrast women dominated the blouse-making segment of the industry.[17] A small number of shirtwaist workers, including Lemlich, joined the ILGWU in 1906. They formed Local 25, the Ladies' Waist and Dress Makers' Union, a branch that served the primarily female shirtwaist and dressmakers. Local 25 had between thirty-five and forty members at the end of 1906; at the time there were about five hundred blouse factories employing more than forty thousand workers. The men who ran the ILGWU weren't very supportive of Local 25; they viewed the women with distrust, assuming they were in the workforce only until they got married. Men also saw the women as competitors who drove down wages because women were usually paid less for performing the same tasks. Local 25 survived by making an alliance with middle- and upper-class activists in the Women's Trade Union League (WTUL).[18]

The WTUL was founded by William English Walling, a socialist, a former resident of Hull House, and the son of a Louisville millionaire, and Mary Kenney O'Sullivan, a bookbinder, union organizer, and also a former resident of Hull House and Denison House. Walling worked as a factory inspector of the State of Illinois in 1900–1901; he moved to New York in 1902 to become a resident of University Settlement. That same year women organized a boycott of the high prices of beef at kosher butchers. The protest gave Walling the idea of forming an organization for women; he was also motivated to form the WTUL because of reports that poor young women were being forced into prostitution to support themselves and their families. He figured that the higher wages that would result from unionization would help reduce the incidence of this "white slavery." He went to England to study female trade unions that had been organized since the early 1870s to push for shorter working hours and better working conditions.

The formation of the WTUL was announced in 1903 at the annual meeting of the American Federation of Labor being held in Boston. Prominent settlement workers in Boston, Chicago, and New York, such as Robert Woods, Vida Scudder, Jane Addams, Lillian Wald, Helena Dudley, Mary McDowell, Mary Morton Kehew, and the sisters Mary Dreier and Margaret Dreier Robins, took on leadership roles in the organization. There were also prominent society ladies among its membership, such as Eleanor Roosevelt and Helen Taft, daughter of President William Taft. The organization's goal was to form a coalition of middle- and upper-class women who would collaborate with working-class women to help organize female factory workers into unions. In addition to O'Sullivan, working-class activists such as Leonora O'Reilly, Rose Schneiderman, and Clara Lemlich participated actively in the organization and eventually took on leadership roles. At the beginning Jane Addams and Lillian Wald served as officers of the WTUL.

The WTUL experienced growing pains, but eventually two wealthy sisters, Mary Dreier of New York and Margaret Dreier Robins of Chicago, took over and stabilized it. However, working-class women felt the middle- and upper-class allies were too domineering and that they themselves should run the organization. By 1909 the WTUL was on solid ground, and throughout that summer its members could be found supporting strikers and unfurling their banners outside factory doors at closing time in New York's garment district.[19]

Lemlich utilized her skill as a draper (drapers cut patterns and molded the cloth over a dressmaker's dummy to get the fit of the garment right; few women were drapers) to get hired at different factories. By the time she organized the Leiserson strike she had already organized three strikes in as many years. She led a ten-week wildcat strike at Weisen and Goldstein in 1907 to protest the pressure being placed on workers to speed up production. She also led a strike against Gotham shirtwaist factory in 1908 to protest the company's policy of firing better paid men and replacing them with lower paid women.[20]

As labor unrest increased, so did the growth of strikebreaking agencies. Using thugs, criminals, gangsters, and "detectives" to break strikes became a profitable enterprise. Strikebreaking companies signed contracts with factory owners. The Greater New York Detective Agency courted owners in the summer of 1909 by sending letters to the leading shirt manufacturers promising to "furnish trained detectives to guard life and property, and, if necessary, furnish help of all kinds, both male and female, for all trades." Detective agencies promised to provide agents to spy on workers, thugs to disrupt the picket lines and beat up workers, and replacement workers to keep the factories operating.[21]

Striking was in full swing by the middle of the summer. In July two hundred workers at Rosen Bros. went on strike after demanding a 20 percent wage increase. The owners refused to negotiate with the strikers; instead they hired strikebreakers to attack the picketers. The police usually sided with the owners and strikebreakers when they were called in to quell disturbances; strikers were usually jailed for their actions while the strikebreaking ruffians went unpunished. Despite the forces arrayed against them Rosen Bros. workers remained on strike for nearly a month. Eventually the company capitulated, giving workers the raise they demanded and rehiring them as the busy season approached. The success of the Rosen Bros. workers emboldened other garment workers. In August almost seven thousand young women (most of them teenagers) making scarves and ties walked out of two hundred sweatshops for a month. The workers demanded an end to work in cramped, stuffy bedrooms and cellars, the locations of many of the sweatshops. More than fifteen hundred tailors who operated buttonhole machines also struck briefly in August 1909. Though only a few hundred workers were dues-paying members of Local 25 in the summer of 1909, when the local organized a rally more than two thousand people attended. This made organizers believe that more large-scale collective activities were feasible.[22]

Blanck and Harris employed various tactics to prevent a strike at Triangle. They used an "insider contractor" system, wherein owners allotted space on the assembly line to contractors, who in turn were responsible for hiring workers to operate the machinery. The contractors were paid a lump sum based on the number of blouses completed each payday; they then paid those working for them. The result was that the wage scale was not the same for workers who did the same type and quantity of work. Contractors could also skim the workers' wages in whatever way they pleased. When the owners shortchanged the contractors, the contractors shortchanged the workers. It was the shortchanging of contractors, who were thrown out of the factory when they expressed their anger that triggered a 1908 walkout. The strike began spontaneously late on a Saturday evening and was over by Monday morning.[23]

Blanck and Harris also mobilized against the unionization efforts by establishing an in-house union at the factory called the Triangle Employees Benevolent Association. (All the officers of Triangle's union were Blanck and Harris's relatives.) They also fired employees who belonged to independent unions. Despite these efforts Triangle was hit by a much more organized and long-lived strike in fall 1909. In late September the WTUL and Local 25 held a secret meeting at Clinton Hall on the Lower East Side. Among those attending were about 150 Triangle workers. Blanck and Harris heard about the gathering and sent company spies to observe their workers. The next day the owners stomped onto the factory floor, stopped work, and began promoting their in-house union. They also announced that workers who joined other unions would be terminated. Triangle workers interested in independent unions did not back down. A day later workers arrived at the Asch Building to find the factory shuttered; newspapers carried ads seeking new workers. This prompted the Triangle workers to go on strike.[24]

The Triangle owners did not allow workers to picket peacefully outside the factory. In the early morning hours of October 4 about a dozen young women were picketing the factory when a posse of prostitutes and pimps from the Bowery and replacement workers walked purposefully toward the factory. When the former Triangle workers pleaded with the strikebreakers not to cross the line the prostitutes attacked the picketers. The women scuffled while the pimps kept tabs on how their recruits were doing. Eventually the pimps joined the melee and helped administer a thrashing to the thoroughly overwhelmed picketers. When cops from the Mercer Street Station (the "penitentiary precinct," located about two blocks away) finally arrived, they simply rounded up the bruised and battered picketers and hauled them off to jail. The thugs from the Bowery were not charged. Anna Held arrived for picket line duty as the picketers were being taken to jail. When she asked what had happened she was arrested. Scenes such as this were repeated throughout the strike. The Jefferson Market magistrate's court, located across Washington Square, was busy with strikers trying to plead their case and being fined by the judges. Like Clara Lemlich, Joe Zeinfield—the chair of the Triangle strike committee—was badly beaten by a gang of hired hooligans; he needed more than thirty stitches to close the gashes on his face. Plainclothes police also monitored the picket lines on Harris's and Blanck's behalf.[25]

About a month into the Triangle strike the upper-class women of the WTUL began walking the picket lines alongside the workers. The police were reluctant to arrest the rich ladies at first, but changed their minds when the aristocrats refused to leave. The arrest of Mary Dreier, president of the WTUL and daughter of a German-born merchant, on November 4 made headlines. Dreier's arrest came two days after the election that swept a host of corrupt Tammany officials from office. No one wanted to deal with the publicity that came from the images of upper-class women being assaulted by strikebreakers and police and then being incarcerated afterward, so the police were told to lay off the strikers—at least for a while.[26]

While activists agitated for a general strike the union stalled. Local 25 called a strike meeting in late October, but realizing they had almost no funds and were unprepared for such a move, quickly cancelled it. In the meantime Blanck and Harris made tactical moves of their own to counter the union's actions. Sensing that there might be a general strike Blanck and Harris took the lead in helping manufacturers to prepare for that eventuality. They invited a *New York Times* reporter to tour the Triangle factory to show that production had continued despite the strike. Blanck made it clear that he had no intention of recognizing the union. "All of this trouble," he said, "is over this union business. We did not recognize it and we do not intend to. We told the girls that we were willing to listen to any complaints and to receive any suggestions from our employes themselves, but we had to draw the line on three or four east side gentlemen stepping into tell us how to run our business. It is an outrage the way the girls who have remained loyal—and they are the great majority of our force—have been treated by these people [the strikers] and their sympathizers."[27]

The Triangle owners sent a letter to other blouse manufacturers urging them to join the Employers Mutual Protection Association that they were forming. The letter stated, "You are aware of the agitation that is now going on in our shops; our satisfied workers are being molested and interfered with, and the so-called union is now preparing to call a general strike. In order to prevent this irresponsible union in gaining the upper hand . . . let us know as soon as you possibly can if you will be willing to form and join an EMPLOYERS MUTUAL PROTECTION ASSOCIATION."[28]

On November 22, 1909, Local 25 called a meeting to discuss a general strike. Thousands of garment workers squeezed into the Great Hall of Cooper Union, located four blocks east of Triangle Waist. More than one hundred speakers were crammed onto the stage as the overflow crowds filled nearby halls. For about two hours speaker after speaker, including the headliner, Samuel Gompers, founder of the American Federation of Labor, urged caution. Gompers said, "I have never declared a strike in all my life, I have done my share to prevent strikes. [However,] there comes a time when *not* to strike is but to rivet the chains of slavery upon our wrists. . . . I say, friends, *do not enter too hastily*. But when you can't get the manufacturers to give you what you want, *then* strike! And when you strike, then let the manufacturers know that you are on strike!"[29]

After Gompers's speech there was no call for a strike vote and the meeting was winding down. Three months after leading the ongoing strike at the Leiserson factory and ten weeks after receiving a vicious beating, Clara Lemlich decided to take matters into her own hands. Just before the next speaker took the stage, Lemlich leaped to the microphone and interrupted the proceedings. The twenty-three-year-old Lemlich (some accounts list her as two years older) said, "I have listened to all speakers, I have no further patience for talk, as I am one of those who feels and suffers from the things pictured. I move that we vote on a general strike." Bedlam ensued. After the moderator restored calm and cautioned attendees as to the outcome of their actions, he called for a second to Lemlich's motion. As organizers hastened to inform those in the overflow halls of the strike vote, it was approved in one hall after another.[30]

The next morning workers reported to their jobs as usual. Then about fifteen thousand walked off the job as the union scrambled to cope with the massive general strike they had on their hands. Strikers made their way to local meeting halls, where they made a list of their strike demands: no night work, shorter working hours, wages arranged by a committee, better treatment from the owners. The union demanded a 20 percent pay increase, a fifty-two-hour workweek, recognition of the union as the bargaining agent for all shirtwaist workers (that is, a closed shop), pay for overtime work, and advance notice of layoffs.[31]

The Triangle owners and other manufacturers responded to the strike by hiring Italian women as replacement workers; between 6 and 10 percent of the strikers were Italians. By the second day roughly twenty thousand workers were on strike. Then the cutters union decided to honor the strike. This was significant since the cutters were primarily male and the best paid workers in the garment industry; cutters were also highly skilled, so manufacturers could not replace them easily. More than five hundred factories were affected by the strike. More than seventy owners of smaller factories capitulated within the first forty-eight hours. Workers from those factories returned to work having secured a 20 percent pay increase, a fifty-two-hour work week, and a closed shop.[32]

Blanck and Harris were perturbed not only by the scale of the strike but by the hasty surrender of many garment manufacturers. They and twenty more leading factory owners called an emergency meeting at the Broadway Central Hotel to develop counterstrategies. They resolved to form a manufacturers' association, resist the strikers, refuse to meet their demands, and try to crush the union. Blanck and Harris had survived a long strike by hiring strikebreakers to disrupt picket lines and beat up union organizers, making friends with the police, and providing incentives for replacement workers. Because the Triangle owners could prove to their colleagues that it was possible to weather a strike, others manufacturers looked to them for direction. The following day an even larger group of manufacturers met at the Hoffman House Hotel, where representatives from about one hundred companies signed a "no surrender" declaration. The stampede to settle was halted, and the policy of dealing with strikers by using brute force, thuggery, psychological and ideological warfare, and ignoring the union went into effect.

During the strike the Triangle owners sought to keep nonstriking workers happy by allowing them to dance in the factory on their lunch break; the owners even offered weekly prizes to the best dancers. In the meantime they made plans to move their operations outside of Manhattan to escape the radicalism that had captured the imagination of the workers. They opened a branch factory in Yonkers and planned to advertise in the conservative Yiddish *Morgen Journal* for new workers willing to work in the far-away suburbs. The Yonkers factory operated for ten days before strikers went there and picketed; it was closed down on December 10. Workers in the Yonkers factory had been led to believe that their workplace would not be picketed. As a result of the pickets and fights that broke out on the picket lines, many of the women commuting to Yonkers stopped showing up for work.[33]

Strikers carried picket signs reading "We Are Striking for Human Treatment."[34] They were beaten and arrested by the police. The judges were not very sympathetic to the strikers; when a group appeared before Magistrate Willard Olmsted he told them, "You are on strike against God and nature, whose prime law it is that man shall earn his bread in the sweat of his brow."[35] Another judge likened the strikers to prostitutes, saying, "You picketers are without shame. . . . You are no better than streetwalkers."[36]

The strikers drew the attention of aristocratic women other than those affiliated with the WTUL. Alva Smith Vanderbilt Belmont, Anne Morgan (J. P. Morgan's daughter), and President Taft's daughter became interested in the plight of the shirtwaist workers and in the strike. Energetic young women from Wellesley, Bryn Mawr, Vassar, Smith, and Barnard also eagerly lent their help and support. Some left college to participate in strike activities, while others moved in with shirtwaist workers. Sympathizers who remained on campus organized a variety of activities in support of the strikers, such as a boycott of blouses made in nonunion factories. They also planned to start a cooperative factory staffed by strikers; the students at Wellesley pledged to buy the first one thousand blouses made by the co-op. Yet despite this coming together of upper- and middle-class women to aid the cause of working-class women, a schism threatened the fragile coalition. While women such as Alva Belmont wanted to use the shirtwaist strike as a medium to push a wider array of feminist causes, longtime WTUL activists wanted to keep the focus on the factory workers' demands. Not to be dissuaded, Belmont organized a rally at the Hippodrome on December 6 that was attended by seven thousand people.[37]

Anne Morgan also got involved. In an interview with the *New York Times* she said, "We can't live our lives without doing something to help them. . . . Fifty-two hours a week seems little enough to ask."[38] On December 15 she and her friends invited strike leaders to attend a luncheon at the exclusive Colony Club that was attended by 150 members. The strikers told their stories, then club members passed a hat for donations, collecting $1,300 ($26,690 in 2005 dollars) in a matter of minutes. Alva Belmont's fundraising efforts resulted in several wealthy women, including the wife of the railroad magnate Collis Huntington, the wife of the tin czar Warner Leeds, and John Jacob Astor's daughter, donating to the strikers.[39]

Faced with a general strike garment manufacturers were no longer averse to hiring blacks as strikebreakers. Racial equality was not an agenda item that received much attention in Local 25. Like the charities and settlements, unions tended to ignore blacks and other minorities in their organizing efforts. When strike organizers heard that blacks were being hired, Elizabeth Dutcher of the WTUL and two shirtwaist workers went to a meeting being held by blacks at the AME Zion Church in Brooklyn to try to persuade them not to take jobs in the shirtwaist companies. The blacks reminded Dutcher and her companions of the racism of the labor movement and in Local 25. Dutcher and her delegation urged the blacks not to act as strikebreakers, but they also urged the labor movement to be more open about the entry of blacks into the trades and the movement.[40]

One of the perks manufacturers provided for some replacement workers was to transport them to and from work in automobiles. Wealthy activists countered by loaning their cars and chauffeurs to the strikers to bring them to the picket lines and to help them organize more efficiently. The cars made a motor parade down Fifth Avenue and around the garment district; Anne Morgan also rented seven taxis for the parade.[41] As time passed, the police grew impatient with the strikers and with the rich women trying to help them. Alva Belmont was arrested on the picket line and spent a night in jail. Her arrest brought even more publicity to the strike when she announced that she would put up her house to bail out all the strikers who had been arrested. As reports of rough treatment of female strikers filtered out, organizers convinced a group of prominent businessmen to observe the picket lines to monitor police behavior. Strikers' efforts were also buoyed when about fifteen thousand shirtwaist makers walked off the job in Philadelphia after learning that New York manufacturers planned to shift their blouse-making operations to their city. Replacement workers in New York began walking out of the factories to join the strikers. To top it off, the Triangle Waist Company had to shut down briefly. Before it was all over, an estimated thirty thousand workers were on strike.[42]

Unfortunately the fragile coalition between the WTUL, Local 25, and the wealthy women recently converted to the cause cracked and fell apart. There were several reasons for this. First, WTUL activists who had been longtime supporters of Local 25 and the working-class activists were resentful of the high-profile activities of the newly converted supporters and the attention-grabbing headlines they had a knack for generating. There were also tensions around whether the strike should be framed as a labor dispute or expanded and framed in terms of women's suffrage. Some were critical of the event held at the Colony Club, where strikers were invited to have lunch with the wealthy members. One critic described the lunch as a gathering in which a "bejeweled, befurred, belaced and begowned audience" contrasted sharply with the "ten wage slaves, some of them mere children," who were invited in to tell their stories. The donations were also scoffed at. Some of the longtime activists felt that, overall, the involvement of the wealthy women had a damaging effect on the strike.[43]

As December drew to a close many of the small and midsize manufacturers reached settlements with the strikers, leaving only the large factories as holdouts. Around Christmas time the owners made a proposal: they agreed to an arbitration panel, a pay raise, shorter working hours, and no punishment of workers who were union members. They did not agree to the closed shop system. Local 25 leaders rejected the offer; they refused to bargain if Local 25 was not recognized as the bargaining unit of the workers and if the manufacturers did not agree to a closed shop. The decision split the coalition and was denounced by some newspapers. Some progressive activists thought the manufacturers' proposal was reasonable and that the decision to reject it was an indication that the union was being unduly influenced by radical factions. A rally at Carnegie Hall on January 2, 1910, was the event that finally splintered the coalition. While strikers were elated by the fiery social justice speech of Morris Hillquit's, who was the leader of the Socialist Party of America, the rich activists from elite mansion districts were shocked and troubled by what they heard, so much so that the next day Anne Morgan issued a statement that criticized Hillquit and ceased her activities on behalf of the strikers. Her withdrawal from the scene meant that much of the mainstream press stopped covering the strike. Some WTUL board members were also distressed by what they saw as growing radicalism. A proposal was floated to join forces with Morgan to form an organization to rival the WTUL.[44]

The strike continued but never regained the momentum it had leading up to Christmas. The new year signaled the approach of the busy shirtwaist season, so the owners were anxious to settle. At the same time the union was gearing up for the cloak makers strike and was eager to settle the current strike as well. As a result in February Local 25 accepted the proposal they had rejected in December. The major shirtwaist companies, such as Triangle, Leiserson, and Bijou, eventually rehired the striking workers at higher wages. Although manufacturers promised to recognize the union, they did so only in the sense that they did not prohibit their employees from joining; thus they effectively resisted closed shops. On February 8, 1910, the union called off the strike.[45]

THE FIRE

The Triangle fire broke out thirteen months later. The image of women and young girls, literally ablaze, leaping from the ninth-story windows with their flaming skirts billowing in the wind horrified the country. The fire broke out around 4:40 p.m., just as workers were preparing to leave for the day. Though the Triangle Company had a no-smoking policy, higher status skilled workers such as the cutters routinely ignored the posted signs. Small fires had broken out in the factory during working hours before, but workers got these under control by dousing the flames with water that was always kept in the red fire buckets located around the factory. The factory manager, Samuel Bernstein, recalled at least three such fires being put out before they could spread.[46]

This time it would be different. The fire broke out in Isidore Abramowitz's wooden cutting bin. The bins under each cutter's wooden table contained hundreds of pounds of scrap cloth and tissue paper. All told there were thousands of pounds of cloth and paper sitting in such bins on the eighth floor. Louis Levy, a rag trader, testified at the trial of Harris and Blanck that he had been purchasing scraps of cloth from the Triangle owners for three or four years. He said the bins were cleaned about six times per year. The last time he emptied the scraps, on January 15, 1911, he had removed 2,252 pounds of materials. Because it was the busy season at least a ton of scraps was in the bins under the tables. It was into Abramowitz's bin that someone tossed a match or cigarette butt on the evening of Saturday, March 25, 1911.

About 180 people worked on the floor that day. The fire was discovered just as they were preparing to leave. Abramowitz was one of the first to grab a red fire pail and begin dousing the flames. Blanck reported that there were a hundred pails scattered around the factory. These were of little help. The fire mushroomed in size and spread rapidly. Other cutters grabbed buckets and began pouring water onto the flames, but within seconds the fire was out of control. Some workers grabbed the fire hose stored in the stairwell, but no water issued forth because the water tank on the roof was empty. Panicked workers tried to leave the factory through the narrow Greene Street exit, but because the guards inspected workers' bags daily the exit was designed to let only one worker through at a time. Some rushed to the elevators and the fire escape. There was a huge pileup at these doors because they were designed to swing inward and were difficult to open with scores of workers pressed up against them. The Washington Place doors were kept locked; it took workers some time to find a key to open the eighth-floor door; they were never able to unlock the ninth-floor door.[47]

In the meantime the bookkeeper placed a call to the tenth floor to warn them of the fire. The Triangle phone system was set up so that a call could be made from the eighth floor directly to the tenth floor's switchboard operator, but not from the eighth to the ninth floor. Once a call was made to the tenth floor it should have been relayed to the ninth floor, but the panic-stricken operator who took the call dropped the phone and never relayed the call to the ninth floor.[48]

Ida Schwartz described the panic inside the building and her frantic search for an exit: "When I heard that cry [of 'fire'] I almost fainted. I ran out of the dressing-room to the Greene Street door. A crowd was about that, so I went to the fire-escape, but that was crowded, and then I rushed to the Washington Place door where I saw Meyer Utal, a machinist, Abe Renowitz, and Jake Kleine, who are all dead now, and they were trying to open the door."[49]

Outside the Asch Building people went about their business on a balmy spring afternoon. Parkgoers lounged in nearby Washington Square, members of New York's upper crust sipped afternoon tea in their mansions, and New York University students attended classes or studied in the library. Frances Perkins of the National Consumers' League was having tea with Mrs. Gordon Norris, who lived in one of the brownstones on the north side of Washington Square. High-rises had engulfed much of the Washington Square neighborhood, but owners of the brownstones on the north side of the square held their ground and managed to retain a genteel residential enclave. Perkins lived on nearby Waverly Place. When the wail of fire engines interrupted their conversation, both women rushed outside to find out what the commotion was about; when they saw the smoke they, like others on the street, followed it to the Asch Building.[50] William Gunn Shepherd, a newspaper reporter, happened to be in Washington Square at around 4:45 p.m. when he noticed a puff of smoke emanating from the Asch Building. He too made his way toward the building. There pedestrians heard the sound of breaking glass and looked up to see bodies, some ablaze, hurtling through the air. Shepherd wrote

> I learned a new sound—a more horrible sound than description can picture. It was the thud of a speeding, living body on a stone sidewalk. Thud-dead. Thud-dead. Thud-dead. Thud-dead. . . . Up in the [ninth] floor girls were burning to death before our very eyes. . . . Down came the bodies in a shower, burning, smoking—flaming bodies, with disheveled hair trailing upwards. . . . I looked upon the heap of dead bodies and I remembered these girls were shirtwaist makers. I remembered their great strike of last year in which these same girls had demanded more sanitary conditions and more safety precautions in these shops. These dead bodies were the answer.[51]

A *Washington Post* writer also described the scene:

> Pedestrians going home through Washington place to Washington Square at 4:50 were scattered by the whiz of something rushing through the air before them; there was a horrible crash on the pavement, and a body flattened on the flags.[52]

The *Chicago Sunday Tribune* carried an article with the following description:

> She climbed to the sill, stood in black outline against the light, hesitating, then, with a last touch of futile thrift, slipped her chatelaine bag over her wrist and jumped. Her body went whirling downward through the woven wire glass of a canopy to the flagging below. Her sisters who followed, flamed through the air like rockets.[53]

The fire burned for only half an hour, but it was devastating. Because workers tried to put out the fire themselves the first alarm was not called in until 4:45. Onlookers too sent in alarms around the same time. As many as seventy workers, as well as Harris and Blanck, were on the tenth floor when they were alerted to the blaze. Harris immediately took control of the situation and began directing workers to the roof and helping others to safety. Once before he had shown this same calm demeanor while putting out a small fire in one of the bins. Blanck, who had his five- and twelve-year-old daughters and their governess with him, panicked. When Eddie Markowitz, the chief shipping clerk, went to the tenth floor he found a terror-stricken Blanck gripping his daughters. Markowitz, who had gone to rescue his order book, grabbed little Mildred from Blanck's arms, pulled him by the coat, and led them and the governess to the Greene Street stairs and up to the roof. Samuel Bernstein, who went to the ninth floor, could not enter because the fire was too intense; he continued up to the tenth floor, where he helped to lead workers to the Greene Street stairs and up onto the roof of the building.[54]

Most of the workers on the eighth and tenth floors escaped, but many on the ninth floor died when they were unable to force open the locked Washington Place door. At the time of the fire the labor law stated, "All doors leading in and to any such factory shall be so constructed as to open outwardly where practicable, and shall not be locked, bolted or fastened during work hours." Moreover the flimsy rear fire escape of the building collapsed, killing several and cutting off one of the avenues of escape. The fire escape stopped twenty-two feet above the ground, and directly below it was a basement skylight; several people died when they crashed through it. There was another problem with the fire escape: the windows leading to it opened outward, so once workers pried the locked windows open, they blocked the escape from those behind them.

Fire trucks arrived quickly, the first only two minutes after the first alarm was sent. Eight engines answered the call, but they were hampered by the dead bodies sprawled all over the sidewalk and ladders that reached only to the sixth floor. For a few minutes firemen tried to save jumpers by using fire nets. Onlookers also tried to catch people by using tarpaulins. But the nets and tarpaulins broke when workers jumped into them and the strategy was soon abandoned.[55]

An estimated 250 people were on the ninth floor that evening. When the fire broke out some of the workers, finding the Washington Place door locked and the Greene Street exit too crowded to get through, took the stairs to the tenth floor. From there they made their way to the roof and were pulled to safety by New York University law students who stretched ladders from the university building to the Asch Building. The students rescued about 150 people.[56]

Other workers who tried to use the elevators on the lower floors couldn't get into them. Fourteen-year-old Eva Harris recalled her horror as she tried to escape the eighth floor by elevator, only to watch it pass her by: "The girls crushed against the doors [of the elevator] could see the cars going up. . . .

Some of the girls were clawing the doors and screaming 'Stop! Stop! For God's sake stop!'" The elevator operators eventually came back and picked up some of the workers, but according to Harris, by then "some of the girls had already jumped."[57]

One woman deliberated before she jumped. Leon Stein, author of *The Triangle Fire*, describes how the woman inched her way toward the window ledge. The flames engulfed her, but as the crowd below watched she slowly opened her handbag and pulled out a few bills and a handful of coins, then tossed the money out the window. As the bills floated down and the coins hit the sidewalk with a ringing sound, she jumped.[58]

Despite the historical significance of the fire there has not been an accurate count of the dead or a reliable list of victims. The most often cited numbers of dead are 146 and 147, though the total is sometimes higher and sometimes lower. Some sources claim that all but twenty-three of those killed were women. Fifty-four people died jumping to the sidewalk, about fifty died inside the factory, and another nineteen died in the elevator shaft. Approximately two dozen people died when they fell from the collapsing fire escape.[59]

The bodies were stored in a makeshift morgue at Charities Pier (also known as Misery Lane). One hundred thousand people lined up to view the remains the following day; three hundred thousand people braved heavy rains to march in a mass funeral procession on April 5, 1911.[60]

THE RELIEF EFFORT

Immediately after the fire the mayor William Gaynor gave $100 to relief efforts and called on people to help out. Though most of the Triangle workers were not unionized, Local 25, the WTUL, the United Hebrew Trades, the Workmen's Circle, and the Jewish *Daily Forward* formed a Joint Relief Committee on March 28 to help collect and administer funds for the victims of the fire. The Charity Organization Society also administered relief funds on behalf of the Red Cross. The WTUL was in charge of visiting victims and investigating their condition. The Joint Relief Committee and the Charity Organization Society reached an agreement whereby the latter would oversee the cases of nonunion workers, while the Joint Relief Committee would deal with cases involving union members. The Joint Relief Committee buried victims, paid wages until victims recovered from injuries, found them new jobs, dispensed money to victims' families, sent remittances overseas, helped family members living overseas migrate to the United States, and placed children orphaned by the fire in boarding schools and paid for their expenses. Some of the injured were sent to the countryside to recuperate.[61]

The public gave generously to the Triangle Fire Fund. By March 30 the fund exceeded $50,000. Andrew Carnegie contributed $5,000. Other prominent businessmen such as John Jacob Astor and M. C. D. Borden contributed a thousand dollars each. Mrs. Jefferson Seligman contributed $500 and Goldman, Sachs and Company contributed $250. Anne Morgan and Mrs. J. P. Morgan each contributed $100, and the Metropolitan Life Insurance Company contributed $2,500 to be used for special relief of families of policyholders. Performing groups such as the White Rats of America and the Associated Actresses of America each contributed $200. Alva Belmont and Anne Morgan reemerged to help the workers; they organized a benefit at the Metropolitan Opera House in April 1911 that raised $8,350. Funds were collected by the Red Cross, the *New York Times*, Local 25, and the Independent Order of B'nai B'rith. Numerous benefit performances were also organized to help relief efforts. By May 1912 the fund had reached $120,000 ($2.46 million in 2005 dollars).[62]

THE TRIAL

Seven indictments of manslaughter were filed against Isaac Harris and Max Blanck on April 11, 1911. They were arrested and released on $25,000 bail. However, the Triangle owners didn't accept culpability. They hired Max D. Steuer, one of the best and most expensive trial attorneys in the city, at a fee of about $10,000 each. Like the defendants, Steuer was an Eastern European Jewish immigrant; he lived in a Lower East Side tenement on Henry Street and had worked in the garment factories when he first immigrated to New York. Steuer called a bevy of witnesses to contradict and whittle away at the prosecution's case, argued by Charles Bostwick. He hammered away at the fact that the Triangle factory had passed inspection less than a month before the fire. He also effectively raised doubt as to whether the doors were locked.[63] Some workers who testified on behalf of the defendants acknowledged that the Triangle owners had increased their wages before they testified; some saw their weekly wages jump from $18 to $25 (from $356 to $495 in 2005 dollars); others got more modest increases.[64]

Harris and Blanck argued that the labor law did not require that all doors in a factory be kept unlocked during working hours. They argued that if one door was kept open, which was the case in the Triangle factory, then they were in compliance with the law. The court disagreed with this interpretation of the law, but the question of whether the owners had padlocked the door, ordered them padlocked, or knew they were padlocked at the time of the fire remained unanswered.

The magistrate presiding over the case, Judge Thomas C. T. Crain, helped Blanck and Harris's case by issuing very specific instructions that made it difficult for jurors to vote for a conviction. In order to convict jurors had to find, without a reasonable doubt, that not only were the doors locked, but that Blanck and Harris knew that they were locked at the time the fire broke out on March 25. The jury of twelve men, several of whom were salesmen or fellow shirtwaist manufacturers, deliberated for about two hours before reaching a verdict. The owners were acquitted of manslaughter charges on December 27, 1911.[65]

It was typical at the time for factories to split the risk of insurance among different companies. Triangle was insured by forty-two different companies; Blanck and Harris settled their claims with forty-one of them. The Royal Insurance Company Limited balked at paying the $5,000 claim made against them; it was suspicious enough of the blaze to conduct its own independent investigation, but Blanck and Harris sued the company to collect on the claim. In all Triangle claimed to have lost $203,585 in the fire; Royal Insurance claimed that the amount was more like $174,750. With their unflappable attorney, Max Steuer, by their side, Harris and Blanck collected almost the full amount they were insured for: $194,750.

The relatives of the victims did not fare as well. Though victims' families sued the Triangle owners, nothing was ever collected from Harris and Blanck. Steuer also defended them against these suits. In 1914 civil suits brought by the families of twenty-three of the victims were settled by an insurance company with payments of $75 ($1,400 in 2005 dollars).[66]

Soon after the trial some of the jurors spoke out. Victor Steinman, a shirtwaist manufacturer, reported that he regretted voting to acquit Harris and Blanck. He said that at the time of voting he was confused; he believed that the Washington Street door was locked, but because he couldn't be certain that Harris and Blanck knew it was locked at the precise moment of the fire he did not register a guilty vote.[67]

THE FACTORY INVESTIGATING COMMISSION

Immediately after the fire the WTUL and the ILGWU demanded an investigation. Some called for a new trial. Though Local 25 was a small fledgling organization and the ILGWU was a relatively weak union at the beginning of the strike season of 1909, since then the ILGWU had signed on thousands of new members and had weathered two significant strikes. In addition to the shirtwaist strike of 1909–1910 the union had organized the cloak makers' two-month strike of 1910, when sixty thousand workers walked of the job.

The cloak makers were successful in getting their demands met; they got a fifty-four-hour week, overtime pay, ten legal holidays, compulsory arbitration, and the establishment of joint labor-management boards. Out of this agreement came the "Protocol of Peace," drafted by the prominent Boston lawyer Louis Brandeis, that established a mechanism for resolving workplace conflicts without resorting to strikes. The union had more power and used it to push for an official investigation of the fire. William Randolph Hearst and other prominent people also began agitating for an investigation. Hearst used his newspaper, the *American*, to promote that agenda and even created his own commission to investigate the tragedy. The Fifth Avenue Association also pushed for an investigation of the fire and increased inspection of factories in the city.[68]

Harris and Blanck weren't the only ones blamed for the tragedy. Some critics noted that if the first wave of firefighters who arrived on the scene had had axes they might have been able to break down the doors and possibly save more lives. Others pointed the finger at the New York Building Department, charging negligence for allowing the factory to operate without adequate safety precautions.[69]

On June 30, 1911, Governor John A. Dix signed a law creating the New York Factory Investigating Commission, consisting of five legislators, two union activists, and two corporate representatives. The Commission began holding hearings on factory conditions immediately. It inspected 228 factories between 1911 and 1912 and found that more than half of the approximately eleven thousand workers in those factories worked above the sixth floor, beyond the reach of the fire hoses. The Commission also noted that the New York Labor Department was unaware of the existence of many of the factories: while investigating the cloak and suit industry, the Joint Board of Sanitary Control found that 30 percent of the shops inspected were not listed in the Labor Department's records. The Commission also found that between 50 and 75 percent of the fires occurring at the time were preventable if simple and inexpensive precautions were taken. After collecting testimonies from thousands of witnesses from 1911 to 1914 the Factory Investigating Commission made more than sixty recommendations. These included enforcing the no-smoking policies inside factories, cleaning up and removing scraps and rubbish, using metal receptacles for flammable wastes, holding fire drills, requiring that buildings taller than seven stories have sprinklers, mandating a fifty-four-hour work week for female factory workers, regulating the use of dangerous machinery, and updating the fire-alarm system. New standards were also set for lighting and ventilation. Factories had to install bathrooms, and greater restrictions were placed on piecework. Fifty-six of the Commission's recommendations were eventually adopted; the institution of a minimum wage for women and children was the major piece of legislation that failed to pass. Illinois, Massachusetts, and other states also appointed factory investigating commissions.[70]

THE FIFTH AVENUE ASSOCIATION AND FACTORY REGULATION AFTER THE FIRE

One outcome of the investigation was the institution of strict limits on the number of workers a factory could have. The factory occupancy law that took effect on February 1, 1914, also regulated the width of the stairways in buildings and specified a minimum of thirty-six square feet per person in the factories. On February 5 the Fifth Avenue Association released a report after inspecting 130 factories; the Association found varying levels of compliance with the new law and predicted that the occupancy limits and regulations regarding stairway width would call for such drastic reduction in the factory workforce that manufacturers would move their operations elsewhere. The report noted that factories hired a minimum of 130 workers; under the occupancy law the average number of workers would be eighty-seven. Twenty-three factories employing 2,685 people were in full compliance or would be soon; eighty-nine factories employing 11,500 people would have to reduce their workforce by 50 to 70 percent to be in compliance, plus make radical structural alterations to their buildings to meet the new standards. Eleven factories employing two thousand workers had moved or were planning to move to between Sixth and Eighth Avenue. Seven factories employing five hundred people would have to close. The report added that the Association opposed the policy of increasing by 50 percent the number of workers in buildings with automatic sprinklers. Copies of the report were sent to the New York Department of Labor and to the superintendent of the Building Department.[71] Robert Grier Cooke, president of the Association, argued that the restrictions would result in "a better class of trade—the kind that Fifth Avenue has and that we want it to have in the future. . . . Fifth Avenue is a tremendous asset for a high-class businessman. . . . We want the right kind of business. We want the right kind of business building."[72]

The aesthetics of the loft buildings and the fiscal soundness of the enterprises operating within them were also at issue. Lofts, constructed between 1870 and 1930, were often cheaply built and financially insecure. The buildings tended to be long and narrow, so despite having large windows, lofts had long, dark corridors. The structure was usually sturdy enough to store heavy loads, and each structure was rented floor by floor. The plain and unpartitioned interiors sported high ceilings. Lofts ranged in height from seven to over twenty stories. Some stood vacant for years; once vacant, it was difficult to convert the building to other uses.[73]

In addition to helping to pass building height restrictions and zoning laws that would limit the number of lofts constructed on Fifth Avenue, the Fifth Avenue Association used other tactics to limit the spread of factories in the neighborhood. Property owners refused to lease their buildings to garment manufacturers.[74] Merchants in the Association collaborated with each other to boycott manufacturers who made their garments above Thirty-fourth Street. In addition, bankers refused to finance the construction of lofts on Fifth Avenue. Walter Stabler, controller of Metropolitan Life, one of the largest financiers of building construction in the city, announced at the 1916 membership drive of the Fifth Avenue Association that Metropolitan and other large financial institutions had created a policy of refusing "loans on loft buildings in the Fifth Avenue region above Thirty-fourth Street." Another strategy was to place restrictions on the use of existing loft buildings. The Association supported the policy stipulating that only one-quarter of buildings that were more than two stories high could be used for manufacturing purposes.[75] The strategies were effective. A census of factories in the area showed that between 1914 and 1915 the number of workers in the Fifth Avenue zone decreased from 73,000 to 51,476.[76]

THE TRIANGLE COMPANY AFTER THE FIRE

After the fire the Triangle Waist Company continued to operate, but at a new location at Fifth Avenue and Sixteenth Street, not far from where the old factory was located. Blanck was the president of the new company and Harris the secretary. After several years they moved their operations again, farther uptown to West Thirty-third Street. Though the Triangle Waist Company never gained the prominence it once had in the shirtwaist industry, continued run-ins with the law kept the owners, particularly Blanck, in the spotlight.[77]

Max Blanck was arrested again in September 1913 on charges of locking his Fifth Avenue factory door during working hours. Charges were brought by Walter J. Dugan, an inspector with the Bureau of Fire Prevention. The factory employed about 150 women. Once again Steuer defended Blanck, arguing that the kind of lock used on the door was approved by the state's Labor Department. Chief Justice Isaac Russell found Blanck guilty but charged him only $20 and apologized for having to levy any fine against him at all, saying, "You seem to have shown every disposition to abide by the law, and I find that Labor Bureau Inspectors passed your locks regularly before this inspection by an agent of the Bureau of Fire Prevention. It seems to have been a case of a failure of the two sets of Inspectors to agree as to what constitutes a dangerous lock under the law."[78]

In December of that year the Fire Prevention Bureau visited the company at the request of the owners, who wanted to show off the new locks they had installed on their fire exits. However, inspectors noticed that the shop floor was littered with cuttings from shirtwaists, a large amount of flammable material was sitting in wicker baskets, and a pile of paper six feet deep stood in one corner.[79]

The negative publicity continued for Triangle. In 1914 the company was caught stitching fake Consumers' League labels into its garments. The Consumers' League label was the official seal certifying that garments were manufactured under good workplace conditions. Blanck defended the action by arguing that the factory's reputation was unfairly tarnished. Neither the company nor its owners were indicted or fined for these later violations.

Triangle struggled on until the end of the First World War. Around 1920 the two partners split. Blanck continued to run the Normandy Waist (later Normandie Waist) Company, while Harris opened a tailor's shop named Harris Waist Shop. After 1925 both men seem to have faded from the business scene.[80] The Asch Building is now a New York University science lab. On March 23, 2003, the building was officially designated a historic landmark. A plaque, commemorating the fire, was also unveiled.[81]

FACTORY INSURANCE AND FIRES

The 1911 fire was not the first at Triangle. Between April 1902 and March 1911 the company had filed claims and collected insurance for nine different fires of unknown origin. Yet surprisingly reports from more than a thousand workers at various factories revealed that conditions at Triangle were better than many. As at Triangle, the doors of many factories were locked during work hours. A traveling salesman interviewed by the *New York Times* reported that nine-tenths of the factories locked their doors while workers were inside: "I am a salesman and have sold waists, skirts, coats . . . for different manufacturers in this city, and I assure you without bias that nine-tenths of the manufacturers lock the doors of the factories during working hours." The WTUL received information that some factories had no fire escapes and workers were unaware of where the emergency exits were or how to get to the roof of the building.[82]

Though Triangle may not have been the worst factory, its fire prevention scheme was not up to the standards employed in the safest factories. From the 1880s on New England cotton mills were equipped with automatic sprinklers, firewalls, and fireproof doors. Commercial buildings in Philadelphia were being built with fireproof stairways by 1911. None of these innovations were widely used in Manhattan's factories. Of more than a thousand factories surveyed in 1910 only one garment factory had sprinklers. One reason Manhattan's loft buildings did not have firewalls or doors was that they weren't originally designed to be used as factories. The insurance companies also played a role in the laissez-faire attitude factory owners had regarding safety. Instead of lowering premiums for taking preventive measures, the insurance companies charged higher and higher premiums to repeat offenders. Though the Triangle owners submitted claims for several fires they were still able to buy large insurance policies on their factories.[83]

In fact some wonder whether some of the fires at Triangle weren't deliberately set, and whether fires deliberately set to collect large insurance premiums weren't a calculated business strategy throughout the garment industry. The first of Triangle's many fires broke out around 5 a.m. on April 5, 1902, shortly after the factory moved to the Asch Building. By the time the fire department got the alarm and arrived at the building the contents of the factory had already burned. Another early-morning fire broke out at the factory on November 1. Blanck and Harris collected about $32,000 from both fires. Both fires occurred at the end of the twice-yearly busy season for the garment industry. If owners did not make careful estimates, they could be left with excess inventory at the end of the busy cycle; such inventory was difficult to sell and very costly. Considering that Blanck and Harris had moved from a small shop to the much larger and more expensive Washington Square space the previous year, they might have been financially strained. Under such circumstances carefully set fires could be lucrative. Indeed fire broke out in Harris and Blanck's Diamond Waist Company in a nearby Mercer Street loft in the predawn hours one April day in 1907. Three years later an after-hours fire broke out at Diamond, also in April. All the insurance claims were paid.[84]

Collier's magazine published a series of articles on the prevalence of suspicious fires in the garment industry, noting that there were more fires in New York than in all the major European cities combined. The Factory Investigating Committee appointed by the governor determined that the per capita loss from fires in the United States was higher than in six leading European countries: $2.51 in the United States compared to thirty-three cents in Europe. The *Collier's* articles found a relationship between macro-level trends in the fashion industry and the occurrence of fires. When Parisian salons rejected feathered hats, within a month there were fires in three American feather factories. Similarly in 1910, soon after Paris declared it wanted plain women's dresses, a rash of fires ravaged New York's braid and embroidery factories. Triangle wasn't the only shirtwaist factory that burned in 1911. That year Paris frowned on the shirtwaist; by the end of the year one insurance company had paid claims for fires at ten shirtwaist factories. In the three years prior to 1911 the same company paid claims on only six shirtwaist factory fires. However, the fires were forcing insurers to change how they insured companies. After the feather factory fires insurers began canceling the policies of feather companies. As the shirtwaist factories burned in 1911 a large insurer began canceling those policies.[85]

Triangle owners insured their businesses heavily. In 1909, the year the shirtwaist industry was paralyzed by a wave of strikes, Harris and Blanck tried to increase their fire insurance policy by a substantial amount. One insurance company insisted on inspecting the factory and hired a fire-prevention expert, P. J. McKeon of Columbia University. He was concerned about the locked doors, the overcrowding, and the large number of workers toiling so far above ground. He recommended sprinklers for the building and regular fire drills. He also recommended H. F. J. Porter to conduct fire drills, but though Porter contacted Triangle to offer his services the owners never hired him.[86]

One of the busy seasons for shirtwaist companies wound down at the end of March. At the time of the fire Triangle was insured for $200,000, much more than the factory was actually worth. The owners bought this large policy less than a year after coming off a crippling strike and at a time when Parisian attitudes toward shirtwaists were having a dampening effect on the market. This information raises the question of what might have happened at Triangle in the absence of the deadly accidental fire.[87]

Gender, Class, and Ethnicity

The Triangle case study illustrates how women, ignored by the male-dominated unions, were systematically exploited in the workplace. Male unionists resented the presence of women in the workplace, and it wasn't until women asserted themselves that they were grudgingly admitted into unions and allowed to organize. The union's shortsightedness played right into the hands of male owners and supervisors of factories, who devalued women's work by paying them less than men. Factory owners were also more reluctant to promote women. As already mentioned, at Triangle men held the supervisory positions and higher paying jobs while women were often classified as trainees for lengthy periods. Though the factory owners deliberately sought out Eastern European Jewish women to staff the factory because of their needlework skills, these women were underpaid once they gained employment in the factory. Women were frequently taunted and humiliated on the job. Their purses were searched daily and some of the women were strip-searched. The strikebreakers were predominantly female.

The Triangle owners capitalized on Jewish pan-ethnic solidarity to recruit and maintain its workforce. Other ethnic groups, such as Italians, were hired as strikebreakers to keep labor unrest in check. The workforce was almost exclusively white; blacks were hired only as strikebreakers. Despite the fact that the owners themselves had been exploited in sweatshops, they did not hesitate to exploit their workforce, then blame them for their plight. So on the one hand the Triangle owners provided job opportunities to new immigrants who could not speak English and who would have had a difficult time finding work, but on the other they worked these immigrants long and hard for abysmal wages.

The class struggles that erupted between the Triangle owners and the workers pitted rising immigrants on the make like Horatio Alger against less successful or less ambitious poorer immigrants. Blanck, Harris, Leiserson, and other immigrants struck it rich within a couple of decades of landing in America. These parvenus were caught between two worlds: they were no longer of the class of tenement dwellers who dragged themselves to the factories each day to labor over the machines, yet they were reliant on the tenement dwellers to generate their wealth; at the same time they were not of the patrician class that dictated tastes, finances, and political life in the city. Indeed as they were fighting to suppress their workforce by trying to keep the unions out, they were also engaged in a battle with the elites in the Fifth Avenue Association who did not want garment factories to mar the aesthetic appeal of their upscale neighborhoods. It is clear from their actions that members of the Association did not consider Blanck, Harris, Leiserson, and others of their ilk to be part of their inner circle. The fact that Blanck and Harris were wealthy, had slowly inched their way uptown with their factories, and had established their homes uptown meant little to the Association. The kind of business the Triangle owners engaged in and the type of workers they hired were offensive to members of the Fifth Avenue Association, and they campaigned vigorously against the establishment of such factories in and around Fifth Avenue. In some ways the struggle between the Fifth Avenue Association and the Triangle owners was an intraclass one. All the players were wealthy, yet the distinction between old wealth and new wealth took on significance during the struggle. While many of the members of the Association were aristocrats, factory owners such as Blanck and Harris were still aspiring to be accepted into the upper class. Their ethnicity also set them apart from the patricians. They were neither American-born nor Western European gentry. In aristocratic circles people of Eastern European ancestry were still looked at askance.

Other Industrial Accidents

Factory fires continued for some time after Triangle. For instance, a similar fire ravaged a five-story Bowery factory in April 1912, killing one worker. Workers trapped in the building jumped out of windows.[88] Many workers have been killed or injured on the job over the past two centuries. During the nineteenth century, when steam power was being introduced into American factories and transportation systems, there were a number of accidents. After 1,770 people were killed and 990 injured in steamboat accidents between 1825 and 1848 a federal Steamboat Inspection Service was formed in 1852. This did not stop the horrendous accidents, however. Thirty-two people were killed when the steamer *Seawanhaka* burned in New York Harbor on June 28, 1880. Even more horrendous was the devastating fire on the steamboat *General Slocum* on June 15, 1904, that killed between one thousand and eleven hundred people.[89]

The federal inspection of steam engines did not extend to the factories. After the Massachusetts Chief Justice Lemuel Shaw developed the doctrine of "assumption of risk" in 1842 states adopted it to resolve legal claims. Under the doctrine workers were said to assume the risks of the job in return for their wages; ergo they and their families were not entitled to further compensation in the event of job-related injury or death. Thus in 1850, when a steam boiler exploded in a New York City building, killing sixty-seven workers and injuring fifty others, there was no recourse under the law. The bodies were so badly burned and disfigured that a "funeral for the unrecognized" was held. Hundreds of well-wishers donated $28,000 for the victims and their families. A similar accident killing nineteen occurred in Hartford in 1854.[90]

In 1908 the federal government estimated that seventeen thousand workers were killed in industrial accidents that year. Of the 38 million employed men and women in 1913, approximately twenty-three thousand were killed on the job and seven hundred thousand were injured (see Table 13). Between 1907 and 1912 industrial accidents were blamed for nearly 10 percent of deaths among males; industrial accidents accounted for 23 percent of the deaths among miners, 49 percent among electric linemen, and 72 percent among gunpowder makers.[91]

Table 13 Estimates of Fatal Industrial Accidents in the United States, 1913

Male Workers in Various Employment Sectors	Number of Workers	Number of Fatal Industrial Accidents	Rate per Thousand
Metal mining	170,000	680	4.00
Coal mining	750,000	2,625	3.50
Fisheries	150,000	450	3.00
Navigation	150,000	450	3.00
Railroad	1,750,000	4,200	2.40
Electric (light and power)	68,000	153	2.25
U.S. Navy and Marine Corps	62,000	115	1.85
Quarrying	150,000	255	1.70
Lumber industry	531,000	797	1.50
U.S. Army	73	109	1.49
Building and construction	1,500,000	1,875	1.25
Transportation	686,000	686	1.00
Street railway	320,000	320	1.00
Watchmen, police, firefighting	200,000	150	0.75
Telephone and telegraph (including linemen)	245,000	123	0.50
Agriculture (including forestry and animal husbandry)	12,000	4,200	0.35
Manufacturing (general)	7,277,000	1,819	0.25
All other employed males	4,678,000	3,508	0.75
Total employed males	*30,760,000*	*22,515*	*0.73*
Total employed females	*7,200,000*	*540*	*0.075*

Source: *Industrial Accident Statistics*, Bulletin 157, U.S. Department of Labor, Bureau of Labor Statistics, 1915, 5.

Worker Health and Safety and Product Substitution

During the nineteenth century thousands of workers died or became ill because of job-related diseases: stone cutters died of lung disease; cigar and tobacco makers contracted heart and respiratory ailments; hat makers developed nerve disorders from inhaling the mercury used to treat fur and felt; and women ingesting radium while painting numbers on clock faces developed radiation poisoning. Alice Hamilton, a doctor who lived at Hull House for about ten years, was among the first to undertake studies of occupational diseases such as carbon monoxide poisoning, pneumonia, rheumatism, and phosphorus-related illness in match-factory workers, known as phosphorus necrosis or "phossy jaw."[92] Phosphorus necrosis could cause the deterioration of the jaw and loss of the eyes.

In 1908 Hamilton joined forces with John Andrews, executive secretary of the American Association for Labor Legislation, to launch a campaign to force factories to use an inexpensive substitute (sesquisulphide) for white phosphorus, the cause of the debilitating phossy jaw. Though doctors began reporting cases of phossy jaw as early as 1851 in America, practically nothing was done about the disease until 1909, when the U.S. Bureau of Labor published a study investigating the incidence of the disease. The study was conducted primarily by Irene Osgood and John Andrews. By the time the American study was released, French scientists had developed sesquisulphide as a substitute for white phosphorus. By 1908 the disease had all but disappeared in European match factories. Eventually the Diamond Match Company, which owned the U.S. rights to sesquisulphide, made the process available to other match companies free of royalties. In 1912 Congress passed the Esch-Hughes bill, which made it uneconomical for companies to continue using white phosphorus because of taxation and regulatory requirements.[93]

Hamilton and others also examined occupational health hazards arising from the industrial use of lead, arsenic, brass, zinc, and cyanide. Between 1910 and 1911 Hamilton and her colleagues visited 314 factories to identify the sources of lead poisoning and its effects on workers. In addition to epidemiological information Hamilton found widespread evidence of job blackmail, meaning that workers were afraid to report their health problems for fear of losing their jobs.[94]

WHITE CITY, WHITE LEAD: A FATAL ATTRACTION

In 1893 Chicago planned to host the Columbian Exposition. The exposition represented for the city a new Chicago that had arisen from the ashes of the Great Fire. Designers of the site planned a gleaming white, marble-like finish for the buildings. The original intent was to cover the buildings with "staff," an inexpensive combination of plaster of Paris and jute fiber, to create the desired effect. However, staff did not remain white for long in the windy, smoky city. Consequently the fair's four hundred buildings were painted with white lead paint. The Hall of Manufactures alone was coated with sixty tons of paint.

The workers paid a severe price to create the futuristic white city celebrated by fairgoers.[95] It wasn't long before activists took note of the workers' debilitation and began to educate the public about the dangers of lead and push for legislation to protect workers and the public.

Though some nineteenth-century reformers were quick to blame workers' illness on lifestyle, hospital records of patients show that many of those illnesses stemmed from exposure to industrial poisons. For instance, in 1887 a Philadelphia doctor, David Denison Steward, reported sixty-four cases of lead poisoning among his patients. Eight of the cases were fatal, and most of the victims were children under the age of fourteen. All the deaths were attributable to eating buns made with yellow lead chromate coloring. The tainted buns were traced to two bakers who used the less expensive lead chromate (or "extract of egg") instead of eggs in their batter. Other doctors reported lead poisoning from contaminated drinking water, canned or foil-wrapped food, lead-based ointments on nursing mother's breasts or faces, lead nipple shields, and sugar of lead ingested for medicinal purposes.[96]

The workplace was a major source of exposure to lead. Records of the German Hospital in Newark in 1888 showed that 7.8 percent of the male patients, some of whom worked at a local smelter, had lead poisoning, 3.4 percent had pneumonia, 5.9 percent had phthisis, 8.8 percent suffered from other lung diseases, 1 percent had mercury poisoning, 9.8 percent had typhoid fever, 15.6 percent had had a work-related accident, and 0.5 percent suffered from alcoholism. Though Newark had one of the highest incidences of lead poisoning, surveys done at other hospitals between 1888 and 1900 uncovered other hotspots of lead poisoning. Three percent of the male workers admitted to St. Joseph's Hospital in Tacoma, Washington, in 1900 suffered from lead poisoning, as did 3.1 percent of those admitted to Pennsylvania Hospital in Philadelphia in 1896 and 1.5 percent of those admitted to Massachusetts General Hospital in Boston in 1896.[97]

Responding to the growing concern over occupational accidents, illness, and death, President Theodore Roosevelt established the U.S. Bureau of Labor in 1902 as a part of the new Department of Commerce. The Bureau engaged leading industrial medicine experts to join the staff and to investigate factory conditions. Alice Hamilton was one of the physician investigators hired. Hamilton visited the Wetherill white lead factory in Philadelphia in May 1911. Wetherill, a company that began corroding lead in 1804, was the first large-scale producer of white lead in America. Though the factory was fitted with hoods and blowers for dust collection, the room in which workers separated out the corroded white lead was extremely dusty. On the third floor, where the dried lead was packaged for shipment, heaps of white lead stretched from wall to wall. The Italian, Polish, and Hungarian workers breathed the dust freely. The foreman, who had worked at the plant for thirty-eight years, insisted that there was no evidence of lead poisoning among the workers, but Hamilton's inquiry of local doctors and area hospitals unearthed twenty-seven cases of lead poisoning among Wetherill workers in the previous sixteen months. Prior to becoming a member of the Bureau of Labor investigative team Hamilton had studied the lead industry in Illinois, where she found 578 cases of lead poisoning in workers at 77 facilities.[98] In her nationwide study of the lead industry, Hamilton reported 358 cases (22 percent) of lead poisoning among 1,600 workers in white lead paint factories and 1,769 cases (24 percent) among 7,400 workers in lead smelting and refining facilities.[99]

Alice Hamilton corresponded with the factories she inspected about the dangers of lead, the installation of protective equipment, providing protective gear for workers, and the need for medical care of the workers. Hamilton also called for a ban on the interior use of white lead paint.[100]

Some in the lead industry resisted reforms. For instance, Webster Wetherill responded to Hamilton's assessment of his factory by saying that the lead Hamilton saw lying on the factory floor was "hard and lumpy and so little dust could possibly blow off" the piles. He thought little or no dust blew off the grinding floor either.[101]

When workers realized that there were few safeguards in the workplace to protect them many began to purchase insurance. Between 1870 and 1900 the number of workers taking out insurance policies increased from 11,000 to 3.5 million. These policies were tailored to specific kinds of jobs. The policies paid death benefits and sometimes paid the wages of a policyholder who could not work because of illness. They rarely paid for medical care. Unions responded to the problem by developing benefit plans for workers. Granite cutters, whose exposure to silica dust resulted in a high death rate from pulmonary diseases, had the first union in the country to develop a death benefit plan.[102]

CHANGING LABOR PRACTICES

In 1906 Richard Ely and other reformers founded the American Association for Labor Legislation, an organization committed to researching work-related illness, and in so doing brought needed attention to the issue of occupational health and safety. The organization sought legislative solutions to occupational hazards and was instrumental in drafting and promoting state labor laws prior to the First World War; it was also involved in the campaign to eradicate phossy jaw. The creation of the U.S. Department of Labor signaled a new era in workplace health and safety. As the government took the issue of workplace injury more seriously, the number of factory inspections increased, from 300 in 1907 to 425 in 1911.[103]

Despite the resistance of industries to protecting workers, state workers compensation laws began to push them in that direction. Before 1909 injured workers had to settle for the compensation their employers gave them, or they had to go to court to prove that their employer was responsible for their injuries. This was a daunting task for workers, and most could not afford court costs. However, by 1915 twenty-three states guaranteed workers that their employers would pay compensation as long as the workers could prove their injuries were sustained in the workplace. This change prompted many factories to hire their own doctor to monitor the health of workers. It wasn't long, though, before workers and unions began complaining that company physicians were abusing their position by using medical screening exams to weed out union sympathizers and to unfairly prevent people from being hired or keeping their jobs.[104]

Workers compensation had a serious flaw that employers exploited. The laws were designed to compensate for specific injuries suffered on the job; they all but ignored the chronic conditions typical of occupational diseases that arose over a long period of time and whose symptoms and causes might be hard to detect. In the first decades of the twentieth century fewer than ten states required employers to compensate workers who developed occupational diseases; as late as 1946 only thirteen states compensated for occupational diseases. Employers who were reluctant to pay compensation claims were quick to use the courts to argue that occupational diseases such as lead poisoning, silicosis, and asbestosis were not job injuries and were therefore not covered by workers compensation laws. In other instances, workers who moved from one job to another had difficulty making a convincing argument that long-term illnesses were work-related injuries eligible for compensation under the law.[105]

State-level oversight of occupational health was a low-level priority until the 1930s. Though many believed that progress would be made if state and local governments got involved, few states focused on the issue until the Social Security Act allocated federal funds for it. Before 1935 only five state health departments (Connecticut, Maryland, Mississippi, Ohio, and Rhode Island) examined occupational health, and even then, these activities were limited. Only Massachusetts and New York had industrial hygiene units within their state labor departments. By 1936 there were seventeen industrial hygiene units in state and local health departments, and by 1938 that number had risen to twenty-six. The budget for these units totaled $750,000.[106]

Sweatshops, Fair Wages, and Child Labor

Despite being exposed to similar dangers in the workplace, women and children doing the same work as men were paid less than men. The meat-processing industry furnishes an example of this. In 1890 the highest wages went to skilled butchers, who earned between $1,200 and $1,500 ($24,630 to $30,800 in 2005 dollars) annually; the average packinghouse worker earned $615 ($12,600) annually (nationwide, factory workers earned an average of $427 per year). Of the 17,500 people employed in the packinghouses that year, 405 were children sixteen or younger. The children earned an average of $187 ($3,800) per year, and the 487 women earned $299 ($6,100). More specifically, female clerks earned an average of $423 ($8,700) and male clerks earned $1,100 ($22,500). Women worked in a variety of jobs in the packinghouses; they sewed bags around hams (men did this task too), laundered smocks and aprons, labeled jars and cans, chopped cooked meat, and stuffed cans and guided them through the soldering machines.[107]

Packinghouse workers began to unionize in the 1860s and to campaign for the eight-hour day in the early 1880s, but, as was the case in the garment industry, unionization efforts bypassed women. Activists from Hull House and Northwest Settlement House initiated efforts to investigate the packinghouses. They also campaigned to establish a minimum wage for women and helped to organize the women and children into unions. To do this activists collaborated with the National Consumers' League (founded in 1899) and the WTUL. In addition to collective bargaining and strikes, unions sought to address occupational health and safety through a network of hospitals they established. The activists brought national attention to the issues, which in turn helped their unionization efforts. The National Consumers' League focused on the occupational hazards women faced in the workplace. The WTUL and the League identified the most oppressive work sites and targeted them for consumer boycotts and mobilization drives. While the WTUL focused on unionizing women, the League saw itself as an environmental, worker, and consumer protection organization; it researched the effects of a number of industrial poisons on workers and consumers, including benzol, carbolic acid, radium in watches, and tetraethyl lead in gasoline.[108]

Building on the work of the settlement houses the Workers' Health Bureau, organized in 1921 by female reformers, linked public health, the environment, and working conditions with trade union activism. The Bureau was an effective worker advocacy organization that established and supervised union health departments, conducted occupational and environmental research, and disseminated its findings to union members. The organization dissolved in 1928 after the American Federation of Labor and other leading trade unions withdrew their support. As the Great Depression engulfed the country unions moved away from the broader agenda of working conditions and environment to focus on wage issues, so the Bureau became less relevant to their concerns.[109]

As early as 1813 Connecticut had passed a law requiring some education and "attention to morals" for indentured children. As the nineteenth century drew to a close Progressive Era reformers and labor activists took up the issue of child labor, which working-class activists identified as problematic. In 1889 a pamphlet titled *Our Toiling Children* argued that child labor should be eliminated and education should be mandatory in America.[110] The Hull House activist Florence Kelley demonstrated that child labor was a widespread practice in which children were toiling long hours under dangerous conditions instead of being schooled.[111]

In 1880 there were 1,118,000 children between the ages of ten and fifteen in the industrial labor force. Children as young as four worked in the textile industry, and as young as eight as miners. In Connecticut alone there were ten thousand child laborers between the ages of ten and fifteen; sixty-eight hundred of them worked in factories. Though state law mandated that children under fifteen could not be employed for more than ten hours per day or fifty-eight hours per week, an 1885 investigation found that this law was routinely violated. In an effort to curtail the use of child laborers Chicago passed similar laws in 1881, and in 1889 and 1891 the State of Illinois strengthened those restrictions. This resulted in a sharp decline in the use of child laborers. Though the city expanded in size between 1880 and 1890, the census reported a 50 percent decline in the number of children who were working. Although twenty-four states had compulsory education laws, only Massachusetts enforced them; six states had no enforcement powers at all. To make matters worse, the school system was inadequate to serve all the children. For instance, in Philadelphia there were ten thousand more children than the schools could accommodate, and Albany had twelve thousand spots for thirty-six thousand children.[112]

Unlike Chicago, in Philadelphia a study of child labor showed that there

Table 14 Percentage of Labor Force Participation of 12- to 17-Year-Old White Youths in Philadelphia, 1860 and 1900

	1860			1900		
Age	At Home	In School	At Work	At Home	In School	At Work
Males						
12	18	80	2	12	79	9
13	25	72	3	10	67	23
14	43	49	8	12	43	45
15	44	43	23	10	29	61
16	37	21	42	11	16	73
17	26	10	64	10	11	79
Females						
12	22	75	3	16	81	4
13	31	65	4	20	60	20
14	40	52	8	24	45	31
15	43	34	23	23	28	49
16	55	15	30	27	15	58
17	48	8	44	29	9	62

Source: Walter Licht, *Getting Work: Philadelphia, 1840–1950* (Philadelphia: University of Pennsylvania Press, 1992), 20–21.

was a growing trend toward using teenagers in the workforce as the nineteenth century wore on. Figures from 1880 show that while the average annual income for adult male workers was $441 ($8,700 in 2005 dollars), the average household expenses were $564 ($11,200). Consequently children were put to work to help meet the family's expenses. The data show the rate of labor force participation for white youngsters. Boys were being drafted into the workforce at a phenomenal rate. Table 14 shows that 23 percent of thirteen-year-old boys were working in 1900, as were 45 percent of fourteen-year-olds, 61 percent of fifteen-year-olds, and more than 70 percent of older teenage boys.[113]

White girls were also drafted into the labor force at increasing rates during the period of study. Twenty percent of thirteen-year-old girls worked in 1900, as did 31 percent of fourteen-year-old girls, and 49 percent of fifteen year olds. More than half of teenage white girls older than fifteen worked in 1900.[114]

The data for the labor force participation of black youth are a bit more sketchy and unpredictable. Ten percent of thirteen-year-old black boys and 6 percent of black girls of a similar age worked in 1900. Half of the fifteen-year-old black boys and 56 percent of black girls of the same age worked in

1900.[115] This pattern may be the result of increased European immigration displacing blacks from industrial jobs.

Though Pennsylvania enacted a compulsory school attendance law in 1895 requiring children eight to thirteen to spend at least sixteen hours per week in school, many children worked instead. In 1903 Philadelphia's Central Textile Workers Union launched a campaign to improve working conditions, especially for those under fifteen. At the time children comprised more than 25 percent of the city's textile workforce. The union wanted to reduce the work week from sixty to fifty-five hours without any reduction in pay, arguing that such a reduction would be healthier for children. After weeks of fruitless negotiations the union called a strike on May 29 and more than ninety thousand textile workers walked off the job, shutting down 636 textile businesses in the city.[116]

Mary Harris, known as "Mother Jones," having convinced the strikers to make child labor the central issue in the strike, helped to organize a rally on June 17 in which fifty thousand adults were led by seven thousand children in a protest march. Two weeks later Harris organized a march of child textile workers from Philadelphia to Oyster Bay, Long Island, where President Theodore Roosevelt had a home. The children left Philadelphia on July 7 and reached Oyster Bay on August 1, but when they got there Roosevelt was unavailable to meet with them. Though the strike dragged on for three months after the march, the issue of child labor remained on the national agenda. The actions in Philadelphia also led to the formation of the Pennsylvania Child Labor Committee, which lobbied for child labor legislation.[117]

In 1905 the compulsory school attendance bill was modified to require full-school-year attendance for children under fourteen. However, child labor continued to be a problem. On December 9, 1906, a photography exhibit of child laborers sponsored by the Consumers' League of Philadelphia opened at Horticultural Hall. More than twenty-five thousand people viewed the photographs. The Consumers' League launched a campaign against child labor that labeled Philadelphia as "The Greatest Child Employing City" in the country. The group put pressure on politicians to curb the practice of child labor by instituting compulsory school attendance laws and inspection and certification of factories.[118]

Florence Kelley used photographs of pallid, maimed, and deformed children in her campaign to raise the minimum working age to sixteen and to make it mandatory for children to attend school until then. Factory inspectors and truant officers were deployed to ensure compliance. Kelley argued that poor families, forced to send their children to work in the factories, actually became poorer as the children competed directly with adults for jobs and were paid less for their labor. Thus the abolition of child labor was one way to protect the living wage. By 1913 Kelley and other children's advocates had forced factories to stop employing children under the age of fourteen.[119]

fourteen

THE INDUSTRIAL WORKPLACE

Air Pollution

While reformers made great strides in removing garbage from streets, developing sewage systems, and delivering clean water to urbanites, the air remained foul. Air pollution was a nuisance and a health hazard in cities. Burning coal caused enormous pollution problems as industrial facilities disgorged heavy black smoke. In particular bituminous coal filled the air with sulfuric smoke and particulate matter; soot coated and corroded buildings and statues, irritated the eyes and lungs, and resulted in increased incidences of tuberculosis and pneumonia.[1] Pittsburgh, the epicenter of the iron and steel industry, was so polluted it earned the moniker "the Smoky City." In 1791 an observer complained that the city was "kept in so much smoke as to affect the skin of the inhabitants."[2] Half a century later a visitor to Pittsburgh wrote

> A dense cloud of darkness and smoke, visible for some distance before [a traveler] reaches it, hides the city from his eyes until he is in its midst; and yet half this volume is furnished by household fires, coal being the only fuel of the place. . . . The whole city [is] under the influence of steam and smoke. The surface of the houses and streets are so discolored as to defy the cleansing power of water, and the dwellings are preserved in any degree of neatness, only by the unremitting labors of their tenants, in morning and evening ablutions. The very soot partakes of the bituminous character of the coal, and falling—color expected—like snowflakes, fastens on the face and neck, with a tenacity

which nothing but the united agency of soap, hot water, and the towel can overcome.[3]

By the early 1820s St. Louis, a city that rivaled Pittsburgh for smokiness, had depleted its supply of wood and turned to burning soft coal found in nearby southern Illinois. In 1822 civic leaders launched an unsuccessful campaign to teach residents how to burn coal with less smoke. The following year the *Missouri Republican* described the smoke as so dense "as to render it necessary to use candles at mid-day," and during the 1840s an anonymous letter to the editor warned that the city could become "an emporium for disease" because of the smoke.[4] Smoke was a problem in New England too. Henry Adams noted how rapidly industrial facilities were gobbling up and transforming rural areas: "Tall chimneys reeked smoke on every horizon, and dirty suburbs filled with scrap-iron, scrap-paper, and cinders formed to the setting of every town."[5]

SMOKE ABATEMENT ORDINANCES

Chicago too had serious air pollution problems. Despite the city's windy locale, by the 1880s the smoke was so thick it was difficult to see across the street. In 1881 activists succeeded in getting the Chicago City Council to pass an air pollution law that declared, "[The] emission of dark smoke from the smokestack of any boat or locomotive or from any chimney anywhere within the city shall be [considered] a public nuisance." The ordinance authorized a municipal inspection force to patrol the streets in search of violators. The ordinance was challenged, but the Illinois Supreme Court upheld it in 1892. Though some industrialists resisted smoke abatement measures St. Louis and Pittsburgh passed similar laws.[6]

Civic groups and chambers of commerce insisted on government intervention, and in 1904 the U.S. Bureau of Mines began investigating the burning of bituminous coal and smokeless fuel that would reduce air pollution in the cities. The study found that the level of smoke in cities was affected by the amount of public pressure to reduce pollution, the existence of smoke abatement ordinances and a smoke ordinance department, the number of personnel assigned to smoke departments, topography and climate, type of fuel used, level of industrialization, type of industries, and volume of steamboat and railroad traffic. Where there was strong public support and organized efforts to reduce smoke cities made the greatest progress toward smoke abatement.[7] Cities with large numbers of brick kilns, heating and annealing furnaces, puddling furnaces, and locomotive and steamboat boilers had a greater smoke problem. In Chicago locomotives produced 43 percent and special furnaces accounted for 12.5 percent of the smoke emitted in the city.[8]

Despite the passage of smoke abatement ordinances in some cities, municipalities such as Minneapolis, Louisville, Pittsburgh, St. Louis, and Chicago continued to be so smoky that it was dark at midday and visibility was limited to less than a city block. The haze caused traffic accidents, illness, and death. Urban foresters in Cleveland and St. Louis reported that the smoke killed trees and corroded buildings.[9]

As air conditions worsened, three groups of activists intensified efforts to reduce the pollution: businessmen concerned about smoke damage to their property, engineers who envisioned a technological solution, and grassroots activists concerned about health as well as passing and enforcing antismoke laws. The most vocal antismoke activists were women's clubs such as the Women's Club of Cincinnati, the Wednesday Club of St. Louis, and the Ladies Health Protective Association of Pittsburgh. The goals of the antismoke movement dovetailed with the goals of the sanitary reform movement: the rhetoric of "civic motherhood" and "municipal housekeeping" urged women to be the keepers and protectors of family, home, and community virtue and pride. Thus women became active in the movement because they believed it endangered the health of their families and detracted from the cleanliness and beauty of their communities.[10]

Business leaders, concerned with the rising costs of air pollution, began to estimate the cost of smoke damage. St. Louis suffered an estimated $6 million ($123 million in 2005 dollars) in smoke damage in 1906. Five years later smoke damage cost Cleveland $6 million, Cincinnati $8 million ($158 million), and Chicago $50 million ($990 million). Businessmen's groups such as the Citizens' Smoke Abatement Association of St. Louis and the Chicago Society for the Prevention of Smoke (founded in 1892 to help reduce smoke in time for the Columbian Exposition) were founded and led by the cities' business magnates. Groups such as the Smoke Prevention Association of America (renamed the Air Pollution Control Association), founded in 1907, were formed to help reduce air pollution.[11]

In 1912 the Bureau of Mines found that 12 of 240 cities with a population under fifty thousand and 17 of 60 cities with a population of between fifty thousand and two hundred thousand had smoke abatement ordinances. The larger cities were more likely to have smoke abatement ordinances: of the twenty-eight cities with a population over two hundred thousand, twenty-three had smoke abatement ordinances. Of the five that did not have ordinances, three of those cities relied heavily on oil and did not have a smoke problem. New Orleans was one of the cities without a smoke abatement ordinance. The other two, San Francisco and Seattle, relied heavily on oil.[12]

The Chicago smoke ordinance became the standard of the time, and several cities copied aspects of the program. Chicago established a Department of Smoke Inspection whose head was appointed by the mayor and the city council. The smoke inspector appointed deputy smoke inspectors. An eight-member smoke commission was also formed. Plans to build new power plants or renovate old ones had to be approved by the smoke inspector before construction could begin, and owners had to obtain a permit before altering chimneys or furnaces. The fee for inspection ranged from one to three dollars. The ordinance also prohibited the emission of dense smoke from any stack for more than six minutes in any hour. Violators risked being fined $10 to $100 for each violation. Any inspector taking bribes for inspections could be fined as much as $100.[13]

Asbestos

The asbestos industry is another in which corporate callousness led to workers being exposed to hazardous working conditions; consequently large numbers of workers were killed or injured on the job. Though the asbestos industry provided medical care for workers, workers were placed at risk because they lacked access to independent medical care.

When corporations had to pay workers compensation for job-related injuries and illnesses, they hired their own company doctors, compelled workers to see these doctors, and kept medical records of the workers. The case of asbestos helps us to understand how the shift to industrial medicine affected workers' health and placed some at risk.

Johns-Manville, the world's largest asbestos company, was founded in 1858 in New York as the H. W. Johns Manufacturing Company. The company had mines in Canada and the United States. The firm's twenty-one-year-old founder, Henry Ward Johns, patented inventions for roofing and insulation products. Asbestos was used as a heat insulator as early as 1866, and asbestos cement was introduced about 1870. Johns died of phthisis pulmonitis (believed to be asbestosis) in 1898.[14]

Exposure to asbestos occurs in a wide variety of operations, from the mining and processing of ore to its use in house construction, ship- and boat building, insulation, pipefitting, steamfitting, boilermaking, and textiles.[15] Asbestos consumption rose steadily in the United States, from less than 100,000 tons in 1912 to 800,000 tons in the 1970s, from which point it has declined.

Since the beginning of the twentieth century an estimated 30 million tons of asbestos have been used and disposed of in the environment.[16] Despite the enthusiasm for the product, by 1918 the dangers of asbestos were apparent enough that U.S. and Canadian insurance companies stopped selling life insurance policies to asbestos workers. By the 1920s descriptions of the often fatal respiratory disease asbestosis appeared regularly in the medical literature. Asbestosis, which causes scarring of the lung, respiratory problems, and heart failure, is similar to black lung and other chronic chest diseases that afflict miners and others who work with minerals and fibers.[17]

SCIENTIFIC STUDIES OF ASBESTOS

The hazards of asbestos have been known for a long time. According to Pliny, when the Romans found that slaves working with asbestos became disabled and often died from breathing difficulties, they made the slaves wear transparent bladder skins over their mouths to reduce the amount of dust they inhaled. Not much more was written on the topic until 1897, when a Viennese physician wrote about the health problems of asbestos weavers and their families. A year later the Lady Inspectors of Factories wrote about the lung diseases found among asbestos workers.[18]

In 1906 and 1907 work done by M. Auribault and testimony by Dr. H. Montague documented the link between asbestos and lung disease.[19] In 1924 the *British Medical Journal* published an article on fibrosis and the inhalation of asbestos, and in 1927 the journal published an article on pulmonary asbestosis. W. E. Cooke, the author of the articles, described the lung scarring evident in a female textile worker who wove asbestos for twenty years.[20] Other studies examining the hazards of asbestos appeared in the scientific literature during the 1930s. In 1933 and 1934 American and British journals published studies on pneumoconiosis and pulmonary asbestosis.[21] The American asbestos companies and their insurers also began to conduct studies of asbestos in 1929. In 1932 they examined workers at the Johns-Manville factory in Manville, New Jersey; 327 of 1,140 (29 percent) had pneumoconiosis.[22] Dr. Anthony J. Lanza, director of the Institute of Industrial Medicine at New York University and assistant medical director at Metropolitan Life, began studying the effects of asbestos on workers. Lanza, whose research was sponsored by Raybestos-Manhattan, Johns-Manville, and their insurance company Metropolitan Life, found that 67 of 126 workers he examined suffered from asbestosis. Vandiver Brown, attorney for Johns-Manville, and other representatives of the companies exerted strong editorial control over what Lanza published. Brown asked that Lanza not report negative findings, and Lanza complied. Lanza published his altered results in the *Journal of the American Medical Association* in 1935 and an edited volume on asbestosis in 1938. Nine asbestos companies also funded an animal study conducted by Dr. Leroy U. Gardner at Saranac Laboratories in Saranac, New York, in 1936 that showed that changes could be detected in the lungs of guinea pigs within a year of exposure to asbestos dust. Gardner died in 1946 before he could make an official report of his findings.[23] The U.S. Public Health Service took an interest in asbestos when it conducted a study of American textile workers in the South in 1938. The report urged the elimination of exposure to hazards in the textile factories, including asbestos.[24]

The first large-scale study of American asbestos insulation workers was conducted in 1945. Researchers who examined workers at navy shipyards on the East Coast found only three cases of asbestosis; they concluded that coating naval vessels with asbestos was a relatively safe operation. However, because 95 percent of the workers studied had been on the job less than ten years, researchers underestimated the potential for such workers to develop asbestosis, which typically does not manifest for ten to twenty years after exposure.[25] In 1947 the American Conference of Governmental Industrial Hygienists recommended a threshold limit of 5 parts per million (ppm) per cubic foot of air for asbestos dust. That threshold limit was reduced to 2 ppm in 1968.[26]

Studies conducted during the 1950s and 1960s also pointed to the dangers of asbestos. In 1955 R. Doll, a British epidemiologist, found that asbestos textile workers experienced rates of lung cancer deaths ten times higher than other textile workers.[27] In a landmark study Selikoff, Churg, and Hammond examined the prevalence of asbestosis in insulation workers in 1965. As Table 15 shows, 542 of the 1,117 workers examined (48.5 percent) had asbestosis. The study also showed that workers were more likely to develop asbestosis and manifest advanced stages of the disease the longer they were exposed to asbestos.[28]

After noticing that fifteen of seventeen patients employed at an asbestos company in New Jersey had developed asbestos-related lung diseases, Selikoff and Hammond sought to get the employment records of workers in the asbestos industry. When they were denied access they contacted the New York and New Jersey locals of the International Association of Heat and Frost Insulators and Asbestos Workers and conducted a cohort study of the union workers employed in the asbestos industry. Because the union kept records of work histories to administer death benefits, the researchers were able to start with a cohort of 632 workers in 1943. By 1973, 444 of the 632 workers were dead, 198 (44.6 percent) from cancer and 37 from asbestosis.[29]

The number of Americans age fifteen and over dying of asbestosis increased

Table 15 Results of X-Ray Examinations of Asbestos Insulation Workers

Onset of Exposure (Years of Exposure)	Sample Size	Percentage Normal	Percentage Abnormal	Asbestosis (Grade)		
				1	2	3
40 years or more	121	5.8	94.2	35	51	28
30–39	194	12.9	87.1	102	49	18
20–29	77	27.2	72.8	35	17	4
10–19	379	55.9	44.1	158	9	0
0–9	346	89.6	10.4	36	0	0
Total	*1,117*	*51.5*	*48.5*	*366*	*126*	*50*

Source: Irving J. Selikoff, Jacob Churg, and E. Cuyler Hammond, "The Occurrence of Asbestosis among Insulation Workers in the United States," *Annals of the New York Academy of Sciences* 132, no. 12 (1965): 147.

from fewer than 100 in 1968 to 1,265 in 1999. There were 10,914 asbestosis deaths between 1990 and 1999 alone. White males were far more likely to die from asbestosis than others; more than 96 percent of those dying from asbestosis during 1990–99 were males, and 93 percent were white. Most (98 percent) were fifty-five and older. Residents in California, Pennsylvania, New Jersey, Florida, Washington, and Virginia accounted for almost half of the asbestosis cases. In 1998 and 1999 asbestosis surpassed coal workers' pneumoconiosis as the most prevalent form of death from lung disease.[30] See Table 16 for the industries with the highest number of workers dying from asbestosis and Table 17 for the occupations in which workers were most likely to die from asbestosis.

Table 16 shows that almost one-fourth of all the people who died of asbestosis worked in the construction industry. This helps to account for the racial and gender disparity in the disease since the construction trade is predominantly male and white. Six percent of the afflicted worked in ship- and boat-building jobs, and about 4 percent worked in the chemical industry. Table 17 shows that laborers working in certain occupations within these industries were very vulnerable to contracting asbestosis. Plumbers, pipefitters, and steamfitters were the occupations with the greatest number of asbestosis deaths; 8 percent of the asbestos deaths occurred in people in these occupations.

CLAIMS AND LITIGATION

Asbestos workers have been filing claims since the 1920s and some have received compensation since the 1930s. The first American asbestos claim was

Table 16 Asbestosis: Most Frequently Recorded Industries on Death Certificates, U.S. Residents Age 15 and Over, Selected States and Years, 1990–1999

Industry	Number of Deaths	Percentage
Construction	702	24.6
Ship- and boat building and repairing	171	6.0
Industrial and miscellaneous chemicals	124	4.3
Railroads	89	3.1
Miscellaneous nonmetallic and stone products	75	2.6
General government, not elsewhere classified	71	2.5
Blast furnaces, steelworks, rolling and finishing mills	67	2.3
Unspecified manufacturing industries	61	2.1
Electric light and power	55	1.9
Elementary and secondary schools	53	1.9
All other industries	1,286	45.0
Industry not reported	105	3.7
Total	*2,859*	*100.0*

Source: National Institute for Occupational Safety and Health, *National Occupational Research Agenda: Priorities for the 21st Century* (Washington: U.S. Department of Health and Human Services, National Institute for Occupational Safety and Health, 2002), 9.

filed in 1927 by a foreman in the weaving department of an asbestos company. The Massachusetts Industrial Accident Board awarded the man compensation for disability from occupational lung disease. Lawsuits were filed against Johns-Manville (renamed Schuller International Group in 1992) as early as 1929, when workers claimed disability as a result of lung disease. In 1933 the company's board of directors approved a settlement in asbestosis lawsuits filed by eleven workers in New Jersey.[31] A total of $35,000 was paid to settle that suit. Throughout the 1930s scores of lawsuits were filed against asbestos companies; workers won settlements in some of these cases. One hundred and two cases against Johns-Mansville went to trial, and the average settlement per case was $25,000.[32] Even workers who did not handle asbestos directly got sick and filed suit. In 1937 Mr. Wetzel, a former bookkeeper at the National Asbestos Manufacturing Company, filed a lawsuit against the New Jersey company to recover damages for lung disease. As a letter written by an employee of the United States Gypsum Company commenting on strategies to defeat Wetzel's claims shows, by the time the Wetzel case was filed, companies were focused

Table 17 Asbestosis: Most Frequently Recorded Occupations on Death Certificates, U.S. Residents Age 15 and Over, Selected States and Years, 1990–1999

Industry	Number of Deaths	Percentage
Plumbers, pipefitters, and steamfitters	238	8.3
Managers and administrators, not elsewhere classified	129	4.5
Electricians	125	4.4
Carpenters	120	4.2
Insulation workers	108	3.8
Laborers, except construction	95	3.3
Supervisors, production occupations	85	3.0
Welders and cutters	78	2.7
Janitors and cleaners	74	2.7
Truck drivers	66	2.3
All other occupations	1,639	57.3
Occupation not reported	102	3.6
Total	*2,859*	*100.00*

Source: National Institute for Occupational Safety and Health, *National Occupational Research Agenda: Priorities for the 21st Century* (Washington: U.S. Department of Health and Human Services, National Institute for Occupational Safety and Health, 2002), 9.

on developing strategies for winning most of the suits alleging that workers contracted lung diseases because of their work with asbestos and gypsum. As the number of asbestos suits being filed mounted, Dr. Kenneth Wallace Smith, medical director of Johns-Mansville, made a recommendation to the company in 1949 that it not inform workers about their asbestos-related illnesses.[33]

A steady stream of asbestos suits were filed throughout the 1940s and 1950s, but from the 1960s onward asbestosis got increased attention. Interest was generated by the growing number of cases and the discovery of documents from several asbestos companies that raised questions about the behavior of corporate executives and their insurance agents. One high-profile case was *Borel v. Fibreboard I* in the late 1960s. Clarence Borel, an insulation worker from 1936 to 1969, discovered he had asbestosis in 1969. He testified that at the end of each workday his clothes were coated with dust: "You just move them just a little and there is going to be dust, and I blowed this dust out of my nostrils by handfuls at the end of the day, trying to use water too, I even used Mentholatum in my nostrils to keep some of the dust from going down in my throat, but it is impossible to get rid of all of it. Even your clothes just stay dusty continually unless you blow it off with an air hose." Borel acknowledged that he knew for years that inhaling the asbestos dust was bad for him, but he did not know it could kill him. He and fellow asbestos workers thought the dust "dissolve[d] when it hit your lungs."[34] He claimed that he was not supplied with a respirator during his early work years. Though respirators were available on some jobs later on, insulation workers usually were not required to wear them and had to make a special request if they wanted to use one. Borel stated that the respirators were uncomfortable, particularly in hot weather: "You can't breathe with the respirator." He argued that none of the respirators available to him prevented the inhalation of asbestos dust. He told the court that he sometimes wore a wet handkerchief over his nose to prevent the inhalation of asbestos.[35]

At first doctors diagnosed Borel with pleurisy. In 1964 a doctor examined him and cautioned him to stay away from asbestos dust. In January 1969 he was hospitalized and a lung biopsy was performed; it was then that he was diagnosed with pulmonary asbestosis. In February 1970, when Borel's right lung was removed, doctors found that he also had mesothelioma. Borel died before his case went to trial (his widow was substituted as plaintiff).[36]

Borel filed suit on October 20, 1969, against eleven corporations. So how much and when did the plaintiff and defendant know about the dangers of asbestos? The plaintiff alleged that the defendants failed to take reasonable precautions to warn him of the dangers of using asbestos; failed to inform him about safe and sufficient protective apparel and equipment and safe methods of using the products; failed to test asbestos products to ascertain what hazards might arise from using them; and failed to remove the products from the market once the hazards were known. The trial court submitted the case to a jury, which found that all the defendants except Pittsburgh and Armstrong negligent, but that none were grossly negligent. Borel was found to be contributorily negligent.[37]

The defendants appealed the jury's decision on the grounds that the danger of asbestos was not foreseeable at the time Borel used the products. They claimed that the dangers were not known until 1968. They argued further that because of the long latency period of the disease, Borel would have contracted the disease long before this date. Borel countered this argument by stating that asbestosis is a cumulative disease; thus even his most recent exposure in 1968 could have contributed to his overall condition. Besides, there was ample evidence in the scientific literature and numerous studies done of asbestos workers that pointed to the hazards of asbestos,[38] and though the defendants claimed that they did not know of the dangers of asbestos until the late 1960s, they had been receiving and settling workers' asbestos-related claims since the late 1920s.

Borel argued that the evidence indicated that none of the defendants ever tested its products to determine how they affected insulation workers. Neither did the defendants try to determine if the exposure of insulation workers to asbestos dust exceeded the threshold limit values recommended by the American Conference of Governmental Industrial Hygienists, or whether those standards were accurate or reliable.

The court concluded that the dangers from inhaling asbestos dust was foreseeable to the defendants at the time the products were being sold. The court also found that the products were unreasonably dangerous and that the defendants had an obligation to warn the user. The defendants countered that since the dangers of the asbestos were obvious, they had no obligation to warn workers. The court, considering this argument disingenuous, found in favor of Borel.[39]

The defendants in the Borel case were Owens-Corning Fiberglass Corporation, Standard Asbestos Manufacturing and Insulating Company, Unarco Industries, Eagle-Picher Industries, Combustion Engineering, Fibreboard Paper Products Corporation, Johns-Manville Products Corporation, Pittsburgh Corning Corporation, Philip Carey Corporation, Armstrong Cork Corporation, and Ruberoid Corporation. The first four corporations settled before trial, paying a combined $20,902.20. The trial court ruled in favor of Combustion Engineering because the plaintiff could not show that he was ever exposed to any product manufactured by the company. The remaining six defendants went to court. Borel's claim against them amounted to a total of $58,534.04. The defendants argued that the judgment was too high and that the amount claimed was $46,669.98, not the amount set by the jury.[40]

Pittsburgh Corning and Armstrong Cork contended that Borel was not exposed to their product until after 1962 and 1966, respectively. However, the court ruled that because of the latency period of asbestos and the disease's individual characteristics the ill effects of asbestos might not manifest for five to ten years after exposure. The court also ruled that though Borel was contributorily negligent (because he did not wear a respirator all the time), his behavior did not constitute a misuse of the product. Indeed the evidence suggests that the defendants gave no instructions to insulation workers to use respirators or warned them of the dangers of asbestos. Though Borel was negligent in not wearing a respirator, he used the asbestos exactly as it was intended. Therefore his action did not absolve the asbestos companies of their responsibilities.[41]

The Borel case was back in court in 1974. Three of the defendants, Johns-Manville, Fibreboard, and Ruberoid, claimed they had warned users of their products. Johns-Manville began placing warning labels on packages in 1964, and Fibreboard and Ruberoid placed warning labels on their products beginning in 1966. All three warning labels read, "This product contains asbestos fiber. Inhalation of asbestos in excessive quantities over long periods of time may be harmful. If dust is created when this product is handled, avoid breathing the dust. If adequate ventilation control is not possible wear respirators approved by the U.S. Bureau of Mines for pneumoconiosis producing dusts."[42]

The court noted that the warning label did not indicate the gravity of the risk, that is, that exposure to asbestos dust could cause asbestosis, mesothelioma, and other cancers. The mild suggestion that the product may be harmful did not adequately convey the severity of the danger. The court also pointed out that given the nature of their work, insulation workers cannot avoid breathing asbestos dust. The respirators that Borel and other insulation workers were given were ineffective. The court concluded that the cautions placed on the products were not warnings because they did not communicate the gravity of the risk to the users of the product.[43]

THE ASBESTOS "PENTAGON PAPERS"

The discovery of documents, the so-called asbestos Pentagon papers or Sumner Simpson letters, in 1974 in a vault in a Raybestos factory in Stratford, Connecticut, buttressed workers' claims that companies knew about the dangers of asbestos and failed to warn or protect them. The asbestos Pentagon papers, written between 1933 and 1945, included correspondence between asbestos company senior executives, lawyers, physicians, consultants, and the companies insuring Johns-Manville, Raybestos-Manhattan (renamed Raymark in 1982), and other asbestos companies. The letters were filed by Marguerite Garvey, secretary of Sumner Simpson, president of Raybestos from 1929 to 1948, and implicated the companies in a coverup. The papers were released to a New Jersey lawyer representing an asbestos worker in 1977.[44]

The asbestos Pentagon papers had an immediate impact on the courts. In 1978 South Carolina Circuit Court Judge James Price, reviewing the documents, said the correspondence showed "a pattern of denial and disease and attempts at suppression of information which is highly probative." He stated further, "[The correspondence] reveals written evidence that Raybestos-

Manhattan and Johns-Manville exercised an editorial prerogative over the publication of the first study of the asbestos industry which they sponsored in 1935. [The correspondence] further reflects a conscious effort by the industry in the 1930s to downplay, or arguably suppress, the dissemination of information to employees and the public for fear of the promotion of lawsuits."[45] With the discovery of the papers and the public outrage that ensued, asbestos workers were much more likely to have the courts rule in their favor and receive larger settlements, at least in the short term.

BANKRUPTCY PROTECTION

A flood of asbestos claims and lawsuits has been filed since the 1970s, leading major corporations to file for bankruptcy protection. In 1982 Johns-Manville filed for bankruptcy under Chapter 11 of the Federal Bankruptcy Act. The company took this step to limit its liability from what it thought would be an estimated fifty-two thousand lawsuits amounting to more than $2 billion. The company emerged from bankruptcy protection in 1988 and established the Manville Trust to compensate victims.[46] However, asbestos companies woefully underestimated the number of lawsuits that eventually would be filed. According to a study conducted in 2005 by the RAND Institute for Civil Justice, about 730,000 people have filed asbestos-related claims, amounting to more than $70 billion, and an additional 50,000 to 75,000 new cases will be filed each year.

Three funds were established in the 1980s to compensate asbestos workers. The Manville Trust, the most significant of the three, disbursed $3.3 billion to injured workers through the middle of 2005. Originally only about three hundred companies were the targets of lawsuits, but as funds have been depleted asbestos injury suits have expanded to target eighty-four hundred firms involved in all aspects of asbestos production, sales, and handling. At least seventy-three companies have filed for bankruptcy through 2004 as a result of litigation. These bankruptcies have also resulted in the loss of between fifty thousand and sixty thousand jobs; each displaced worker has lost between $25,000 and $50,000 in income and an average of $8,300 in retirement funds.[47]

Asbestos claims are still pouring in. Johns-Manville (which was bought by Warren Buffett's firm, Berkshire Hathaway, in December 2000 for $2.2 billion) reported that the number of claims filed against the company for the first three quarters of 2001 increased by 55 percent over the same period in 2000. During the first nine months of 2001 the trust received 69,500 new claims; this compares to 59,200 claims filed for all of 2000.[48] During the 1990s estimates of the ultimate cost of asbestos hovered around $40 billion, but figures released by Tillinghast-Towers Perrin, a management consulting and actuarial firm, in 2001 estimated that the cost will be between $55 billion and $65 billion. The company estimated that about 39 percent of the cost would be borne by asbestos defendants, 31 percent by overseas insurance companies, and 30 percent by American insurers. They also estimated that there was a shortfall of as much as $33 billion in funds to cover litigation costs. Given these sobering predictions, it is not surprising that victims are making haste to file claims while the asbestos trusts are seeking ways to deny claims or limit payouts to victims. In 2001 Manville Trust was paying victims five cents for every dollar claimed to ensure that money was left in the $2 billion trust to cover future claims. In the third quarter of 2001 the trust paid an average of $2,400 per claim; during the same period in 2000 the trust paid an average of $4,000 per claim. The trust also slashed payments to dying cancer victims to $10,000 or less; during the late 1980s such victims were being compensated hundreds of thousands of dollars each.[49] Once suits are settled, claimants receive about 42 percent of the settlement; about 27 percent goes to the plaintiff's attorney's fees and 31 percent goes to defense costs.[50]

Asbestos defendants question the upsurge in the number of claims being filed and are investigating what they suspect are fraudulent claims. Beginning in September 2005 the Manville Trust ceased making payments of claims based on reports submitted by nine doctors and three X-ray screening companies that conducted or verified tens of thousands of claims (the doctors and companies are under investigation). Manville Trust's subsidiary, Claims Resolution Management Corporation, has begun investigating claimants submitting both asbestos and silica lawsuits. Claims Resolution Management has found that 5,174 of 8,629 silica claimants had already filed asbestos injury claims with the company. It is unlikely that so many people could have suffered exposure to both silica and asbestos. The Claims Resolution Management investigation and other probes have discovered that the X-ray screening companies recruit people and examine them to see if they have symptoms of silicosis and asbestosis. After the exams the companies file two reports simultaneously diagnosing those screened with silicosis and asbestosis.[51]

Everyone agrees that something should be done about asbestos compensation. The U.S. Supreme Court has struck down class action settlements on two occasions and has pushed Congress to resolve the impasse through legislative action.[52] The latest attempt by Congress to deal with asbestos claims took the form of Senate Bill 852, called the Fairness in Asbestos Injury Resolution Act. The bill was intended to take most of the asbestos litigation cases out of the courts and resolve them through a no-fault administrative process overseen by the Office of Asbestos Disease Compensation in the Department of Labor. The bill proposed the creation of a $140 billion fund to pay off claims. Companies currently named as defendants in asbestos cases would contribute $90 billion to the fund, insurers would be responsible for $46 billion, and bankruptcy trusts would contribute an additional $4 billion.[53]

The Fairness in Asbestos Injury Resolution Act has been controversial, and two government reports issued in 2005 inflamed the debate. In August the Congressional Budget Office published a report estimating that in its first ten years (2006–2015) the program would cost $70 billion, yet estimates of revenues from asbestos defendants and the consolidation of existing asbestos trust funds would amount to only $63 billion over the same period. The program was projected to receive about $140 billion in funds from defendants over a thirty-year period; however, the Congressional Budget Office estimated that claims amounting to between $120 billion and $150 billion might be submitted for the next fifty years.[54] A few months later, in November, the General Accountability Office issued a report stating that federal victims' compensation programs such as the Fairness in Asbestos Injury Resolution Act tend to last longer than initially intended and cost more than estimated. For instance, the Black Lung Program (financed by an excise tax on coal) was estimated to cost about $3 billion and had an end date of 1976, but between 1969 and 2004 the program, which compensates miners disabled with pneumoconiosis and other work-related illnesses, cost $41 billion.[55]

A number of conservative and progressive groups opposed the Fairness in Asbestos Injury Resolution Act. The Association of Trial Lawyers of America opposed the bill on the grounds that it robbed victims of their right to due process and full compensation and provided a $20 billion windfall to the largest asbestos companies, which would actually end up paying less into the asbestos fund than it is estimated they would pay in the absence of the bill. Opponents point to USG Corp.'s announcement weeks before the Senate vote on the bill that after almost five years in bankruptcy the company had a reorganization plan that would create a $4 billion trust fund to pay asbestos claimants. However, if Congress passed the bill, USG's payment into the federal asbestos trust fund would be capped at $900 million. The bill was also opposed by groups such as the Asbestos Disease Awareness Organization, the Coalition for Asbestos Reform, the Environmental Working Group, the AFL-CIO, USAction, and Public Citizen. A bipartisan group of senators, taxpayer rights groups, and insurers also objected; these groups raised concerns about the inadequacy of funds to cover current and future claims, the $20 billion reduction in contributions from major asbestos companies, and the rights of victims to get fair treatment under the proposed system. In February 2006, when the bill was being debated in Congress, about 150,000 asbestos claimants sent letters to Congress opposing it. On February 14 the bill failed to pass the Senate.[56]

As the stalemate drags on, state courts have become more proactive in addressing asbestos and silica liability and claims. Courts in Illinois, Maryland, Massachusetts, Minnesota, New York, Virginia, and Washington have established what is being referred to as "inactive dockets" in which cases involving claimants who cannot prove their asbestos claims through a physical manifestation of asbestos-related disease are placed on an inactive list. If such claimants develop asbestos-related diseases in the future their claims will be activated. Florida, Georgia, Ohio, Texas, and Mississippi have passed laws requiring claimants to satisfy medical criteria before filing a claim. Texas and Virginia have laws regarding case consolidation and laws requiring a stricter connection between claimants and the venues in which they can file claims.[57]

Steel Cities

BLACK RAIN AND HOT BALLS OF FIRE

Pollution and industrial accidents proved to be an ongoing fact of life for many workers and residents living near steel mills, factories, and similar facilities. Steel mills were egregious polluters and the communities around them suffered the consequences. Air pollution pervaded steel cities such as Gary, Indiana, South Deering, Illinois, and Youngstown, Ohio, in the 1940s and 1950s: "Steel production permeated the environment of Gary. . . . Every evening the mills presented viewers with a display of giant torches, erupting sparks, and massive factories engraved against a glowing red sky. Day and night, black and red smoke wafted through the atmosphere while oils, greases, and chemicals streaked across rivers and lakes. For those who lived and worked in Gary, pollution was inescapable."[58] In nearby Southeast Chicago the community also suffered:

> Viewed from . . . [the] privileged altitude [of the Chicago Skyway], the area below seems a maze of smokestacks and train tracks, dull maroon buildings, and occasional splotches of green parkland. Vast steel mills dwarf the soldierly rows of houses encrusted with decades of

> industrial grime. . . . [Once] the Skyway was built in 1956 . . . travelers . . . [were] lofted high above the nitty-gritty of life on . . . [the] streets. Through car windows shut tight against the acrid air, they see Chicago's steel mills merging with those of Whiting and Gary to form an immense, continuous industrial chain. . . .
>
> Commercial Avenue is also the main artery linking the disjointed mill neighborhoods. Running south, it wends its way through Slag Valley, a residential islet marked off by an enormous expanse of slag heaps, piles of scrap metal, and into South Deering, home of Wisconsin Steel. . . . [South Deering's] streets are poorly paved; its frame houses sag with the weight of almost a century; its air reeks of sulfur dioxide. Wisconsin Steel occupies more acreage than all of the homes, schools, churches, and stores combined.[59]

Similarly, since the 1940s Struthers, a city close to the Youngstown steel mills, wrestled with the highly acidic "black rain" that fell over the town, caused by pollutants from the mill's smokestacks mixing with vapors in the higher atmosphere to produce a very corrosive precipitation. One resident describes the pollution permeating the working-class communities adjacent to the steel mills:

> Our home . . . like many in the area, was covered with an imitation red-brick aluminum siding. After years of black rain the siding had grayed terribly. My father would have to hose the house down every few days to wash away the rainfall. On days when the weather was good, it was very common to walk across the driveway and feel your sneakers swishing over tiny particles of graphite, spewed across the neighborhood by mill blast and open hearth furnaces. . . . [When] the morning air was heavy with dew, captured within the dampness was a porridge-like sooty substance. You could reach up and feel the stuff floating in the air. . . . [Throughout the year] "you were wiping black rain off the hood" [of your car]. . . . I would occasionally sleep at my grandparents' home . . . in Youngstown. Their house sat on a hill, just a short hop from Republic Steel's Stop 5 entrance. The plant had four monstrous blast furnaces and fifteen open hearth furnaces continuously running through three daily shifts. My second-floor bedroom included a window that opened up onto a rooftop. . . . I would sometimes crawl out onto the roof at night to get a better view of the "fire." The fire was actually hot balls of orange gasses shooting into the air . . . [releasing] an explosion of sound and gasses into the air.[60]

Fred Mancini recalled growing up in West Aliquippa, Pennsylvania, in the shadows of Jones and Laughlin's steel mill: "Growing up in West Aliquippa was dirty and filthy. My mother got up in the morning, she had to sweep the porch and wipe the windows off, everyday."[61] John Hoerr described growing up in McKeesport, Pennsylvania, where the steel mills were like

> enormous steaming vessels, clanging and banging, spouting great plumes of smoke, and searing the sky with the Bessemer's reddish orange glow. . . . Noondays were often as dark as night . . . when inversions trapped great clouds of smoke close to earth, and the downtown sidewalks were so thick with ferruginous dust from the open hearth and Bessemer furnaces, that they gave off a metallic sheen. Smoke seemed to seep out of the very pores of the mill buildings. Every morning housewives all over town put on babushkas and swept clouds of dust off their front porches.[62]

In the 1940s Sparrows Point, Maryland, a company town and the home of Bethlehem Steel, was dominated by the hulking buildings of the mill and smokestacks belching white smoke. Just behind the mill were rows of three-storey homes housing five thousand people. The mill so dominated the lives of residents that they sometimes described the diurnal rhythms in terms of plant operations: "The sun always rose over the blast furnaces and set over the plate mill." One woman described the pollution from Sparrows Point:

> When the open hearths were built, the dirt that came from there was red. Then they put in a terrific large blast furnace and if that gave off when the wind was blowing into town, you got the black. It was just like little beads—little pebbles. At home you were sweeping and hosing off all the time. You never opened a window there—you had the air conditioners and the windows were caulked. But as clean as you kept the place, you could still wipe that dirt off the windowsills or whatever was around. When it came to washing and hanging out clothes, you'd have to look to see which way the wind blew. If it came from the open hearth, you were going to get red clothes, so you didn't hang out. If it came from the east side, the clothes would be black.[63]

Bethlehem Steel's Dundalk mill, also located in Maryland, was just as irksome to Dundalk residents. The company town housed both steel and shipyard workers and the residents referred to the fine coating of red dust that covered everything as "gold dust." Remarking on the dust, one foreman told a steelworker, "Remember, as long as that's there, you're working."[64]

WORKER HEALTH AND SAFETY

Steelmills provided many jobs but they were dangerous places to work—workers were exposed to dangerous substances in the air and water or suffered serious or fatal injuries on the job. The challenges of living adjacent to and working in a steel mill are exemplified by life in Gary and surrounding towns. The lakeshore town of Gary grew into one of the most important centers of steel production in the country. In 1889 Standard Oil established a refinery in Whiting that eventually became the largest refinery in the world. Inland Steel built a plant in neighboring East Chicago in 1901, and in 1906 U.S. Steel began building its Gary plant. U.S. Steel was formed in 1901 when J. P. Morgan purchased the Carnegie holdings in steel; by 1918 U.S. Steel owned 145 steel plants that accounted for half of the steel produced in America.[65]

U.S. Steel occupied 3,700 acres (about eight square miles) of Gary's lakeshore. The company, Gary's largest employer, operated an integrated steel mill that produced coke, iron, and steel. Gary Works produced an average of twenty thousand tons of iron each day. The plant drew its water from Lake Michigan, and during the 1970s the mill discharged about 775 million gallons of polluted water into the lake and the Grand Calumet River (which flows into the lake) each day. The U.S. Steel facility discharged wastes through five outfalls into the lake and fourteen outfalls into the river. During the 1970s the mill discharged an average of 180 pounds of phosphorus, 325 pounds of phenol, 3,100 pounds of cyanide, 3,400 pounds of fluorides, 5,100 pounds of ammonia, 82,000 pounds of chlorides, and 180,000 pounds of sulfates into the water each day. The waste water also contained heavy metals such as iron, manganese, and chromium.[66] Nearby steel mills were major polluters also. Wisconsin Steel emitted carbon monoxide into the air and polluted the water with cyanide, iron, and lead. In 1973 a consultant estimated that it would cost more than $150 million to bring Wisconsin Steel into compliance with the Clean Air and Clean Water Acts.[67]

Though conditions at U.S. Steel were dangerous and unhealthy most of Gary's residents worked there. In 1967 the company employed about 65 percent of Gary's workforce. At their peak U.S. Steel, Inland Steel, and Standard Oil employed fifty thousand people in Gary and nearby towns. Hundreds of coke ovens operated simultaneously. In 1992, for instance, U.S. Steel operated 268 ovens, each of which was capable of holding fifteen to twenty-five tons of coal. Each coke oven battery is a collection of ten to one hundred ovens operated together.[68] Visibility in the mill was so limited by particulate matter that at times workers had to ring bells, clang shovels, and tie themselves together with rope to communicate and navigate through the haze. One worker reported spitting up particles of coke three years after leaving U.S. Steel.[69]

Some parts of the coking cycle are particularly dangerous. Pushing coke from the ovens into quenching cars can be a major source of pollution. If the coke is not properly prepared airborne pollutants are emitted. Quenching hot coke with cold water also releases water vapor and particulate matter. Coke emissions may contain over ten thousand gaseous compounds, vapors, and particulates, including benzene, toluene, isomers of xylene, cyanide compounds, naphthalene, phenol, and other known or suspected carcinogens.[70]

At Bethlehem Steel's Johnstown, Pennsylvania, plant workers reported that the mill was so dusty that they "were still coughing two or three hours after [they] left work" each day.[71] Fred Mancini of Jones and Laughlin's steel plant in Aliquippa described the effects of the mill:

> It was dirty and filthy, it was dirty, you wouldn't believe it, when you took a shower. I scrubbed my hair every night when I took a shower. I'd get in bed with my wife and she'd say that I stink from the Byproducts. I'd come home and would get a white towel and wipe my face and black would come off. . . . We finished midnight like on a Thursday . . . and we didn't have to go back till Tuesday. By the time you went back Tuesday the black was out of your nose. . . . It just kept coming out, you breathed it. You breathed those fumes. . . .
>
> If you would just hit a beam—all that dust fell right on you if you hit against the wall it just went all over your clothes. . . . They finally decided that this stuff was cancer causing and they give us yellow clothes and respirators. But that wasn't heck. . . . We wasn't allowed to take our clothes home and wash them anymore. They gave us these yellow clothes, just a jacket and pants. . . . So we had to actually keep our clothes there at work. You wouldn't take them home to wash, they were too dirty. . . . When I come home I would put my clothes on the porch and I shook them out because there were cockroaches in the lockers . . . you didn't want to get in your house.[72]

Similar conditions were reported at the mills in South Deering and Southeast Chicago.

The mills were operational twenty-four hours a day to keep them hot and the rolls turning; laborers worked twelve-hour shifts, seven days a week. In the summer the ambient temperature in parts of the mill exceeded 100 degrees

Fahrenheit; it was bitterly cold in the winter. The air was laden with dust and vapors and workers had short lifespans. In 1906 forty-six men were killed at South Works alone.[73]

At the steel mill in Sparrows Point, Maryland, a worker said, "The heat was the worst part. You'd never see a guy bare-chested like they show in the paintings 'cause he'd be toasted it was so hot."[74] A worker described the tin mills at Sparrows Point:

> Working conditions in the old tin mills were hot, dirty, noisy, just about unbearable. The heat was so bad that the men had to wear long woolen underwear even through the summertime to keep the heat off their bodies. They wore wooden clog shoes because the floors got so hot that the leather shoes would just eat up on them. For the most part, there was some spell time between heats, but the average guy worked—the perspiration through his underwear and outerwear.
>
> The men had to take the steel out of a furnace with a pair of hand tongs and then feed it into the tin mills. The tongs would get so hot that they would have to wrap rags round their hands to keep their hands from blistering. In order to cool the tongs, the men would put them in buckets of water. The tongs weighed about 45 pounds—along with the sheet it weighed about 70 pounds. The men had no cleaning facilities, no washroom facilities, so consequently they brought their own piece of toilet tissue and washed their hands in the same water that they cooled the tongs off on at the end of the shift.[75]

The coke ovens were the dirtiest, most dangerous parts of the plant. A steelworker described the coke works at U.S. Steel in Youngstown: "The first time I worked there I think I lasted three hours. Man, it was hot down there. The fumes, the heat, and everything else were terrible. . . . I left work that day and said I didn't care if I didn't get paid [I wouldn't go back]. . . . Lots of summer help came to work in the coke plant. They always had a big turnover there." Accidents and injury were a constant in mill workers' lives. Some suffered hearing loss from the noise in the mills. A Youngstown steelworker recalled what happened to a coworker on the job: "One of the ugliest [accidents I saw] was when this guy walked over the catwalk over the straightening rolls, and he got drilled by a pipe through the right kidney. It went in about a foot. It was a four and a half inch pipe, but because it was so hot it cauterized. He stayed like this for two hours with his insides partially hanging out."[76] A worker at Wisconsin Steel described similar dangers: "You've gotta work in a rolling mill to appreciate [the dangers]. What you don't want is for a bar to land on you because they go right through you like a knife through butter."[77] Working on the catwalks was particularly dangerous:

> [In] the casting department . . . the temperature is usually over 100 degrees. Overhead cranes swing ladles filled with 200 tons of molten steel across the vast room, jets of cold water shoot into contact with boiling steel, and the hydrogen gas that feeds the reheating ovens is always in danger of exploding. If the machinery jams, hundreds of tons of steel may be ruined. . . . One crew member has to shove a lance over and over again into the boiling metal, while standing on two half-inch-thick sheets of steel a hundred feet from the ground. While shoving the lance, he pushes forward closer and closer to the mouth of the ladle until he's on the verge of falling in."[78]

A worker at Sparrows Point elaborates: "When a man fell into a ladle of hot metal, the ladle, the metal, everything was buried in the ground." Another worker explained how the machinery could easily mangle a worker: "I passed a form on a stretcher under a blanket. It didn't look right. It didn't look like a person lying there and yet again it did in a way. When I got to my job I saw they were sweeping a man up with a broom and a shovel and putting it in a shopping bag. So what I had seen was half the form. . . . The man had been torn up by a hot ladle car. He had gone to dodge the charging machine on the platform and he ran right into the hot ladle car. At that time they had the motor out in front of the car and that's what ground him up. . . . The worst thing was I knew the guy. . . . Quiet guy, around 40 years old."[79]

Blacks in Steel: Split Labor Markets and Labor Castes

THE GREAT NORTHERN MIGRATION

Between 1870 and 1900 many blacks moved from one part of the South to another and from rural to urban areas within the South to take advantage of job opportunities and to follow the expanding cotton cultivation. But then a number of social, economic, and environmental factors drove them out of the South altogether. Escalating racial violence (intimidation, assaults, lynchings, imprisonment); Jim Crow laws; the economic and labor market structure (low wages, confinement to the secondary labor market, constrained job opportunities); and environmental disasters (the boll weevil, devastating rains and floods that ravaged cotton crops in 1915 and 1916 in Louisiana, Mississippi,

Alabama, Georgia, and Florida) pushed blacks to move. At the same time a less hostile racial climate and expanding economic opportunities in the North pulled blacks to migrate.[80]

The Great Northern Migration marks one of the greatest redistributions of population in American history. The rate was slow yet steady before 1910, but increased dramatically after that. Between 1870 and 1910 535,000 blacks left the South for the North; in the next forty years 3.5 million followed suit. At the peak of the migration blacks left the South at the rate of sixteen thousand each month.[81] In the North they were used as strikebreakers or to alleviate labor shortages created by the dramatic drop in European immigration resulting from the passage of restrictive immigration legislation and the First World War.[82]

The demographic change in the North was profound and rapid. In the early twentieth century only about 4.3 percent of blacks born in the South moved to the North; by 1950, 20.4 percent had relocated from the South to the North. Between 1910 and 1920 the black population increased by 66 percent in New York, 148 percent in Chicago, 611 percent in Detroit, and 500 percent in Philadelphia. In Chicago fifty thousand blacks moved into the Black Belt.[83] Between 1920 and 1930 fifteen thousand blacks from Mississippi, Arkansas, Alabama, Tennessee, and Georgia moved to Gary to work in the steel mills. The migration flow slowed somewhat during the Depression, but during the 1940s another twenty thousand blacks moved to the city. By the end of the Second World War sixty-five hundred blacks worked at U.S. Steel. In 1950 three-quarters of all black males in the city worked in the industrial sector.[84]

UNIONS AND STRIKEBREAKING

As was the case in the garment industry, technological advances and the deskilling of jobs that enabled the development of large integrated mills and meatpacking facilities made it easier for employers to substitute one unskilled worker for another and to replace striking workers. As the skilled craftsmen who worked with a product from start to finish gave way to legions of unskilled workers doing precise repetitive tasks, factories increased the number of workers they hired.[85] Such operations were more prone to labor unrest as workers objected to low wages, long hours, and the speed-up of assembly lines.

Unions commonly discriminated against blacks, barring them from joining and objecting to their being hired in many types of industrial jobs. Corporate executives, recognizing the shortsightedness of racially exclusive unionization, used race as an effective divide-and-conquer strategy. They hired blacks to break 141 strikes between 1847 and 1929. Most of these strikes were in the North; in only twelve instances were blacks used to break strikes south of Virginia.[86]

Whatley argues that racism made blacks willing to break strikes, especially when the strikebreaking action was directed at unions that had a history of discrimination against them. The railroad brotherhoods excluded blacks by a constitutional clause; the Sons of Vulcan (the Amalgamated Association of Iron and Tin Workers) barred blacks from joining until black strikebreakers helped to defeat the union during the 1875 strike in Pittsburgh. Blacks were also banned from the building trade unions in Chicago until blacks broke their strike in 1900. Thus strikebreaking resulted in blacks gaining access to higher paying jobs and the unions.[87] For a long time union leaders failed to understand what motivated blacks to break strikes or to see how the actions of unions played a role in black strikebreaking. They saw black strikebreaking in terms of blacks "taking" jobs from whites, interfering with their gains, and hindering white progress. In 1901 John Mitchell, president of the United Mine Workers, characterized black strikebreaking: "I know of no element that is doing more to create disturbance in mining circles than the system of importing colored labor to take white men's place."[88] In 1905 Samuel Gompers argued, "If the colored man continues to lend himself to the work of tearing down what the white man has built up, a race hatred far worse than any ever known will result. Caucasian civilization will serve notice that its uplifting process will not be interfered with in any way."[89]

This is not to say that blacks and other minorities were averse to unionizing. In 1924 the NAACP sent a letter to the AFL:

> Negro labor in the main is outside the ranks of organized labor, and the reason is, first, that white union labor does not want black labor, and secondly, black labor has ceased to beg admission to the union ranks because of its increasing value and efficiency outside of unions.
>
> We face a crisis in inter-racial labor conditions; the continued and determined race prejudice of white labor, together with the limitation of immigration, is giving black labor tremendous advantage. The Negro is entering the ranks of semi-skilled and skilled labor and he is entering mainly and necessarily as a "scab." He will soon be in a position to break any strike when he can gain economic advantage for himself.[90]

Yet whites continued to exclude blacks from unions. Between 1916 and 1920 there were 18,633 strikes in America. In the peak year of 1919 more than 4.16 million workers went on strike. However, as late as 1925 less than 1 percent of the approximately 5 million African American workers in the country were unionized. In fact prominent organizations such as the Urban League and the NAACP actively promoted black strikebreaking as a means of gaining access to industrial jobs. During the 1919 steel strike, for instance, corporations recruited 40,000 black strikebreakers. About 3, 500 were recruited to work in Homestead, Pennsylvania; 2,000 in Buffalo; 5,000 in Youngstown; and 8,000 in Chicago. Hence the racial homogeneity of the unions and the antipathy unionists felt toward blacks encouraged both split labor market dynamics and black strikebreaking. The Interchurch World Movement reported that the effective use of black strikebreakers was one of the main reasons the union lost the 1919 strike.[91]

At first blacks were primarily recruited from the South and used to break strikes in the North. However, after 1910 many corporations recruited blacks from the large urban centers in which the mills were located or in nearby northern towns or states. It should be noted that before 1910 only a few industries—iron, steel, coal mining, meatpacking, railroad industries, and longshoremen—used blacks to break strikes. Whatley explains that these were industries that had extensive southern branches that trained and employed large numbers of African American workers who could be called upon to break strikes. After 1910 about 61 percent of the strikes using black strikebreakers occurred in these industries.

The use of black strikebreakers sometimes resulted in violence as white laborers tried to prevent blacks from crossing picket lines, attacked them, or tried to run them out of town. The 1919 meatpackers strike in Chicago escalated into a race riot in July. In October a race riot broke out in Gary during the steelworkers strike. The police, militia, and federal troops were sent in to quell the violence.[92]

When was it most expedient for corporations to use black strikebreakers? Whatley found that black strikebreakers were deployed most frequently when the level of European immigration was low and when there were also many strikes to be broken. Thus in the 1890s, a period of many strikes and falling European immigration rates, twenty-nine strikes were broken by African Americans. Black strikebreaking again peaked during the First World War, which coincided with the passage of restrictive immigration legislation. This suggests that blacks were used as substitutes for white immigrant strikebreakers.[93]

There was another reason corporations used black strikebreakers: their presence increased the likelihood that the strike would turn violent and that the government would intervene, and government intervention usually favored the corporation. In some instances the police, the state militia, or the national guard were dispatched to escort and protect strikebreakers or guard industrial facilities.[94] Scholars have developed a violence-instigation thesis to explain the deployment of tactics by the civil rights movement that is applicable here. As the argument goes, when one set of actors in a conflict takes action to provoke violent or repressive responses from opponents, in this case striking white laborers attacking black strikebreakers, the likelihood of government action increases.[95]

Cliff Brown studied union organizing in Chicago and Gary between 1917 and 1919 and found that when unions, such as the Gary steelworkers, completely excluded blacks, the antagonism between the two was already so high that corporations didn't need to resort to manipulating racial hostilities to generate animus. However, when unions, such as the meatpackers in Chicago, attempted to organize multiracial coalitions and develop solidarity with black workers, corporations, seeing this as a threat, adopted divide-and-rule strategies that promoted the instigation of violence between blacks and whites.[96]

While black strikebreakers were often characterized as ignorant, rural folks duped into strikebreaking, the situation was more complicated. An analysis of black strikebreakers indicates that some were urban, many were already skilled industrial workers at the time they were recruited, and many knew they were being recruited to break strikes when they decided to accept job offers. Their prior work experience in southern factories helped to make them effective strikebreakers. The corporate practice of transferring men who had already broken strikes in one location to other locations to break strikes points to the fact that blacks knew that they were being hired as strikebreakers. It also points to the agency of black strikebreakers: taking a strikebreaking job was a calculated strategy for many. In some cases companies recruited blacks from the very cities in which the strikes were occurring. For instance, in the 1921 Chicago stockyard strike corporate owners opened an employment office in the heart of the city's black community. During the 1919 steel strike eight thousand black strikebreakers were recruited from Chicago. It was reported that so many blacks lined up for the jobs that five hundred had to be turned away.[97]

Table 18 shows the increase in the number of blacks hired in the steel industry in ten northern cities between 1910 and 1920. In 1910 there were 1,572 blacks working in the steel industry in these cities. This number rose to 20,762 by 1920. There were about 125,000 black steelworkers nationwide in 1920.[98]

The incidence of black strikebreaking dwindled after the Wagner-Connery Act, passed in 1935, which required employers to rehire striking workers after

Table 18 Increase in the Number of Black Steelworkers in Ten Northern Cities, 1910–1920

	1910		1920	
Cities	Number	Percentage	Number	Percentage
Buffalo	6	0.1	672	6.1
Chicago	37	0.1	3,540	13.0
Cleveland	38	0.2	3,323	14.1
Columbus	329	7.9	1,210	33.2
Detroit	4	0.1	1,887	12.9
Gary	105	0.9	2,060	13.7
Newark	11	0.6	488	7.8
Philadelphia	464	7.2	2,572	13.8
Pittsburgh	576	1.9	4,810	16.1
Toledo	2	0.1	200	6.9
Total	*1,572*		*20,762*	

Source: Cliff Brown, "Racial Conflict and Split Labor Markets," *Social Science History* 23, no. 3 (1998): 322.

labor disputes were settled.[99] Both the NAACP and the National Urban League objected to the Act, claiming, "While we deplore the necessity of strikebreaking . . . it is a weapon left to the negro worker whereby he may break the stranglehold that certain organized labor groups have utilized in preventing him complete absorption into the American labor market."[100]

Some unions adopted a strategy of excluding unskilled workers from their ranks.[101] Since most blacks were classified as unskilled laborers, this meant they were excluded from the unions on the basis not only of their race, but of their class. This only began to change when whites realized that multiracial organizing could strengthen the unions.

RACIAL DIFFERENCES IN JOB CLASSIFICATION AND WAGES

The discrimination did not stop once blacks were hired into the factories. As Table 19 shows, it was common practice to hire blacks into the unskilled segment of the labor force. This practice led to the development of labor castes. Pressed Steel Car hired only a small number of blacks, and that number did not change between 1916 and 1917. Half were working as unskilled laborers. A. M. Byers hired two hundred black workers in 1917, 60 percent of whom

Table 19 Blacks Employed at Selected Iron and Steel Plants in Pittsburgh in 1917

Company	Employed before 1916	Employed in 1917	Percentage Unskilled Labor in 1917
Jones and Laughlin	400	1,500	100
National Tube	100	250	100
Pittsburgh Forge and Iron	0	75	100
American Steel and Wire	25	25	100
Oliver Iron and Steel	0	50	100
Carnegie Steel	1,500	4,000	95
Lockhart Iron and Steel	0	160	95
Crucible Steel	150	400	90
Clinton Iron and Steel	25	25	75
Carbon Steel	50	200	75
A. M. Byers	0	200	60
Pressed Steel Car	25	25	50
Total	*2,275*	*6,910*	—

Source: Adapted from Cliff Brown, "Racial Conflict and Split Labor Markets," *Social Science History* 23, no. 3 (1998): 323. See also Abraham Epstein, *The Negro Migrant in Pittsburgh* (Pittsburgh: University of Pittsburgh Press, 1918), 31.

were categorized as unskilled laborers. Between 75 and 100 percent of the black laborers working in the remaining ten companies were classified as unskilled laborers in 1917.[102]

Policies at companies such as U.S. Steel illustrate how castes and patterns of inequality arose and were maintained. The company had a policy of assigning blacks to the most dangerous jobs. Their tasks included heavy lifting and exposure to grease, toxic chemicals, and fumes, carcinogenic substances, dangerous acids, flammable liquids, and hot tar. Most blacks worked in the coke plant, the dirtiest section of the mill. Temperatures in the coke ovens hovered at 2,000 degrees. The company-issued insulated clothing provided minimal relief for the mill hands working directly on top of the coke batteries. These workers often developed circulatory problems; workers who fed coal into the ovens were usually covered in soot. The constant crushing of coal filled the air with particulate matter. U.S. Steel instituted a promotion system in the 1940s that trapped blacks in the coke plant. Seniority was calculated on a unit-by-unit basis, and because the coke plant was deemed an independent unit workers there could not amass enough seniority points to transfer to other jobs. In addition the managers at the mill withheld from blacks information about job openings in other areas of the plant. In the unlikely event that black workers heard about and applied for work in other areas they were denied permission to transfer. Some black applicants were given unusually difficult tests, while others were simply told they did not qualify for the jobs.[103]

A similar kind of job stratification could be found at Bethlehem Steel's Sparrows Point mill, where African Americans were given the least skilled, hottest, and dirtiest jobs. Most of the managers and workers who lived in Sparrows Point were white, native born, and of English, Irish, Welsh, and German ancestry; many came from rural Pennsylvania, Maryland, and neighboring southern states. The African American workers were from the rural South.[104] One worker described how difficult it was for blacks to get jobs at the plant and how they were discriminated against:

> I was hired at the mill in the 1940s. When I applied, all the blacks were put into a large auditorium. We were asked our age, where we were from, and what we expected out of the company in the future. If you were known to have a high school background, they wouldn't hire you. If you were from Baltimore, they were reluctant to hire you. You had to tell them that you had a job on a farm and no more than a grade school education, then they'd hire you.
>
> The company would look at a black's size, his physical fitness. My foreman made it quite clear that he wanted blacks with strong backs and no education. I almost didn't get hired because I had a book in my hand that I had brought to read on my way down on the streetcar. Prior to the unions, in the early 1940s and 1950s, blacks were hired mainly for the hot, hard, physical labor, the "CB's department"—the "colored boys' department." That's where you stayed. If you tried to transfer, you were given a termination notice. Wages were much lower, much lower than for white workers. Job classes ran from 1 to 32. Black workers usually stayed around job class 1 and no higher than a job class 4.[105]

These observations are borne out by statistics. A comparison of hourly wages paid to black and white workers in ten industries in Pittsburgh in 1925 showed that black workers composed a small segment of the industrial workforce and, in most instances, were paid less than white workers doing the same job.[106] Table 20 shows that this trend was still common in 1940. A comparative study of the wages of male workers found that blacks in twenty-one northern states

Table 20 Comparison of Weekly Wages of Black and White Male Workers in Industries in Northern States, 1940

	Blacks		Whites		
Industry	Wages	*n*	Wages	*n*	Wage Ratio
Meatpacking	$23.76	79	$26.00	788	0.91
Chemicals	23.92	32	27.84	976	0.86
Paper products	20.88	26	24.82	1423	0.84
Auto	27.22	178	33.1	3640	0.82
Steel	25.11	166	30.71	3382	0.82
Electrical machinery	22.73	10	28.79	1607	0.79
Trucking	17.96	38	24.84	1308	0.72
Water transportation	19.57	42	27.41	460	0.71
Construction	19.52	249	28.04	6692	0.70
Street transportation	21.50	40	31.55	1186	0.68
Railroad	21.41	238	32.17	4355	0.66
Textile	15.35	24	28.61	1386	0.54

Source: Adapted from Thomas N. Maloney, "Degrees of Inequality," *Social Science History* 19, no. 1 (1995): 35.

were systematically paid less than whites in the twelve industrial sectors examined. Blacks came closest to parity in the meatpacking industry, earning 91 percent of the wages whites earned. Blacks earned 82 percent of the wages whites earned in the steel industry. Though the wages for blacks and whites were highest in the auto industry, blacks earned only 82 percent of the wages whites earned. Blacks fared worst in the textile industry, where they earned only slightly more than 50 percent of the wages earned by white workers.[107]

Table 21 shows that in 1940 blacks were more likely than whites to be assigned to the lowest job category in the twelve industrial sectors studied in northern industries. In every industrial sector except meatpacking and trucking blacks were more likely to be working as laborers than in the two other job categories, operatives and skilled workers, combined. Even then, 48 percent of black meatpackers and 50 percent of black truckers were laborers. In the worst-case scenario 87 percent of the blacks in railroad jobs and 85 percent of those in street transportation were laborers. Less than 10 percent of black workers were in skilled positions in meatpacking, chemicals, railroad, street transportation, trucking, and water transportation.

Table 21 Distribution of Black and White Workers in Selected Northern Industries, 1940

Industry	Skill Level	Black	White	Ratio
Meatpacking	Laborer	.482	.317	1.52
	Operative	.506	.559	.905
	Skilled	.012	.125	.096
Water transportation	Laborer	.767	.487	1.57
	Operative	.209	.397	.526
	Skilled	.023	.116	.198
Chemicals	Laborer	.546	.322	1.70
	Operative	.394	.434	.908
	Skilled	.061	.244	.250
Paper products	Laborer	.462	.245	1.89
	Operative	.423	.546	.775
	Skilled	.115	.210	.548
Steel	Laborer	.667	.332	2.01
	Operative	.226	.335	.675
	Skilled	.107	.334	.320
Construction	Laborer	.708	.304	2.33
	Operative	.075	.099	.758
	Skilled	.217	.597	.363
Railroad	Laborer	.873	.284	3.07
	Operative	.082	.293	.280
	Skilled	.045	.423	.106
Trucking	Laborer	.500	.154	3.25
	Operative	.500	.801	.624
	Skilled	0	.045	0
Auto	Laborer	.525	.129	4.07
	Operative	.332	.523	.635
	Skilled	.144	.349	.413
Electrical machinery	Laborer	.600	.147	4.08
	Operative	.300	.484	.620
	Skilled	.100	.369	.271
Textile	Laborer	.208	.045	4.62
	Operative	.667	.728	.916
	Skilled	.125	.226	.553
Street transportation	Laborer	.850	.087	9.77
	Operative	.125	.733	.171
	Skilled	.025	.181	.138

Source: Adapted from Thomas N. Maloney, "Degrees of Inequality," *Social Science History* 19, no. 1 (1995): 36.

In contrast, relatively low percentages of whites were classified as laborers. The one aberration was water transportation, where 49 percent of whites and 77 percent of blacks were classified as laborers. Otherwise, less than 35 percent of whites in the remaining industries were laborers. In two industries, street transportation and textiles, less than 10 percent of whites were laborers. While relatively few blacks were in skilled positions in all but one industry, 60 percent of whites were in skilled positions in construction, and 42 percent were in skilled positions in railroad jobs.

The worst ratios were seen in street transportation, where blacks were almost ten times more likely to be classified as laborers than whites. Blacks were also about five times more likely to work as laborers in the textile industry and four times more likely to work as laborers in electrical machinery and auto industries. Conversely no blacks were listed as skilled in the trucking industry and were only 10 percent as likely as whites to work in skilled positions in meatpacking and 11 percent as likely to work in skilled positions in the railroad industry.

In the steel industry blacks earned 82 percent of the wages white workers earned. Furthermore, blacks were twice as likely to be classified as laborers: 67 percent of black steelworkers were laborers compared to 33 percent of white steelworkers. Twenty-three percent of blacks were operatives compared to 34 percent of whites. Only 11 percent of blacks were in skilled positions in the steel industry, while 33 percent of whites occupied such positions.

RACE AND THE WORK ENVIRONMENT

Black workers in steel mills did not wait for the 1970s studies linking work near coke batteries with increased rates of lung and kidney cancer to tell them that this work was endangering their health. During the 1940s and 1950s the union filed many petitions and grievances protesting the heat, smoke, and dust in the coke plant. As early as 1935 thirty-five black steelworkers sued U.S. Steel on the grounds that they had contracted pneumoconiosis, silicosis, and tuberculosis as a result of smoke inhalation on the job. Workers assigned to this unit tried to transfer out as quickly as possible. However, most blacks weren't allowed to.[108]

White privilege helped white workers regardless of class or skill; white workers benefited from the permanent placement of black workers in coke plants and other dangerous parts of the industrial operations. As long as the company, unions, and white workers kept this arrangement, whites would not be subjected to the hazards of the most dangerous work for long. For blacks and other minorities reversing their exclusion from whole classes of jobs, improving degraded work environments, and ridding the workplace of institutionalized racism were as much a part of their quest for improved quality of life as cleaner air, protective gear, and improved safety conditions. Black workers understood that in order to improve their position racist hiring and promotion policies had to be halted with as much urgency as improving the physical conditions in the work environment. So while both black and white workers pushed for protective gear, better ventilation, and a safer work environment, black workers interjected demands for racial equality into the discourse. Black unionists and workers therefore had to oppose the wishes of the white unionists in demanding that the racist system of occupational apartheid be dismantled.

To press their demands during the 1940s black workers at U.S. Steel, recognizing that their work in the coke plant was crucial to the operation of the whole mill, started pressuring the company to open up other positions in the mill to blacks. When the company refused, blacks walked off the job. Black union members used the strike tactic several times a year, succeeding in opening up jobs in many areas of the plant. Despite these gains, however, the union worked in concert with the company to block many efforts by blacks to gain racial equality on the job. Because U.S. Steel and the union used the same hiring and promotion policies in all the plants, African Americans developed a national strategy. Black workers across the country sued both the company and the Steelworkers Union. As a result in 1974 all three parties signed a consent decree establishing new promotion guidelines. According to the new policy, seniority was calculated based on time spent at the company rather than in a particular unit. But racial equality was slow in coming. Four years after the decree was signed blacks still constituted 90 percent of the coke plant workers, while relatively few blacks worked in sections of the plant with the best working conditions.[109]

RACE AND HOUSING

Gary was originally a company town laid out to replicate the bureaucratic hierarchy of the mill and reflect the ideological thinking of its owner. Steel mill owners were not overly concerned with aesthetics or residents' access to the lakefront when they designed the town, hence they spread out the mill along the waterfront, effectively blocking workers from gaining access to the water for recreational purposes. The best houses were reserved for the highest paid employees. These were managers and supervisors; all were native born of Western European ancestry. Their houses were built closest to the plant so that they had an easy walk to work. Unfortunately for these employees, their close proximity to the mill meant they got the worst pollution. Hence the workers exposed to the least pollution inside the factory were subject to the worst pollution at home. As one moved down the hierarchy, workers were housed in smaller, less well-built homes farther away from the mill. When the workforce was all white, Eastern Europeans lived in the worst housing on the southern fringe of the city. Just after the First World War, when blacks and a small number of Hispanics moved to the city, they lived among the Eastern European immigrants on the Southside. There blacks lived in the worst housing. As better housing opportunities materialized, whites moved to all-white neighborhoods, leaving blacks, who could not get housing elsewhere, in the southern section of the city, called Midtown. In 1950 97 percent of Gary's black population lived in Midtown, an area covering two square miles. Realtors refused to show blacks homes in other neighborhoods, and city authorities rejected their applications for public housing in neighborhoods outside of Midtown. As Midtown's population increased from thirty-eight thousand in 1950 to sixty-two thousand in 1960 housing conditions deteriorated rapidly. In addition housing construction did not keep pace with demand; the number of new dwelling units increased by 33 percent while the population increased by more than 61 percent. The ecology of the neighborhood and the quality of life deteriorated also. Garbage accumulated in the streets, and abandoned cars and vacant lots contributed to neighborhood blight. Rats and other vermin ran rampant throughout the neighborhood. According to the 1960 census, one in five homes lacked plumbing. Housing code violations were pervasive, and banks refused to lend money for home improvements in the area.[110]

Sparrows Point was another company town laid out to replicate the bureaucratic, racial, and ethnic hierarchy in the factory. The Maryland Steel Company originally built the plant and town, and in 1916 Charles Schwab, the head of Bethlehem Steel, acquired Maryland Steel. The mill's workforce was about a third black and another third recent immigrants. Employees got housing according to their salary, rank, ethnic origin, and race on streets numbered A to K. The typical lot was 125 feet deep.[111]

About eight hundred homes had been constructed by 1904. The size of the houses and the front porches got progressively smaller to indicate the declining status of the worker. The general manager's house was the only residence on A Street; it was near the beach. Superintendents and other upper-level managers lived on B and C Streets in spacious, detached frame houses on lots twenty-eight feet wide. These homes cost between $2,500 and $3,100 to build; many had six or seven rooms (three to four bedrooms). The business district, D Street, separated upper management from foremen and white skilled laborers, who lived on E and F Streets. These homes were snug duplex cottages on lots twenty-eight feet wide, built at a cost of $1,350 for brick and $1,100 for frame construction. There were also several hundred row houses for skilled workers on lots that were fourteen feet wide. The black residents were separated from the white residents by Humphreys Creek, where G Street would have been located. A bridge eight hundred feet long connected both sides of town. The black section of town contained H through K Streets. Two-room bungalows with outhouses were constructed for blacks here, but as housing became scarce these were used to house white immigrants. Many black workers ended up renting bunks in shanties originally intended for temporary housing. One barracks housing blacks was surrounded by a high barbed-wire fence with a guard at the gate. Another set of barracks was inside the steel mill yard, completely surrounded by iron works. Each barracks contained ten one-room shanties; each room was ten feet by fourteen feet and had two double bunks. Each shanty cost $50 to construct.[112]

STEEL IN DECLINE

The largest geographic concentration of steel mills in the United States is located in the Great Lakes region: 46 percent of the steel mills and about 80 percent of the country's steel-making capacity is located in New York, Pennsylvania, Ohio, Indiana, Illinois, and Michigan. Most of the integrated mills were established in this region because of the proximity to water and good transportation routes. But during the 1980s American steel production declined and many mills closed. Between 1977 and 1992 the industry lost more than 61 percent of its employees and 58 percent of its facilities; the number of facilities decreased from 504 to 247, and wages fell dramatically. However, while the large integrated mills were shutting down, the number of minimills in operation doubled (the minimills use electric arc furnaces to melt scrap and other materials to make steel products). Minimills tend to operate with a small, nonunionized workforce receiving lower wages.[113]

Approximately 570,000 workers were employed in the industry in 1979; this number had declined to about 262,700 by 1991. Prior to the 1980s steelworkers' wages were 45 percent higher than the wages of other workers, but between 1979 and 1999 their wages fell while the wages of other workers remained relatively flat. Though steelworker wages are still slightly higher than that of other workers in general, they are now lower than that of workers in durable goods industries and on par with the wages of other workers in the manufacturing sector.[114]

Seeking Environmental Reforms through the Unions

After the Great Depression workers placed more emphasis on workplace hazards. They also began to pay attention to and highlight unsafe environmental conditions in their communities. Eventually laborers and community residents began to collaborate on both issues. Working-class activism became more commonplace in the Second World War era as industrial expansion resulted in unprecedented levels of pollution inside and outside the workplace.

By the 1940s working-class people who had once relied on a variety of religious, ethnic, and social service organizations to help them assimilate began relying more heavily on the unions and government to help them in their economic and social advancement. Though the strikes were deadly in Youngstown, South Chicago, and other places, participation in the labor strikes of the 1930s gave Gary's steelworkers the confidence to challenge U.S. Steel and their own union; exasperated with dogmatic supervisors, limited job security, low pay, and hazardous working conditions, thirteen steelworkers met in a tavern basement in 1933 to begin unionizing. Workers at Sparrows Point began to organize in the early 1930s; they met secretly in the storefront space of the Amalgamated Association of Iron, Steel, Metal and Tin Workers. Steelworkers met in secret because workers sympathetic to the union were harassed or fired by Bethlehem Steel. However, the need for secrecy stymied organizing efforts, and because the Amalgamated Association did not commit enough resources to the effort the organizing drive floundered without producing a union.

After the passage of the Wagner-Connery Act in 1935 workers had some protection. The Congress of Industrial Organizations formed the Steel Workers Organizing Committee to unionize steelworkers. It sent organizers to mills in and around Pittsburgh, Chicago, and Gary, and in 1936 Gary's steelworkers joined up. This time around, steel companies across the country relied less on black strikebreakers and more on intimidation to dissuade workers from unionizing. Some companies hired the Pinkerton National Detective Agency to spy on, intimidate, and harass union members. Despite threats from mill management, labor organizers got the support of workers, and in 1937 U.S. Steel recognized the union and signed a contract covering 375,000 workers in 142 companies. The Steel Workers Organizing Committee organized the Baltimore area steel mills, Sparrows Point, Dundalk, and Highlandtown in the 1940s. Between 1940 and 1942 Sparrows Point steelworkers met regularly at O'Connors Liquor Store and Bar, a few doors from where steelworkers had met secretly in the 1930s, to organize the unionization drives at Sparrows Point and Highlandtown.[115]

Despite the successful unionizing efforts at Gary, South Chicago was wracked by a bitter strike at Republic Steel's East Side Plant in 1937. Approximately twenty thousand workers struck in May of that year when their demands for a union and wage increases were not met. At the time of the strike there were twenty-five thousand workers at the plant, eight thousand of whom were black. All but about twenty-five of the black workers went on strike with their white coworkers. Republic Steel strengthened its police force in preparation for the strikes: 370 men carried 552 pistols, 64 rifles, 245 shot guns, 143 gas guns, 58 gas billy clubs, 232 nightsticks, 2,707 gas grenades, and 178 billy clubs.

On May 30 a crowd estimated to be between one thousand and five thousand protesters marched toward the mill gate. The police ordered them to disperse. As the protestors turned to leave, some marchers reportedly threw rocks and a tree branch at the police, who retaliated by firing two hundred shots at the fleeing marchers; ten were killed, thirty sustained wounds that left them permanently disabled, twenty-eight were hospitalized, and another thirty received emergency room medical treatment. None of the marchers who was killed had frontal wounds; seven were shot in the back and three in the side. What became known as the Memorial Day Massacre shocked the country.[116] An eyewitness gave an account of the events:

> I was standing within a few feet of a neighbor of mine (Lee Tisdale), a black man just like me. He had a wife and five children. The oldest was only ten. An automobile rolled up with three policemen. One got out with a machine gun; I looked back but he didn't. They shot him in the back six times, and then drove off. Ten union members were killed. . . . There were more than a hundred beaten to the ground. They were hauling people off in ambulances for over two hours. I was clubbed over the head and still have the scar to show it. I saw a woman that I know knock down four Chicago policemen with a nine bar before they knocked her down and beat her.[117]

Later that year striking steelworkers were also killed in Canton, Ohio, and Beaver Falls, Pennsylvania.

There is a long history of corporations responding to unionization efforts with violence directed at labor organizers and workers. During the 1919 steel strike, three thousand police deputies were brought to McKeesport to suppress strikers. An estimated twenty-five thousand armed men employed by the steel companies patrolled plants in Pennsylvania from Pittsburgh to Clairton. Bethlehem Steel made life difficult for strikers by evicting them from company houses and foreclosing on their mortgages. Picketers were arrested and public gatherings barred. The U.S. Army occupied the city of Gary. In all, twenty-two people were killed in the 1919 campaign.[118]

THE RIGHT TO REFUSE HAZARDOUS WORK, AND THE CATCH

Though some workers sought to improve environmental conditions through their unions, in Gary and other parts of the country unions de-emphasized workplace and community environmental issues. As the limits of the union strategy became clear, Gary's workers formed independent working-class environmental groups to tackle the issues of concern to them. They challenged management on a variety of issues related to health and safety and at times resorted to insurgent actions. Workers sometimes ignored the formal grievance procedures laid out in the 1937 contract and sometimes staged wildcat strikes. This was not unique to Gary: across the country workers used the wildcat strike as a strategy to mediate shop floor conflicts and call attention to hazardous work conditions, which their contracts did not address.[119] At Wisconsin Steel, for instance, workers used wildcat strikes and sick-outs to force management to improve shop floor conditions. One union officer recalled, "If the spark testers had a gripe, I would tell them, 'Call in sick, you all got the flu.' The company would threaten to fire each one who didn't show up for work. So I said, 'Go ahead, go ahead.' And nobody was ever fired."[120]

However, workers had to exercise caution when participating in wildcat and sympathy strikes. In 1978 Eugene Goldenfeld, a steelworker, was suspended for thirty-eight days from the Gary U.S. Steel plant for refusing to cross the picket line that had been set up at the plant by the Elgin, Joliet, and Eastern Railroad employees, who operated the tracks, offices, and switching equipment at Gary Works. Goldenfeld argued in court that he had a right to honor the picket lines of other workers and that he should not be disciplined, but U.S. Steel argued that Goldenfeld's sympathy strike violated his union contract, which had a no-strike clause. The court agreed with U.S. Steel that the no-strike clause constituted a clear waiver of the employee's right to engage in a sympathy strike.[121]

The only recourse some workers had under their contract was the right to refuse unsafe work, but this clause did not address the issue of improving work conditions.[122] An individual or small group choosing to refuse unsafe work could undermine collective bargaining, as they could be fired and replaced by others willing to do the work.[123] However, a few strategically placed workers refusing to work could hold an entire shop floor hostage if they controlled crucial parts of the mechanism, the incapacitation of which rendered the rest of the factory inoperable. Such strategies tended to illicit participation of all or most of the workers on the floor. A crane operator at the Youngstown steel plant explained, "I could shut the whole mill down if I didn't move those ingots. . . . The process depended on me and gave me an advantage."[124]

UNION CONTRACTS AND WORKPLACE SAFETY

Because union contracts did not compel a company to fix, eliminate, or reduce hazards to workers, steelworkers resorted to strikes to get managers to improve working conditions. This was particularly true of the very hazardous coke plant at Gary's U.S. Steel plant, where shop floor leaders struck many times over working conditions. In 1953, 450 coke plant workers staged a two-day walkout because management ignored their complaints about warped furnace doors and lack of protective clothing. Workers in the sheet and tin mill struck over the fumes from the enameling process. Once workers walked off the job, management responded quickly by sending engineers to install exhaust fans at the site.[125] Workers also used the health and safety committees to force management to provide shielding devices such as ear plugs, uniforms, helmets, and shoes to protect workers against the noise, heat, and chemicals. Sometimes management's response was limited; for example, in the 1940s when coke plant workers demanded protective clothing, they received only goggles, shoes, gloves, and masks.[126] Despite the fact that the company was reluctant to furnish safety clothing and equipment to workers, U.S. Steel filed and got tax exemptions for the purchase of these items.[127]

Workers had good reason to be concerned about their health. Maps produced by the National Cancer Institute showing the geographic distribution of cancer mortality rates between 1950 and 1969 and from 1970 to 1994 show excess rates of cancer in people living in industrialized areas, particularly in the vicinity of petrochemical plants. People in the Northeast, along the southern edge of the Great Lakes (including Gary), and along the Mississippi Delta were strongly affected. Lake County, Indiana, where Gary is located, was among the counties with some of the highest cancer mortality rates in the country. Some of these cancers might be due to exposure within the factories and contamination of communities adjacent to hazardous facilities.[128] Table 22 lists some common cancers that occur in various occupations.

It was becoming clear that workers could not rely solely on the Steelworkers Union to advocate environmental reforms for them. The union, having adopted the strategy of business unionism, still focused on enhancing job growth and lobbying for higher wages and increased benefits, and were not overly aggressive in criticizing the company for fear that the antagonistic ap-

Table 22 Occupational Exposure, Cancers, and Carcinogens

Type of Cancer	Carcinogen	Occupation
Liver cancer	Arsenic, vinyl chloride	Tanners, smelters, vineyard workers, plastics workers
Nasal cavity and sinus cancer	Chromium, isopropyl, oil, nickel, wood and leather dusts	Glass, pottery, and linoleum workers, nickel smelters, mixers and roasters, electrolysis workers, wood, leather, and shoe workers
Lung cancer	Arsenic, asbestos, silica, chromium, coal products, iron oxide, mustard gas, nickel, petroleum, ionizing radiation, bischloromethylether	Vintners, miners, quarry workers, asbestos users, textile workers, insulation workers, masons, tanners, smelters, glass and pottery workers, coal tar and pitch workers, iron and steel foundry workers, electrolysis workers, retort workers, radiologists, radium dial painters, chemical workers, construction workers, plumbers, pipefitters, steamfitters, ship- and boat builders, boilermakers, nonmetallic mineral and stone products workers, petroleum refining, electric light and power workers, mechanical engineers
Bladder cancer	Coal products, aromatic amines	Asphalt, coal tar, and pitch tar workers, gas stokers, still cleaners, dyestuffs users, rubber workers, textile dyers, paint manufacturers, leather and shoe workers
Bone marrow	Benzene, ionizing radiation	Benzene, explosives, and rubber cement workers, distillers, dye users, painters, radiologists

Sources: P. Cole and M. B. Goldman, "Occupation," in *Persons at High Risk of Cancer*, edited by J. F. Fraunmeni Jr. (New York: Academic Press, 1975), 167–84; Samuel S. Epstein, *The Politics of Cancer* (New York: Anchor Press, 1979), 32, 78; National Institute for Occupational Safety and Health, *The Work-Related Lung Disease Surveillance Report*, Publication 2003-111, U.S. Department of Health and Human Services, National Institute of Occupational Safety and Health, 2002, xxiii–xxix; National Institute for Occupational Safety and Health, *Health Effects of Occupational Exposure to Respirable Crystalline Silica* (Washington, D.C.: U.S. Department of Health and Human Services, National Institute for Occupational Safety and Health, 2002), 5.

Table 23 Percentage of Union Contracts Containing Workplace Health and Safety Clauses

Worker Safety	Year							% Increase
	1957	1961	1966	1971	1979	1983	1987	
Any Safety Clause								
Manufacturing	69	71	69	71	87	87	89	20
Nonmanufacturing	38	48	48	52	73	72	77	39
General Duty Clause								
Manufacturing	43	39	43	48	58	64	66	23
Nonmanufacturing	12	20	19	28	36	38	43	31
Joint Safety Committee								
Manufacturing	31	34	35	38	55	57	62	31
Nonmanufacturing	12	14	18	19	24	26	27	15
Ongoing Physical Exam	12	14	10	17	22	23	22	10

Source: James Robinson, *Toil and Toxics: Workplace Struggles and Political Strategies for Occupational Health* (Berkeley: University of California Press, 1991), 53.

proach would undermine their bargaining power. For a brief period in the 1960s the unions did support environmental reforms, although even then they did not seek to initiate the reforms but merely supported some of the campaigns initiated by the middle class. For example, once the city of Gary formed the Air Pollution Advisory Board, labor requested and obtained a seat on the board. However, local unions sometimes undercut each other's attempts at environmental reforms. In March 1965 the Gary Works local pledged an all-out battle to institute an air pollution reduction program. But when U.S. Steel proposed a pollution-abatement plan for its facilities in Gary that same year, the president of the sheet and tin mill local was one of the first to advise the mayor to reject the proposal on the grounds that it contained too many loopholes.[129]

Still, some unions realized that they had to put the issue of worker health and safety back on the agenda and get health and safety clauses written into union contracts. Table 23 shows the percentage of union contracts that included worker health and safety measures in their contracts for the thirty-year period spanning 1957 to 1987. In 1957 69 percent of all contracts in the manufacturing sector had safety clauses; thirty years later 89 percent did. Though less common in the nonmanufacturing sector, contracts with health and safety clauses increased by 39 percent over the same period. There was a 31 percent increase in the number of contracts in the manufacturing sector that had joint safety committees and a 15 percent increase in the nonmanufacturing sector.[130]

WORKER DISCONTENT AND INCREASED ENVIRONMENTAL ACTIVISM

Despite the fact that unions were making progress on occupational health issues, some workers were dissatisfied about the pace of progress. Toward the end of the 1960s 14,000 of the 90 million workers in the country were dying accidentally each year. Injuries resulted in about 2 million lost workdays annually. In addition, there were 390,000 new cases of occupational illnesses and 100,000 deaths annually stemming from occupational diseases.[131] Consequently unions were pressured by their members to promote worker health and safety more aggressively. Miners for Democracy ousted corrupt officials of the United Mine Workers and exposed the union's refusal to press management on the problem of black lung disease, and insurgent teamsters confronted union leaders about truck safety. As a result in 1973 union members seeking election to leadership positions focused on workplace and community hazards.[132]

Mirroring the national trend, in the late 1960s some union members in Gary openly expressed their dissatisfaction with the union's handling of worker health and safety issues. Having participated in working-class environmental groups seeking to reduce pollution and increase access to recreational facilities, steelworkers wanted to improve working conditions. Recognizing the limits of pursuing environmental reforms through the unions, the steelworkers sought other options. To this end, discontented steelworkers collaborated with local scientists on occupational health initiatives. After establishing contacts with an industrial chemist who offered his services for pollution research, the workers established a lab to test the air and water. Out of this partnership, the Calumet Environmental and Occupational Health Committee (CEOHC) was formed in 1972. The CEOHC investigated health hazards at U.S. Steel and filed complaints with the federal Occupational Safety and Health Administration (OSHA).[133]

Through public education campaigns, the CEOHC heightened workers' awareness of the hazards that abounded in the factory. Workers began seeking technical assistance and filing complaints. For example, at U.S. Steel's pickling plant workers complained of exposure to ammonium chloride fumes. The company argued that the workers did not face any risks and conducted air quality tests that supported its position. However, the CEOHC taught workers how to conduct their own tests and file OSHA complaints. They found that the ammonium chloride levels in the plant were extremely high. In addition the CEOHC pressured federal inspectors to examine the problems in the Gary area.[134]

At the end of 1972 some of the unionists who had been working with the CEOHC formed a new group, the Workers for Democracy (WFD, modeled after Miners for Democracy). The thirty-member steelworker environmental group focused on worker health and safety within the factory; they continued to file OSHA reports without the consent or approval of the union leadership. In December 1972 the WFD requested OSHA inspections of the coke oven batteries, blast furnaces, sintering plant, and plate mill. An OSHA representative toured the coke plant and found that the workers there inhaled levels of coal tar fumes far in excess of legal limits. U.S. Steel was issued a citation and ordered to remedy the situation within a year. Taking advantage of a provision in the Clean Air Act that permitted citizen lawsuits, the WFD then sued U.S. Steel for excessive air pollution. In addition the WFD submitted a set of demands regarding coke oven controls to the steel companies and the U.S. Environmental Protection Agency. In some ways their demands followed the standard union line: distributing hazardous work among more laborers in an effort to reduce the exposure of each individual. But the WFD also requested the replacement of faulty oven doors, the use of clean water for quenching hot metals and coal, continuous medical surveillance of workers, and procedural changes in the baking of coke.[135]

OSHA, THE UNIONS, AND WORKPLACE HAZARDS

In the 1970s the union, feeling pressure from the WFD, began to use OSHA provisions to help improve health and safety conditions at the plant.[136] Nationwide, unions started to include in their contracts environmental protection, safety equipment, the right to refuse hazardous work, and hazard payment clauses. Table 24 shows that from 1970 to 1980 there was a 40 percent increase in the number of manufacturing contracts that had environmental protections. In both the manufacturing and the nonmanufacturing sectors the number of contracts including safety equipment increased; this is particularly true in the nonmanufacturing sector, where inclusion of such clauses increased from 35 percent of all contracts in 1970 to 49.5 percent a decade later. The number of manufacturing contracts with the right to refuse hazardous work had been climbing slowly since 1974, yet only 26 percent of contracts had such provisions in 1980. Although the manufacturing sector was more likely to have

Table 24 Percentage of Union Contracts Containing Environmental Protection Clauses

Safety	1970	1971	1972	1973	1974	1975	1976	1978	1980
Environmental Protection									
Manufacturing	—	11.7	12.6	10.0	11.5	14.0	14.0	15.1	16.4
Nonmanufacturing	—	7.9	4.6	4.7	4.7	5.7	6.7	5.9	6.3
Safety Equipment									
Manufacturing	47.6	52.3	51.7	51.8	52.8	56.3	56.5	58.0	58.7
Nonmanufacturing	35.0	35.5	42.9	43.8	47.0	48.0	46.1	51.0	49.5
Joint Safety Committee									
Manufacturing	39.6	39.2	36.5	39.0	38.2	40.6	44.7	47.6	55.0
Nonmanufacturing	16.7	18.2	15.8	15.3	16.6	16.7	16.3	18.1	19.9
Right to Refuse Hazardous Work									
Manufacturing	—	—	—	—	14.8	16.7	16.6	17.3	26.3
Nonmanufacturing	—	—	—	—	25.0	26.1	24.7	26.5	24.9
Hazard Pay									
Manufacturing	—	9.4	10.7	11.6	11.0	9.6	10.2	10.0	9.8
Nonmanufacturing	—	30.3	29.4	28.8	28.7	31.0	29.8	29.8	31.2

Source: James Robinson, *Toil and Toxics: Workplace Struggles and Political Strategies for Occupational Health* (Berkeley: University of California Press, 1991), 55.

hazardous work, only 10 percent of manufacturing contracts and 33 percent of nonmanufacturing contracts had hazard pay provisions.[137] The demand for safety equipment and safety committees were the two clauses that were most frequently written into union contracts.

Environmental Hazards in the Community

POLLUTION

Despite all the concern about environmental issues in the factory, some of the earliest working-class environmental groups in Gary were not formed inside the factory, nor did they deal with pollution and hazards in the workplace. To combat pollution in the community a white working-class environmental group, the Calumet Community Congress (CCC), was formed. Like the WFD, the CCC saw itself as a radical alternative to organized labor and the entrenched political structures of the city.[138] While the WFD focused on improving conditions inside the factory, the CCC worked outside, organizing community groups in an effort to build a united working-class alliance against Gary's polluters.

Members of the CCC organized themselves around their shared experiences of immigration from Europe, their quest for upward mobility, their feeling of abandonment by organized labor, and their desire for enhanced political power. This strategy helped the CCC to recruit members from a wider circle of people than just steelworkers. Industrial pollution became the group's central issue.[139]

The first meeting of the CCC was held on December 6, 1970; it attracted about fifteen hundred people. Elected delegates from labor unions, Latino organizations, suburban improvement groups, churches, student groups, and women's groups participated. These community organizations would form a pressure group that would break the stronghold that U.S. Steel had over Gary. The group was hailed as a new model for building an effective coalition of activists to challenge corporate polluters and shape environmental policies in local communities; activists and organizations all over the country were eager to emulate the CCC. Dozens of observers, most of whom were from white ethnic organizations wanting to develop similar organizations in Baltimore, Newark, Providence, Chicago, Cleveland, Detroit, and Philadelphia, attended that first meeting.[140]

Many of the activists who organized the CCC received training at the renowned community organizer Saul Alinsky's Industrial Areas Foundation Training Institute in Chicago. Critics of the CCC such as Democrat John Krupa, a dominant figure in county politics, capitalized on the presence of Alinsky-trained activists to refer to the group as "atheistic and Communistic."[141]

At the time the CCC was created, about 80 percent of the residents of Gary were black. They stood a good chance of being left out of environmental decision making if they did not become more proactive in organizing the community. Black organizers planned to start a parallel organization and so monitored the CCC proceedings carefully. One CCC leader, James Wright, met with Obediah Sims, a black organizer, to discuss future organizing. Delegates to the CCC meeting agreed on a ten-point action plan to combat pollution and political corruption, seek tax reform, and improve access to open space.[142]

The CCC also got the attention of national politicians. Senators Edmund Muskie of Maine, Edward M. Kennedy of Massachusetts, and Vance Hartke of Indiana sent telegrams endorsing the organization. The consumer advocate

Ralph Nader invited the CCC to join him in investigating pollution at U.S. Steel. Local churches funded the CCC. Similar groups were formed in Chicago, Pittsburgh, Newark, New York, Baltimore, and Cleveland. A national group, the Task Force on Urban Problems, was also launched.[143]

The CCC was organized around the following issues: flooding, pollution of the storm sewers, lack of access to fishing piers for working-class anglers, and steel mill emissions. The organization enjoyed enthusiastic community support at first. However, within a few days of the formation of the group U. S. Steel was on the offensive, describing its leaders as communists intent on destroying the American way of life. The conflict between U.S. Steel and the CCC escalated quickly as the group organized a direct-action campaign against the corporation. The CCC called attention to the company's environmental violations and sponsored a trip to the suburban home of the U.S. Steel superintendent, J. David Carr, who called the group communist. There the thirty-six CCC members demanded to speak with Carr but were told that he wasn't home. They distributed a flyer with Carr's picture over a caption reading, "Wanted: Pollution Outlaw."[144]

The CCC did not spare Gary's second largest polluter, the Northwest Indiana Public Service Company (NIPSCO, the electric company). At the time NIPSCO's trucks were dumping tons of ash from the settling pits of their coal-fired plant into a marsh on the eastern edge of its complex. Earth-moving equipment then spread the ash over the marsh. In 1971 NIPSCO built a mile-long earthen dike to contain the flow in the marsh. Beyond the dike lay Cowles Bog, the site of Henry Cowles's pioneering turn-of-the-century studies on the principles of ecological plant succession. The naturally acidic Cowles Bog is a registered national landmark and a part of the Indiana Dunes National Lakeshore. When water backed up behind the dike NIPSCO breached the dike and the alkaline ash-contaminated water drained into the bog.[145] The CCC distributed flyers blaming NIPSCO for a dangerous gas explosion in Glen Park, a Gary suburb, and for the degradation of Lake Michigan because of its hot water releases. In 1972 CCC protests convinced public regulators to reduce NIPSCO's requested rate increase. When NIPSCO announced in 1971 that it would build a nuclear power plant opposition from activists helped to stall the process; ten years later and after NIPSCO had spent $206 million on the plant the company announced it was abandoning plans to build the facility.[146]

Despite successfully increasing community awareness of pollution, the CCC received internal and external criticism for its controversial, confrontational, and militant style. Business, labor, and the Democratic Party leadership criticized the group, they lost the support of churches and many community groups, and their membership declined. In response a splinter group, the Calumet Action League, was formed in 1973. Led by one of the original founders of the CCC, the League claimed that the CCC had taken its attack on industry too far and assured supporters that it would take a more moderate approach. Soon after the formation of the League the CCC disbanded.[147]

INCINERATORS AND LANDFILLS

Working-class environmental groups sprang up in other mill towns. Irondalers Against the Chemical Threat (I-ACT) had about forty core activists who mobilized two hundred or more supporters regularly to protest industrial contamination in the communities of South Deering and Hegewisch on the Indiana-Illinois border. In 1982 I-ACT joined a coalition of other groups trying to stop the siting of a 289-acre toxic dump that would have been located a few blocks from South Deering's grammar school. The owner, Waste Management, Inc., already operated two landfills in the area. Activists claimed that the soil and water in the neighborhood was already highly contaminated with heavy metals. Waste Management ignored the citizens' protests and refused to meet with representatives of I-ACT. In desperation I-ACT supporters blocked a nearby Waste Management landfill in Burnham. On two different occasions they formed a human chain across the entrance of the facility, effectively preventing any vehicles from entering or leaving. Police removed and arrested protesters the second time. I-ACT joined forces with activists in the more upscale community of Hegewisch, located four miles south of South Deering, who had been fighting the siting of landfills in the area since 1976. In 1983 activists also tried to stop a chemical treatment plant (located halfway between South Deering and Hegewisch) from burning PCBs. However, despite intense community opposition Waste Management got permits to build its facility. With its legal options exhausted, I-ACT tried to seek a political solution. Though the community worked with their alderman, they had grown frustrated with his support for the proposed facility, even after residents led him on a tour of the proposed site. On August 24, 1983, activists invited the newly elected mayor of Chicago, Harold Washington, to meet with them. Five hundred people crammed into St. Kevin's Church to hear Washington, the first African American mayor of Chicago, say that he was investigating their complaints and that the permits issued to Waste Management had not gone through the proper procedures. After the meeting a temporary ban was placed on construction of landfills in the area until new standards were put in place to protect the water and develop effective buffer zones.[148]

ENVIRONMENTAL ACTIVISM IN GARY'S BLACK COMMUNITY

As in many cities across the country, Gary's blacks linked the struggle for civil rights with the struggle for environmental equality. They fought for access to beaches, parks, and other public recreational facilities. Beginning in the late 1950s labor activists, NAACP representatives, and community residents decided to take action to integrate the city's beaches and parks.[149]

Richard Hatcher, Gary's first African American mayor, spoke to a group of local white middle-class environmental activists from the group Community Action to Reverse Pollution in May 1970. He tried to articulate some of the differences between the black environmental ideology and agenda and that of middle-class whites. He argued that blacks did not share white environmentalists' definition of ecology; for blacks, the relevant environmental issues were poor sanitation, overcrowding, substandard housing, and vermin: "The mothers of poor babies must consider, in their planning for the night, that in their environment there are rats which may bite their children; that there are roaches which crawl over and spoil any food left unprotected." Hatcher rejected what he considered to be the false dichotomy between environmental change and social justice. He argued that a nation wealthy enough to send spaceships to the moon could certainly combat poverty and pollution if the will existed. In his view the causes of poverty and the causes of pollution stemmed from the same root: the "technological and financial powers" that created the problems in the first place, then showed little interest in solving them.[150]

Hatcher's environmental justice perspective resonated with blacks because African Americans have always understood environmental problems in the broader context of social, political, and economic inequalities. Despite the fact that blacks in Gary were better off than many blacks in the South and in other northern cities, persistent racism blocked upward mobility and sparked a movement for social change. Hence workplace hazards and the condition of the physical environment in black communities were components of local civil rights campaigns. During the 1940s black activism and master frames emphasized job discrimination; during the 1950s the focus was on integrating facilities. In the 1960s blacks focused on gaining a greater share of political power, and environmental activism became a more explicit and prominent part of that agenda.[151]

Hatcher used his office as a platform from which to launch an attack on pollution and campaign to heighten blacks' awareness of environmental issues. He emphasized air quality in his speeches, press conferences, and public appearances, dispelling the myth among blacks that pollution was "a white issue." As Hatcher saw it, blacks ought to be concerned about pollution and other environmental issues. At the 1972 National Black Political Convention he urged black activists from across the nation to be wary of their relationship with major corporations that degraded the environment and exploited their workers. In that same year in Gary the African American newspaper *Info* reported the results of a poll that showed that pollution, crime, and drugs were among the top concerns of its readership. African Americans began working to enact environmental changes through a network of existing community organizations.[152]

As public concern over environmental degradation increased blacks began to challenge the polluters. Black residents of Ivanhoe Gardens in Western Gary took on Bucko Construction, an asphalt manufacturer, in the 1970s. During the 1940s and 1950s the neighborhood surrounding the Bucko plant was predominantly white, but the racial composition changed in the 1960s when whites moved away to cleaner neighborhoods and blacks from Midtown moved into the area. Although the company installed dust-collecting devices, dust still filled the air, making it difficult to breathe. Some residents blamed their respiratory ailments on the foul air. In 1972 a local precinct representative circulated a petition among residents demanding that the Gary Air Pollution Control Division attend to the problem. The city responded by persuading the company to overhaul its pollution control system.[153]

Midtown residents were also active on the environmental front. One of Lyndon Johnson's Great Society programs, Model Cities, helped residents of low-income areas facilitate social, economic, and physical improvements in their neighborhoods. Hatcher secured $13 million for the Model Cities program that served Midtown and adjoining neighborhoods. The Model Cities Residents Committee fought blight and air pollution and organized tree-planting initiatives and trash-collection drives.[154]

Theoretical Analysis

THE EFFECTIVENESS OF WORKING-CLASS ACTIVISM

The coalition of working-class activists and their middle- and upper-class allies who confronted the shirtwaist industry and urged thousands of workers to strike used a rhetoric of entitlement in their campaigns. They experienced cognitive liberation; that is, women who were once fearful of losing their jobs began to see and frame their plight in terms of the wider labor market dynamics. They recognized that they were in a split as well as a dual labor market in which they were paid less than men for doing the same work and were deliberately kept at the lowest end of the wage scale. They framed their message in terms of unjust treatment on the job and called for fair play at the hands of their employers, police, detectives, and the courts. They developed a salient identity around dignified womanhood and independence that resonated with many women across the country. Middle- and upper-class activists found they had something in common with working women: they too sought independence. By organizing campaigns and strikes, working-class female garment workers successfully transformed their image to that of intelligent, competent, fearless, worthy contributors to the economy and their households. Through campaigns such as these they helped to influence the social construction of feminism. For these women, feminism wasn't only about getting the vote and liberating oneself from the constraints of the home; it was about the expectation that one should be treated with dignity regardless of class, gender, or ethnicity, and that equality also meant equity in wages.

While political forces (city hall, corporations, police, and judges) were arrayed against the activists, striking garment workers garnered public support by manipulating the media more effectively than their opponents did. The working-class activists showed that petite, peaceful, demurely dressed, white working women were being beaten by cops and hired thugs. This was frequently interspersed with images of well-known aristocratic women, including the daughters of industrial moguls, being dragged off to jail for helping poor women. Working-class women were somewhat successful by aligning themselves with wealthy women who could mobilize both monetary resources and influential people to aid the cause. The working-class female activists also aligned themselves with male-dominated labor unions that were gaining political influence at the time.

Though activists sought fundamental changes in the system, their fragile coalition with upper-class women could not withstand the radical politics expressed by some working-class activists. They ended up capitulating to the factory owners without winning the major concessions they fought for. (Male garment workers won some of these the following year.) After the disastrous Triangle fire activists focused on seeking change through government policies and legislation.

The factory owners may not have enjoyed positive press coverage, but they got the workers to return as a largely nonunionized workforce and for wages the owners set. The factory owners used a variety of countertactics to help their cause. They hired strikebreakers to intimidate, assault, and replace striking workers. They also used conservative Jewish women as well as Italian women to replace the more radical Jews of Eastern European ancestry whom they relied on before the strikes. Unlike other industries the shirtwaist industry did not import black men to replace striking workers. This might be due to the fact that the shirtwaist owners did not have branches of their operations in the South.[155]

Like the garment industry, the steel industry operated with split and dual labor markets. However, unlike the shirtwaist industry, when steel companies were threatened with strikes and unionization drives they moved aggressively to recruit and use blacks to break the strikes. As was the case in the shirtwaist industry, the steel industry used force to intimidate workers. Both industries instigated violence against workers to attract government intervention. Neither the Steelworkers Union nor the ILGWU organized black workers. Blacks were used quite effectively by steel corporations to cripple strikes until the passage of the Wagner-Connery Act and when the Steelworkers Union finally began to organize across racial lines.

During the early twentieth century activists' efforts to curb air pollution focused less on mass mobilization and more on enacting government regulations. Middle-class activists sought incremental reforms in the system as they focused on getting businesses to change their practices and the government to set standards, regulate behavior, and monitor compliance. In the second half of the century working-class activists became more involved in fighting air and water pollution inside and outside of industrial complexes. During the 1960s and 1970s middle- and working-class environmental activists in Gary had trouble bridging the class divide to form lasting environmental coalitions. This was not unique to Gary. Middle-class residents, having fled polluted cities, focused their activism on preservation, limiting or prohibiting development, promulgating zoning laws to maintain the integrity of their communities, limiting access to recreational facilities, and promoting slow growth. While these issues and the way they were framed resonated with middle-class activists, the framing either threatened the working class or was not very salient to them. Furthermore, issues that were foremost on the minds of working-class people, such as air and water pollution in industrial facilities and neighboring communities, workplace hazards, occupational health, and inadequate recreational facilities, were either de-emphasized or missing from middle-class environmental discourses. Thus middle- and working-class groups either ignored or opposed each other or had great difficulty working together.[156]

This being the case, working-class environmental groups in Gary seemed to operate in fits and starts. While the Calumet Community Congress began as a popular group with an injustice frame that resonated with many community residents and institutions, the group fell apart a few years after its founding.[157] Plagued by internal ideological rifts, the inability to develop an effective master frame to attract and sustain a mass following, and disagreements over strategies the organization could not maintain its cohesion or support base. Some of the CCC leaders seem to have misread the extent to which their supporters would back certain types of radical direct-action tactics. As McAdam argues, movement leaders have to ensure that the collective identity and frame they articulate is aligned with those of their supporters.[158] If the leaders get too far ahead of their supporters, that is, if they push for radical actions when supporters are not willing or able to go along, and are not successful in communicating their goals and frames to their supporters, they will lose support. The CCC's support base did not want to engage in certain kinds of radical, direct actions and the organization did not take the time to communicate the goals of and need for such actions to their supporters. Furthermore, they lost ground when critics used their actions to criticize and frame the CCC in unflattering terms. The group was ill-equipped to handle the rapid countermobilization that was organized, and they were not successful at countering or diffusing the tactics and negative framings of their critics.[159]

The Buffington Pier Community Coalition was another short-lived working-class environmental group. Though the group began with a very focused message of gaining access to fishing piers that resonated well with its supporters, this group, a loosely knit aggregation of Gary's West Side residents, did not develop the institutional infrastructure and did not mobilize the resources needed to fight prolonged environmental battles. Consequently, the group disbanded after its first defeat. The Calumet Environmental and Occupational Health Committee and the Workers for Democracy were more effective and long-lived. These groups had very focused frames: workplace safety, a clean environment, and worker health. They worked within established institutional frameworks: the workplace, OSHA, and to some extent the unions. They had a clearly defined target group and support base: workers in the factories. Consequently they had an easier time framing their message so that individual workers could align their identities with the new collective identity being formed. These groups basically borrowed a salient identity, that of a steelworker, and expanded it to include an environmental identity.[160] Though these groups opposed the union's handling of safety issues, they had some support from the union; that is, the union acted as a safety net that protected them from arbitrary firings or other kinds of reprisals that might be directed at workers for their activism.

Irondalers Against the Chemical Threat was successful in building and maintaining a coalition between activists for several years. The group also had the support of the church. Some of the activists were former steelworkers displaced by plant closings. This organization constructed a frame focused on industrial contamination of local communities. They targeted corporations that, though powerful, weren't industrial juggernauts in the communities. They used a mixture of legal, political, and direct-action strategies, and they resorted to radical action only after exhausting legal and political channels. The group used their environmental activism to build solidarity, boost morale, and build the confidence of local residents in small towns shellshocked from the closing of Wisconsin Steel and other facilities. In this respect I-ACT was more successful than the CCC in building a community group to fight pollution. Capitalizing on the political opportunity that presented itself with the election of Harold Washington as mayor of Chicago, I-ACT used the new mayor to call attention to problems in their community and highlight the neglect of their community by political leaders.

Because they had little contact with middle-class environmental groups, some working-class groups foundered when they hit snags. This occurred because the working-class groups were not able to draw on the experience, resources, and networks of the middle-class groups to help them with their campaigns. In contrast, middle-class environmental groups in Gary not only allied themselves to national groups, but were able to tap the resources of well-established groups to aid their causes.[161] These well-established environmental groups drew upon decades of experience in preserving open space, challenging industrial development, opposing pollution, and developing proactive zoning policies.

The asbestos industry exposed thousands of workers to hazardous working conditions. Workers have relied heavily on the courts to adjudicate their conflicts; they have also focused on securing passage of government legislation and securing compensation for their illnesses. Today hundreds of grassroots environmental groups exist in white working-class communities across the country. Similarly, African Americans, Native Americans, Latinos, and Asians have formed hundreds of environmental justice groups all over America. In addition to the traditional working-class issues discussed earlier, ethnic minority environmental groups focus on environmental inequalities, such as the disproportionate number of noxious industrial complexes, hazardous waste facilities, and pollution located in communities of people of color. Some of the white working-class groups are also organized around environmental justice issues. Hence coalitions of minorities and whites have joined forces to work together to improve environmental conditions in their communities.[162]

National polls about environmental quality found that throughout the 1960s increasing numbers of people considered air and water pollution to be serious issues in their communities. In 1970 69 percent of the respondents felt this way about air pollution and 74 percent felt the same about water pollution.[163] Seventy percent of respondents to a different poll thought there was "a lot" or "some" air pollution in their community.[164] In 1972 61 percent of respondents were concerned about reducing water pollution, and 60 percent had concerns about reducing air pollution. Four years later, 57 percent of respondents were concerned about reducing water pollution and 55 percent were concerned about reducing air pollution.[165] It is fair to say that persistent working-class activism around workplace health and safety and environmental hazards has helped to catapult these issues to national prominence. Today toxic substances and environmental hazards are among the top environmental issues in the country.

CONCLUSION

Three centuries ago America was on the brink of urbanization. Since then cities have proliferated and undergone tremendous changes. The urban population grew rapidly and cities expanded outward and upward to accommodate them. Cities also became industrialized. During the eighteenth century and the nineteenth, urban population growth quickly outpaced cities' abilities to cope with the demands placed on them. Rural out-migration fueled the growth of many European and American cities, but American cities also grew because of massive immigration from Europe. Though European cities were forced to wrestle with the consequences of overcrowding earlier than American cities, American cities had a new wrinkle to contend with: in addition to the class tensions that characterized European cities, American cities such as New York had a much wider range of racial and ethnic groups crammed together.

American urbanization was accompanied by enormous social stratification and escalating class, racial, and ethnic tensions. Ethnic and racial groups that had little or no prior contact with each other found themselves stacked on top of each other in deplorable housing, competing fiercely for horrid low-wage jobs, drinking each other's wastes from contaminated water sources, and sifting through each other's trash to eke out a living. At the same time people were trying to understand and cope with their new life circumstances in this volatile mix, cities seemed to be exceeding their ability to accommodate large numbers of people. Consequently environmental issues such as sewage and garbage disposal, clean water, decent and affordable housing, navigable roads, and clean air became urgent problems that cities had to deal with. Resolving these problems required a massive outlay of cash, a centralized planning system, civic governance, and forethought. Cities were woefully lacking in these, and as a result were frequently ravaged by epidemics and wracked by social unrest. Cities lurched from one crisis to another as civic leaders tried to figure out how to make them work.

From the beginning cities had religious, political, economic, and policy elites who played the role of civic leaders and reformers. These reformers covered the ideological spectrum from conservative to progressive. While they were more inclined to frame problems in moral and religious terms at first, it became evident that the resolution of the problems required more than religious reform. Increasingly reformers came to realize that many of the problems were environmental in nature and that their resolution required an environmental approach. Hence by the middle of the nineteenth century environmental activists were very involved in reform work.

The cities have been at the forefront of environmental activism for a long time. Urban environmental activists embarked on initiatives to improve sanitation, waste disposal practices, and public health; provide clean water and recreational opportunities; develop safe and affordable housing and comprehensive zoning; and create a legacy of urban parks and open space. In short, urban activists laid the foundation for many of the environmental campaigns and strategies that were used by wilderness and wildlife activists later on.

Today American cities, like many in Europe, are more racially and ethnically diverse than ever. Class as well as racial and ethnic tensions still pose a challenge to stability. At the same time, civic leaders are stronger, the infrastructure for governance is more powerful, social control is more institutionalized, and civil society is more developed. Urbanites are more tolerant of each other's differences, and there is greater understanding of how to achieve the common good. In addition to stronger governments cities now have better physical infrastructure to serve the populace and are now far more capable of responding to social, political, and economic crises than they were even a century ago. City leaders spend more time forecasting and planning so that urban centers can anticipate and prepare for crises rather than simply react to them.

Modern cities rely on environmental professionals in local, regional, state, and federal agencies to set, monitor, and enforce regulations and plan and develop environmental contingencies. Environmental agencies have budgets to undertake initiatives. Though millions of environmental activists from thousands of environmental organizations still play important roles in policy making the context in which they operate has changed. During the nineteenth century early environmental activists operated in a social and political context in which the activists themselves diagnosed the problems and prescribed the solutions. There was a tremendous vacuum in leadership in government on these issues. Activists often had no regulations or guidelines to support their cause or guide them. Governments were slow to identify and solve problems, and there was little or no institutional infrastructure for dealing with environmental affairs. Things have changed. While government entities are still targets of action for environmental activists and sometimes act slowly, contemporary activists interact with existing government agencies that are charged with overseeing a broad range of environmental concerns.

When eighteenth- and nineteenth-century urbanites consumed water mixed with their neighbors' wastes, the results were devastating outbreaks of diseases. Today American cities are cleaner and the epidemics of the past have all but vanished. Thanks to dramatic improvements in sanitation and a more sophisticated understanding of how diseases are transmitted, people in industrialized countries drink water already used and disposed of by their upstream neighbors without triggering outbreaks of deadly epidemics.

But American cities still face significant problems. Though municipalities no longer dump their trash and sewage in open gutters along the sides of city streets, water pollution is a significant problem in some areas. Today the water is more likely to be contaminated with pesticides and other toxic chemicals than with fecal matter, entrails, or trash. Solid waste disposal is still a problem as many cities run out of landfill space and citizens oppose new facilities.[1] The politics of locating waste dumps and other noxious facilities in cities follows a familiar pattern. As early as the seventeenth century noxious facilities expelled from white neighborhoods were relocated to black communities. Such facilities were allowed to foul the air and water with impunity. In contemporary times there is an increased likelihood that such facilities are located in poor, inner-city neighborhoods inhabited by blacks and other minorities.[2]

While cities try to resolve their waste disposal problems by shipping their wastes to poor, rural, out-of-state (often minority) communities or developing countries, most people realize that these are not long-term viable or sustainable options.[3] Though yellow fever and cholera no longer pose a threat to American urbanites, cancers and other illnesses have supplanted them. In short, the air, water, solid waste disposal systems, housing infrastructure, transportation systems, and public open space are still issues that environmental reformers wrestle with in the twenty-first century.

In general, housing conditions have improved dramatically and laws are in place to protect tenants. However, neighborhoods are still segregated, and there are still pockets of substandard housing that would make nineteenth-century reformers cringe. Most notably the public housing projects that reformers fought long and hard for became uninhabitable in the 1990s. Some of the same conditions that reformers fought to eradicate a century ago—low wages, limited skills, high unemployment, limited employment opportunities, low educational attainment—still plague residents of housing projects. Contemporary housing reformers are pinning their hopes on programs like HOPE (Home Ownership and Opportunity for People Everywhere) VI, which is replacing dilapidated public housing with mixed-income developments to which some former residents of the housing projects have access. Residents of HOPE VI developments have an option to rent or buy.[4]

At the other end of the spectrum the wealthy have not lost their desire to cloister themselves in exclusive residential enclaves. In bustling downtown areas they live in well-guarded condominium complexes and luxury high-rises. In the ruralized suburbs idealized by the likes of Downing, the well-to-do build gated communities or sprawling mansions on rapidly disappearing farmland. The quaint suburbs of the mid-nineteenth century such as Olmsted's Riverside have been enveloped by cities, becoming either a part of the city proper or an inner-ring suburb. Those wishing to bask in the solace of the long, lazy days of summer that Emerson cherished have to drive much farther into the countryside to find such bucolic surroundings.

Cities still struggle to control land use. Such groups as the Fifth Avenue Association were effective in influencing early comprehensive zoning laws and in maintaining the character of Upper Fifth Avenue. The avenue still has the feel of an exclusive mixed commercial and residential district. Chicago's "Magnificent Mile" on Michigan Avenue has also been successful in capturing the essence of this type of urban neighborhood.

Workplace health and safety are still a challenge. Though we have made great strides in understanding the causes of occupational illnesses and reducing workplace hazards, many workers are still being killed or injured on the job. On average, 9,000 U.S. workers sustain disabling injuries on the job each day. In 2004 there were 4.3 million nonfatal injuries and 249,000 job-related illnesses in private industry workplaces; there were 4.4 million nonfatal injuries and illnesses in 2003, a rate of 4.8 cases per 100 equivalent full-time workers. In 2003 the rate was 5.0 per 100. Data from NIOSH show that annually $145 billion is spent on injuries and $26 billion on diseases. The number of work-related fatalities has been declining steadily since 1992, but the 2004 figures showed a 2 percent increase over 2003 figures. In 2004 there were 5,703 fatal work injuries in the United States; the number was 5,575 in 2003. Ninety percent of work-related fatalities occur in private industry. Fatal work injuries occurred at a rate of 4.1 per 100,000 workers; this is a slight increase over the rate of 4.0 recorded for 2002 and 2003. The highest rate recorded since 1992 was 5.3 per 100,000 in 1994.[5]

To respond to these issues environmental activism has grown tremendously over the past century. For several decades this was a heavily middle-class movement, but the movement has grown more diverse in the past three decades.[6] Around the time large numbers of low-income grassroots environmental groups were mobilizing against toxins and other environmental hazards during the 1980s, there was widespread public perception that some of these hazardous conditions posed a large threat to individuals. For instance, in 1989, 69 percent of the respondents in a national poll felt threatened by the disposal of hazardous wastes; 65 percent by contamination of underground water supplies; 60 percent by the pollution of lakes, rivers, and oceans; 60 percent by food additives and pesticides; 58 percent by pollution caused by business and industry; and 52 percent by vehicular pollution. All the figures showed an increase over the 1987 figures.[7] Moreover 82 percent of the respondents in a 1988 Roper poll thought severe air and water pollution would create a "serious problem" twenty-five to fifty years hence.[8] In 1990, 55 percent of respondents thought pollution would be a "very serious threat," and 46 percent thought the quality of their drinking water had worsened in the past five years.[9] In 1998, 53 percent of the respondents in a CBS poll thought the environment would worsen over the next century, while only 15 percent thought it would improve.[10] More than 66 percent of respondents thought that hazardous waste disposal; contamination of ground water supplies; pollution of lakes, rivers, and oceans; and general air pollution posed a "large threat" to the environment.[11]

Despite the large number of environmental groups operating in working-class communities and the increasing effectiveness of those groups in taking action to improve environmental conditions in poor communities, those communities are still threatened by industrial pollution, degraded air and water, hazardous wastes, chemical spills, accidental toxic releases, and explosions, to name a few. Throughout the country contaminated sites abound. More than thirty-five thousand hazardous waste sites have been evaluated by the Environmental Protection Agency since Superfund came into existence in 1980. In 2006, 970 Superfund sites had already been cleaned up, and scores more were in the process of being cleaned up. Yet 1,300 sites still remain on the National Priorities List, a list of the most hazardous sites. It is estimated that it will take an average of $35 million to clean up each site. Superfund sites are found in each state, but New Jersey, Pennsylvania, and California have the largest number. About 57 percent of the sites are related to various industries and businesses, 28 percent are landfills or waste dumps, 11 percent are government sites, and about 4 percent are related to mining operations.[12]

Cities are still trying to keep pace with the demand for adequate access to open space. In the past 150 years urban park administrators have come full circle in the way they think about acquisition of land for and the funding of open space. While earlier government entities considered it their role to fund and administer public parks, since the 1970s there has been a shift away from this position. Increasingly, local governments are slashing park budgets and looking to the private sector to finance public parks. Once again this raises serious questions about access and equity. One hopes that public awareness of these trends will stimulate vigorous debates that will help provide some answers about the nature of public goods such as urban parks and the role of government in safeguarding these goods.

Tempting though it was to write a simpler book, one written solely from a classic environmental perspective, the more I researched the topic, the greater my sense that such a book would do a grave injustice to the topic. The environmental challenges cities face are inextricably connected to their demographic context, social classes, labor market dynamics, and politics. I hope that this account of how American cities confronted and dealt with environmental problems will provide some deeper insights into the problems, broaden our perspective, and help us approach contemporary problems with a greater awareness and understanding of the issues.

NOTES

Introduction

1. For example, see Roderick Nash, *Wilderness and the American Mind*, 3rd ed. (New Haven: Yale University Press, 1982); Steven Fox, *The American Conservation Movement: John Muir and His Legacy* (Madison: University of Wisconsin Press, 1981); Max Oelschlaeger, *The Idea of Wilderness: From Prehistory to the Age of Ecology* (New Haven: Yale University Press, 1991).
2. Stephan Thernstrom, Ann Orlov, and Oscar Handlin, *Harvard Encyclopedia of American Ethnic Groups* (Cambridge, Mass.: Belknap Press, 1980), 869.
3. R. A. Schermerhorn, *Comparative Ethnic Relations: A Framework for Theory and Research* (New York: Random House, 1970); Michael Banton, "United States Racial Ideology as Collective Representation," *Ethnic and Racial Studies* 10, no. 4 (1987); Howard M. Bahr, Bruce A. Chadwick, and Joseph A. Strauss, *American Ethnicity* (Lexington, Mass.: D. C. Heath, 1979).
4. See, for example, Andrew Hurley, *Environmental Inequalities: Class, Race, and Industrial Pollution in Gary, Indiana, 1945–1980* (Chapel Hill: University of North Carolina Press, 1995).
5. Pierre van den Berghe, *Race and Racism: A Comparative Perspective* (New York: Wiley, 1978), 191–93; Banton, "United States Racial Ideology."
6. Louis Wirth, *The Problem of Minority Groups*, edited by R. Linton (New York: Columbia University Press, 1945), 347–49.
7. C. Wagley and M. Harris, *Minorities in the New World* (New York: Columbia University Press, 1967), 10.
8. Harold J. Abramson, *Ethnic Diversity in Catholic America* (New York: Wiley, 1973); Milton M. Gordon, *Human Nature, Class, and Ethnicity* (New York: Oxford University Press, 1978); Werner Sollors, *Beyond Ethnicity, Consent, and Descent in American Culture* (New York: Oxford University Press, 1986); Thernstrom et al., *Harvard Encyclopedia of American Ethnic Groups*, 869; Edna Bonacich, "A Theory of Ethnic Antagonism: The Split Labor Market," in *Social Stratification: Class, Race, and Gender in Sociological Perspective*, edited by David Grusky (Boulder: Westview Press, 1994), 475;

Adalberto Aguirre Jr. and Jonathan H. Turner, *American Ethnicity: The Dynamics and Consequences of Discrimination*, 2nd ed. (New York: McGraw-Hill, 1998), 2; Michael Omi and Howard Winant, *Racial Formation in the United States: From the 1960s to the 1990s*, 2nd ed. (New York: Routledge, 1994), 9–48.

9. Banton, "United States Racial Ideology," 1.
10. P. I. Rose, *The Subject Is Race* (New York: Oxford University Press, 1968); Earl W. Count, *This Is Race: An Anthology Selected from the International Literature on the Races of Man* (New York: Schuman, 1950); Nancy D. Fortney, "The Anthropological Concept of Race," *Journal of Black Studies* 8, no. 1 (1977): 35–54.
11. Banton, "United States Racial Ideology"; Barbara Fields, "Ideology and Race in American History," in *Region, Race, and Reconstruction*, edited by J. M. Kousser and J. M. McPherson (New York: Oxford University Press, 1982), 151; Jeffrey Prager, "American Political Culture and the Shifting Meaning of Race," *Ethnic and Racial Studies* 10, no. 1 (1987): 75.
12. Van den Berghe, *Race and Racism*, 237–38; Michael Banton, *The Idea of Race* (London: Tavistock, 1977); Ernst Mayr and Peter Ashlock, *Principles of Systematic Zoology*, 2nd ed. (New York: McGraw-Hill, 1991).
13. Robert E. Park and Ernest W. Burgess, *Introduction to the Science of Sociology* (Chicago: University of Chicago Press, 1921), 140–41.
14. Edward Wilson and William Brown, "The Subspecies Concept and Its Taxonomic Application," *Systematic Zoology* 2 (1953): 97–111; Frank Livingstone, "On the Nonexistence of Human Races," in *The Concept of Race*, edited by Ashley Montagu (New York: Collier Books, 1964), 46–60.
15. See, for example, Kwame Appiah, *In My Father's House* (New York: Oxford University Press, 1992); Michael Banton, "Analytical and Folk Concepts of Race and Ethnicity," *Racial and Ethnic Studies* 2, no. 2 (1979): 127–38.
16. Van den Berghe, *Race and Racism*, 237–38; Banton, *The Idea of Race*. It should be noted that Robin Andreasen and other scholars still believe that race can be classified biologically at the same time it is perceived to be a social construct, and that both views of race are compatible. See Robin O. Andreasen, "Race: Biological Reality or Social Construct?," *Philosophy of Science* 67 (2000): s653–s666.
17. Omi and Winant, *Racial Formation*, 14–15.
18. See also Thomas F. Gossett, *Race and the History of an Idea in America* (Dallas: Southern Methodist University Press, 1963); Rose, *The Subject Is Race*; Madison Grant, *The Passing of a Great Race* (New York: Charles Scribner's Sons, 1916); Allen Chase, *The Legacy of Malthus* (New York: Knopf, 1977); Daniel J. Kelves, *In the Name of Eugenics: Genetics and the Uses of Human Heredity* (New York: Knopf, 1985).
19. Omi and Winant, *Racial Formation*, 14–16; Robert E. Park, "Race and Culture," in *The Collected Papers of Robert E. Park*, vol. 1, edited by Everett Hughes et al. (Glencoe, Ill.: Free Press, 1950); William Peterson, "Concepts of Ethnicity," in *Ethnicity: Theory and Experience*, edited by Nathan Glazer and Daniel P. Moynihan (Cambridge, Mass.: Harvard University Press, 1975).
20. The race relations cycle was posited by Park. See Park and Burgess, *Introduction to the Science of Sociology*, 735; Park, "Race and Culture."

21. Park and Burgess, *Introduction to the Science of Sociology*, 735; Park, "Race and Culture"; Oscar Handlin, *Boston's Immigrants: A Study of Acculturation* (Cambridge, Mass.: Harvard University Press, 1941); Oscar Handlin, *The Uprooted: The Epic Story of the Great Migrations That Made the American People* (Boston: Little, Brown, 1951); Milton Gordon, *Assimilation in American Life: The Role of Race, Religion, and National Origins* (New York: Oxford University Press, 1964); W. Lloyd Warner and Leo Srole, *The Social Systems of American Ethnic Groups* (New Haven: Yale University Press, 1945); Irving L. Child, *Italian or American? The Second Generation in Conflict* (New Haven: Yale University Press, 1943); Gunnar Myrdal, *An American Dilemma* (New York: Harper and Row, 1944).
22. Horace Kallen, *Culture and Democracy in America* (New York: Boni and Liveright, 1924); Andrew Greeley, *Why Can't They Be Like Us? America's White Ethnic Groups* (New York: Dutton, 1971); Nathan Glazer and Daniel P. Moynihan, *Beyond the Melting Pot: The Negroes, Puerto Ricans, Jews, Italians, Irish of New York City* (Cambridge, Mass.: MIT Press, 1970).
23. Alejandro Portes and Robert D. Manning, "The Immigrant Enclave: Theory and Empirical Examples," in *Social Stratification: Class, Race, and Gender in Sociological Perspective*, edited by David Grusky (Boulder: Westview Press, 1994), 509–10; Omi and Winant, *Racial Formation*, 14–16; Park, "Race and Culture."
24. Omi and Winant, *Racial Formation*, 16–23.
25. William Julius Wilson, *The Declining Significance of Race* (Chicago: University of Chicago Press, 1978); C. West, *Race Matters* (Boston: Beacon Press, 1993); C. Willie, *Caste and Class Controversy on Race and Poverty* (Dix Hills, N.Y.: General Hall, 1989); J. Durant Thomas Jr. and Kathleen H. Sparrow, "Race and Class Consciousness among Lower- and Middle-Class Blacks," *Journal of Black Studies* 27, no. 3 (1997): 334–51.
26. Aguirre and Turner, *American Ethnicity*, 12; E. Ellis Cashmore, "Prejudice," in *Dictionary of Ethnic and Race Relations*, edited by E. Ellis Cashmore (London: Routledge, 1988), 22; Beverly Daniel Tatum, *"Why Are All the Black Kids Sitting Together in the Cafeteria?" and Other Conversations about Race* (New York: HarperCollins, 1999), 7–9.
27. Michael Banton, "Discrimination: Categorical and Statistical," in *Dictionary of Ethnic and Race Relations*, edited by E. Ellis Cashmore (London: Routledge, 1988), 79.
28. Robert K. Merton, "Discrimination and the American Creed," in *Discrimination and National Welfare*, edited by R. H. MacIver (New York: Harper and Row, 1949), 99–126. See also Joseph F. Healey, *Race, Ethnicity, Gender, and Class: The Sociology of Conflict Group Change*, 2nd ed. (Thousand Oaks, Calif.: Pine Forge, 1998), 7–12.
29. Aguirre and Turner, *American Ethnicity*, 13.
30. Tatum, *"Why Are All the Black Kids Sitting Together,"* 11–12.
31. Michael Banton and Robert Miles, "Racism," in *Dictionary of Ethnic and Race Relations*, edited by E. Ellis Cashmore (London: Routledge, 1988), 247.
32. Stokely Carmichael and Charles Hamilton, *The Politics of Liberation in America* (New York: Penguin, 1967); Michael Banton, "Institutional Racism," in *Dictionary of Ethnic and Race Relations*, edited by E. Ellis Cashmore (London: Routledge, 1988), 146.

33. Richard Dyer, "White," *Screen* 29 (1988): 44–64; Jane Flax, *Disputed Subjects* (New York: Routledge, 1993); J. H. Katz and A. E. Ivey, "White Awareness: The Frontier of Racism Awareness Training," *Personnel and Guidance Journal* 55 (1992): 485–88; Michael Omi and Howard Winant, "On the Theoretical Status of the Concept of Race," in *Race, Identity, and Representation in Education*, edited by C. McCarthy and W. Crichlow (New York: Routledge, 1993); Avtar Brah, "Difference, Diversity and Differentiation," in *"Race," Culture, and Difference*, edited by J. Donald and A. Rattansi (London: Sage, 1992), 126–45; Judith Lorber, *Paradoxes of Gender* (New Haven: Yale University Press, 1994); Betsy Lucal, "Oppression and Privilege: Toward a Relational Conceptualization of Race," *Teaching Sociology* 24 (July 1996): 245–55.
34. Tatum, *"Why Are All the Black Kids Sitting Together,"* 3–17; David Wellman, *Portraits of White Racism* (New York: Cambridge University Press, 1977); Peggy McIntosh, "White Privilege: Unpacking the Invisible Knapsack," *Independent School* 49, no. 2 (1990): 31–35; Omi and Winant, *Racial Formation*, 69–76.
35. McIntosh, "White Privilege," 31–35.
36. Tatum, *"Why Are All the Black Kids Sitting Together,"* 3–17; Wellman, *Portraits of White Racism*; McIntosh, "White Privilege," 31–35; Omi and Winant, *Racial Formation*, 69–76.
37. Lerone Bennett Jr., *The Shaping of Black America: The Struggles and Triumphs of African-Americans, 1619 to the 1990s* (New York: Penguin Books, 1993); St. Clair Drake and Horace R. Cayton, *Black Metropolis: A Study of Negro Life in a Northern City* (1945; Chicago: University of Chicago Press, 1993); Eugene D. Genovese, *Roll, Jordan, Roll: The World the Slaves Made* (New York: Vintage Books, 1972); Paula Giddings, *When and Where I Enter: The Impact of Black Women on Race and Sex in America* (New York: Bantam Books, 1984); Hurley, *Environmental Inequalities*; Aldon Morris, *The Origins of the Civil Rights Movement: Black Communities Organizing for Change* (New York: Free Press, 1984); William M. Tuttle, *Race Riot: Chicago in the Red Summer of 1919* (New York: Atheneum, 1970).
38. Dorceta E. Taylor, "The Rise of the Environmental Justice Paradigm: Injustice Framing and the Social Construction of Environmental Discourses," *American Behavioral Scientist* 43, no. 4 (2000): 508–80.
39. For discussions of white privilege, see McIntosh, "White Privilege," 31–35; Tatum, *"Why Are All the Black Kids Sitting Together,"* 3–17; Wellman, *Portraits of White Racism*.
40. Edward K. Spann, *The New Metropolis: New York City, 1840–1857* (New York: Columbia University Press, 1981), 40–41; Richard Bradley, "Nativism and Anglo-Saxonism," in *The Encyclopedia of New England*, edited by Burt Feintuch and David H. Watters (New Haven: Yale University Press, 2005), 388–89.
41. Marilyn Halter and Robert L. Hall, "Ethnic and Racial Identity," in *The Encyclopedia of New England*, edited by Burt Feintuch and David H. Watters (New Haven: Yale University Press, 2005), 330–37; Bradley, "Nativism and Anglo-Saxonism," 388–89.
42. Bonacich, "A Theory of Ethnic Antagonism," 474–76.
43. Marvin Harris, *Patterns of Race in the Americas* (New York: Walker, 1964); Bonacich,

"A Theory of Ethnic Antagonism," 478; Michael Reich, "The Economics of Racism," in *Social Stratification: Class, Race, and Gender in Sociological Perspective*, edited by David Grusky (Boulder: Westview Press, 1994), 469–74.

44. Emilie J. Hutchinson, *Women's Wages* (New York: AMS Press, 1968), 59–61; H. G. Heneman and Dale Yoder, *Labor Economics* (Cincinnati: Southwestern, 1965), 543–44.
45. Bonacich, "A Theory of Ethnic Antagonism," 479.
46. Oliver C. Cox, *Caste, Class, and Race* (New York: Modern Reader, 1948), 411.
47. Michael J. Piore, "The Dual Labor Market: Theory and Implications," in *Social Stratification: Class, Race, and Gender in Sociological Perspective*, edited by David Grusky (Boulder: Westview Press, 1994), 359.
48. Reich, "The Economics of Racism," 469; Thomas S. Moore, *The Disposable Workforce: Worker Displacement and Employment Instability in America* (New York: Aldine de Gruyter, 1996).
49. William Thompson, *An Inquiry into the Principles of the Distribution of Wealth Most Conducive to Human Happiness* (1824; New York: A. M. Kelley, 1963); Thomas Hodgskin, *Labour Defended against the Claims of Capital* (1825; New York: A. M. Kelley, 1963); Karl Marx and Friedrich Engels, *Communist Manifesto*, edited by Robert Tucker (1848; New York: Norton, 1978); Erik Olin Wright, "Race, Class, and Income Inequality," in *Classes, Power, and Conflict: Classical and Contemporary Debates*, edited by Anthony Giddens and David Held (Berkeley: University of California Press, 1982); Benjamin B. Ringer and Elinor R. Lawless, *Race, Ethnicity, and Society* (New York: Routledge, 1989).
50. Max Weber, *Economy and Society 1*, edited by Guenther Ross and Claus Wittich (New York: Bedminster, 1968); Anthony Giddens, *The Class Structure of Advanced Societies* (New York: Harper and Row, 1973); F. Parkin, *Class Inequality and Political Order* (New York: Praeger, 1971); Frank Parkin, *Class, Inequality, and Political Order* (London: Tavistock, 1974); Frank Parkin, *Marxism and Class Theory* (New York: Columbia University Press, 1979); Wright, "Race, Class, and Income Inequality."
51. W. Lloyd Warner, "American Class and Caste," *American Journal of Sociology* 9 (1936): 524–31; W. Lloyd Warner, *Social Class in America* (New York: Harper and Row, 1960); Talcott Parsons, *The Social System* (Glencoe, Ill.: Free Press, 1951); Ralf Dahrendorf, *Class and Class Conflict in Industrial Society* (Palo Alto, Calif.: Stanford University Press, 1959), 139; Gerhard Lenski, *Power and Prestige* (New York: McGraw-Hill, 1966), 95.
52. Pierre Bourdieu, *Distinction: A Social Critique of the Judgment of Taste* (London: Routledge and Kegan Paul, 1984).
53. For discussions of nineteenth-century American class structure, see Robert L. Macieski, "Cities and Suburbs," in *The Encyclopedia of New England*, edited by Burt Feintuch and David H. Watters (New Haven: Yale University Press, 2005), 191; Eric Homberger, *Mrs. Astor's New York: Money and Social Power in a Gilded Age* (New Haven: Yale University Press, 2002), 2.
54. Macieski, "Cities and Suburbs," 191.

55. Michael Spector and John Kitsuse, "Social Problems: A Reformulation," *Social Problems* 20 (1973): 145–59; Bert Klandermans, "The Social Construction of Protest and Multiorganizational Fields," in *Frontiers in Social Movement Theory*, edited by Aldon D. Morris and Carol McClurg Mueller (New Haven: Yale University Press, 1992), 78; John A. Hannigan, *Environmental Sociology: A Social Constructionist Perspective* (New York: Routledge, 1995), 32–33.
56. Joel Best, *Images of Issues: Typifying Contemporary Social Problems* (New York: Aldine de Gruyter, 1989), 252; Hannigan, *Environmental Sociology*, 33.
57. Carol McClurg Mueller, "Building Social Movement Theory," in *Frontiers in Social Movement Theory*, edited by Aldon Morris and Carol McClurg Mueller (New Haven: Yale University Press, 1992), 92: 19–20; Pamela E. Oliver and Gerald Marwell, "Mobilizing Technologies for Collective Action," in *Frontiers in Social Movement Theory*, edited by Aldon D. Morris and Carol McClurg Mueller (New Haven: Yale University Press, 1992), 251–72; Mayer N. Zald, "Culture, Ideology, and Strategic Framing," in *Comparative Perspectives on Social Movements: Political Opportunities, Mobilizing Structures, and Cultural Framings*, edited by Doug McAdam, John D. McCarthy, and Mayer N. Zald (Cambridge: Cambridge University Press, 1996), 267–68.
58. Best, *Images of Issues*, 250; Spector and Kitsuse, "Social Problems," 145–59; Hank Johnston, Enrique Larana, and Joseph Gusfield, "Identities, Grievance, and New Social Movements," in *Social Movements: Perspectives and Issues*, edited by Steven M. Buechler and F. Kurt Cylke Jr. (Mountain View, Calif.: Mayfield, 1997), 286; William A. Gamson, Bruce Fireman, and Steven Rytina, *Encounters with Unjust Authority* (Homewood, Ill.: Dorsey Press, 1982); William A. Gamson, "The Social Psychology of Collective Action," in *Frontiers in Social Movement Theory*, edited by Aldon D. Morris and Carol McClurg Mueller (New Haven: Yale University Press, 1992), 53–76.
59. Joel Best, "Rhetoric in Claims-Making," *Social Problems* 34, no. 2 (1987): 101–21.
60. William A. Gamson and David S. Meyer, "Framing Political Opportunity," in *Comparative Perspectives on Social Movements: Political Opportunities, Mobilizing Structures, and Cultural Framings*, edited by Doug McAdam, John D. McCarthy, and Mayer N. Zald (Cambridge: Cambridge University Press, 1996), 275–90.
61. N. Rafter, "Claims-Making and Socio-Cultural Context in the First U.S. Eugenics Campaign," *Social Problems* 35 (1992): 27.
62. P. R. Ibarra and J. I. Kitsuse, "Vernacular Constituents of Moral Discourse: An Interactionist Proposal for the Study of Social Problems," in *Reconsidering Social Constructionism: Debates in Social Problems Theory*, edited by J. A. Holstein and G. Miller (New York: Aldine de Gruyter, 1993).
63. See Hannigan, *Environmental Sociology*, 35–36.
64. Hannigan, *Environmental Sociology*, 36–37; Roger E. Kasperson, "The Social Amplification of Risk: Progress in Developing an Integrative Framework," in *Social Theories of Risk*, edited by Sheldon Krimsky and Dominic Golding (Westport, Conn.: Praeger, 1992), 152–78; Ortwin Renn, "The Social Arena Concepts of Risk Debates," in *Social Theories of Risk*, edited by Sheldon Krimsky and Dominic Golding (Westport, Conn.: Praeger, 1992), 179–96.

65. Examples of other influential social movements are the civil rights movement, the antinuclear movement, the peace movement, the women's movement, the labor movement, and the temperance movement.

66. John D. McCarthy and Mark Wolfson, "Consensus Movements, Conflict Movements, and the Cooptation of the Civic and State Infrastructure," in *Frontiers in Social Movement Theory*, edited by Aldon Morris and Carol McClurg Mueller (New Haven: Yale University Press, 1992), 275; Gamson and Meyer, "Framing Political Opportunity," 283.

67. Gary T. Marx and Doug McAdam, *Collective Behavior and Social Movements: Process and Structure* (Englewood Cliffs, N.J.: Prentice-Hall, 1994), 11.

68. Gamson, "The Social Psychology of Collective Action," 68; Ralph H. Turner and Lewis M. Killian, *Collective Behavior* (Englewood Cliffs, N.J.: Prentice-Hall, 1987); David A. Snow and Robert D. Benford, "Master Frames and Cycles of Protest," in *Frontiers in Social Movement Theory*, edited by Aldon Morris and Carol McClurg Mueller (New Haven: Yale University Press, 1992), 136–38.

69. D. McAdam and D. Snow, "Social Movements: Conceptual and Theoretical Issues," in *Social Movements: Readings on Their Emergence, Mobilization, and Dynamics*, edited by D. McAdam and D. Snow (Los Angeles: Roxbury, 1997), xix–xx; D. Aberle, *The Peyote Religion among the Navaho* (Chicago: Aldine, 1966).

70. Steven M. Buechler and F. Kurt Cylke Jr., *Social Movements: Perspectives and Issues* (Mountain View, Calif.: Mayfield, 1997), 58–63; G. T. Marx and McAdam, *Collective Behavior*, 81–82. See also William Kornhauser, *The Politics of Mass Society* (Glencoe, Ill.: Free Press, 1959).

71. Mancur Olson, *The Logic of Collective Action, Public Goods, and the Theory of Groups* (Cambridge, Mass.: Harvard University Press, 1965). For critiques of Olson's rational choice theory, see Myra Marx Feree, "The Social Psychology of Collective Action," in *Frontiers in Social Movement Theory*, edited by Aldon D. Morris and Carol McClurg Mueller (New Haven: Yale University Press, 1992), 29–52; Mueller, "Building Social Movement Theory," 3–25; Debra Friedman and Doug McAdam, "Collective Identity and Activism: Networks, Choices and the Life of a Social Movement," in *Frontiers in Social Movement Theory*, edited by Aldon D. Morris and Carol McClurg Mueller (New Haven: Yale University Press, 1992), 156–73.

72. Mueller, "Building Social Movement Theory," 6.

73. Anthony R. Oberschall, *Social Conflict and Social Movements* (Englewood Cliffs, N.J.: Prentice-Hall, 1973); William A. Gamson, *The Strategy of Protest* (Homewood, Ill.: Dorsey Press, 1975); John D. McCarthy and Mayer N. Zald, "Resource Mobilization and Social Movements: A Partial Theory," *American Journal of Sociology* 82 (1977): 1212–41; Charles Tilly, "Getting It Together in Burgundy," *Theory and Society* 4 (1977): 479–504.

74. Mueller, "Building Social Movement Theory," 6–7.

75. Feree, "The Social Psychology of Collective Action," 29; Bert Klandermans, "Mobilizing and Participation: Social Psychological Expansions of Resource Mobilization Theory," *American Sociological Review* 49 (1984): 583–84; Jean L. Cohen, "Strategy or

Identity: New Theoretical Paradigms and Contemporary Social Movements," *Social Research* 52, no. 4 (1985): 668; William A. Gamson and Andre Modigliani, "Media Discourse and Public Opinion on Nuclear Power," *American Journal of Sociology* 95 (1989): 1–38; David A. Snow, Burke Rochford Jr., Steve Worden, and Robert A. Benford, "Frame Alignment Processes, Micro-Mobilization, and Movement Participation," *American Sociological Review* 51 (1986): 464–81; D. A. Snow and Benford, "Master Frames and Cycles of Protest"; Alberto Melucci, *Nomads of the Present: Social Movements and Individual Needs in Contemporary Society* (London: Hutchinson Radius, 1989); Bert Klandermans and Dirk Oegema, "Potentials, Networks, Motivations and Barriers," *American Sociological Review* 52 (1987): 519–31; Doug McAdam and Ronnelle Paulsen, "Specifying the Relationship between Social Ties and Activism," in *Social Movements: Readings on Their Emergence, Mobilization, and Dynamics*, edited by Doug McAdam and Donald Snow (Los Angeles: Roxbury, 1997), 145–57; Gamson et al., *Encounters with Unjust Authority*; Doug McAdam, *Political Process and the Development of Black Insurgency: 1930–1970* (Chicago: University of Chicago Press, 1982).

76. Peter Eisinger, "The Conditions of Protest Behavior in American Cities," *American Political Science Review* 67 (1973): 11–28.
77. McAdam, *Political Process*, 36–59.
78. James Rule and Charles Tilly, "Political Process in Revolutionary France: 1830–1832," in *1830 in France*, edited by John M. Merriman (New York: New Viewpoints, 1975), 41–85; Craig Jenkins and Charles Perrow, "Insurgency of the Powerless: Farm Worker Movements (1946–1972)," *American Sociological Review* 42, no. 2 (1977): 249–68; Charles Tilly, *From Mobilization to Revolution* (Reading, Mass.: Addison Wesley, 1978); Sidney Tarrow, "Struggling to Reform: Social Movements and Policy Change During Cycles of Protest," Western Societies Program Occasional Paper No. 15, New York Center for International Studies, Cornell University, Ithaca, N.Y., 1983.
79. Doug McAdam, "Conceptual Origins, Current Problems, Future Directions," in *Comparative Perspectives on Social Movements: Political Opportunities, Mobilizing Structures, and Cultural Framings*, edited by Doug McAdam, John D. McCarthy, and Mayer Zald (Cambridge: Cambridge University Press, 1996), 23–40. See Charles D. Brockett, "The Structure of Political Opportunities and Peasant Mobilization in Central America," *Comparative Politics* 23, no.3 (April 1991), 253–74; Hanspeter Kreisi, Ruud Koopmans, Jan Willem Duyvendak, and Marco G. Guigni, "New Social Movements and Political Opportunities in Western Europe," *European Journal of Political Research* 22 (1992): 219–44; Sidney Tarrow, *Power in Movement: Social Movements, Collective Action, and Mass Politics in the Modern State* (Cambridge: Cambridge University Press, 1994); Dieter Rucht, "The Impact of National Contexts on Social Movement Structures: A Cross-Movement and Cross National Comparison," in *Comparative Perspectives on Social Movements: Political Opportunities, Mobilizing Structures, and Cultural Framings*, edited by Doug McAdam, John D. McCarthy, and Mayer N. Zald (Cambridge: Cambridge University Press, 1996), 205–26.
80. McAdam, *Political Process*; Frances Fox Piven and Richard A. Cloward, *Poor People's Movements: Why They Succeed, How They Fail* (New York: Vintage Books, 1979);

Frances Fox Piven and Richard A. Cloward, "The Structuring of Protest," in *Social Movements: Perspectives and Issues*, edited by Steven M. Buechler and F. Kurt Cylke Jr. (Mountain View, Calif.: Mayfield, 1997), 326–42.

81. Klandermans and Oegema, "Potentials, Networks, Motivations, and Barriers," 519–31.

82. R. E. Petty and J. T. Cacioppo, *Attitudes and Attitude Change: The Social Judgment-Involvement Approach* (Dubuque, Iowa: W. C. Brown, 1981); Klandermans and Oegema, "Potentials, Networks, Motivations, and Barriers," 519–31; McAdam and Paulsen, "Specifying the Relationship between Social Ties and Activism," 146.

83. Martien Briet, Bert Klandermans, and Frederike Kroon, "How Women Become Involved in the Women's Movement of the Netherlands," in *The Women's Movements of the United States and Western Europe: Consciousness, Political Opportunities, and Public Policy*, edited by Mary Fainsod Katzenstein and Carol McClurg Mueller (Philadelphia: Temple University Press, 1987); Luther P. Gerlach and Virginia H. Hine, *People, Power, and Change: Movements of Social Transformation* (Indianapolis: Bobbs-Merrill, 1970); Max Heirich, "Changes of the Heart: A Test of Some Widely-Held Theories of Religious Conversion," *American Journal of Sociology* 83 (1977): 653–80; Doug McAdam, "Recruitment of High-Risk Activism: The Case of Freedom Summer," *American Journal of Sociology* 82 (1986): 64–90; Anthony Orum, *Black Students in Protest: A Study of the Origins of the Black Student Movement*, Rose Monograph Series (Washington, D.C.: American Sociological Association, 1972); David A. Snow, *The Nichiren Shosu Buddhist Movement in America: A Sociological Examination of Its Value Orientation, Recruitment Efforts, and Spread* (Ann Arbor: University of Michigan Microfilms, 1976); David A. Snow, Louis Zurcher, and Sheldon Eckland-Olson, "Social Networks and Social Movements," *American Sociological Review* 45 (1980): 787–801; Donald Von Eschen, Jerome Kirk, and Maurice Pinard, "The Organizational Substructure of Disorderly Politics," *Social Forces* 49 (1971): 529–44; Louis A. Zurcher and R. George Kirkpatrick, *Citizens for Decency: Antipornography Crusades as Status Defense* (Austin: University of Texas Press, 1976); Charles D. Bolton, "Alienation and Action: A Study of Peace Group Members," *American Journal of Sociology* 78 (1972): 537–61; McAdam and Paulsen, "Specifying the Relationship between Social Ties and Activism," 146–47.

84. Anthony R. Oberschall, "Loosely Structured Collective Conflicts: A Theory and an Application," in *Research in Social Movements, Conflict, and Change*, edited by Louis Kreisberg (Greenwich: JAI Press, 1980), 45–68; McAdam, "Recruitment of High-Risk Activism"; Orum, *Black Students in Protest*; Edward J. Walsh and Rex H. Warland, "Social Movement Involvement in the Wake of a Nuclear Accident: Activists and Free Riders in the TMI Area," *American Sociological Review* 48 (1983): 764–780; McAdam and Paulsen, "Specifying the Relationship between Social Ties and Activism," 146–47.

85. Elaine Sharp, "Citizen Perceptions of Channels for Urban Service Advocacy," *Public Opinion Quarterly* 44 (1980): 362–76; Steven Finkel, "Reciprocal Effects of Participation and Political Efficacy: A Panel Analysis," *American Journal of Political Science* 29,

no. 4 (1985): 891–913; Arthur Neal and Melvin Seeman, "Organizations and Powerlessness: A Test of the Mediation Hypothesis," *American Sociological Review* 29 (1964): 216–26; Cynthia Woolever Sayre, "The Impact of Voluntary Association Involvement on Social-Psychological Attitudes," paper presented at the American Sociological Association, New York City, August 1980; Steven Craig, "Efficacy, Trust and Political Behavior," *American Politics Quarterly* 7, no. 2 (1979): 225–39; Ronelle D. Paulsen, "Educational, Social Class and Participation in Collective Action," *Sociology of Education* 64, no. 2 (1991): 96–110; S. L. Sutherland, *Patterns in Belief and Action: Measurement of Student Political Activism* (Toronto: University of Toronto Press, 1981); Eva F. Travers, "Ideology and Political Participation among High School Students: Changes from 1970 to 1979," *Youth and Society* 13, no. 3 (1982): 327–52.

86. Eisinger, "The Conditions of Protest Behavior," 123; Sidney Verba and N. Nie, *Participation in America: Political Democracy and Social Equity* (New York: Harper and Row, 1972); Arthur Miller, Edie Goldenberg, and Lutz Erbring, "Type-set Politics: Impact of Newspaper on Public Confidence," *American Political Science Review* 73 (1979): 67; Julian Rotter, "Generalized Expectancies for Internal versus External Control of Reinforcements," *Psychological Monographs* 80 (1966): 1–28; Philip Converse, "Change in the American Electorate," in *The Human Meaning of Social Change*, edited by Angus Campbell and Philip Converse (New York: Russell Sage, 1972); George Balch, "Multiple Indicators in Survey Research: The Concept 'Sense of Political Efficacy,'" *Political Methodology* 1 (1974): 1–44; Neal and Seeman, "Organizations and Powerlessness"; Melvin Seeman, "Alienation and Engagement," in *The Human Meaning of Social Change*, edited by Angus Campbell and Philip Converse (New York: Russell Sage, 1972).
87. Sharp, "Citizen Perceptions," 362–76.
88. Gamson and Meyer, "Framing Political Opportunity," 285–86.
89. Sheldon Stryker, "Identity Salience and Role Performance: The Relevance of Symbolic Interaction Theory for Family Research," *Journal of Marriage and the Family* 30 (1968): 558–64; Sheldon Stryker, "Symbolic Interactionism: Themes and Variations," in *Social Psychology: Sociological Perspectives*, edited by Morris Rosenberg and Ralph Turner (New York: Basic, 1981), 23–24.
90. McAdam and Paulsen, "Specifying the Relationship between Social Ties and Activism," 145–57; G. J. McCall and J. L. Simmons, *Identities and Interactions*, 2nd ed. (New York: Free Press, 1978); Morris Rosenberg, *Conceiving the Self* (New York: Basic Books, 1979).
91. Petty and Cacioppo, *Attitudes and Attitude Change*; Klandermans and Oegema, "Potentials, Networks, Motivations, and Barriers," 519–31; McAdam and Paulsen, "Specifying the Relationship between Social Ties and Activism," 146.
92. Erving Goffman, *Frame Analysis* (Cambridge, Mass.: Harvard University Press, 1974); D. A. Snow and Benford, "Master Frames and Cycles of Protest," 133–55; D. A. Snow et al., "Frame Alignment Processes, Micro-Mobilization, and Movement Participation," 464–81; Turner and Killian, *Collective Behavior*, 242; Piven and Cloward, *Poor People's Movements*, 12; Barrington Moore, *Injustice: The Social Bases of Obedience and*

Revolt (White Plains, N.Y.: M. E. Sharpe, 1978), 88; McAdam, *Political Process*, 51; Gamson, "The Social Psychology of Collective Action," 68; Gamson and Meyer, "Framing Political Opportunity," 285.

93. Gamson, "The Social Psychology of Collective Action," 68; Turner and Killian, *Collective Behavior*; D. A. Snow and Benford, "Master Frames and Cycles of Protest," 136–38; D. A. Snow et al., "Frame Alignment Processes, Micro-Mobilization, and Movement Participation," 464–81.
94. Gamson, "The Social Psychology of Collective Action"; Robert B. Zajonc, "Feeling and Thinking: Preferences Need No Inferences," *American Psychologist* 35 (1980): 151–75.
95. Edward J. Walsh, "Resource Mobilization and Citizen Protest in Communities around Three Mile Island," *Social Problems* 29 (1981): 1–21.
96. D. A. Snow et al., "Frame Alignment Processes, Micro-Mobilization, and Movement Participation," 464–81; John D. McCarthy, "Pro-Life and Pro-Choice Mobilization: Infrastructure Deficits and New Technologies," in *Social Movements in an Organizational Society*, edited by Mayer Zald and John McCarthy (New Brunswick, N.J.: Transaction, 1987), 49–66; Gamson, "The Social Psychology of Collective Action," 60.
97. D. A. Snow et al., "Frame Alignment Processes, Micro-Mobilization, and Movement Participation," 464–81; Goffman, *Frame Analysis*.
98. Milton Rokeach, *The Nature of Human Values* (New York: Free Press, 1973); Robin Williams, *American Society: A Sociological Interpretation* (New York: Knopf, 1970); D. A. Snow et al., "Frame Alignment Processes, Micro-Mobilization, and Movement Participation," 464–81.
99. Gamson et al., *Encounters with Unjust Authority*; McAdam, *Political Process*; Piven and Cloward, *Poor People's Movements*; Ralph H. Turner, "The Theme of Contemporary Social Movements," *British Journal of Sociology* 20 (1969): 390–405; Myra Marx Feree and Frederick D. Miller, "Mobilization and Meaning: Toward an Integration of Social Psychological and Resource Perspectives on Social Movements," *Sociological Inquiry* 55 (1985): 36–61; Louis A. Zurcher and David A. Snow, "Collective Behavior: Social Movements," in *Social Psychology, Social Perspectives*, edited by Morris Rosenberg and Ralph Turner (New York: Basic Books, 1981), 447–82; Tamotsu Shibutani, "On the Personification of Adversaries," in *Human Nature and Collective Behavior: Papers in Honor of Herbert Blumer*, edited by Tomatsu Shibutani (Englewood Cliffs, N.J.: Prentice-Hall, 1970), 223–33; Turner and Killian, *Collective Behavior*; Bert Klandermans, "The Expected Number of Participants, the Effectiveness of Collective Action, and the Willingness to Participate: The Free-Riders Dilemma Reconsidered," paper presented at the annual American Sociological Association meeting, Detroit, August 1983; Klandermans, "Mobilizing and Participation," 583–600; Oberschall, "Loosely Structured Collective Conflicts," 45–88; M. Olson, *The Logic of Collective Action*; Bruce Firemen and William Gamson, "Utilitarian Logic in the Resource Mobilization Perspective," in *The Dynamics of Social Movements*, edited by Mayer Zald and John D. McCarthy (Cambridge, Mass.: Winthrop, 1979), 8–44; Pamela E. Oliver, "If You

Don't Do It, Nobody Else Will: Active and Token Contributions to Local Collective Action," *American Sociological Review* 49 (1984): 601–10; D. A. Snow et al., "Frame Alignment Processes, Micro-Mobilization, and Movement Participation," 464–81.

100. D. A. Snow et al., "Frame Alignment Processes, Micro-Mobilization, and Movement Participation," 464–81.
101. McAdam, *Political Process*, 129–31; Friedman and McAdam, "Collective Identity and Activism," 162–63.
102. D. A. Snow and Benford, "Master Frames and Cycles of Protest," 138–41; Harold H. Kelley, *Causal Schemata and the Attribution Process* (Morristown, N.J.: General Learning Press, 1972); C. W. Mills, "Situated Actions and Vocabularies of Motive," *American Sociological Review* 5 (1940): 404–13.
103. D. A. Snow and Benford, "Master Frames and Cycles of Protest," 138–41.
104. Ibid.
105. David Snow and Robert D. Benford, "Ideology, Frame Resonance, and Participant Mobilization," *International Social Movement Research* 1 (1988): 197–217; Walter R. Fisher, "Narration as a Human Communication Paradigm: The Case of Public Moral Argument," *Communication Monographs* 51 (1984): 1–23.
106. Best, "Rhetoric in Claims-Making"; Gamson and Meyer, "Framing Political Opportunity," 275–90; Rafter, "Claims-Making and Socio-Cultural Context in the First U.S. Eugenics Campaign," 27; Ibarra and Kitsuse, "Vernacular Constituents of Moral Discourse"; Hannigan, *Environmental Sociology*, 35–36.
107. See D. E. Taylor, "The Rise of the Environmental Justice Paradigm," 508–80 for further discussion of submerged frames.
108. Paul DiMaggio and Walter W. Powell, "The Iron Cage Revisited: Institutional Isomorphism and Collective Rationality in Organizational Fields," *American Sociological Review* 48 (1983): 147–60.
109. Amos H. Hawley, "Human Ecology," in *International Encyclopedia of the Social Sciences*, edited by David L. Sills (New York: Macmillan, 1968), 328–37.
110. Michael T. Hannan and John H. Freeman, "The Population Ecology of Organizations," *American Journal of Sociology* 82 (1977): 929–64.
111. There are two kinds of isomorphisms: competitive and institutional. See ibid. for a discussion of competitive isomorphism, an approach that focuses on system rationality, market competition, niche change, and fitness measures. Institutional isomorphism is discussed in Rosabeth Moss Kanter, *Commitment and Community* (Cambridge, Mass.: Harvard University Press, 1972), 152–54. Howard E. Aldrich, *Organization and Environments* (Englewood Cliffs, N.J.: Prentice-Hall, 1979) examines the forces pushing organizations in a field toward understanding each other and accommodation with each other and with the outside world.
112. Paul J. DiMaggio, "State Expansion and Organizational Fields," in *Organizational Theory and Public Policy*, edited by Richard H. Hall and Robert E. Quinn (Beverly Hills, Calif.: Sage, 1983); Paul J. DiMaggio, "Structural Analysis of Organizational Fields: A Blockmodel Approach," *Research in Organizational Behavior* 8 (1986): 335–70; DiMaggio and Powell, "The Iron Cage Revisited."

113. Paul J. DiMaggio, "Constructing an Organizational Field as a Professional Project: U.S. Art Museums," in *New Institutionalism in Organizational Analysis*, edited by Walter Powell and Paul J. DiMaggio (Chicago: University of Chicago Press, 1991), 267.
114. Ibid., 267.
115. Aldrich, *Organization and Environments*; Jeffrey Pfeffer and Gerald Salancik, *The External Control of Organizations: A Resource Dependence Perspective* (New York: Harper and Row, 1978); Peter M. Blau and W. Richard Scott, *Formal Organizations* (San Francisco: Chandler, 1962); W. M. Evan, "An Organizational-set: Toward a Theory of Interorganizational Relations," in *Approaches to Organizational Design*, edited by J. D. Thompson (Pittsburgh: University of Pittsburgh Press, 1966), 173–88; Howard E. Aldrich and Peter V. Marsden, "Environments and Organizations," in *Handbook of Sociology*, edited by Neil J. Smelser (Beverly Hills, Calif.: Sage, 1988); Charles Perrow, *Complex Organizations: A Critical Perspective*, 3rd ed. (New York: Random House, 1986); W. Richard Scott, *Organizations: Rational, Natural, and Open Systems*, 2nd ed. (New York: Wiley, 1987); David Whetten, "Interorganizational Relations," in *Handbook of Organizational Behavior*, edited by Jay Lorsch, Howard E. Aldrich, and Jeffrey Pfeffer (Englewood Cliffs, N.J., 1987); Howard E. Aldrich and Jeffrey Pfeffer, "Environments of Organizations," *Annual Review of Sociology* 2 (1976): 79–105.
116. Walter W. Powell and Paul J. DiMaggio, *The New Institutionalism in Organized Analysis* (Chicago: University of Chicago Press, 1991), 67; James G. March and Michael Cohen, *Ambiguity and Choice in Organizations* (Bergen: Universitetsforlaget, 1976); Richard M. Cyert and James G. A. March, *Behavioral Theory of the Firm* (Englewood Cliffs, N.J.: Prentice-Hall, 1963); Magali Sarfatti Larson, *The Rise of Professionalism: A Sociological Analysis* (Berkeley: University of California Press, 1977), 49–52; Randall Collins, *The Credential Society* (New York: Academic Press, 1979); Charles Perrow, "Is Business Really Changing?," *Organizational Dynamics* (Summer 1974): 31–44; Marco Orru, Nicole Woolsey Biggart, and Gary G. Hamilton, "Organizational Isomorphism in East Asia," in *The New Institutionalism in Organizational Analysis*, edited by Walter Powell and Paul DiMaggio, 361–89.
117. Powell and DiMaggio, *The New Institutionalism in Organizational Analysis*, 67; Perrow, "Is Business Really Changing?," 31–44.
118. See DiMaggio, "Constructing an Organizational Field as a Professional Project," 275–76.
119. Ibid., 287; Powell and DiMaggio, *The New Institutionalism in Organizational Analysis*, 64–65; DiMaggio and Powell, "The Iron Cage Revisited."
120. Pfeffer and Salancik, *The External Control of Organizations.*
121. Ibid.
122. Ibid.
123. M. N. Czudnowski, *Political Elites and Social Change* (DeKalb: Western Illinois University Press, 1983); Harold D. Lasswell, "Study of Political Elites," in *Harold D. Lasswell on Political Sociology*, edited by Dwaine Marvick (Chicago: University of Illinois Press, 1977); C. Wright Mills, "The Power Elite," in *Social Stratification: Class, Race, and Gender in Sociological Perspective*, edited by David B. Grusky (Boulder: Westview

Press, 1994), 161–69; Kenneth Prewitt and Alan Stone, *The Ruling Elites: Elite Theory, Power, and American Democracy* (New York: Harper and Row, 1973), 84–85; Anthony Giddens, "Elites and Power," in *Social Stratification: Class, Race, and Gender in Sociological Perspective*, edited by David Grusky (Boulder: Westview Press, 1994), 170–74.

124. Edward Alsworth Ross, *Social Control: A Survey of the Foundations of Order* (New York: Macmillan, 1901). See also Paul Boyer, *Urban Masses and Moral Order in America, 1820–1920* (Cambridge, Mass.: Harvard University Press, 1978), 224–28.

125. Works of European theorists include A Jamison, R. Eyerman, and J. Cramer, *The Making of the New Environmental Consciousness: A Comparative Study of the Environmental Movements in Sweden, Denmark, and the Netherlands* (Edinburgh: University of Edinburgh Press, 1990); A. Jamison, *The Making of the New Environmental Movement in Sweden* (Lund: Research Reports, Department of Sociology, 1987); H. Kitschelt, "Political Opportunity Structures and Political Protest," *British Journal of Political Science* 16, no. 1 (Spring 1986): 57–86; Wolfgang Rudig, "Peace and Ecology Movements in Western Europe," *West European Politics* 11, no. 1 (January 1988): 26–39; Rucht, "The Impact of National Contexts," 205–26; Hanspeter Kreisi, "The Organizational Structure of New Social Movements in Political Context," in *Comparative Perspectives on Social Movements: Political Opportunities, Mobilizing Structures, and Cultural Framings*, edited by Doug McAdam, John D. McCarthy, and Mayer N. Zald (Cambridge: Cambridge University Press, 1996), 152–84; Paulo Donati, *Citizens and Consumers: The Ecology Issue and the 1970s Movement in Italy* (Chestnut Hill, Mass.: Boston College, Department of Sociology, 1988). Those of American theorists include J. O'Brien, "Environmentalism and Mass Movement: Historical Notes," *Radical America* 17 (1983): 2–3; Ron Eyerman and Andrew Jamison, "Environmental Knowledge as an Organizational Weapon: The Case of Greenpeace," *Social Science Information* 2 (1989): 99–118; Walsh and Warland, "Social Movement Involvement in the Wake of a Nuclear Accident," 764–80.

126. Doug McAdam, John D. McCarthy, and Mayer N. Zald, "Introduction: Opportunities, Mobilizing Structures, and Framing Processes—Toward a Synthetic," in *Comparative Perspective on Social Movements: Political Opportunities, Mobilizing Structures, and Cultural Framings*, edited by Doug McAdam, John D. McCarthy, and Mayer N. Zald (Cambridge: Cambridge University Press, 1996), 1–20.

127. Briet et al., "How Women Become Involved in the Women's Movement of the Netherlands"; Gerlach and Hine, *People, Power, and Change*; Heirich, "Changes of the Heart"; McAdam, "Recruitment of High-Risk Activism"; Orum, *Black Students in Protest*; D. A. Snow, *The Nichiren Shosu Buddhist Movement*; D. A. Snow et al., "Social Networks and Social Movements"; Von Eschen et al., "The Organizational Substructure," 529–44; Zurcher and Kirkpatrick, *Citizens for Decency*; Bolton, "Alienation and Action"; McAdam and Paulsen, "Specifying the Relationship between Social Ties and Activism," 146–47.

128. McAdam, "Conceptual Origins, Current Problems, Future Directions," 23–40.

1. The Evolution of American Cities

1. Louis P. Tremante, "Farming," in *The Encyclopedia of New York City*, edited by Kenneth T. Jackson (New Haven: Yale University Press, 1995), 390–91.
2. Thomas Janvier, *In Old New York* (1894; New York: St. Martin's Press, 2000), xvi.
3. Edwin G. Burrows, and Mike Wallace, *Gotham: A History of New York City to 1898* (New York: Oxford University Press, 1999), 41–43; John S. Abbott, *Peter Stuyvesant: The Last Dutch Governor of New Amsterdam* ([1873] Whitefish, Mont.: Kessinger Publishing, 2004), 125.
4. Burrows and Wallace, *Gotham*, 41, 43; James E. Mooney, "Stuyvesant, Peter," and Craig D. Bida, "Paving," both in *The Encyclopedia of New York City*, edited by Kenneth T. Jackson, 1133–34, 886–87.
5. Burrows and Wallace, *Gotham*, 43; Charles E. Rosenberg, *The Cholera Years: The United States in 1832, 1849, 1866* (Chicago: University of Illinois Press, 1987), 17.
6. Michael R. Corbett, "Meatpacking," in *The Encyclopedia of New York City*, edited by Kenneth T. Jackson, 745–46.
7. Burrows and Wallace, *Gotham*, 43–44; Donald J. Cannon, "Firefighting," in *The Encyclopedia of New York City*, edited by Kenneth T. Jackson, 408–12.
8. Burrows and Wallace, *Gotham*, 46.
9. Burrows and Wallace, *Gotham*, 44, 46.
10. Burrows and Wallace, *Gotham*, 44–45; Clara J. Hemphill and Raymond A. Mohl, "Poverty," in *The Encyclopedia of New York City*, edited by Kenneth T. Jackson, 932–34.
11. Burrows and Wallace, *Gotham*, 156–57.
12. In 1730, when the population of the city had reached 8,622 (7,045 whites and 1,577 blacks), the richest 10 percent of the population, about 140 merchants and landowners, held almost half of the taxable wealth. Burrows and Wallace, *Gotham*, 88, 144; Kenneth T. Jackson and David S. Dunbar, *Empire City: New York through the Centuries* (New York: Columbia University Press, 2002), 18.
13. L. J. Krizner and Lisa Sita, *Peter Stuyvesant: New Amsterdam and the Origins of New York* (New York: Rosen Publishing Group, 2002), 1–85; Burrows and Wallace, *Gotham*, 72–85; Janvier, *In Old New York*, 25; Corbett, "Meatpacking," 745–46.
14. Burrows and Wallace, *Gotham*, 84–85; Margaret Latimer, "Bowling Green," and Laura Lewison, "Lawn Bowling," both in *The Encyclopedia of New York City*, edited by Kenneth T. Jackson, 132, 657; Abigail Franks, to Napthali Franks, 1734, in *Letters of the Franks Family (1733–1748)*, edited by Leo Hershkowitz and Isidore S. Meyer (Waltham, Mass.: American Jewish Historical Society, 1968), 27.
15. Janvier, *In Old New York*, xvi, 25, 31.
16. Ibid., 47, 50, 53–54; Jackson and Dunbar, *Empire City*, 19.
17. Michael Sletcher, *New Haven: From Puritanism to the Age of Terrorism* (Charleston: Arcadia, 2004), 65; Keith N. Morgan and Richard M. Candee, "Architecture," Joseph S. Wood, "Village Commons and Greens," and Robert J. Imholt, "New Haven, Conn.," all in *The Encyclopedia of New England*, edited by Burt Feintuch and David H. Watters, 66, 618–19, 237; Thomas J. Farnham, "New Haven, 1638 to 1690," in *New Haven: An Illustrated History*, edited by Floyd Shumway and Richard Hegel, 12, 14;

Leland M. Roth, *A Concise History of American Architecture* (Boulder: Westview Press, 1980), 22–23.

18. Farnham, "New Haven," 12–13; Sletcher, *New Haven*, 84, 110.
19. Steven H. Corey, "Waste Management," in *The Encyclopedia of New England*, edited by Burt Feintuch and David H. Watters, 620–21.
20. Ibid. Oliver Ayer Roberts, *History of the Military Company of Massachusetts, Now Called, the Ancient and Honorable Artillery Company of Massachusetts: 1637–1888* (Boston: A. Mudge and Son, 1895), 406.
21. Elizabeth McNulty, *Boston Then and Now* (San Diego: Thunder Bay Press, 1999), 5; Sam Bass Warner, "A Brief History of Boston," and Nancy S. Seaholes and Amy Turner, "Diagramming the Growth of Boston," both in *Mapping Boston*, edited by Alex Kreiger, David Cobb, and Amy Turner (Cambridge, Mass.: MIT Press, 2001), 3, 6, 16–17; Corey, "Waste Management," 620–21.
22. S. Lawrence Dingman, "Water Supply," in *The Encyclopedia of New England*, edited by Burt Feintuch and David H. Watters, 625–26.
23. Alex Kreiger, "Experiencing Boston: Encounters with the Places on the Maps," in *Mapping Boston*, edited by Alex Kreiger, David Cobb, and Amy Turner, 156; Ronald Dale Karr, "Boston Common," in *The Encyclopedia of New England*, edited by Burt Feintuch and David H. Watters, 212–13.
24. Macieski, "Cities and Suburbs," 192.
25. Mary Maples Dunn and Richard S. Dunn, "The Founding, 1681–1701," in *Philadelphia: A 300-Year History*, edited by Russell F. Weigley, 1–10.
26. Ibid., 10; Roth, *A Concise History of American Architecture*, 23–24.
27. Dunn and Dunn, "The Founding, 1681–1701," 11, 14–16.
28. David Paul Nord, "Readership as Citizenship in Late-Eighteenth-Century Philadelphia," in A *Melancholy Scene of Devastation: The Public Response to the 1793 Yellow Fever Epidemic*, edited by J. Worth Estes and Billy G. Smith (Canton, Mass.: Science History Publications/USA, 1997), 21; Dunn and Dunn, "The Founding, 1681–1701," 11–16, 25; Edwin B. Bronner, "Village into Town, 1701–1746," in *Philadelphia: A 300-Year History*, edited by Russell F. Weigley, 44.
29. Dunn and Dunn, "The Founding, 1681–1701," 11; Bronner, "Village into Town," 61. There were fires in Philadelphia, but not on the same scale as in New York, Boston, and other cities.
30. Dunn and Dunn, "The Founding, 1681–1701," 20; Bronner, "Village into Town," 58–59; Harry M. Tinkcom, "The Revolutionary City, 1765–1783," in *Philadelphia: A 300-Year History*, edited by Russell F. Weigley, 151.
31. Theodore Thayer, "Town into City, 1746–1765," in *Philadelphia: A 300-Year History*, edited by Russell F. Weigley, 102.
32. Sherry H. Olson, *Baltimore: The Building of an American City* (Baltimore: Johns Hopkins University Press, 1997), 1, 10, 18–19, 29, 34, 36.
33. Macieski, "Cities and Suburbs," 191.
34. See City Homes Association, *Tenement Conditions in Chicago* (Chicago: City Homes Association, 1901); Elizabeth Blackmar, *Manhattan for Rent: 1785–1850* (Ithaca, N.Y.: Cornell University Press, 1989), 14–43; Spann, *The New Metropolis*; Amy Bridges,

A City in the Republic: Antebellum New York and the Origins of Machine Politics (New York: Cambridge University Press, 1984); Boyer, *Urban Masses*, 4–11; John F. Bauman, "Introduction: The Eternal War of the Slums," in *From Tenements to the Taylor Homes: In Search of an Urban Housing Policy in Twentieth-Century America*, edited by John F. Bauman, Roger Biles, and Kristin M. Szylvian (University Park: Pennsylvania State University Press, 2000), 5; S. H. Olson, *Baltimore*, 37.

35. Sherrill D. Wilson and Larry A. Greene, "Blacks," in *The Encyclopedia of New York City*, edited by Kenneth T. Jackson, 112–15.
36. A bushel is approximately one basketful. Burrows and Wallace, *Gotham*, 32–33; S. D. Wilson and Greene, "Blacks," 112–15; Leslie M. Harris, *In the Shadow of Slavery: African Americans in New York City, 1626–1863* (Chicago: University of Chicago Press, 2002), 233–24.
37. Burrows and Wallace, *Gotham*, 32–33.
38. S. D. Wilson and Greene, "Blacks," 112–15; Harris, *In the Shadow of Slavery*.
39. Tyler Anbinder, *Five Points: The 19th-Century New York City Neighborhood That Invented Tap Dance, Stole Elections, and Became the World's Most Notorious Slum* (New York: Free Press, 2001), 14–15; Burrows and Wallace, *Gotham*, 32–33; Carol Groneman, "Collect," and Eric A. Goldstein and Mark A. Izeman, "Water," both in *The Encyclopedia of New York City*, edited by Kenneth T. Jackson, 250, 1244–46; Janvier, *In Old New York*, 53–55; Jackson and Dunbar, *Empire City*, 17. In 1640 war broke out between Indians and whites; hostilities continued for several years. See Jackson and Dunbar, *Empire City*, 28.
40. Goldstein and Izeman, "Water," 1244–46.
41. Anbinder, *Five Points*, 14–15; Groneman, "Collect," 250; Corbett, "Meatpacking," 745–46.
42. Anbinder, *Five Points*, 15–16, 19.
43. Wallace and Burrows, *Gotham*, 475–76; Jacob Riis, *How the Other Half Lives* (1890; New York: Penguin, 1997), 14, 20; Corbett, "Meatpacking," 745–46.
44. Homberger, *Mrs. Astor's New York*, 33; David Sinclair, *Dynasty: The Astors and Their Times* (London: J. M. Dent, 1983), 178–79.
45. Sletcher, *New Haven*, 11, 50–54; Judith A. Schiff, "The Social History of New Haven," in *New Haven: An Illustrated History*, edited by Floyd Shumway and Richard Hegel, 96; J. Ritchie Garrison, "Agriculture," and Pamela Snow, "Agriculture in the Precolonial and Colonial Eras," both in *The Encyclopedia of New England*, edited by Burt Feintuch and David H. Watters, 4–5, 23–24.
46. Macieski, "Cities and Suburbs," 191; Paul Albert Cyr, "New Bedford, Mass.," and J. Stanley Lemons, "Newport, R.I.," both in *The Encyclopedia of New England*, edited by Burt Feintuch and David H. Watters, 236–37, 237–38.
47. John H. Ellis, *Yellow Fever and Public Health in the New South* (Lexington: University of Kentucky Press, 1992), 22, 61; Craig E. Colten, "Basin Street Blues: Drainage and Environmental Equity in New Orleans, 1890–1930," *Journal of Historical Geography* 28, no. 2 (2002): 237–38, 242; E. C. Richard, *Louisiana: An Illustrated History* (Baton Rouge: Louisiana Public Broadcasting, 2003), 7.
48. Colten, "Basin Street Blues," 238–40, 245–48, 250.

49. Ibid., 251; *Land Development Company of Louisiana v. City of New Orleans*, 1926, 13 F. 2d 898, U.S. Dist., July 8; *Land Development Company of Louisiana v. City of New Orleans*, 1927, 17 F. 2d 1016, U.S. App., April 1.
50. *Buchanan v. Warley*, 1917, 245 U.S. 60, 38 S. Ct. 16, 62 L. Ed. 149, November 5.
51. Ellis, *Yellow Fever and Public Health in the New South*, 142.
52. Ibid., 130–31, 142–43.
53. Ibid., 22.
54. S. H. Olson, *Baltimore*, 49.
55. John Radford, "Race, Residence and Ideology: Charleston, South Carolina in the Mid-Nineteenth Century," *Journal of Historical Geography* 2 (1976): 329–46; John Kellog, "Negro Clusters in Postbellum South," *Geographical Review* 67 (1977): 313–21; Colten, "Basin Street Blues," 237, 239–42.
56. Colten, "Basin Street Blues," 240.
57. Paul A. Gilje, "Riots," in *The Encyclopedia of New York City*, edited by Kenneth T. Jackson, 1006–8; Sletcher, *New Haven*, 30–31.
58. Gilje, "Riots," 1006–8.
59. Burrows and Wallace, *Gotham*, 473.
60. Gilje, "Riots," 1006–8.
61. Anbinder, *Five Points*, 7–12.
62. Ibid.
63. Gilje, "Riots," 1006–8.
64. Ibid.; Boyer, *Urban Masses*, 69; Spann, *The New Metropolis*, 67–91, 205–41; Blackmar, *Manhattan for Rent*, 175; H. W. Brands, *T. R.: The Last Romantic* (New York: Basic Books, 1997), 15.
65. Gilje, "Riots," 1006–8.
66. Sletcher, *New Haven*, 113; Schiff, "The Social History of New Haven," 101.
67. Sletcher, *New Haven*, 58, 62; Dorothy Ann Lipson, "From Puritan Village to Yankee City, 1690 to 1860," in *New Haven: An Illustrated History*, edited by Floyd Shumway and Richard Hegel, 42.
68. Halter and Hall, "Ethnic and Racial Identity," 330–37; Bradley, "Nativism and Anglo-Saxonism," 388–89; Joseph Edgar Chamberlin, *The Boston Transcript: A History of Its First Hundred Years* (Manchester, N.H.: Ayer Publishing, 1969), 37–47.
69. Macieski, "Cities and Suburbs," 198.
70. Bronner, "Village into Town," 60.
71. W. E. B. Du Bois, *The Philadelphia Negro: A Social Study* (1899; Philadelphia: University of Pennsylvania Press, 1996), 28.
72. Ibid., 29–30; Nicholas B. Wainwright, "The Age of Nicholas Biddle, 1825–1841," in *Philadelphia: A 300-Year History*, edited by Russell F. Weigley, 295–96.
73. Elizabeth M. Geffen, "Industrial Development and Social Crisis, 1841–1854," in *Philadelphia: A 300-Year History*, edited by Russell F. Weigley, 337–38, 353; Burrows and Wallace, *Gotham*, 473.
74. S. H. Olson, *Baltimore*, 98–100.
75. Ellis, *Yellow Fever and Public Health in the New South*, 23; Gerald M. Capers, *The*

Biography of a River Town: Memphis, Its Heroic Age (Chapel Hill: University of North Carolina Press, 1939), 181, 177–78.

76. Ellis, *Yellow Fever and Public Health in the New South*, 23.

2. *Epidemics, Cities, and Environmental Reform*

1. John Duffy, *The Sanitarians: A History of American Public Health* (Urbana: University of Illinois Press, 1990), 38; Ellis, *Yellow Fever and Public Health in the New South*, 6, 31; Margaret Humphreys, *Yellow Fever and the South* (Baltimore: Johns Hopkins University Press, 1992), 2, 18–19; Kenneth R. Foster, Mary F. Jenkins, and Ana Coxe Toogood, "Introduction to the 1993 Edition," in *Bring Out Your Dead: The Great Plague of Yellow Fever in Philadelphia in 1793*, edited by John Harvey Powell (1949; Philadelphia: University of Pennsylvania Press, 1993), xiii.
2. J. Worth Estes, "Introduction: The Yellow Fever Syndrome and its Treatment in Philadelphia, 1793," and Billy G. Smith, "Comment: Disease and Community," both in *A Melancholy Scene of Devastation: The Public Response to the 1793 Philadelphia Yellow Fever Epidemic*, edited by J. Worth Estes and Billy G. Smith (Canton, Mass.: Science History Publications/USA), 2–7, 149; Humphreys, *Yellow Fever and the South*, 6; Powell, *Bring Out Your Dead*, xx.
3. Ellis, *Yellow Fever and Public Health in the New South*, 31; B. G. Smith, "Comment: Disease and Community," 150; Susan E. Klepp, "Appendix 1: 'How Many Precious Souls are Fled?' The Magnitude of the 1793 Yellow Fever Epidemic," in *A Melancholy Scene of Devastation: The Public Response to the 1793 Philadelphia Yellow Fever Epidemic*, edited by J. Worth Estes and Billy G. Smith (Canton, Mass.: Science History Publications/USA, 1997), 168–70.
4. Powell, *Bring Out Your Dead*, xvii, 11–12, 93; Foster et al., "Introduction to the 1993 Edition," ix; Nord, "Readership as Citizenship in Late-Eighteenth-Century Philadelphia," 21; Sally F. Griffith, "A Total Dissolution of the Bonds of Society: Community Death and Regeneration in Mathew Carey's *Short Account of the Malignant Fever*," in *A Melancholy Scene of Devastation: The Public Response to the 1793 Philadelphia Yellow Fever Epidemic*, edited by J. Worth Estes and Billy G. Smith (Canton, Mass.: Science History Publications/USA, 1997), 45.
5. Humphreys, *Yellow Fever and the South*, 18–19; Foster et al., "Introduction to the 1993 Edition," ix, xii, xviii; P. M. Kollock, "Notes on the Epidemic Fever of 1854," *Southern Medical and Surgical Journal* 11 (1855): 469–70; Powell, *Bring Out Your Dead*, 1–2, 13–14; Nord, "Readership as Citizenship in Late-Eighteenth-Century Philadelphia," 24; Michael McMahon, "Beyond Therapeutics: Technology and the Question of Public Health," in *A Melancholy Scene of Devastation: The Public Response to the 1793 Yellow Fever Epidemic*, edited by J. Worth Estes and Billy G. Smith (Canton, Mass.: Science History Publications/USA, 1997), 102.
6. McMahon, "Beyond Therapeutics," 104; Powell, *Bring Out Your Dead*, 91; Mathew Carey, *A Short Account of the Malignant Fever, Lately Prevalent in Philadelphia: With a Statement of the Proceedings That Took Place on the Subject in Different Parts of the United*

States (Philadelphia: Mathew Carey Publishing, 1793); B. G. Smith, "Comment," 150–51.

7. Steven Johnson, *The Ghost Map: The Story of London's Most Terrifying Epidemic—and How It Changed Science, Cities, and the Modern World* (New York: Riverhead Press, 2006), 86, 101, 111–36.
8. Powell, *Bring Out Your Dead,* 20–21, 39; Martin S. Pernick, "Politics, Parties, and Pestilence: Epidemic Yellow Fever in Philadelphia and the Rise of the First Party System," in *A Melancholy Scene of Devastation: The Public Response to the 1793 Philadelphia Yellow Fever Epidemic,* edited by J. Worth Estes and Billy G. Smith (Canton, Mass.: Science History Publications/USA, 1997), 122.
9. Matthew Clarkson, letter to the Philadelphia city commissioners, August 22, 1793, in *Federal Gazette,* August 30, 1793. Clarkson sent another letter reiterating these ideas on August 27; see Matthew Clarkson, letter to the Philadelphia city commissioners, August 27, *Federal Gazette,* August 30, 1793.
10. Powell, *Bring Out Your Dead,* 39.
11. Elizabeth Carpenter Piechocinski, *The Old Burying Ground: Colonial Park Cemetery* (Savannah, Ga.: Oglethorpe Press, 1999), 1–2.
12. Pernick, "Politics, Parties, and Pestilence," 121–22; Powell, *Bring Out Your Dead,* 4–6.
13. Griffith, "A Total Dissolution of the Bonds of Society," 50–54; Phillip Lapsansky, "'Abigail, a Negress': The Role of the Legacy of African Americans in the Yellow Fever Epidemic," in *A Melancholy Scene of Devastation: The Public Response to the 1793 Philadelphia Yellow Fever Epidemic,* edited by J. Worth Estes and Billy G. Smith (Canton, Mass.: Science History Publications/USA, 1997), 64–68, 71; Absalom Jones and Richard Allen, *A Narrative of the Proceedings of the Black People, during the Late Awful Calamity in Philadelphia in the Year 1793; and A Refutation of Some Censures, Thrown upon Them in Some Late Publications* (Philadelphia: William W. Woodward at Franklin's Head, 1794), 4, 10; Julie Winch, *Philadelphia's Black Elite: Activism, Accommodation, and the Struggle for Autonomy, 1787–1848* (Philadelphia: Temple University Press, 1988), 10, 15–16; Foster et al., "Introduction to the 1993 Edition," x–xi; Powell, *Bring Out Your Dead,* 56–57, 64–65, 94–96; McMahon, "Beyond Therapeutics," 99. Rush's sister died of yellow fever; Rush himself caught the fever but survived.
14. Jones and Allen, *A Narrative of the Proceedings of the Black People,* 4.
15. Ibid., 14; Powell, *Bring Out Your Dead,* 98–99.
16. Nord, "Readership as Citizenship in Late-Eighteenth-Century Philadelphia," 26; Lapsansky, "'Abigail, a Negress,'" 67–68; Jones and Allen, *A Narrative of the Proceedings of the Black People,* 4, 7, 15; Winch, *Philadelphia's Black Elite,* 16.
17. Some blacks, especially those who had migrated to the city from the West Indies, as well as some whites, were immune to the epidemic because of prior exposure to the fever. According to one estimate the fever struck 11 percent of blacks and 14 percent of whites. Powell, *Bring Out Your Dead,* 95; Lapsansky, "'Abigail, a Negress,'" 66–68.
18. Powell, *Bring Out Your Dead,* 94–95, 97, 99; Carey, *A Short Account of the Malignant Fever,* 10, 78. Not only did Carey's account of the epidemic, which sold over ten thousand copies within months of its publication, ignore the positive contributions

of blacks during the epidemic, but he accused blacks of extortion and looting. Jones and Allen, *A Narrative of the Proceedings of the Black People*, 6–8, 21–23. Jones and Allen wrote their essay to rebut Carey's account of the epidemic and to respond to widespread accusations of wrongdoing. The quote regarding the poisoning of wells is in Nord, "Readership as Citizenship in Late-Eighteenth-Century Philadelphia," 26. For excerpts from Elizabeth Drinker's diary, see Elaine Forman Crane, *The Diary of Elizabeth Drinker* (Boston: Northeastern University Press, 1991), 112–23.

19. Powell, *Bring Out Your Dead*, 47, 93–94; Carey, *A Short Account of the Malignant Fever*.
20. Janvier, *In Old New York*, 139–44; Homberger, *Mrs. Astor's New York*, 72–73.
21. Copied from Janvier, *In Old New York*, 140.
22. Homberger, *Mrs. Astor's New York*, 72–73.
23. Copied from ibid., 73.
24. Janvier, *In Old New York*, 140.
25. Corey, "Waste Management," 620–21.
26. Humphreys, *Yellow Fever and the South*, 2, 4–5, 18.
27. Ibid., 5; Ellis, *Yellow Fever and Public Health in the New South*, 41–43, 46–47, 57.
28. Ellis, *Yellow Fever and Public Health in the New South*, 46–50.
29. Ibid., 41–43.
30. Humphreys, *Yellow Fever and the South*, 49; Duffy, *The Sanitarians*.
31. Piechocinski, *The Old Burying Ground*, 11–13.
32. Humphreys, *Yellow Fever and the South*, 6–7; Ellis, *Yellow Fever and Public Health in the New South*, 31.
33. Powell, *Bring Out Your Dead*, 14, 61.
34. Ibid., 22–23. For example, see *Federal Gazette*, September 13, 1793.
35. Doctors feared that the incessant tolling of the bells to signal the death of new victims was causing undue stress among the population. See Powell, *Bring Out Your Dead*, 43.
36. "Preventative against the Raging Yellow Fever," *Federal Gazette*, September 13, 1793.
37. Powell, *Bring Out Your Dead*, 46.
38. Humphreys, *Yellow Fever and the South*, 60–61, 89; Ellis, *Yellow Fever and Public Health in the New South*, 43.
39. Humphreys, *Yellow Fever and the South*; C. E. Rosenberg, *The Cholera Years*, 8–10.
40. Chad Ludginton, "Quarantine Riots," in *The Encyclopedia of New York City*, edited by Kenneth T. Jackson (New Haven: Yale University Press, 1995), 965–66; Harris, *In the Shadow of Slavery*, 23–24.
41. Humphreys, *Yellow Fever and the South*, 5; Ellis, *Yellow Fever and Public Health in the New South*, 31; Foster et al., "Introduction to the 1993 Edition," xii–xiii; Powell, *Bring Out Your Dead*, xix; B. G. Smith, "Comment," 149.
42. Humphreys, *Yellow Fever and the South*, 49.
43. Powell, *Bring Out Your Dead*, 43; 86–87; Pernick, "Politics, Parties, and Pestilence," 126–27; Jacqeulyn C. Miller, "Passions and Politics: The Multiple Meanings of Benjamin Rush's Treatment for Yellow Fever," in *A Melancholy Scene of Devastation: The Public Response to the 1793 Philadelphia Yellow Fever Epidemic*, edited by J. Worth

Estes and Billy G. Smith (Canton, Mass.: Science History Publications/USA), 86–87. Genêt was the young French minister to the United States who tried to push Americans to sympathize with French revolutionaries. He recruited Americans to fight for France as seamen on French privateers and as soldiers raiding British Canada and Spanish Louisiana. He also established French maritime prize courts in the United States. Contemptuous of American sovereignty, Genêt publicized his views in sympathetic newspapers. France eventually recalled him. See Richard Miller, "The Federal City, 1783–1800," in *Philadelphia: A 300-Year History*, edited by Russell F. Weigley, 190.

44. Powell, *Bring Out Your Dead*, 243–44; Carey, *A Short Account of the Malignant Fever*.
45. McMahon, "Beyond Therapeutics," 98–105; Edgar P. Richardson, "The Athens of America, 1800–1825," in *Philadelphia: A 300-Year History*, edited by Russell F. Weigley, 226–30.
46. Humphreys, *Yellow Fever and the South*, 54.
47. Foster et al., "Introduction to the 1993 Edition," xiv.
48. Alan Clendenning, "5 Yellow Fever Deaths in Brazil," Associated Press, January 16, 2008.
49. Johnson, *The Ghost Map*, 33–34, 111–36; Richard J. Evans, "Epidemics and Revolutions: Cholera in Nineteenth-Century Europe," *Past and Present* 120 (August 1988): 123–46.
50. C. E. Rosenberg, *The Cholera Years*, 2–4, 37; Howard Markel, "Cholera," in *The Encyclopedia of New York City*, edited by Kenneth T. Jackson (New Haven: Yale University Press, 1995), 219; Anbinder, *Five Points*, 23; Christopher Wills, *Yellow Fever Black Goddess: The Coevolution of People and Plagues* (Cambridge, Mass.: Perseus, 1996), 111–12; Johnson, *The Ghost Map*, 37; Evans, "Epidemics and Revolutions," 127.
51. C. E. Rosenberg, *The Cholera Years*, 17–18, 21–23.
52. Most likely the disease was introduced to the city by an incoming ship. Ibid., 25–26, 30.
53. Ibid., 30.
54. Ibid., 17–18, 43–45, 58–59.
55. Though there were some similarities in the way European and American elites framed the disease during the 1831–32 pandemic, the poor on both continents responded differently. In Russia, Austria-Hungary, Prussia, and Paris the poor felt that the ruling class was trying to eliminate them by poisoning them with the disease. As a result riots broke out, castles were ransacked, and government officials were murdered. In Britain doctors were attacked in numerous cholera riots by the poor, who thought the doctors were killing them off to get corpses to practice on in medical schools. In Paris and other European cities the disease was also framed in racial terms as a scourge of the "uncivilized" peoples of Asia being visited on the culturally and biologically superior races of Europe. The American poor did not perceive things this way; neither did they riot. Ibid., 41–42; François Delaporte, *Disease and Civilization: The Cholera in Paris, 1832*, translated by Arthur Goldhammer (Cambridge, Mass.: MIT Press, 1986); Catherine J. Kudlick, *Cholera in Post-Revolutionary Paris: A Cultural History* (Berkeley:

University of California Press, 1996); Evans, "Epidemics and Revolutions," 127–28, 136–39.

56. *Mercury* (New York), July 18, 1832.

57. John Pintard, *Letters from John Pintard to His Daughter Liza Noel Pintard Davidson 1816–1833*, edited by Dorothy C. Barck (4 vols.; New York: New-York Historical Society, 1941), 4:68–69, 73, 75–76; James E. Mooney, "John Pintard," in *The Encyclopedia of New York City*, edited by Kenneth T. Jackson (New Haven: Yale University Press, 1995), 906.

58. C. E. Rosenberg, *The Cholera Years*, 25, 29, 32; Markel, "Cholera," 219.

59. C. E. Rosenberg, *The Cholera Years*, 37, 56; Ellis, *Yellow Fever and Public Health in the New South*, 30.

60. Markel, "Cholera," 219; Ellis, *Yellow Fever and Public Health in the New South*, 30.

61. C. E. Rosenberg, *The Cholera Years*, 5.

62. Ibid., 23, 26–28, 68.

63. *Evening Post* (New York), July 3, 1832.

64. C. E. Rosenberg, *The Cholera Years*, 28, 32–33; Vincent Seyfried, "Far Rockaway," in *The Encyclopedia of New York City*, edited by Kenneth T. Jackson (New Haven: Yale University Press, 1995), 391–92.

65. C. E. Rosenberg, *The Cholera Years*, 37.

66. Ibid., 88–89.

67. Johnson, *The Ghost Map*, 59, 68–79.

68. John M. Snow, *Snow on Cholera* (New York: Commonwealth Fund, 1936); C. E. Rosenberg, *The Cholera Years*, 6; Wills, *Yellow Fever Black Goddess*, 112–15; Johnson, *The Ghost Map*, 68–79, 104–8.

69. Johnson, *The Ghost Map*, 159–88.

70. Ibid., 98–99. For more on Pacini, see "Who First Discovered *Vibrio Cholera*?," on the Web page of the UCLA School of Public Health's Department of Epidemiology, http://www.ph.ucla.edu/EPI/snow/firstdiscoveredcholera.html (accessed on November 11, 2006).

71. The outbreak of 1873 was confined to the Mississippi Valley. C. E. Rosenberg, *The Cholera Years*, 3–4. *Vibrio cholerae* was renamed in 1965 *Vibrio cholerae Pacini 1854* in honor of Pacini's work. Johnson, *The Ghost Map*, 213; see also "Who First Discovered *Vibrio Cholera*?"

72. Wills, *Yellow Fever Black Goddess*, 118.

73. Johnson, *The Ghost Map*, 36–37, 40.

74. C. E. Rosenberg, *The Cholera Years*, 112–13.

75. Ibid., 113; Burrows and Wallace, *Gotham*, 477.

76. Burrows and Wallace, *Gotham*, 477.

77. By the late 1850s the pigs were located north of Eighty-sixth Street. Ibid., 786; C. E. Rosenberg, *The Cholera Years*, 113.

78. Burrows and Wallace, *Gotham*, 786–87; Corbett, "Meatpacking," 745–46.

79. Humphreys, *Yellow Fever and the South*, 46–47; Ellis, *Yellow Fever and Public Health in the New South*, 1.

80. James C. Mohr, *Plague and Fire: Battling Black Death and the 1900 Burning of Honolulu's Chinatown* (New York: Oxford University Press, 2005), 7, 9; Myron Echenberg, "Pestis Redux: The Initial Years of the Third Bubonic Plague Pandemic, 1894–1901," *Journal of World History* 13, no. 2 (2002): 429–49; Norman E. Cantor, *In the Wake of the Plague: The Black Death and the World It Made* (New York: Free Press, 2001); Burl Burlingame, "Plague on Our Shores: Dark Days," *Honolulu Star-Bulletin*, January 24, 2000; Thomas W. McGovern, M. D. Faad, and Arthur M. Friedlander, "Plague," in *Medical Aspects of Chemical and Biological Warfare*, edited by Frederick R. Sidell, Ernest T. Takafuji, and David R. Franz (Washington: Office of the Surgeon General, 1997), 480–82.
81. Paul D. Hoeprich, M. Colin Jordan, and Alan R. Ronald, *Infectious Diseases: A Treatise of Infectious Processes* (Philadelphia: Lippincott, 1994), 1302–12; Mohr, *Plague and Fire*, 8; "Observations on Plague," *Journal of the American Medical Association* 288, no. 2 (1902): 143; McGovern et al., "Plague," 482.
82. Marilyn Chase, *The Barbary Plague: The Black Death in Victorian San Francisco* (New York: Random House, 2003), 188–91; Mohr, *Plague and Fire*, 12; "The Plague Pandemic," *Journal of the American Medical Association* 288, no. 2 (1902): 143; McGovern et al., "Plague," 482, 486–87; Y. Han, D. Zhou, X. Pang, Y. Song, L. Zhang, J. Bao, Z. Tong, X. Zhang, J. Wang, "Microarray Analysis of Temperature-Induced Transcriptome of Yersinia Pestis," *Microbiology and Immunology* 488, no. 11 (2004): 791–805.
83. Public Broadcasting Service, "A Science Odyssey: People and Discoveries, Bubonic Plague Hits San Francisco," 1998, WGBH, Boston.
84. Mohr, *Plague and Fire*, 13.
85. "Stop the Asiatics," *Austin's Hawaiian Weekly*, January 20, 1900, 1; "The Grim Visitor," *Austin's Hawaiian Weekly*, December 16, 1899, 2; "Filth in Honolulu Described, Plague Feared," *Polynesian*, October 15, 1859, 2; Mohr, *Plague and Fire*, 14; L. F. Alvarez, "Disinfectant: Dr. Alvarez says it is Useless," *Hawaiian Gazette*, December 22, 1899.
86. Mohr, *Plague and Fire*, 30; M. J. Keeling and C. A. Gilligant, "Metapopulation Dynamics of Bubonic Plague," *Nature* 407 (2000): 903–906; Burlingame, "Plague on Our Shores: Dark Days"; McGovern et al., "Plague," 498. One passenger on the *Nippon Maru* died of the plague before reaching Japan. After the *Nippon Maru* left Honolulu it was quarantined in San Francisco Bay. The bodies of two passengers who jumped overboard trying to escape quarantine were found to be infected. A third passenger, a young girl, died aboard the ship of a mysterious illness and was buried at sea. Another ship, the *Manchuria*, could have also been the source of the plague in Honolulu. "Nippon Maru is Infected. Glands of the Ship's Drowned Sailors Contain Bubonic Bacilli—No Danger of Infection on Land," *New York Times*, June 30, 1899; "Nippon Maru's Sailors Drown. Two Bodies from Quarantined Ship found in San Francisco Bay," *New York Times*, June 29, 1899; "Surgeon General Wyman says Bubonic Disease will not Spread," *New York Times*, June 27, 1899.
87. Mohr, *Plague and Fire*, 32–34; Myron Echenberg, *Plague Ports: The Global Urban Impact of Bubonic Plague, 1894–1901* (New York: New York University Press, 2006), 195; "Epidemic is dying out," *Pacific Commercial Advertiser*, September 17, 1895, 1;

"Epidemic is about ended, says PCA," *Pacific Commercial Advertiser*, September 20, 1895, 1.

88. Mohr, *Plague and Fire*, 34, 37–38, 56; Echenberg, *Plague Ports*, 196.
89. Mohr, *Plague and Fire*, 56; editorial, *Independent*, January 3, 1900, 4; editorial, *Independent*, December 13, 1899, 3; "Today Honolulu Throws off Four Months' Shackles," *Hawaiian Gazette*, May 1, 1900, 2–3.
90. O. A. Bushnell, *Gifts of Civilization* (Honolulu: University of Hawaii Press, 1993), 181–82.
91. Mohr, *Plague and Fire*, 56.
92. Sheldon Watts, *Epidemics and History: Disease, Power, and Imperialism* (New Haven: Yale University Press, 1998).
93. Mohr, *Plague and Fire*, 56–58, 64, 66; John H. Wilson, "All Work on Sewer Outfall has Stopped," *Pacific Commercial Advertiser*, January 3, 1900, 3.
94. Mohr, *Plague and Fire*, 14, 59, 62–63; "Census by Citizens' Committee Sets Honolulu Population at More than 40,000," *Pacific Commercial Advertiser*, January 31, 1900, 3; Echenberg, *Plague Ports*, 186.
95. Mohr, *Plague and Fire*, 59; Echenberg, *Plague Ports*, 186; Alvarez, "Disinfectant," *Hawaiian Gazette*, December 22, 1899, 1; Burl Burlingame, "Plague on Our Shores: City at War," *Honolulu Star-Bulletin*, January 31, 2000.
96. Mohr, *Plague and Fire*, 65–66; Echenberg, *Plague Ports*, 195–96; Burl Burlingame, "Plague on Our Shores: False Hope," *Honolulu Star-Bulletin*, January 25, 2000.
97. Mohr, *Plague and Fire*, 66–67; Echenberg, *Plague Ports*, 197; Burlingame, "Plague on Our Shores: City at War."
98. Mohr, *Plague and Fire*, 67–68, 76, 79; Burlingame, "Plague on Our Shores: False Hope," and "Plague on Our Shores: City at War"; C. B. Wood, "Plague on Maui. Seven Deaths Reported by Dr. Wood," *Hawaiian Gazette*, February 13, 1900, 5.
99. Mohr, *Plague and Fire*, 69, 72–74; "He Hoopalaleha no Ona Waiwai" [Land Owners Remain Indifferent], *Ke Aloha Aina*, December 23, 1899, 5.
100. Mohr, *Plague and Fire*, 73–74; Echenberg, *Plague Ports*, 197.
101. Mohr, *Plague and Fire*, 77–79.
102. "Today Honolulu Throws off Four Months' Shackles," *Hawaiian Gazette*, May 1, 1900; "Pioo ke Kulanakauhale—Paa Hou no Alanui I ke Kiai ia e no Koa—Hookapu ia na Poe o ka Apna e Hoomalu Mua ia Aole e Puke Iwaho—Eia no ka Ma'I ke Pahola mai nei" [City in Panic—The Streets Once Again Closed and Guarded by Troops—People within the Quarantined Area Restricted from Leaving—The Disease Spreads], *Ke Aloha Aina*, December 30, 1899, 3; Mohr, *Plague and Fire*, 78–81.
103. Mohr, *Plague and Fire*, 86–87, 90.
104. Ibid., 89–90; Jerome Platt, Maurice Jones, and Arleen Platt, *The Whitewash Brigade: The Hong Kong Plague of 1894* (London: Dix Noonan Webb, 1998), 77–78.
105. "Advertiser Runs Front Page Sketch of Burning of Buildings in Chinatown," *Pacific Commercial Advertiser*, January 2, 1900, 1, 4; "Burning of Plague Infested Houses Described," *Pacific Commercial Advertiser*, January 5, 1900, 7; "Is the City Quarantined?," *Hawaiian Gazette*, January 5, 1900, 4; Mohr, *Plague and Fire*, 91–92.
106. "Laweia no ka Homalu ana" [Taken under Quarantine], *Independent*, January 11, 1900,

4; Dr. Kobayashi, "Japanese MD, Mobbed by Irate Japanese in Quarantine during Plague Epidemic," *Pacific Commercial Advertiser*, January 10, 1900, 2; John F. Colburn, "Property Owner Objects to Burning of Chinatown during Plague," *Pacific Commercial Advertiser*, January 10, 1900, 2; Mohr, *Plague and Fire*, 92–95; Burlingame, "Plague on Our Shores: False Hope."

107. Mohr, *Plague and Fire*, 99, 114–15, 118.

108. "He Ku Maoli i Ka Walohia" [Sadly . . .], and "Na Make o ke Kulanakauhale nei" [Death Toll in the City], both in *Ke Aloha Aina*, January 6, 1900, 4, 7.

109. "He Oiaio Anei hoi Keia?" [Is This True?], *Ke Aloha Aina*, January 20, 1900, 2; Mohr, *Plague and Fire*, 115–16.

110. Saito Miki to Clifford B. Wood, Honolulu Board of Health, January 17, 1900, Hawaii State Archives; Clifford B. Wood to Saito Miki, Honolulu Board of Health, January 17, 1900, Hawaii State Archives; "Story Cites Special Consideration Given French Woman in Quarantine, Hard Treatment for Japanese," *Pacific Commercial Advertiser*, February 1, 1900, 3; Mohr, *Plague and Fire*, 116–17.

111. PCA Advocates Burning of Entire Chinatown," *Pacific Commercial Advertiser*, January 16, 1900, 4; Mohr, *Plague and Fire*, 120–22, 157.

112. Echenberg, *Plague Ports*, 200–201, 204.

113. Mohr, *Plague and Fire*, 125–32; Echenberg, *Plague Ports*, 201–202; Burl Burlingame, "Plague on Our Shores: The Great Chinatown Fire," *Honolulu Star-Bulletin*, February 1, 2000; "Great Chinatown Fire as a Result of Plague," and "Destroyed by Chinatown Fire: Good Description of," both in *Pacific Commercial Advertiser*," January 22, 1900, 1, 8.

114. Mohr, *Plague and Fire*, 132–33, 136–41, 145–46; Echenberg, *Plague Ports*, 202–3; Department of Land and Natural Resources, "Honolulu Responds to the Plague," 2005, State Historic Preservation Division; Burlingame, "Plague on Our Shores: City at War" and "Plague on Our Shores: The Great Chinatown Fire."

115. Mohr, *Plague and Fire*, 148–50; "Many Haole Companies Subscribe to United Chinese Society Fund to Feed Plague Victims," *Pacific Commercial Advertiser*, March 7, 1900, 10.

116. Mohr, *Plague and Fire*, 144, 157–58, 161, 164–68; Honolulu Board of Health, letter to the Board of Health of Sydney, New South Wales, January 31, 1900, Hawaii State Archives; Burlingame, "Plague on Our Shores: The Great Chinatown Fire"; "200 Hawaiians Destitute from Plague," *Pacific Commercial Advertiser*, March 15, 1900, 12.

117. Mohr, *Plague and Fire*, 190–93; Echenberg, *Plague Ports*, 204; "U.S. Government Declines Responsibility to Pay Fire Claims for Chinatown Fire," *Pacific Commercial Advertiser*, October 11, 1900, 5.

118. Echenberg, *Plague Ports*, 204; "Today Honolulu Throws off Four Months' Shackles," *Hawaiian Gazette*, May 1, 1900; 2–3; "Attend Mass Rally to Protest Grievances from Chinatown Fire," and "Band with Japanese in Protest against the Court of Claims," both in *Pacific Commercial Advertiser*, April 9, 1900, 6; "Building will Probably Begin Again on the Old Plague Site," *Pacific Commercial Advertiser*, July 12, 1900, 3; "Chair-

man of the Chinese Citizens' Committee Blames Board of Health for the Plague," *Pacific Commercial Advertiser*, January 23, 1900, 6.

119. Honolulu Board of Health, letter to the Board of Health of Sydney, New South Wales, January 31, 1900, Hawaii State Archives; Mohr, *Plague and Fire*, 171–72.

120. Alfred D. Morris, "The Epidemic That Never Was: Yellow Fever in Hawaii," *Hawaii Medical Journal* 54 (November 1995): 781–84; Wood, "Plague on Maui. Seven Deaths Reported by Dr. Wood," *Hawaiian Gazette*, February 13, 1900.

121. M. Chase, *The Barbary Plague*; Mohr, *Plague and Fire*, 171–72; Public Broadcasting Service, "A Science Odyssey."

122. Paul J. Edelson, "Quarantine and Social Inequality," *Journal of the American Medical Association* 290 (2003): 2874; G. B. Risse, "'A Long Pull, a Strong Pull, and All Together': San Francisco and the Bubonic Plague, 1907–1908," *Bulletin of Historical Medicine* 66 (1992): 260–62; C. McClain, "Of Medicine, Race, and American Law: The Bubonic Plague Outbreak of 1900," *Law and Social Inequality* 13 (1988): 447–513; Mohr, *Plague and Fire*, 171–72; Public Broadcasting Service, "A Science Odyssey."

123. J. L. Caten and L. Kartman, "Human Plague in the United States: 1900–1966," *Journal of the American Medical Association* 205, no. 6 (1968): 81–84; M. Chase, *The Barbary Plague*.

124. Mohr, *Plague and Fire*, 198–200, 203.

125. Alexandra Minna Stern and Howard Markel, "International Efforts to Control Infectious Diseases, 1851 to the Present," *Journal of the American Medical Association* 292 no. 12 (2004): 1474–79.

126. Edelson, "Quarantine and Social Inequality," 2874; Echenberg, *Plague Ports*, 198–200.

3. Wealthy Urbanites

1. Homberger, *Mrs. Astor's New York*, 1, 3–4.
2. Mooney, "Stuyvesant, Peter," 1133–34.
3. Pamela Plakins Thornton, "Gentlemen Farmers," in *The Encyclopedia of New England*, edited by Burt Feintuch and David Watters (New Haven: Yale University Press, 2005), 42; Homberger, *Mrs. Astor's New York*, 5.
4. Janvier, *In Old New York*, 192–97; Gilbert Tauber, "Lispenard Swamp," in *The Encyclopedia of New York City*, edited by Kenneth T. Jackson (New Haven: Yale University Press, 1995), 679.
5. *Manual of the Corporation of the City of New York* (New York: D. T. Valentine, 1865), 583–84.
6. Copied from Janvier, *In Old New York*, 194–97.
7. Ibid., 197–98; Gilbert Tauber "Ranelagh Garden," in *The Encyclopedia of New York City*, edited by Kenneth T. Jackson, 679, 985–86. Quit-rents are a kind of levy or land tax imposed on freehold land by the government or its agents. See Dunn and Dunn, "The Founding, 1681–1701," 17. There is some uncertainty about the time of Rutgers's death. Tauber, "Ranelagh Garden," writes that he died in 1746.

8. Jan Seidler Ramirez, "Greenwich Village," in *The Encyclopedia of New York City*, edited by Kenneth T. Jackson, 506–9; Janvier, *In Old New York*, 85–86, 89–90; Jackson and Dunbar, *Empire City*, 19.
9. In 1784 the State Commission of Forfeiture confiscated and auctioned off De Lancey's property. Blackmar, *Manhattan for Rent*, 34–35; Burrows and Wallace, *Gotham*, 178.
10. Burrows and Wallace, *Gotham*, 178.
11. Trinity decided to lease its landholdings in 1734 for a period not to exceed forty years (eventually the church began to grant sixty-three- and ninety-nine-year leases). Before 1770 the church leased all its lots for a £2 annual ground fee for the first seven years, £3 a year for the next seven years, and a £4 annual fee for the remaining years of the contract. The leases stipulated that tenants remove any structures they constructed on the property within ten days of the expiration of the lease. Trinity's lease fees were low compared to others. Small lots near the wharves ranged from £10 to £20 annually, while house lots ranged from £50 to £200 per year. Consequently artisans, shopkeepers, and other small businessmen leased church lots. Ibid.; Blackmar, *Manhattan for Rent*, 30–35; Ramirez, "Greenwich Village," 506–9.
12. Blackmar, *Manhattan for Rent*, 33; Burrows and Wallace, *Gotham*, 178.
13. Burrows and Wallace, *Gotham*, 178–79; Ramirez, "Greenwich Village," 506–9; Elaine Weber Pascu, "Henry Rutgers," and Hilda Reiger, "Chelsea," both in *The Encyclopedia of New York City*, edited by Kenneth T. Jackson; Janvier, *In Old New York*, 85–88, 117; New York Historical Society, 2008, https://www.nyhistory.org/; Richard Schermerhorn, *Schermerhorn Genealogy and Family Chronicles* (New York: Tobias A. Wright Publisher, 1914), 149–61; "The Rhinelander Family, An Old Colonial Fortune Scraps from the Family History," *New York Times*, June 23, 1878, 2; "Roger Morris Park, New York City Department of Parks and Recreation, 2008, http://www.nycgovparks.org; Cynthia Kierner, "Philip Livingston," in *American National Biography* edited by the American Council of Learned Societies (New York: Oxford University Press, 2008); Eugene R. Fingerhut and Joseph S. Tiedmann, *The Other New York: The American Revolution Beyond New York City, 1763–1787* (Albany: SUNY Press, 2005), 23–24.
14. Macieski, "Cities and Suburbs," 193; Thornton, "Gentlemen Farmers," 42; Ronald Dale Karr, "Brookline, Mass.," in *The Encyclopedia of New England*, edited by Burt Feintuch and David H. Watters, 214–15; "John Cabot House" Beverly Historical Society and Museum, http://www.beverlyhistory.org; Henry Cabot Lodge and George Cabot, *Life and Letters of George Cabot* (Cambridge, Mass.: Harvard University Press, 1877); Robert Means Lawrence, *The Descendants of Major Samuel Lawrence of Groton, Massachusetts, with Some Mention of Allied Families* (Boston: Riverside Press, 1904); Robert Sobel, *The Entrepreneurs: Explorations within the American Business Tradition* (New York: Weybright and Talley, 1974), 1–50.
15. Homberger, *Mrs. Astor's New York*, 166–67; James David Glunt and Ulrich Bunnell Phillips, editors, *Florida Plantation Records: From the Papers of George Noble Jones* (St. Louis: Missouri Historical Society, 1927); Sue Burns, *Painting the Dark Side: Art and the Gothic Imagination in Nineteenth-Century America* (Berkeley: University of California Press, 2004), 95–96; "Middleton of South Carolina," *The South Carolina Historical and Genealogical Magazine* 1, no. 1 (January 1900): 228–50.

16. Several of the mansions along Bellevue Avenue are open for public tours. See www.NewportMansions.org.
17. Macieski, "Cities and Suburbs," 193.
18. Homberger, *Mrs. Astor's New York*, 68–69; Burrows and Wallace, *Gotham*, 456–57; Jackson and Dunbar, *Empire City*, 157; "Cornbury to Tryon—1702–1775," *Courts and Lawyers of New York: A History, 1609–1925*, edited by Alden Chester and E. Melvin Williams (Clark, N.J.: Lawbook Exchange, Ltd., 2004), 507–626; Miriam Medina, "Biographical Sketches of Wealthy Men of the Colonial Era in New York," in *A Classification of American Wealth; History and Genealogy of the Wealthy Families of America*, http:www.thehistorybox.com; James V. Marshall, *The United States Manual of Biography and History* (Philadelphia: James B. Smith, 1856), 137–39; *The Colonial Laws of New York from the Year 1644 to the Revolution*, vol. 4 (Albany: J. B. Lyon, 1894), 1016.
19. William E. Dodge, *Old New York: A Lecture* (New York: Dodd, Mead, 1880), 21–22. See also "Annual Dinner of the Chamber of Commerce, Speeches by William E. Dodge, Judge Noah Davis, Mr. A. A. Low, Mayor Havemeyer, Honorable William M. Evarts, and Other Gentlemen," *New York Times*, May 8, 1874, 1.
20. Homberger, *Mrs. Astor's New York*, 70–71.
21. James Kirke Paulding to Sally Hanlon, August 27, 1802, in *The Letters of James Kirke Paulding*, edited by Ralph M. Aderman (Madison: University of Wisconsin Press, 1962), 19. For more on Paulding see Rufus Rockwell Wilson and Irving Stone, *Rambles in Colonial Byways* (Philadelphia: J. B. Lippincott Company, 1900), 169–70.
22. Jeremy Cockloft, "The Stranger at Home; Or a Tour of Broadway," *Salmagundi* 12 (June 27, 1807).
23. Francis J. Grund, *Aristocracy in America* (1839; New York: Harper & Brothers, 1859), 10.
24. Homberger, *Mrs. Astor's New York*, 70–71.
25. Burrows and Wallace, *Gotham*, 456–57; Jackson and Dunbar, *Empire City*, 157.
26. Burrows and Wallace, *Gotham*, 456–57; Jackson and Dunbar, *Empire City*, 157; Homberger, *Mrs. Astor's New York*, 69; Duncan S. Somerville, *The Aspinwall Empire* (Mystic, Conn.: Mystic Seaport Museum, 1983); Dorothie Bobb, "Philip Hone's New York," *American Heritage Magazine* 8, no. 5 (August 1957): 5; Philip Hone, *The Diary of Philip Hone: 1828–1851*, edited by Allan Nevins, vol. 1 (New York: Dodd, Mead and Company, 1927).
27. Homberger, *Mrs. Astor's New York*, 58. The Melvills could not afford to live on Broadway for long and returned to Albany.
28. Frances Milton Trollope, *The Domestic Manners of the Americans* (London: Whittaker, Treacher, 1832).
29. Trollope, *The Domestic Manners of the Americans*.
30. Burrows and Wallace, *Gotham*, 457; C. E. Rosenberg, *The Cholera Years*, 17.
31. Pintard, *Letters*, 2: 220.
32. After the family moved, Mary's father, Amos, decided that Washington Square was too far away from downtown. He moved back to the neighborhood around the Battery, but as immigrants swarmed in and their once fashionable residence became surrounded with boardinghouses, they moved back to Washington Square, then farther

north to Madison Square. Char Miller, *Gifford Pinchot and the Making of Modern Environmentalism* (Washington, D.C.: Island Press, 2001), 38, 49–50.

33. Homberger, *Mrs. Astor's New York*, 72–73.
34. Ibid., 67, 86–87; Kenneth A. Scherzer, "St. John's Park," in *The Encyclopedia of New York City*, edited by Kenneth T. Jackson, 1035.
35. Homberger, *Mrs. Astor's New York*, 80.
36. Scherzer, "St. John's Park," 1035; Burrows and Wallace, *Gotham*, 457–58, 473; Blackmar, *Manhattan for Rent*, 30–32.
37. "St. John's Park," *The Subterranean* 3 (July 26, 1845): 2.
38. Lydia Maria Child, *Letters from New-York* (New York: Charles S. Francis, 1843).
39. Homberger, *Mrs. Astor's New York*, 76–78.
40. Cf. Burrows and Wallace, *Gotham*, 458–59.
41. Homberger, *Mrs. Astor's New York*, 1.
42. Vanderbilt owned stockyards, grain elevators, lumber stores, petroleum tanks, an elevated transit system, a freight railway, train terminals, and the Grand Central Depot. Ibid., 81–83; Scherzer, "St. John's Park," 1035; Burrows and Wallace, *Gotham*, 459, 715; David M. Scobey, *Empire City: The Making and Meaning of the New York City Landscape* (Philadelphia: Temple University Press, 2002), 74–78.
43. Burrows and Wallace, *Gotham*, 715; Janvier, *In Old New York*, 211–16; Peter L. Bernstein, T*he Power of Gold: The History of an Obsession* (New York: John Wiley and Sons, 2000), 274–75.
44. Homberger, *Mrs. Astor's New York*, 57.
45. Ibid., 113–15.
46. Brands, *T. R.*, 4–5.
47. Kreiger, "Experiencing Boston," 154; Alex Kreiger, David Cobb, and Amy Turner, eds., *Mapping Boston* (Cambridge, Mass.: MIT Press, 2001), 190; McNulty, *Boston Then and Now*, 104.
48. McNulty, *Boston Then and Now*, 28, 102, 108, 112, 114; Karr, "Boston Common," 212–13.
49. McNulty, *Boston Then and Now*, 116–17, 124, 126; Douglass Shand-Tucci, *Built in Boston: City and Suburb, 1800–2000* (Cambridge, Mass.: MIT Press, 1999), 19–20; Susanne Klingenstein, "Boston," in *The Encyclopedia of New England*, edited by Burt Feintuch and David H. Watters, 210–12. Today the park is still privately owned by homeowners whose properties overlook the park. The Louisburg Square Proprietors, as they are called, is one of the oldest homeowners associations in America.
50. Thayer, "Town into City," 93–96, 102; Wainwright, "The Age of Nicholas Biddle," 281; Geffen, "Industrial Development and Social Crisis," 332; "Biographical Background" Guide to Pemberton Family Papers, 1741–1789 (Philadelphia: Haverford College Library); Lynne Delehante Destafano, *Charles Norris (1712–1766): Philadelphia Gentleman* (Philadelphia: University of Pennsylvania Press, 1973); R. Kent Newmyer, "Charles Stedman's *History of the American War*," *American Historical Review* 63, no. 4 (July, 1958): 924–34; George B. Tatum and Cortland Van Dyke Hubbard, *Philadelphia Georgian: The City House of Samuel Powel and Some of Its Eighteenth-Century Neighbors* (Middletown, Conn.: Wesleyan University Press, 1976); *Green Mount Cemetery,*

One Hundredth Anniversary, 1838–1938 (Baltimore: Green Mount Cemetery, 1938); R. Kent Lancaster, "Green Mount: The Introduction of the Rural Cemetery into Baltimore," *Maryland Historical Magazine,* 74 (1978): 62–79; "Reminiscences of Baltimore in 1824" *Maryland Historical Magazine* 1 (1906): 113–14; "A Chronicle of Union Square and Environs" (2008), at http://www.union-square.us/us-history.html; Mary Ellen Hayward, Frank R. Shivers, and Richard Hubbard Howland, *Architecture of Baltimore: An Illustrated History* (Baltimore: Johns Hopkins University Press, 2004), 42–43; "Baltimore Attractions: Evergreen House," *New York Times,* January 2, 2009, D2; "About Evergreen," Evergreen Museum and Library (2008), at http://www.museums.jhu.edu/evergreen/.

51. Tinkcom, "The Revolutionary City," 121–22; Wainwright, "The Age of Nicholas Biddle," 292.
52. S. H. Olson, *Baltimore,* 36, 45, 112–13; Elizabeth Fee, Linda Shopes, and Linda Zeidman, *The Baltimore Book* (Philadelphia: Temple University Press, 1991), 24–25.
53. S. H. Olson, *Baltimore,* 114, 167–68.

4. *Social Inequality*

1. Ross, *Social Control.* See also Boyer, *Urban Masses,* 224–28.
2. Parsons, *The Social System.*
3. Ibid.; J. P. Gibbs, *Norms, Deviance, and Social Control: Conceptual Matters* (New York: Elsevier, 1981).
4. Homberger, *Mrs. Astor's New York,* 10, 27.
5. "Worker, Manager, and Society: Responsibility to the Worker," in *Encyclopedia Britannica* (Chicago: Britannica, 2003).
6. Louise A. Breen, "Religion," in *The Encyclopedia of New England,* edited by Burt Feintuch and David H. Watters (New Haven: Yale University Press, 2005), 1282.
7. James A. Morone, "Second Great Awakening," in *The Encyclopedia of New England,* edited by Burt Feintuch and David H. Watters, 1384–85.
8. Burrows and Wallace, *Gotham,* 495.
9. Ibid., 495–96; Lucy Forsyth Townsend "Bethune, Joanna," in *European Immigrant Women in the United States,* edited by Judy Barrett Litoff and Judith McDonnell (Philadelphia: Taylor & Francis, 1994), 21–23.
10. Geffen, "Industrial Development and Social Crisis," 335.
11. Sally K. Ward, Cynthia Duncan, and Jody Grimes, "Poverty," in *The Encyclopedia of New England,* edited by Burt Feintuch and David H. Watters, 242–43.
12. Boyer, *Urban Masses,* 22–33, 34–53; Edwin W. Rice, *The Sunday-School Movement, 1780–1917, and the American Sunday-School Union, 1817–1917* (Philadelphia: American Sunday-School Union, 1917).
13. Burrows and Wallace, *Gotham,* 496–97.
14. Cf. ibid., 497–98.
15. Ibid., 384–85; Alana J. Erickson, "Sunday Schools," in *The Encyclopedia of New York City,* edited by Kenneth T. Jackson (New Haven: Yale University Press, 1995), 1142–43. There are conflicting dates for Ferguson's birth date and the year she opened her

school, although there is consensus that she did run a Sunday School program. Lucy Forsyth Townsend "Joanna Bethune," 21–23; Isaac Ferris, *Semi-Centennial Memorial Discourse of the New-York Sunday-School Union: Delivered on the Evening of the 25th February, 1866, in the Reformed Dutch Church* (New York: J. A. Gray & Green Printers, 1866), 11–12; David W. Wills and Albert J. Raboteau, *African-American Religion: A Historical Interpretation with Representative Documents* (Chicago: University of Chicago Press, 2008).

16. S. H. Olson, *Baltimore*, 64–66; Burrows and Wallace, *Gotham*, 384–85, 499.
17. Boyer, *Urban Masses*, 34–38; Erickson, "Sunday Schools," 1142–43.
18. S. H. Olson, *Baltimore*, 66; Boyer, *Urban Masses*, 34.
19. Malthus was also a curate in the Anglican Church. Thomas Robert Malthus, *Essay on the Principle of Population and a Summary View of the Principle of Population* (1798; New York: Penguin, 1970); Donald Worster, *Nature's Economy: A History of Ecological Ideas*, 2nd ed. (New York: Cambridge University Press, 1994), 149–53.
20. Worster, *Nature's Economy*, 171–73.
21. Daniel Raymond, *Thoughts on Political Economy* (Baltimore: F. Lucas, 1819), part 1, pp. 233, 248–50, 452–53. See also S. H. Olson, *Baltimore*, 68–69.
22. Daniel Raymond, *The Elements of Political Economy* (Baltimore: F. Lucas, 1836).
23. Burrows and Wallace, *Gotham*, 493.
24. Ward et al., "Poverty," 242–43.
25. Burrows and Wallace, *Gotham*, 493–94.
26. Ibid., 494; Boyer, *Urban Masses*, 87–93; Spann, *The New Metropolis*, 82–91.
27. Burrows and Wallace, *Gotham*, 494.
28. Ibid., 494–95; Hemphill and Mohl, "Poverty," 932–34; Anbinder, *Five Points*, 267.
29. Burrows and Wallace, *Gotham*, 495.
30. Ibid.
31. Cf. ibid., 624.
32. Anbinder, *Five Points*, 244; Alana J. Erikson, "Association for Improving the Condition of the Poor," in *The Encyclopedia of New York City*, edited by Kenneth T. Jackson, 61.
33. Anbinder, *Five Points*, 244; Association for Improving the Condition of the Poor, *Annual Report of the New York Association for Improving the Condition of the Poor* (New York: AICP, 1846), 15–20. Hartley was born in England and moved to New York, where he opened a mercantile firm. In addition to being general agent of the Association for Improving the Condition of the Poor, he directed the New York City Temperance Society. Marilyn Thornton Williams, "Robert M(ilham) Hartley," in *The Encyclopedia of New York City*, edited by Kenneth T. Jackson, 531.
34. After Robert Hartley retired from the Association for Improving the Condition of the Poor in 1876 the organization relied less on upper-class men as friendly visitors. In 1879 the Association began employing women as visitors also. Boyer, *Urban Masses*, 87–93; Spann, *The New Metropolis*, 82–91; Anbinder, *Five Points*, 244; Erickson, "Association for Improving the Condition of the Poor," 61.
35. Boyer, *Urban Masses*, 90–91; Spann, *The New Metropolis*, 86–91; Bauman et al., *From Tenements to the Taylor Homes*, 6.

36. Burrows and Wallace, *Gotham*, 778.

37. Ibid., 778–779.

38. Anbinder, *Five Points*, 244.

39. Burrows and Wallace, *Gotham*, 848–49.

40. Corinne T. Field and Marilyn Thornton Williams, "Bathhouses," in *The Encyclopedia of New York City*, edited by Kenneth T. Jackson, 87–88; Erickson, "Association for Improving the Condition of the Poor," 61.

41. The tenement was condemned in 1880 because it was unfit for human habitation. Burrows and Wallace, *Gotham*, 788–89; Anbinder, *Five Points*, 435. See also Riis, *How the Other Half Lives*, 7.

42. Bauman et al., *From Tenements to the Taylor Homes*, 8; Erickson, "Association for Improving the Condition of the Poor," 61; Field and Williams, "Bathhouses," 87–88; Burrows and Wallace, *Gotham*, 788–79; Anbinder, *Five Points*, 435.

43. Ward et al., "Poverty," 242–43.

44. Marilyn Irvin Holt, *The Orphan Trains: Placing Out in America* (Lincoln: University of Nebraska Press, 1992), 42; Stephen O'Connor, *Orphan Trains: The Story of Charles Loring Brace and the Children He Saved and Failed* (Boston: Houghton Mifflin, 2001), 18, 20–22.

45. Jane Allen, "Charles Loring Brace," and Jane Allen, "Children's Aid Society," both in *The Encyclopedia of New York City*, edited by Kenneth T. Jackson, 135, 213; Carolee R. Inskeep, *The Children's Aid Society of New York* (Baltimore: Clearfield Press, 1996), i; O'Connor, *Orphan Trains*, 74.

46. Children's Aid Society, *Eighteenth Annual Report of the Children's Aid Society* (New York: Press of Wynkoop and Ballenbeck, 1870), 18.

47. Children's Aid Society, *Eighteenth Annual Report*, 1871; O'Connor, *Orphan Trains*, 81–93; Homberger, *Mrs. Astor's New York*, 32.

48. Holt, *The Orphan Trains*, 4, 43–45, 82–83; E. C. Richard, *Louisiana*, 7.

49. Though Brace was influenced by Romanticism, he, like his good friend Frederick Law Olmsted, saw pastoral settings as idealized landscapes. Children's Aid Society, *Nineteenth Annual Report of the Children's Aid Society* (New York: Press of Wynkoop and Ballenbeck, 1871), 7–9, 50; Holt, *The Orphan Trains*, 27.

50. Charles Loring Brace, *The Best Method of Disposing of Our Pauper and Vagrant Children* (New York: Wynkoop, Hallenbeck and Thomas, 1859).

51. Children's Aid Society, *Nineteenth Annual Report*, 3.

52. Ibid., 3–5.

53. Ibid., 5–6.

54. O'Connor, *Orphan Trains*, xiii.

55. Charles Loring Brace, "The Little Laborers," *Jacksonville* (Ill.) *Journal*, April 26 and May 24, 1860. Not all children were placed when the trains stopped at a given location. Those who were not selected were put back on the trains. They kept repeating the process until they were eventually taken by a family. See Holt, *The Orphan Trains*; O'Connor, *Orphan Trains*.

56. Boyer, *Urban Masses*, 94–102; Spann, *The New Metropolis*, 88, 270–74; Inskeep, *The Children's Aid Society of New York*, i–ix; O'Connor, *Orphan Trains*, xvii. For more de-

tailed critiques of the emigration program, see O'Connor, *Orphan Trains*, and Holt, *The Orphan Trains*.

57. Holt, *The Orphan Trains*, 2, 64; Children's Aid Society, *Nineteenth Annual Report*, 11.
58. Children's Aid Society, *Eighteenth Annual Report*, 9.
59. Domenica M. Barbuto, *American Settlement Houses and Progressive Social Reform: An Encyclopedia of the American Settlement Movement* (Phoenix: Oryx Press, 1999), 28.
60. Alana J. Erickson, "Fresh Air Fund," in *The Encyclopedia of New York City*, edited by Kenneth T. Jackson, 441; Fresh Air Fund, *Fresh Air Fund Programs* (New York: Fresh Air Fund, 2004).
61. Burrows and Wallace, *Gotham*, 1159; Jane Allen, "Josephine Shaw Lowell," in *The Encyclopedia of New York City*, edited by Kenneth T. Jackson, 695–96; Barbuto, *American Settlement Houses and Progressive Social Reform*, 123–24. For a complete biography of Lowell, see Joan Waugh, *Unsentimental Reformer: The Life of Josephine Shaw Lowell* (Cambridge, Mass.: Harvard University Press, 1997).
62. Burrows and Wallace, *Gotham*, 1159.
63. Riis, *How the Other Half Lives*, 183.
64. Boyer, *Urban Masses*, 147–51; Burrows and Wallace, *Gotham*, 1159.
65. Boyer, *Urban Masses*, 147–51; Burrows and Wallace, *Gotham*, 1159–60.
66. Burrows and Wallace, *Gotham*, 1159–60; Alana J. Erikson, "Charity Organization Society," in *The Encyclopedia of New York City*, edited by Kenneth T. Jackson, 201; Riis, *How the Other Half Lives*, 186–87, 195.
67. Barbuto, *American Settlement Houses and Progressive Social Reform*, 29–30.
68. David E. Nye, *Consuming Power: A Social History of American Energies* (Cambridge, Mass.: MIT Press, 1998), 99; U.S. Bureau of the Census, *Statistical Abstract of the United States, 1900* (Washington: Government Printing Office, 1975), 1–12.
69. Mayer N. Zald and Patricia Denton, "From Evangelism to General Service: On the Transformation of the YMCA," *Administrative Science Quarterly* 8 (June 1963): 214–34; Kathy Peiss, *Cheap Amusements: Working Women and Leisure in Turn-of-the-Century New York* (Philadelphia: Temple University Press, 1986); Boyer, *Urban Masses*, 108–20.
70. Boyer, *Urban Masses*, 112–14; YMCA, *A Brief History of the YMCA* (Chicago: YMCA, 2006), www.ymca.net; Central Connecticut Coast YMCA, *YMCA History* (New Haven: CCC-YMCA, 2006), www.cccymca.org.
71. New York Charity Organization Society, *Charities*, February 24, 1900, 3; City Homes Association, *Tenement Conditions in Chicago*, 145–47; Fee et al., *The Baltimore Book*, 8–9.
72. Slater was born in 1768 and died in 1835. James L. Conrad, "Samuel Slater," in *The Encyclopedia of New England*, edited by Burt Feintuch and David H. Watters, 879; Edward Hugh Cameron, *Samuel Slater: Father of American Manufactures* (Freeport, Maine: B. Wheelright, 1960); Stephen Mihm, "Industrial Revolution—1790–1860," in *The Encyclopedia of New England*, edited by Burt Feintuch and David H. Watters, 826–27.
73. Alexander Hamilton, *Alexander Hamilton's Report on the Subject of Manufactures, Made*

in his Capacity of Secretary of the Treasury, on the Fifth of December 1791, edited by Mathew Carey (1791; Philadelphia: William Brown, 1827).

74. Lowell was born on April 7, 1775, in Newburyport, Massachusetts, into a prominent Boston family. He died in 1817. After his death, one of his sons, Francis Cabot Lowell Jr., joined the company to carry on where his father left off. Mihm, "Industrial Revolution," 826–27; James Beauchesne, "Boston Associates," in *The Encyclopedia of New England*, edited by Burt Feintuch and David H. Watters, 837–38; Caroline Farar Ware, *The Early New England Cotton Manufacture* (Boston: Houghton Mifflin, 1931); "Worker, Manager, and Society: Responsibility to the Worker," and "Francis Cabot Lowell," in *Encyclopedia Britannica* (Chicago: Britannica, 2003).
75. Mihm, "Industrial Revolution," 826–27; Beauchesne, "Boston Associates," 837–38; Thomas Dublin, "Lowell, Mass.," and Susan Alves, "Mill Girls," both in *The Encyclopedia of New England*, edited by Burt Feintuch and David H. Watters, 232–33, 869–870; Christine Lunardini, *What Every American Should Know about Women's History: 200 Events That Shaped Our Destiny* (Holbrook, Mass.: Adams Media, 1977), 41–42, 61–62; "Massachusetts Investigation into Labor Conditions, 1845," House Document No. 50, March 1845.
76. Dublin, "Lowell, Mass.," 232–33; Alves; "Mill Girls," 869–70; Beauchesne, "Boston Associates," 837–38; Mihm, "Industrial Revolution," 826–27.
77. John Avery, "Factory Rules," in *Handbook to Lowell, 1848* (Lowell, Mass.: Hamilton Manufacturing, 1848); "Massachusetts Investigation into Labor Conditions, 1845"; "A Description of Factory Life by an Associationist in 1846," Center of Lowell History, University of Massachusetts, Lowell Libraries, Special Collections, http://library.uml.edu; John Avery, "Boarding House Rules," in *Handbook to Lowell, 1848* (Lowell, Mass.: Hamilton Manufacturing, 1848); "Female Workers of Lowell," *The Harbinger*, 1836.
78. Harriet Hanson Robinson, *Loom and Spindle or Life among the Early Mill Girls* (New York: T. Y. Crowell, 1898), 83–86; Lucy Larcom, "Among Lowell Mill-Girls: A Reminiscence," *Atlantic Monthly* 48 (November 1881). See also "Massachusetts Investigation into Labor Conditions, 1845."
79. Seth Luther, *Address to the Working Men of New England, On the State of Education, and on the Condition of the Producing Classes in Europe and America* (1832; New York: Office of the Working Man's Advocate, 1833), 17–21.
80. "A Week in the Mill," *Lowell Offering* 5 (1845): 217–18. The magazine was renamed the *New England Offering* in 1845. Its last issue was published in 1849.
81. Josephine L. Baker, "A Second Peep at Factory Life," *Lowell Offering* 5 (1845): 97–100.
82. Brian Lombard, *Amoskeag Falls History* (Manchester, N.H.: Amoskeag Falls Management Corporation, November 2000); Public Service of New Hampshire, *Amoskeag Manufacturing Company: The Making of a Mill Town* (Manchester, N.H.: Northeast Utilities, 2006).
83. James P. Hanlan, "Manchester, N.H.," and Tamara K. Hareven, "Amoskeag Manufacturing Company," both in *The Encyclopedia of New England*, edited by Burt Feintuch

and David H. Watters, 233–34, 832–33; Public Service of New Hampshire, *Amoskeag Manufacturing Company*; *Industrial Development in New Hampshire: A Timeline* (Concord: Museum of New Hampshire History, August 8, 2004).

84. "Worker, Manager, and Society: Responsibility to the Worker."
85. See Hurley, *Environmental Inequalities*.
86. Linda Zeidman, "Sparrows Point, Dundalk, Highlandtown, Old West Baltimore: Home of Gold Dust and the Union Card," in *The Baltimore Book: New Views of Local History*, edited by Elizabeth Fee, Linda Shopes, and Linda Zeidman (Philadelphia: Temple University Press, 1991), 176–80.
87. Paul Tedesco, "Industrial New England, 1861–1918," in *The Encyclopedia of New England*, edited by Burt Feintuch and David H. Watters, 827–28.
88. Lunardini, *What Every American Should Know*, 61–62; Lawrence F. Gross, "Textiles," in *The Encyclopedia of New England*, edited by Burt Feintuch and David H. Watters, 881–83; *Industrial Development in New Hampshire*.
89. Luther, *Address to the Working Men of New England*, 36.
90. Anon., "A Week in the Factory," *Lowell Offering*, Series 5, no. 10 (October 1845): 217–18; Lunardini, *What Every American Should Know*, 41–42, 61–62; "Massachusetts Investigation into Labor Conditions, 1845."
91. Statistics of the Lowell Manufactures, January 1, 1835, Library of Congress, Ephemera Collection, portfolio 56, folder 10.
92. Alves, "Mill Girls," 869–70; Lunardini, *What Every American Should Know*, 41–42, 61–62; Juliet H. Mofford, "Women in the Workplace: Labor Unions," *Women's History*, Spring/Summer 1996; "Massachusetts Investigation into Labor Conditions, 1845."
93. "Recruitment of Female Operatives—An Account from the 1840s," *Voice of Industry*, January 2, 1846. The *Voice of Industry* was published by the New England Workingmen's Association; *Industrial Development in New Hampshire*.
94. Herman Melville, "The Paradise of Bachelors and the Tartarus of Maids," *Harper's New Monthly Magazine* 10 (1855): 670–78.
95. "Petition to Massachusetts Legislature for 10-Hour Day," reprinted in *Voice of Industry*, January 15, 1845.
96. "Massachusetts Investigation into Labor Conditions, 1845."
97. Ibid.
98. Ibid.
99. Hanlan, "Manchester, N.H.," 233–34; Hareven, "Amoskeag Manufacturing Company," 832–33; Public Service of New Hampshire, *Amoskeag Manufacturing Company*; *Industrial Development in New Hampshire*.
100. John S. Garner, "Industrial Utopian Communities," in *The Encyclopedia of New England*, edited by Burt Feintuch and David H. Watters, 861.
101. Ibid.
102. Nathaniel Hawthorne, "The Celestial Railroad; or, Modern Pilgrim's Progress. After the Manner of Bunyan . . ." (1845; Boston: J. V. Hines, 1860); Walter Harding, *The Days of Henry Thoreau* (Princeton: Princeton University Press, 1992), 125–26; Robert D. Richardson, *Henry Thoreau: A Life of the Mind* ([1986]; Berkeley: University of California Press, 1996) 101–3.

103. Ralph Waldo Emerson, *Essays* (Boston: Houghton Mifflin, 1883); Henry David Thoreau, *Excursions: The Writings of Henry David Thoreau* (Boston: Houghton Mifflin, 1893); Sherman Paul, *Emerson's Angle of Vision: Man and Nature in the American Experience* (Cambridge, Mass.: Harvard University Press, 1952); Carl Bode, ed., *The Portable Thoreau* (New York: Viking, 1977); Susan Alves, "Literature of Industrialism," and Anne Baker, "Transcendentalist Writers," both in *The Encyclopedia of New England*, edited by Burt Feintuch and David H. Watters, 1004–6, 1032–33.

104. Alves, "Literature of Industrialism," 1004–6; A. Baker, "Transcendentalist Writers," 1032–33; Melville, "The Paradise of Bachelors and the Tartarus of Maids"; Hawthorne, "The Celestial Railroad."

105. Boyer, *Urban Masses*, 154–55; Bauman et al., *From Tenements to the Taylor Homes*, 1–11.

106. Boyer, *Urban Masses*, 156–57; City Homes Association, *Tenement Conditions in Chicago*.

107. The Ladies' Home Missionary Society was founded in 1844. Anbinder, *Five Points*, 245; Lucy Forsyth Townsend, "Bethune, Joanna," 21–23; *New-York Ladies' Home Missionary Society of Methodist Episcopal Church Board Minute Book* (New York: United Methodist Church, July 1, 1851).

108. Anbinder, *Five Points*, 245–46.

109. Ibid., 247–48.

110. Ibid., 250–51.

111. Ibid., 251–52.

112. Ibid., 252, 254–55.

113. Ibid., 256–57.

114. Ibid., 267.

115. Barbuto, *American Settlement Houses and Progressive Social Reform*, viii, 208, 214; Allen F. Davis, *Spearheads for Reform: The Social Settlements and the Progressive Movement: 1890–1914* (New York: Oxford University Press, 1967), 8–12.

116. Barbuto, *American Settlement Houses and Progressive Social Reform*, 208.

117. Ibid., 50–51, 192; Jeffrey Scheuer, "University Settlement," in *The Encyclopedia of New York*, edited by Kenneth T. Jackson, 1217.

118. Barbuto, *American Settlement Houses and Progressive Social Reform*, 51–53, 186–87; David Coit Scudder and Horace Elisha Scudder, *Life and Letters of David Coit Scudder, Missionary in Southern India* (New York: Hurd and Houghton, and Boston: E. P. Dutton and Company, 1864), iii, 1, 379–85.

119. Davis, *Spearheads for Reform*, 8–12, 23; Arthur Mann, *Yankee Reformers in the Urban Age* (Cambridge, Mass.: Belknap Press, 1954), 223; Boyer, *Urban Masses*; Barbuto, *American Settlement Houses and Progressive Social Reform*, 215.

120. Louise Carroll Wade, *Chicago's Pride: The Stockyards, Packingtown, and Environs in the Nineteenth Century* (Urbana: University of Illinois Press, 1987), 31, 219, 225.

121. Davis, *Spearheads for Reform*, 24, 30, 33–36.

122. Allen F. Davis, "Settlement Houses," in *The Reader's Companion to American History*, edited by Eric Foner and Hohn A. Garrity (Boston: Houghton Mifflin, 1991); Barbara Trainin Blank, "Settlement Houses: Old Idea in New Form Builds Communities,"

New Social Worker 5, no. 3 (1998); Reed Ueda, "Settlement Houses," in *The Encyclopedia of New England,* edited by Burt Feintuch and David H. Watters, 246–47.

123. Barbuto, *American Settlement Houses and Progressive Social Reform,* 4–6, 99–100.
124. Mann, *Yankee Reformers,* 119–20, 218–23.
125. Jane Addams, *Forty Years at Hull-House* (New York: Macmillan, 1935); John Kelly, *Leisure* (Englewood Cliffs, N.J.: Prentice Hall, 1996), 154–60; Robert Gottlieb, *Forcing the Spring: The Transformation of the American Environmental Movement* (Washington, D.C.: Island Press, 1993), 47.
126. Wade, *Chicago's Pride,* 289; *The Town of Lake Directory* (Chicago, 1889), 41–46.
127. See Anbinder, *Five Points*; Riis, *How the Other Half Lives.*
128. Davis, *Spearheads for Reform,* 84–85, 91–95; Allan H. Spear, *Black Chicago: The Making of a Negro Ghetto, 1890–1920* (Chicago: University of Chicago Press, 1967), 103; Barbuto, *American Settlement Houses and Progressive Social Reform,* 42; Mary Houlihan, "A Century of Service at Association House," *Chicago Sun-Times,* January 24, 1999, 21.
129. Frederick A. Bushee, "Population" (1898), in *The City Wilderness: A Settlement Study,* edited by Robert Archey Woods (New York: Garrett Press, 1970), 44; *Report of the Committee to Study Conditions of Negroes in the Pleasant Street District* (Boston: Boston Social Union, South End House, 1910); Davis, *Spearheads for Reform,* 95–96.
130. Du Bois, *The Philadelphia Negro.*
131. Du Bois, "The Black North, a Social Study: New York City," part 1, *New York Times,* November 17, 1901, and part 2, *New York Times,* November 24, 1901; Du Bois, "The Black North, a Social Study: Philadelphia," *New York Times,* December 1, 1901; Du Bois, "The Black North, a Social Study: Boston," *New York Times,* December 8, 1901; Du Bois, "The Black North, a Social Study: Some Conclusions," *New York Times,* December 15, 1901.
132. Du Bois, "The Black North, a Social Study: New York City," parts 1 and 2.
133. Du Bois, "The Black North, a Social Study: Philadelphia."
134. Du Bois, "The Black North, a Social Study: Boston."
135. Davis, *Spearheads for Reform,* 96–102, 226–27; City Homes Association, *Tenement Conditions in Chicago,* 156.
136. Spear, *Black Chicago,* 60, 63, 95; "Douglas/Grand Boulevard: The Past and the Promise. Reverend and Mrs. Ranson," Chicago, Chicago Historical Society, 2006.
137. Fannie Barrier Williams (1855–1944) was the wife of S. Laing Williams, a lawyer. She was a prominent national club leader and a journalist who moved to Chicago from New York in 1887. She served on the board of the Frederick Douglass Center, the Phyllis Wheatley Home, and the Illinois Woman's Alliance. Williams was the first African American and woman to serve on the Chicago Library Board. Susan L. Smith, *Sick and Tired of Being Sick and Tired: Black Women's Health Activism in America: 1890–1950* (Philadelphia: University of Pennsylvania Press, 1995), 21–22.

    Ida B. Wells-Barnett was born in Mississippi. She moved to Memphis when she was fourteen and served as a schoolteacher and the editor of the militant black weekly newspaper *Free Speech.* After the newspaper office was torched, she fled Memphis and embarked on a national speaking tour. She spoke at the Columbian Exposition in

Chicago and met Ferdinand L. Barnett, marrying him in 1895. Barnett was a prominent Chicagoan who founded the city's first black newspaper, the *Conservator*, in 1878. Spear, *Black Chicago*, 59–60.

138. Spear, *Black Chicago*, 101–3.
139. Ibid., 104–5.
140. Ibid., 105.
141. S. L. Smith, *Sick and Tired*, 25–26, 153–54; Sandy Date, "Minneapolis Center Plans 'Homecoming,'" *Star Tribune*, June 22, 2001, 1.
142. Davis, *Spearheads for Reform*, 24, 30, 33–36.
143. Ibid., 86–88; Mann, *Yankee Reformers*, 119; Herbert J. Gans, *The Urban Villagers: Group and Class in the Life of Italian-Americans* (New York: Free Press, 1982), 145–53.
144. Davis, *Spearheads for Reform*, 222–25, 229.
145. Ibid., 93, 227, 229.
146. Ibid., 229–31, 236–38.
147. Ibid., 231–32, 235.
148. Ibid., 235.
149. Gottlieb, *Forcing the Spring*, 275.
150. Curtis Lawrence, "Conference Focuses on Settlement Houses," *Chicago Sun-Times*, July 14, 1999, 27; "Chicago Commissioner on Human Relations, Clarence Wood, Named President, CEO of Jane Addams Hull House Association," *Ascribe Newswire*, June 11, 2000; Blank, "Settlement Houses."
151. Blank, "Settlement Houses"; United Neighborhood Centers of America, "About UNCA," Cleveland, 2003.
152. Louis I. Kuslan, "New Haven Industry: A Retrospective View," in *New Haven: An Illustrated History*, edited by Floyd Shumway and Richard Hegel, 92.
153. *New Haven Register*, October 25, 1833.
154. Neil Hogan, *Moments in New Haven Labor History* (New Haven: Greater New Haven Labor History Association, 2004), 18–19.

5. *Data Gathering*

1. See D. A. Snow and Benford, "Ideology, Frame Resonance," 197–217; Fisher, "Narration."
2. Ellis, *Yellow Fever and Public Health in the New South*, 2–3; George Rosen, *A History of Public Health* (New York: M. D. Publications, 1958), 211–20.
3. Ellis, *Yellow Fever and Public Health in the New South*, 4–5; Edwin Chadwick, *Report on an Inquiry into the Sanitary Condition of the Labouring Population of Great Britain* (Edinburgh: University Press, 1842).
4. Johnson, *The Ghost Map*, 190–228.
5. Ellis, *Yellow Fever and Public Health in the New South*, 8–9; Barbara Gutmann Rosenkrantz, *Public Health and the State: Changing Views in Massachusetts* (Cambridge, Mass.: Harvard University Press, 1972), 14–22.
6. Ellis, *Yellow Fever and Public Health in the New South*, 7; John Griscom, *The Sanitary*

Condition of the Laboring Population of New York (New York: Harper, 1845); John H. Griscom, *Anniversary Discourse, before the New York Academy of Medicine. Delivered in Clinton Hall, November 22, 1854* (New York: New Academy of Medicine, 1855).

7. Ellis, *Yellow Fever and Public Health in the New South*, 35.
8. Geoff King, *Mapping Reality: An Exploration of Cultural Cartographies* (New York: St. Martin's Press, 1996), 20–21; Alex Krieger, "Introduction: Revealing a City/Exploring its Maps," in *Mapping Boston*, edited by Alex Krieger and David Cobb with Amy Turner (Cambridge, Mass.: MIT Press, 1999), xi–xii.
9. David Cobb, "Windows to Our Past: Mapping in the Nineteenth Century and Beyond," and Barbara McCorkle, "The Mapping of New England before 1800," both in *Mapping Boston*, edited by Alex Krieger and David Cobb with Amy Turner, 57–58, 23–24; King, *Mapping Reality*, 23; *The Maggiolo Map* (1527), Vesconte De Maggiolo, cartographer, lithographic print, 24 × 71¼ inches, Harvard Map Collection; *The Gastaldi Map* (1556), Giacomo Di Gastaldi, cartographer, woodblock engraving, 10½ × 14¼ inches, private collection, Venice; Paul E. Cohen and Robert T. Augustyn, *Manhattan in Maps: 1527–1995* (New York: Rizzoli International, 1997), 16–95; Naomi Miller, *Mapping Cities* (Seattle: University of Washington Press, 2000), 9.
10. King, *Mapping Reality*, 28–30, 66.
11. Ibid., 61, 63.
12. Sletcher, *New Haven*, 65; *John Brockett's Map of 1641*, New Haven Colony Historical Society; *A Plan of the Town of New Haven for the Year 1748*, by James Wadsworth, Beinecke Library, Yale University.
13. *The Town of Boston in New England* (1722), John Bonner, cartographer, engraving, Boston; Kreiger et al., *Mapping Boston*, 174–75; Dunn and Dunn, "The Founding, 1681–1701," 7–9.
14. Cobb, "Windows to Our Past," 57; Resolves, June 26, 1794, Massachusetts State Archives; Resolves, June 5, 1795, Massachusetts State Archives.
15. *A Map of the City of New York by the Commissioners Appointed by an Act of the Legislature Passed April 3rd, 1807* (1811), John Randel Jr., cartographer, uncolored manuscript on paper, 106 × 30⁷⁄₁₆ inches, New York Public Library, Manuscript and Archives Division; *Randel Farm Map No. 27: The City of New York as Laid out by the Commissioners Appointed by an Act of the Legislature* (ca. 1819–1820), John Randel Jr., cartographer, pen and ink with watercolor on paper, 32 × 20 inches, Manhattan Borough President's Office; Cohen and Augustyn, *Manhattan in Maps*, 102–10.
16. *The Town of Boston in New England*, Bonner (1722).
17. Cholera map, 1832, New York Public Library.
18. Ibid.
19. Citizens' Association, *Report . . . upon the Sanitary Condition of the City* (New York, 1865); *A Chart Showing the Prevalence of Typhus Fever in the City of New York during the Year 1864*, Council of Hygiene, cartographer, Museum of the City of New York.
20. Michael E. Teller, *The Tuberculosis Movement: A Public Health Campaign in the Progressive Era* (Westport, Conn.: Greenwood Press, 1988), 60–61; Karen Olson, "Old West Baltimore: Segregation, African-American Culture, and the Struggle for Equality," in *The Baltimore Book: New Views on Local History*, edited by Elizabeth Fee, Linda

Shopes, and Linda Zeidman (Philadelphia: Temple University Press, 1991), 61; S. H. Olson, *Baltimore*, 325–28.

21. See Mohr, *Plague and Fire*; Department of Land and Natural Resources, "Honolulu Responds to the Plague."
22. Louis Mazzari, "Fire and Firefighting," in *The Encyclopedia of New England*, edited by Burt Feintuch and David H. Watters (New Haven: Yale University Press, 2005), 224–26.
23. Cobb, "Windows to Our Past," 61; Cannon, "Firefighting," 408–12.
24. Thayer, "Town into City," 79; Dunn and Dunn, "The Founding, 1681–1701," 11; Bronner, "Village into Town," 61; Tinkcom, "The Revolutionary City," 151.
25. Cobb, "Windows to Our Past," 61; Cannon, "Firefighting," 408–12.
26. *The Firemen's Guide* (ca. 1834), P. Desobry, cartographer, lithograph, 16 × 19½ inches, New-York Historical Society; *Map of the Croton Water Pipes with the Stop Cocks* (1842), Endicott, cartographer, uncolored lithograph, 13¾ × 10½ inches, New-York Historical Society; Cohen and Augustyn, *Manhattan in Maps*, 116–18; Cobb, "Windows to Our Past," 61; Cannon, "Firefighting," 408–12. Some estimates have the fire burning as many as seven hundred stores and costing as much as $40 million in damage ($887 million in 2005 dollars). See a reprint of Philip Hone's diary in Jackson and Dunbar, *Empire City*, 166.
27. Goldstein and Izeman, "Water," 1244–46; Cohen and Augustyn, *Manhattan in Maps*, 116–18.
28. Cohen and Augustyn, *Manhattan in Maps*, 116–18; *Map of the Croton Water Pipes with the Stop Cocks* (1842); Goldstein and Izeman, "Water," 1244–46.
29. David Garrard Lowe, *The Great Chicago Fire* (New York: Dover, 1979), 1; Richard F. Bales and Thomas F. Schwartz, *The Great Fire of Chicago and the Myth of Mrs. O'Leary's Cow* (Jefferson, N.C.: McFarland and Company, 2005), 16, 20–21; Richard F. Bales, "Did the Cow Do It? A New Look at the Great Chicago Fire," *Illinois Historical Journal* (Spring 1997): 2–24; "The Fire Fiend," *Chicago Tribune*, October 8, 1871, 1; "Origin of the Fire," *Chicago Workingman's Advocate*, October 28, 1871, 1; *Report of the Chicago Relief and Aid Society of Disbursements and Contributions for the Sufferers of the Chicago Fire* (Cambridge, Mass.: Riverside Press, 1874), 8–15.
30. Lowe, *The Great Chicago Fire*, 1; Bales and Schwartz, *The Great Fire of Chicago*, 9–13, 17; *First Special Report of the Chicago Relief and Aid Society* (Chicago: Culver, Page, Hoyne, and Company, 1871), 8–10; N*ew Chicago: A Full Review of Reconstruction, for the Year, by the Journalists of the Chicago Times* (Chicago: Horton & Leonard Printers, 1872), 5–9; Elaine Lewinnek, '"Domestic and Respectable": Suburbanization and Social Control after the Great Chicago Fire," *Iowa Journal of Cultural Studies* 3 (Fall 2003): 20–39.
31. Bales, "Did the Cow Do it?," 2–24; Bales and Schwartz, *The Great Fire of Chicago*, 12–13, 15, 18, 21, 72–73; "The Fire Fiend," 1.
32. Lowe, *The Great Chicago Fire*, 4–5; Bales and Schwartz, *The Great Fire of Chicago*, 21–27, 34; "Houses for the Houseless," *Chicago Times*, October 18, 1871, 2; "Committee on Shelter," *Chicago Tribune*, October 12, 1871, 2; Lewinnek, "Domestic and Respectable," 20–39.

33. Bales and Schwartz, *The Great Fire of Chicago*, "Rebuild the City," *Chicago Tribune*, October 12, 1871, 2; "Devastated Chicago," *New York Times*, October 11, 1871, 1; Elias Colbert and Everett Chamberlin, *Chicago and the Great Conflagration*. (Cincinnati: C. F. Vent, 1871), 230, 251; Frank Luzerne, *Through the Flames and Beyond, or Chicago as It Was and as It Is* (New York: Wells & Company, 1872), 67–68; "Houses for the Houseless," 2.
34. *New Chicago*, 7–12; "Hesing's Mob," *Chicago Times*, January 16, 1872, 1; "The Fire-Bugs," *Chicago Times*, January 17, 1872, 3; "COMMUNISM," *Chicago Tribune*, January 17, 1872, 2; "The North Side Incendiaries," *Chicago Tribune*, January 17, 1872, 2; "Monday Night's Riot," *Chicago Tribune*, January 17, 1872, 2; Lewinnek, "Domestic and Respectable," 20–39; *Report of the Chicago Relief and Aid Society*, 8–20.
35. Karen Swasilak, *Smoldering City: Chicagoans and the Great Fire, 1871–1874* (Chicago: University of Chicago Press, 1995), 44–46; Catherine and Peter O'Leary, "Transcript of Inquiry into Cause of Chicago Fire and Actions of Fire Department Therein," vol. 1 (Chicago: Chicago Historical Society, 1871), 59–60, 246; *Chicago Times*, October 23, 1871; "The Fire," *Chicago Times*, October 18, 1871, 1; *Chicago Evening Journal*, Extra, October 9, 1971, 1; Luzerne, *Through the Flames and Beyond*, 186–96.
36. Bales, "Did the Cow Do It?," 2–24; Bales and Schwartz, *The Great Fire of Chicago*, 51–138; Colbert and Chamberlin, *Chicago and the Great Conflagration*, 202; Harry A. Musham, "The Great Chicago Fire, October 8–10, 1871," in *Papers in Illinois History and Transactions for the Year 1940* (Springfield: Illinois Historical Society, 1941), 94–95; Anthony DeBartolo, "Who Caused the Great Chicago Fire? A Possible Deathbed Confession," *Chicago Tribune*, October 8, 1997, 1; Anthony DeBartolo, "Odds Improve that a Hot Game of Craps in Mrs. O'Leary's Barn Touched off Chicago Fire," *Chicago Tribune*, March 3, 1998, 1.
37. Kreiger et al., *Mapping Boston*, 210; McNulty, *Boston Then and Now*, 48, 68; Macieski, "Cities and Suburbs," 192; Mazzari, "Fire and Firefighting," 224–26.
38. Mazzari, "Fire and Firefighting," 224–26.
39. Cobb, "Windows to Our Past," 60–61.
40. *Dripps I: Map of the City of New-York Extending Northward to Fiftieth Street* (1851), Matthew Dripps, cartographer, uncolored lithograph, 37½ × 78½ inches, private collection; *Dripps II: Map of That Part of the City and Country of New-York North of 50th St.* (1851), Matthew Dripps and H. A. Jones, cartographers, uncolored lithograph, 37¼ × 78½ inches, private collection; untitled map (ca. 1852–54), William Perris, cartographer, 25 × 34¼ inches, New York Public Library, Map Division; Cohen and Augustyn, *Manhattan in Maps*, 124, 128.
41. Cobb, "Windows to Our Past," 58–59.
42. Though Downing drew the design for an "ideal" park for New York City in 1840, this was a purely hypothetical exercise; it was not the map of an actual park site.
43. *Plan of Drainage for the Grounds of the Central Park* (1855), Egbert Ludovicus Viele, cartographer, hand-colored manuscript, 47 × 135¼ inches, New York Municipal Archives; *Sanitary and Topographical Map of the City and Island of New York* (1865), Egbert Ludovicus Viele, cartographer, colored lithograph, 63 × 17½ inches, Library of Congress; *Topographical Atlas of the City of New York, including the Annexed Territory,*

Showing the Original Water Courses and Made Land (1874), Egbert Ludovicus Viele, cartographer, colored lithograph, New York; Cohen and Augustyn, *Manhattan in Maps*, 130–34.

44. Cohen and Augustyn, *Manhattan in Maps*, 137–38; Viele maps, 1874, 1865, 1855; Johnson, *The Ghost Map*, 101–2.
45. Cobb, "Windows to Our Past," 61–62.
46. Dan Beard, *Hardly a Man Is Now Alive: The Autobiography of Dan Beard* (New York: Doubleday, 1939), 266.

6. Sanitation and Housing Reform

1. Powell, *Bring Out Your Dead*, 20–21, 90.
2. David Schuyler, *The New Urban Landscape: The Redefinition of City Form in Nineteenth-Century America* (Baltimore: Johns Hopkins University Press, 1986), 38–43.
3. Ibid., 54, 60–62; *Report of the New York City Sanitary Association* (New York: Academy of Medicine, 1859), 9–10; Council of Hygiene and Public Health—Citizens Association of New York, *Report upon the Sanitary Condition of the City* (New York: Appleton, 1866).
4. Nye, *Consuming Power*, 94–95; Stuart Galishoff, "Triumph and Failure: The American Response to the Urban Water Supply Problem, 1860–1923," in *Pollution and Reform in American Cities, 1870–1930*, edited by M. Melosi (Austin: University of Texas Press, 1980), 35–36; Spann, *The New Metropolis*, 117–38.
5. Spann, *The New Metropolis*, 117, 120; Schuyler, *The New Urban Landscape*, 60; Joel A. Tarr, "The Search for the Ultimate Sink: Urban Air, Land, and Water Pollution in Historical Perspective," *Records of the Columbia Historical Society* 51 (1984): 1–29; Joseph M. Petulla, *American Environmentalism: Values, Tactics, Priorities* (College Station: Texas A&M University Press, 1987), 24; McMahon, "Beyond Therapeutics," 98–105; E. P. Richardson, "The Athens of America, 1800–1825," 226–30.
6. Spann, *The New Metropolis*, 130–34.
7. Catharine Beecher, *Treatise on Domestic Economy for Use of Young Ladies at Home and at School* (Boston: Thomas H. Webb, 1842); Catharine E. Beecher and Harriet Beecher Stowe, *The American Woman's Home* (New York: J. B. Ford, 1869).
8. The term *oecologie* was coined by Ernst Haeckel, a German disciple of Darwin, to denote the emerging field of science that was concerned with the study of the relations of living organisms to the external world, their habitat, and customs. *Oecologie* comes from the Greek word *oikos*, which originally referred to the family household that had conflict, competition, and mutual aid. Worster, *Nature's Economy*, 192.
9. Gottlieb, *Forcing the Spring*, 216–17; Robert Clarke, *Ellen Swallow: The Woman Who Founded Ecology* (Chicago: Follett, 1973).
10. Mildred Chadsey, "Municipal Housekeeping," *Journal of Home Economics* 7 (February 1915): 53–59.
11. Robin W. Doughty, *Feather Fashions and Bird Preservation: A Study in Nature Protection* (Berkeley: University of California Press, 1975), 15.
12. Boyer, *Urban Masses*, 262–63; Martin V. Melosi, *Garbage in the Cities: Refuse, Reform,*

and the Environment, 1880–1980 (College Station: Texas A&M University Press, 1981), 34–35.

13. Waring rented Olmsted's Staten Island farm during the 1850s. He also managed Horace Greeley's Chapaqua farm for two years. Charles E. Beveridge and David Schuyler, *The Papers of Frederick Law Olmstead,* Vol. 3, *Creating Central Park, 1857–1961* (Baltimore: Johns Hopkins University Press, 1983), 12.
14. Melosi, *Garbage in the Cities,* 72–75; Petulla, *American Environmentalism,* 26; Boyer, *Urban Masses,* 263; Beveridge and Schuyler, *The Papers of Frederick Law Olmstead,* Vol. 3, 12.
15. "Down the Drain: Typhoid Fever City," 2003, Chicago Public Library, Digital Collection.
16. Wade, *Chicago's Pride,* 36–39; Edwin C. Larned, letter to the editor, *Chicago Tribune,* December 18, 1864; "Chicago River," editorial, *Chicago Tribune,* December 19, 1864.
17. Wade, *Chicago's Pride,* 39–40, 134–36.
18. Ibid., xi–xiii.
19. City Homes Association, *Tenement Conditions in Chicago,* 181–83.
20. "Down the Drain: Typhoid Fever City."
21. "South Side Dump Menace to City," *Chicago Daily Tribune,* October 27, 1909, 3; Howard E. Wilson, *Mary McDowell: Neighbor* (Chicago: University of Chicago Press, 1928), 140–41, 144; Barbuto, *American Settlement Houses and Progressive Social Reform,* 125–26.
22. H. E. Wilson, *Mary McDowell,* 141–42; "South Side Dump Menace to City," 3; Gottlieb, *Forcing the Spring,* 64; Wade, *Chicago's Pride,* 36.
23. Wade, *Chicago's Pride,* xiv, 358, 367–79.
24. H. E. Wilson, *Mary McDowell,* 146–51; "South Side Dump Menace to City," 3.
25. Gottlieb, *Forcing the Spring,* 61–69; Davis, *Spearheads for Reform,* 85.
26. H. E. Wilson, *Mary McDowell,* 152, 155–57.
27. Ibid., 158–59. See also Sophonisba P. Breckenridge and Edith Abbott, "Housing Conditions in Chicago, Ill.: Back of the Yards," *American Journal of Sociology* 16, no. 4 (January 1911): 434; Robert A. Slayton, *Back of the Yards: The Making of a Local Democracy* (Chicago: University of Chicago Press, 1986); Edith Abbott, *The Tenements of Chicago, 1908–1935* (Chicago: University of Chicago Press, 1936); Wade, *Chicago's Pride.*
28. H. E. Wilson, *Mary McDowell,* 160–62; Gottlieb, *Forcing the Spring,* 611–69; Upton Sinclair, *The Jungle* (New York: Doubleday, 1906); Breckenridge and Abbott, "Housing Conditions in Chicago," 434; Slayton, *Back of the Yards*; Abbott, *The Tenements of Chicago*; Wade, *Chicago's Pride,* xi, xiv.
29. The Wetlands Initiative (Chicago), "Bubbly Creek," 2003.
30. Joel Schwartz, "Tenements," in *The Encyclopedia of New York City,* edited by Kenneth T. Jackson (New Haven: Yale University Press, 1995), 1161–63; Barbuto, *American Settlement Houses and Progressive Social Reform,* 206.
31. J. Schwartz, "Tenements," 1161–63; Riis, *How the Other Half Lives,* 16.
32. Scobey, *Empire City,* 71.
33. Ibid., 72.
34. Riis, *How the Other Half Lives,* 115.

35. Ibid., 115–16.

36. John H. Griscom, "Public Parks vs. Public Health," *New-York Daily Times*, June 30, 1853; Guernsey, letter to the editor, *Journal of Commerce*, June 24, 1851, 2; Blackmar, *Manhattan for Rent*; Spann, *The New Metropolis*, 67, 94, 142; Gottlieb, *Forcing the Spring*, 55–56.

37. City Homes Association, *Tenement Conditions in Chicago*, 12, 88–90, 181–82; Jane Addams, *Twenty Years at Hull-House: With Autobiographical Notes* (New York: New American Library, 1960), 8.

38. Dorothy Gondos Beers, "The Centennial City, 1865–1876," and Nathaniel Burt and Wallace E. Davies, "The Iron Age, 1876–1905," both in *Philadelphia: A 300-Year History*, edited by Russell F. Weigley, 417–70, 494–96.

39. Beers, "The Centennial City," 421–22; Burt and Davies, "The Iron Age," 494–95.

40. Beers, "The Centennial City," 422.

41. Du Bois, *The Philadelphia Negro*, 287, 290, 295.

42. Ibid., 292–93.

43. Ibid., 294.

44. Ibid., 302.

45. Ibid., 304.

46. William Claiborne, "Chicago Ink Deal to Reconstruct Public Housing," *Washington Post*, February 6, 2000, A2; Susan J. Popkin, Victoria E. Gwiasda, Lynn O. Olson, Dennis P. Rosenbaum, and Larry Buron, *The Hidden War: Crime and the Tragedy of Public Housing in Chicago* (New Brunswick, N.J.: Rutgers University Press, 2000); Alex Kotlowitz, *There Are No Children Here* (New York: Doubleday, 1991).

47. Brands, *T. R.*, 4–5.

48. Griscom, *The Sanitary Condition of the Laboring Population of New York* (New York: Harper, 1845); Griscom, "Public Parks vs. Public Health"; John Griscom, "The Importance of Proper Ventilation," *New York Daily Times*, June 30, 1853, 3; Boyer, *Urban Masses*, 234–35.

49. J. Schwartz, "Tenements," 1161–63; Barbuto, *American Settlement Houses and Progressive Social Reform*, 206.

50. J. Schwartz, "Tenements," 1161–63; Barbuto, *American Settlement Houses and Progressive Social Reform*, 207.

51. Allen J. Share, "Dumbbell Tenements," in *The Encyclopedia of New York City*, edited by Kenneth T. Jackson, 348.

52. Barbuto, *American Settlement Houses and Progressive Social Reform*, 207; Anbinder, *Five Points*, 348.

53. Anbinder, *Five Points*, 348.

54. Ibid., 347; Bauman et al., *From Tenements to the Taylor Homes*, xi, 10; City Homes Association, *Tenement Conditions in Chicago*, 23; J. Schwartz, "Tenements," 1161–63.

55. Riis, *How the Other Half Lives*.

56. Davis, *Spearheads for Reform*, 66–67; Spann, *The New Metropolis*, 144–45; Blackmar, *Manhattan for Rent*; Bauman et al., *From Tenements to the Taylor Homes*, 11; Boyer, *Urban Masses*, 234; Robert B. Fairbanks, "From Better Dwellings to Better Neighborhoods: Rise and Fall of the First National Housing Movement," in *From Tenements to*

the Taylor Homes: In Search of an Urban Housing Policy in Twentieth-Century America, edited by John F. Bauman, Roger Bies, and Kristin M. Szylvian (University Park: Pennsylvania State University Press, 2000), 23–30; J. Schwartz, "Tenements," 1161–63; Share, "Dumbbell Tenements," 348.

57. J. Schwartz, "Tenements," 1161–63.
58. Ibid.; Share, "Dumbbell Tenements," 348.
59. Davis, *Spearheads for Reform*, 66; Boyer, *Urban Masses*, 234; Fairbanks, "From Better Dwellings to Better Neighborhoods," 32.
60. "A Plea for the Park Chairs," *New York Times*, July 5, 1901, 5.
61. Welfare Council of New York City, Charity Organization Society of the City of New York, Association of Catholic Day Nurseries, Community Council of Greater New York, *Directory of Social and Health Agencies of New York City* (New York: Columbia University Press, 1918), p. 76.
62. Davis, *Spearheads for Reform*, 71–72; City Homes Association, *Tenement Conditions in Chicago*, 36, 168; "New Laws Urged to End Congestion," *New York Times*, June 4, 1911, 5; "Skyscrapers Are a Growing Menace," *New York Times*, May 27, 1911, XX2.
63. For more on the crisis of legitimacy, see Jürgen Habermas, *Legitimation Crisis* (Boston: Beacon Press, 1975).

7. Conceptualizing Urban Parks

1. See, for instance, Fox, *The American Conservation Movement*, 128–30; Frederick Jackson Turner, *Rediscovering America: John Muir in His Time and Ours* (San Francisco: Sierra Club Books, 1985), 303.
2. John Muir, *Our National Parks* (1901; Boston: Houghton Mifflin, 1981), 362.
3. Frederick Law Olmsted, "Park," *New American Cyclopaedia: A Popular Dictionary of General Knowledge* 2 (1861): 768–75.
4. Alfred Runte, *National Parks: The American Experience* (Lincoln: University of Nebraska Press, 1987), 2.
5. Raymond Williams, *The Country and the City* (New York: Oxford University Press, 1973), 122–25; Roy Rosenzweig and Elizabeth Blackmar, *The Park and the People: A History of Central Park* (Ithaca: Cornell University Press, 1992), 3–4; Elizabeth Barlow Rogers, *Rebuilding Central Park: A Management and Restoration Plan* (Cambridge, Mass.: MIT Press, 1987), 7; George F. Chadwick, *The Park and the Town: Public Landscape in the 19th and 20th Century* (New York: Praeger, 1966), 163–220; Olmsted, "Park"; Olmsted, Vaux, and Company, *Preliminary Report to the Commissioners for Laying Out a Park in Brooklyn, New York: Being a Consideration of Circumstances and Site and Other Conditions Affecting the Design of Public Pleasure Grounds*, January 24, 1866 (Brooklyn, N.Y.: Board of Commissioners of Prospect Park, 1866); Runte, *National Parks*, 2.
6. Kreiger, "Experiencing Boston," 156; S. B. Warner, "A Brief History of Boston," 3; Macieski, "Cities and Suburbs," 188; Boston Parks and Recreation Commission, *Boston Common Management Plan*, December 1990 (Boston: Boston Parks and Recreation Commission), http://www.cityofboston.gov/parks.

7. Karr, "Boston Common," 212–13.
8. Sletcher, *New Haven*, 65, 109–11; Schiff, "The Social History of New Haven," 104.
9. Timothy Dwight, *Travels in New England and New York* (New Haven: S. Converse, 1821).
10. Thomas. J. Campanella, "Elm Street," in *The Encyclopedia of New England*, edited by Burt Feintuch and David H. Watters (New Haven: Yale University Press, 2005), 223; Henry James, *The American Scene* (London: Chapman and Hall, 1907).
11. Thayer, "Town into City," 102; Geffen, "Industrial Development and Social Crisis," 315–16; Burrows and Wallace, *Gotham*, 85; Latimer, "Bowling Green," 132; Lewison, "Lawn Bowling," 657.
12. Laura Wood Roper, *FLO: A Biography of Frederick Law Olmsted* (Baltimore: Johns Hopkins University Press, 1973); Chadwick, *The Park and the Town*; Spann, *The New Metropolis*, 2.
13. Rosenzweig and Blackmar, *The Park and the People*, 4; Rogers, *Rebuilding Central Park*, 7; G. M. Towle, "Boston," in *Picturesque America; Or, The Land We Live In*, edited by William Cullen Bryant (New York: D. Appleton, 1874), 234–35; Olmsted, Vaux, and Company, *Preliminary Report to the Commissioners for Laying Out a Park in Brooklyn*; Macieski, "Cities and Suburbs," 188; Karr, "Boston Common," 212–13.
14. Edward L. Bergman, "Cemeteries," in *The Encyclopedia of New York City*, edited by Kenneth T. Jackson (New Haven: Yale University Press, 1995), 196; Burrows and Wallace, *Gotham*, 102, 156, 185, 606, 890, 838.
15. Groneman, "Collect," 251; Janvier, *In Old New York*, 53–55; Anbinder, *Five Points*, 14–15.
16. Copied from Janvier, *In Old New York*, 59. See also Jackson and Dunbar, *Empire City*, 116.
17. Jackson and Dunbar, *Empire City*, 116.
18. Dorceta E. Taylor, "Central Park as a Model for Social Control: Urban Parks, Social Class and Leisure Behavior in Nineteenth-Century America," *Journal of Leisure Research* 31, no. 4 (1999): 420–77.
19. Charles E. Beveridge and Carolyn Hoffman, *The Papers of Frederick Law Olmsted, Supplementary Series*, Vol. 1: *Writings on Public Parks, Parkways, and Park Systems* (Baltimore: Johns Hopkins University Press, 1997); Beveridge and Schuyler, *The Papers of Frederick Law Olmstead*, Vol. 3; John Alvah Peterson, "The Origins of the Comprehensive City Planning Ideal in the United States, 1840–1911," Ph.D. diss., Harvard University, 1976; Geoffrey Blodgett, "Frederick Law Olmstead: Landscape Architecture as Conservative Reform," *Journal of American History* 62 (1976): 869–89; Galen Cranz, *The Politics of Park Design: A History of Urban Parks in America* (Cambridge, Mass.: MIT Press, 1982).
20. J. A. Peterson, "The Origins of the Comprehensive City Planning Ideal in the United States," 76; Blodgett, "Frederick Law Olmstead"; Beveridge and Hoffman, *The Papers of Frederick Law Olmsted*, Vol. 1.
21. Nye, *Consuming Power*, 33; Herbert G. Guttman, *Work, Culture, and Society in Industrializing America* (New York: Knopf, 1976), 59–62.
22. Rosenzweig and Blackmar, *The Park and the People*, 26; Bridges, *A City in the Repub-*

lic; Spann, *The New Metropolis*, 77–79, 84–85, 144–45; Roy Lubove, "The New York Association for Improving the Condition of the Poor: The Formative Years," *New-York Historical Society Quarterly* 43 (July 1959): 307–27; Andrew Jackson Downing, "A Talk about Public Parks and Gardens," *Rural Essays*, October 1848, 147–53; Ross, *Social Control*.

23. J. A. Peterson, "The Origins of the Comprehensive City Planning Ideal in the United States," 76; Blodgett, "Frederick Law Olmstead"; Beveridge and Hoffman, *The Papers of Frederick Law Olmsted*, Vol. 1; Olmsted, Vaux, and Company, *Preliminary Report upon the Proposed Suburban Village at Riverside, Near Chicago* (New York: Sutton, Bowne and Company, 1868); Frederick Law Olmsted to Charles Loring Brace, December 1, 1853, Library of Congress, Frederick Law Olmsted Papers.
24. Brace and Olmsted grew up in Hartford. Brace's father was a teacher and head of the Hartford Female Seminary. Though his family was not as well off as Olmsted's, Brace had a more sophisticated education than his friend. Brace attended the Yale Divinity School and Union Theological Seminary. Olmsted and Brace became good friends while Brace was at Yale and Olmsted spent time there visiting his brother. Brace's religious background influenced his work with the Children's Aid Society, which he founded in 1853 when he was twenty-seven; he directed the organization for thirty-seven years. Witold Rybczynski, *A Clearing in the Distance: Frederick Law Olmsted and America in the 19th Century* (New York: Touchstone, 1999), 47, 61, 104.
25. Schuyler, *The New Urban Landscape*, 65.
26. Frederick Law Olmsted, *Public Parks and the Enlargement of Towns* (Cambridge, Mass.: Riverside Press, 1870).
27. Burrows and Wallace, *Gotham*, 172–73, 179; Gerard R. Wolfe, "Governors Island," in *The Encyclopedia of New York City*, edited by Kenneth T. Jackson, 493; Janvier, *In Old New York*, 211–12; "Brief History of Governors Island" (New York: Governors Island Preservation and Education Corporation, n.d.), http://www.gipec.org/PDFs/GI History_timeline.pdf.
28. James E. Mooney, "Fox Hunting," and Steven A. Reiss, "Horse Racing," both in *The Encyclopedia of New York City*, edited by Kenneth T. Jackson, 435–36, 557–59.
29. Tauber, "Lispenard Swamp," 985–86; Janvier, *In Old New York*, 208–12, 253–61.
30. Blackmar, *Manhattan for Rent*, 167.
31. Ibid., 164.
32. By 1833 Ruggles had purchased over five hundred building lots, many of them in contiguous clusters. Most of his holdings were heavily mortgaged. Ibid., 166; Kenneth T. Jackson, "Ruggles, Samuel B(ulkley)," in *The Encyclopedia of New York City*, edited by Kenneth T. Jackson, 1027; Burrows and Wallace, *Gotham*, 577.
33. Burrows and Wallace, *Gotham*, 577; Harriet Davis-Kram, "Gramercy Park," in *The Encyclopedia of New York City*, edited by Kenneth T. Jackson, 497.
34. Clement Clarke Moore placed restrictive covenants on the fashionable residences he developed in Chelsea. Blackmar, *Manhattan for Rent*, 166; Spann, *The New Metropolis*, 104–6; Rybczynski, *A Clearing in the Distance*, 40; Burrows and Wallace, *Gotham*, 578.

35. Davis-Kram, "Gramercy Park," 497; Thurman Wilkins, *John Muir: Apostle of Nature* (Norman: University of Oklahoma Press, 1995), 193–95.
36. Burrows and Wallace, *Gotham*, 577–78; Joyce Gold, "Union Square," in *The Encyclopedia of New York City*, edited by Kenneth T. Jackson, 1211.
37. Burrows and Wallace, *Gotham*, 579; Esther Mipaas, "Tompkins Square," in *The Encyclopedia of New York City*, edited by Kenneth T. Jackson, 1190.
38. Burrows and Wallace, *Gotham*, 577–79; Blackmar, *Manhattan for Rent*, 166.
39. Scherzer, "St. John's Park," 1035.
40. Dunn and Dunn, "The Founding, 1681–1701," 1–10; Burrows and Wallace, *Gotham*, 582–83; Olmsted, Vaux, and Company, *Preliminary Report upon the Proposed Suburban Village at Riverside.*
41. See also Rybczynski, *A Clearing in the Distance*, 31; Nash, *Wilderness*; Downing, "A Talk about Public Parks and Gardens," 138–46; Andrew Jackson Downing, "The New York Park," *Rural Essays*, August 1851, 147–53; "The Middle Park," *New York Times*, June 23, 1853, 4; Rosenzweig and Blackmar, *The Park and the People*, 23; George W. Curtis, "Editor's Easy Chair," *Harper's Magazine*, June 1855, 124–25; Spann, *The New Metropolis.*
42. Richard Morris Hunt, *Designs for the Gateways of Southern Entrances of the Central Park, submitted to the Commissioners of the Central Park* (New York: D. Van Nostrand, 1866).
43. C. C. Cook, "New-York Daguerreotyped. Private Residences," *Putman's Monthly* 3 (March 1854): 247–48.
44. Schuyler, *The New Urban Landscape*, 80.
45. Roper, FLO; Cranz, *The Politics of Park Design*; Beveridge and Hoffman, *The Papers of Frederick Law Olmsted*, Vol. 1; Spann, *The New Metropolis*, 214; Rybczynski, *A Clearing in the Distance*, 20.
46. Rogers, *Rebuilding Central Park*, 7–8.
47. Thomas Cole and Winslow Homer were two noted painters of the Hudson River School.
48. Olmsted claimed these works had a profound impact on his life. Roper, FLO; Cranz, *The Politics of Park Design*; Rybczynski, *A Clearing in the Distance*, 61–65; Janvier, *In Old New York*, xvi.
49. Rybczynski, *A Clearing in the Distance*, 78, 80; Roper, FLO; Cranz, *The Politics of Park Design*; Beveridge and Hoffman, *The Papers of Frederick Law Olmsted*, Vol. 1.
50. Olmsted was elected to the Saturday Club in 1883, the year after Emerson's death. Roper, FLO; F. J. Turner, *Rediscovering America*; Lois Wille, *Forever Open, Clear and Free: The Struggle for Chicago's Lakefront* (Chicago: University of Chicago Press, 1972); Rybczynski, *A Clearing in the Distance*, 20.
51. Blodgett, "Frederick Law Olmstead"; Nash, *Wilderness*; Board of Commissioners of the Department of Public Parks, *Description of the Central Park*, second *Annual Report*, year ending May 1, 1859, reprinted in Beveridge and Schuyler, *The Papers of Frederick Law Olmsted*; Cranz, *The Politics of Park Design*; William Cullen Bryant, ed., *Picturesque America; Or The Land We Live In* (New York: D. Appleton, 1872).

52. Nash, *Wilderness*.
53. Edward Waldo Emerson, *Emerson in Concord* (Boston: Houghton Mifflin, 1889), 29; Ralph Waldo Emerson, *Essays* (Boston: Houghton Mifflin Company, 1883), 161–88; Robert D. Richardson Jr., *Emerson: The Mind on Fire* (Berkeley: University of California Press, 1995), 45.
54. R. D. Richardson, *Emerson*, 3–5, 138–42; 206–9; 283. The Emerson home, built in 1828 by J. J. Coolidge, was originally named Coolidge Castle. The Emersons renamed it Bush.
55. Ibid., 245–51; 433–35; Andrew Jackson Downing, *The Fruits and Fruit Trees of America* (1845; New York: John Wiley, 1864).
56. Spann, *The New Metropolis*, 96–97; Rosenzweig and Blackmar, *The Park and the People*, 27–28.
57. Thorstein Veblen, *The Theory of the Leisure Class* (1899; New York: Penguin, 1967); Rosenzweig and Blackmar, *The Park and the People*, 27–28; Blackmar, *Manhattan for Rent*, 145–46; Daniel M. Bluestone, "From Promenade to Park: The Gregarious Origins of Brooklyn's Park Movement," *American Quarterly* 39 (Winter 1987): 529–50; Curtis, "Editor's Easy Chair."
58. For a discussion of public conveyances as undemocratic spaces, see Robin D. G. Kelley, *Race Rebels: Culture, Politics and the Black Working Class* (New York: Free Press, 1996), 56.
59. The last burial occurred on the Green in 1812. Hillhouse reportedly considered building a private cemetery for his family, but dropped the idea in favor of building a public cemetery for the town. Barbara Rotundo, "Cemeteries," in *The Encyclopedia of New England*, edited by Burt Feintuch and David Watters, 217–18; Grove Street Cemetery, "The Grove Street Cemetery," and "History of the Grove Street Cemetery" (New Haven: Grove Street Cemetery, 2006), http://www.grovestreetcemetery.org.
60. Rotundo, "Cemeteries," 217–18; Grove Street Cemetery, "The Grove Street Cemetery," and "History of the Grove Street Cemetery."
61. The first meeting was attended by seven men and a woman. Jacob Bigelow, *A History of Mt. Auburn Cemetery* (1859; Cambridge, Mass.: Applewood Books, 1988), 1–4, 8, 14, 174; Schuyler, *The New Urban Landscape*, 38–40; Rotundo, "Cemeteries," 217–18.
62. Bigelow, *A History of Mt. Auburn Cemetery*, 4–7, 9–10.
63. Schuyler, *The New Urban Landscape*, 38–43; Bigelow, *A History of Mt. Auburn Cemetery*, 11, 28–30, 252; Mount Auburn Cemetery, "Architecture" (Cambridge, Mass.: Mount Auburn Cemetery, 2006).
64. Bigelow, *A History of Mt. Auburn Cemetery*, 252–54; Tamara Plakins Thornton, "Massachusetts Horticultural Society," in *The Encyclopedia of New England*, edited by Burt Feintuch and David Watters, 48–49; Richard P. Carpenter, "Where Travelers Can Rest in Peace before Their Time," *Boston Globe*, July 16, 2006, M3.
65. Schuyler, *The New Urban Landscape*, 38–43; Rybczynski, *A Clearing in the Distance*, 45; Wainwright, "The Age of Nicholas Biddle," 286; Geffen, "Industrial Development and Social Crisis," 316; Macieski, "Cities and Suburbs," 202; R. Kent Lancaster, "Green Mount: The Introduction of the Rural Cemetery into Baltimore," *Maryland Historical Magazine* 74 (1978): 62–79.

66. R. D. Richardson, *Emerson*, 539–40.
67. Joseph Story, "An Address Delivered on the Dedication of the Cemetery at Mount Auburn, September 24th, 1831," reprinted in Bigelow, *A History of Mt. Auburn*, 154–55.
68. Story, "An Address Delivered on the Dedication of the Cemetery at Mount Auburn," 160–62.
69. Jeffrey I. Richman, *Brooklyn's Green-Wood Cemetery: New York Burial Treasure* (Lunenbrug, Vt.: Stinehour Press, 1998), 3–4; Schuyler, *The New Urban Landscape*, 40.
70. Richman, *Brooklyn's Green-Wood Cemetery*, 4.
71. Schuyler, *The New Urban Landscape*, 39.
72. R. D. Richardson, *Emerson*, 539–40.
73. Bergman, "Cemeteries," 196.
74. Ibid.; Edward L. Bergman, "Green-Wood Cemetery," in *The Encyclopedia of New York City*, edited by Kenneth T. Jackson, 509–10; Burrows and Wallace, *Gotham*, 582–83, 639, 719; Richman, *Brooklyn's Green-Wood Cemetery*, 8–10, 16, 33; Carpenter, "Where Travelers Can Rest in Peace before Their Time," M3.
75. Richman, *Brooklyn's Green-Wood Cemetery*, 19.
76. Catharine Maria Sedgwick, *Letters from Abroad to Kindred at Home* (New York: Harper, 1841), 53–54.
77. Bryant was a neighbor and friend of Olmsted. He was very supportive of Olmsted's appointment and tenure at Central Park. See also M. D'Innocenzo, "William Cullen Bryant and the Newspapers of New York," in *William Cullen Bryant and His America: Centennial Conference Proceedings 1878–1978*, edited by S. Brodwin and M. D'Innocenzo (New York: AMS Press, 1983), 39–50; Chadwick, *The Park and the Town*; Schuyler, *The New Urban Landscape*; *New York Mirror*, July 23, 1842; "The Middle Park," *New York Times*, June 23, 1853, 4; Olmsted, "Park."
78. William Cullen Bryant, "A New Park," *New York Evening Post*, July 3, 1844.
79. Caroline Matilda Kirkland, *Holidays Abroad; Or, Europe from the West* (New York: Baker and Scribner, 1849), 93–94.
80. *Democratic Review* 22 (1848): 476.
81. Towle, "Boston," 235–36; C. D. Gardette, "Philadelphia and Its Suburbs," in *Picturesque America; Or, The Land We Live In*, edited by William Cullen Bryant, 42.
82. The park deteriorated so much that the commissioners suggested subdividing and selling the land. In 1867 Olmsted and Vaux were asked to prepare a park plan. It included a monument for the dead soldiers, a saluting ground, athletic fields, walks, seats, and shaded areas. The site was renamed Fort Greene Park. Judith Berck, "Fort Greene," in *The Encyclopedia of New York City*, edited by Kenneth T. Jackson, 428–29; Spann, *The New Metropolis*, 164; Burrows and Wallace, *Gotham*, 854; David Schuyler and Jane Turner Censer, eds., *The Papers of Frederick Law Olmsted*, Vol. 6: *The Years of Olmsted, Vaux and Company* (Baltimore: Johns Hopkins University Press, 1992), 24.
83. Spann, *The New Metropolis*, 164.
84. Schuyler, *The New Urban Landscape*.
85. Rosenzweig and Blackmar, *The Park and the People*, 24.

86. David Schuyler, *Apostle of Taste: Andrew Jackson Downing: 1815–1852* (Baltimore: Johns Hopkins University Press, 1996); Dolores Hayden, *Building Suburbia: Green Fields and Urban Growth, 1820–2000* (New York: Pantheon, 2003), 26, 34.
87. Roper, *FLO*; Bryant, "A New Park"; Downing, "A Talk about Public Parks and Gardens"; Beveridge and Schuyler, *The Papers of Frederick Law Olmstead*, Vol. 3; Andrew Jackson Downing, *A Treatise in the Theory and Practice of Landscape Gardening Adapted to North America with a View to the Improvement of Country Residences* (1841; New York: C. M. Saxton, 1860); Jackson and Dunbar, *Empire City*, 173; J. C. Loudon, *The Suburban Gardener, and Villa Companion* (London: J. C. Loudon, 1838); Hayden, *Building Suburbia*, 26–27, 34; Schuyler, *Apostle of Taste*, 88.
88. Downing, "The New York Park," and "A Talk about Public Parks and Gardens"; Roper, *FLO*; Bryant, "A New Park."
89. Downing, "The New York Park," 147.
90. Andrew Jackson Downing, "The New York Park," *Horticulturist* 8 (August 1851): 345–49.
91. Downing, "The New York Park," 152.
92. Schuyler, *The New Urban Landscape*; Downing, "A Talk about Public Parks and Gardens," 144; Rosenzweig and Blackmar, *The Park and the People*, 5.
93. Downing, "A Talk about Public Parks and Gardens," 144–45; Schuyler, *The New Urban Landscape*; Jackson and Dunbar, *Empire City*, 173; Chadwick, *The Park and the Town*, 181.
94. Cranz, *The Politics of Park Design*; San Francisco Park Commission, *Report*, 1897, 18–19, http://www.sfgov.org.
95. *San Francisco Bulletin*, November, 3, 1900.
96. "Andrew Jackson Downing, Father of American Parks," *Park International*, July 1920, 48; Downing, "A Talk about Public Parks and Gardens," 145.
97. Rosenzweig and Blackmar, *The Park and the People*, 30; Charles E. Beveridge and Paul Rocheleau, *Frederick Law Olmsted: Designing the American Landscape* (New York: Universe, 1998), 16.
98. Cranz, *The Politics of Park Design*.
99. Olmsted, 1865.
100. Cranz, *The Politics of Park Design*.
101. Beveridge and Hoffman, *The Papers of Frederick Law Olmsted*, Vol. 1; George Curtis Waldo Jr., *History of Bridgeport and Vicinity* (New York: S. J. Clarke, 1917), 277–79; *Bridgeport Municipal Register for 1873*, Bridgeport Department of Parks, 117–18, online at City of Bridgeport's Web page; Frederick Law Olmstead and John Charles Olmstead, *Beardsley Park. Landscape Architects' Preliminary Report*, Bridgeport Park Commission, September 1884 (Boston: privately printed, 1884), 3–14; Schuyler, *The New Urban Landscape*.
102. Geffen, "Industrial Development and Social Crisis," 332.
103. Downing's heavily mortgaged estate was sold nine weeks after his death. Beveridge and Schuyler, *The Papers of Frederick Law Olmstead*, Vol. 3; Chadwick, *The Park and the Town*; Cranz, *The Politics of Park Design*; Beveridge and Hoffman, *The Papers of Frederick Law Olmsted*, Vol. 1; Rosenzweig and Blackmar, *The Park and the People*, 30;

Rybczynski, *A Clearing in the Distance*, 101; Hayden, *Building Suburbia*, 35; Schuyler, *Apostle of Taste*, 222.

104. Olmsted, *Public Parks and the Enlargement of Towns.*

8. Elite Ideology

1. "The East River Park," *Journal of Commerce*, June 24, 1851, 2.
2. James Grant Wilson and John Fiske, eds., *Appleton's Cyclopedia of American Biography*, 6 vols. (New York: D. Appleton, 1887–89).
3. The Jones Wood site was located near the present site of the United Nations. Robert Bowne Minturn Jr., *Memoir of Robert Bowne Minturn* (New York: A. D. F. Randolph, 1871), 65–66; "The East River Park," 2; Rosenzweig and Blackmar, *The Park and the People*, 17, 47; "The Middle Park," *New York Times*, June 23, 1853, 4; Spann, *The New Metropolis*, 111–12, 211; Bridges, *A City in the Republic*, 127; Beveridge and Rocheleau, *Frederick Law Olmsted*, 51.
4. Horace Weeks Lyman, *The Prominent Families of New York; Being an Account in Biographical Form of Individuals and Families Distinguished as Representatives of the Social, Professional, and Civic Life of New York City* (New York: Historical Company, 1897), 406; Rosenzweig and Blackmar, *The Park and the People*, 17; Burrows and Wallace, *Gotham*, 791.
5. "The East River Park," 2; Rosenzweig and Blackmar, *The Park and the People*, 17–22; Homberger, *Mrs. Astor's New York*, 10. See Spann, *The New Metropolis*, 212–15, for details on the network ties of the Minturn Circle.
6. "The East River Park," 2.
7. *Post*, June 7 and July 10, 1851; Rosenzweig and Blackmar, *The Park and the People*, 18.
8. Spann, *The New Metropolis*, 161; "The Proposed Great Park," *Journal of Commerce*, June 5, 1851, 2. The *Journal of Commerce* has slightly different estimates of park acreage and the value of park property.
9. Isaac Newton Phelps Stokes, *Iconography of Manhattan Island, 1498–1909* (1915–28; New York: R. H. Dodd, 1967), 968–72; Rosenzweig and Blackmar, *The Park and the People*, 18–19; "The Middle Park," 4; *Post*, August 5, 1850, June 7, 1851.
10. Bryant owned a Long Island estate, and Greeley owned a seven-acre retreat one mile from Jones Wood.
11. Rosenzweig and Blackmar, *The Park and the People*, 20; "The East River Park," 2; "The Proposed Great Park," 2.
12. Rosenzweig and Blackmar, *The Park and the People*, 20–21; Homberger, *Mrs. Astor's New York*, 238; Bryant, "A New Park."
13. Rosenzweig and Blackmar, *The Park and the People*, 20–21, 40–41, 46–47; Homberger, *Mrs. Astor's New York*, 240–41.
14. Rosenzweig and Blackmar, *The Park and the People*, 21–22, 46–47. In the past private property had been taken by eminent domain to build small parks and squares. See Scobey, *Empire City*, 202–3.
15. Spann, *The New Metropolis*, 212–15; Minturn, *Memoir*; Rosenzweig and Blackmar, *The Park and the People*, 21–22, 31; "The East River Park," 2.

16. Bridges, *A City in the Republic*, 2, 40–41, 46–47; 56; Robert Ernst, *Immigrant Life in New York City, 1825–1863* (New York: King's Crown Press of Columbia University Press, 1949), appendix 1; Rosenzweig and Blackmar, *The Park and the People*, 22; Campbell Gibson, "Population of the 100 Largest Cities in the United States: 1790–1990," working paper 27, June 1998 (Washington, D.C.: Population Division, U.S. Bureau of the Census, 1998).
17. "The Middle Park," 4; "The East River Park," 2.
18. See also Rosenzweig and Blackmar, *The Park and the People*, 23; Blackmar, *Manhattan for Rent*, 151–58.
19. Hilary A. Taylor, "Urban Public Parks, 1840–1900: Design and Meaning," *Garden History* 23, no. 2 (Winter 1995): 202.
20. *Post*, July 10, 1851; Richard Amerman to James W. Beekman, February 12, 1850, New York Historical Society, James W. Beekman Papers, 10: 2; Rosenzweig and Blackmar, *The Park and the People*, 33–34; Towle, "Boston," 234–36. Gardette, "Philadelphia and Its Suburbs," 23–46; "The East River Park," 2; Guernsey, "Letter to the Editor," *Journal of Commerce*, June 24, 1851, 2; Rybczynski, *A Clearing in the Distance*, 93.
21. "The Proposed Great Park," 2; Rosenzweig and Blackmar, *The Park and the People*, 30–31.
22. Guernsey, "Letter to the Editor," 2.
23. *Herald*, July 15, 1850; Downing, "The New York Park," 149–50; "The Opponents of a City Park," *New York Times*, June 27, 1853, 2; Rosenzweig and Blackmar, *The Park and the People*, 212.
24. "The East River Park," 2; Rosenzweig and Blackmar, *The Park and the People*, 33–35.
25. Robert Minturn to James W. Beekman, June 23, 1851, and Rufus Prime to James W. Beekman, June 23, 1851, both letters in James W. Beekman Papers, New York Historical Society; Rosenzweig and Blackmar, *The Park and the People*, 33–35.
26. "The Proposed Great Park," 2; Guernsey, "Letter to the Editor," 2; "The Parks," *New York Times*, June 28, 1853, 4; Rosenzweig and Blackmar, *The Park and the People*, 37.
27. Edward Durand and Dana Durand, *The Finances of New York City* (New York: Macmillan, 1898), 190–95; Spann, *The New Metropolis*, 162; Rosenzweig and Blackmar, *The Park and the People*, 37, 39–44.
28. Cohen and Augustyn, *Manhattan in Maps*, 112; *Journal of Commerce*, June 7, 1849; "The Enlargement of the Battery," *Journal of Commerce*, June 5, 1851, 2; Rosenzweig and Blackmar, *The Park and the People*, 37, 39–44; O. B. Bunce, "New York and Brooklyn," in *Picturesque America; Or, The Land We Live In*, edited by William Cullen Bryant (New York: D. Appleton, 1874), 549; Guernsey, "Letter to the Editor," 2; Rybczynski, *A Clearing in the Distance*, 44.
29. *Staats-Zeitung*, August 1, 1851, April 1, 1853, and April 21, 1854; Sean Wilentz, *Chants Democratic: New York City and the Rise of the American Working Class, 1788–1850* (New York: Oxford University Press, 1984); Rosenzweig and Blackmar, *The Park and the People*, 41–42.
30. New York City Board of Aldermen, *Proceedings*, 1849, Vol. 42: 29–33, 98–99, Common Council, New York Public Library, Miscellaneous Collection; *Staats-Zeitung*, August 8, 1851; Rosenzweig and Blackmar, *The Park and the People*, 44.

31. *Commercial Advertiser*, December 2, 1851; Rosenzweig and Blackmar, *The Park and the People*, 44–45.
32. Daniel Tiemann to James W. Beekman, February 23, 1852, James W. Beekman Papers, letter book 2, New York Historical Society; "The Parks," 4; Rosenzweig and Blackmar, *The Park and the People*, 46–49, 60; Blackmar, *Manhattan for Rent*, 175–76.
33. Rosenzweig and Blackmar, *The Park and the People*, 49, 52–53, 81–82.
34. In 1852 the city sold off some of its uptown landholdings, including about eight hundred parcels that fell within the boundaries of Central Park. Given the speculative land boom in the area once the Central Park site emerged as a viable alternative to Jones Wood, it cost the city a large sum to reacquire this land shortly after selling it. Rosenzweig and Blackmar, *The Park and the People*, 80–81.
35. Beveridge and Schuyler, *The Papers of Frederick Law Olmstead*, Vol. 3, 14; Rosenzweig and Blackmar, *The Park and the People*, 81–82; Downing, "The New York Park," 150.
36. Scobey, *Empire City*, 202–3.
37. Ibid.; Rosenzweig and Blackmar, *The Park and the People*, 33, 45–54; Cohen and Augustyn, *Manhattan in Maps*, 117–18.
38. Cohen and Augustyn, *Manhattan in Maps*, 117.
39. *Map of the Croton Water Pipes with the Stop Cocks* (1842).
40. Griscom, "Public Parks vs. Public Health"; *Staats-Zeitung*, July 16, 1853; Spann, *The New Metropolis*, 132, 169; Rosenzweig and Blackmar, *The Park and the People*, 55, 59.
41. *Herald*, December 24, 1858, March 13, 1859; Rosenzweig and Blackmar, *The Park and the People*, 212. See also Spann, *The New Metropolis*, 171.
42. Beveridge and Schuyler, *The Papers of Frederick Law Olmstead*, Vol. 3, 64–65; Rybczynski, *A Clearing in the Distance*, 161–62.
43. Rybczynski, *A Clearing in the Distance*, 162; Cohen and Augustyn, *Manhattan in Maps*, 130–31.
44. Beveridge and Schuyler, *The Papers of Frederick Law Olmstead*, Vol. 3, 16–17; Rybczynski, *A Clearing in the Distance*, 161–64; Cohen and Augustyn, *Manhattan in Maps*, 134.
45. Viele's plan did not receive a single vote. Stung by the rejection and the fact that he was relieved of his Central Park duties, Viele became an enemy of Olmsted and Vaux and was so for the remainder of his life. Olmsted was promoted to the position of architect in chief of Central Park, and Viele was fired after the competition. Vaux did not fare as well as Olmsted after the competition; he was hired as Olmsted's "assistant" and then later as "consulting architect." In 1858 Viele accused both Olmsted and Vaux of plagiarizing his design. All the entrants in the design competition had Viele's original maps and incorporated elements of his work into their plan. Viele sued the city for wrongful dismissal and for compensation for his topographical and park designs. He asked for $2,500 a year in salary for each year since 1858 and $5,000 for the adoption of his plans. In 1864 a jury awarded him $8,625. During the trial Viele argued that the Greensward plan and other competition designs were merely copies of his plan. Beveridge and Schuyler, *The Papers of Frederick Law Olmstead*, Vol. 3, 65, 69; Rybczynski, *A Clearing in the Distance*, 172–73.

46. See also Beveridge and Schuyler, *The Papers of Frederick Law Olmstead*, Vol. 3; Rybczynski, *A Clearing in the Distance*.
47. Beveridge and Schuyler, *The Papers of Frederick Law Olmstead*, Vol. 3; Chadwick, *The Park and the Town*; Frederick Law Olmsted Jr. and Theodora Kimball, *Forty Years of Landscape Architecture: Central Park* (Cambridge, Mass.: MIT Press, 1922), 45–57; S. B. Sutton, introduction to *Civilizing American Cities: A Selection of Frederick Law Olmsted's Writings on City Landscapes*, edited by S. B. Sutton (New York: Da Capo Press, 1997), 2, 4; Rybczynski, *A Clearing in the Distance*, 23, 29, 40; Beveridge and Rocheleau, *Frederick Law Olmsted*, 30; Alan Wallach, "Thomas Cole," and Angela Miller, "Landscape Painting," both in *The Encyclopedia of New England*, edited by Burt Feintuch and David Watters, 136, 152–53.
48. Frederick Law Olmsted to Mariana Griswold Van Rensselaer, June 1893, Frederick Law Olmsted Papers, Library of Congress.
49. Frederick Law Olmsted, *Preliminary Report upon the Yosemite and Big Tree Grove*, 1865, Frederick Law Olmsted Papers, Library of Congress; Beveridge and Hoffman, *The Papers of Frederick Law Olmsted*, Vol. 1; Victoria Post Ranney, Gerard J. Rauluk, and Carolyn F. Hoffman, eds., *The Papers of Frederick Law Olmsted*, Vol. 5: *The California Frontier: 1863–1865* (Baltimore: Johns Hopkins University Press, 1990).
50. Olmsted, *Preliminary Report upon the Yosemite.*
51. Muir, *Our National Parks*, 2.
52. John Muir, *A Thousand-Mile Walk to the Gulf* (1916; Boston: Houghton Mifflin, 1981), 186.
53. Linnie Marsh Wolfe, *John of the Mountains: Unpublished Journals of John Muir* (Boston: Houghton Mifflin, 1938), 367.
54. Edwin Way Teale, *Wilderness World of John Muir* (1954; Boston: Houghton Mifflin, 1982), 96.
55. Clarence C. Cook, *A Description of the New York Central Park* (New York: F. J. Huntington, 1869), 107.
56. Olmsted, Vaux, and Company, *Report upon a Projected Improvement of the Estate of the College of California, at Berkeley, near Oakland*, June 29, 1866, Frederick Law Olmsted Papers, Library of Congress.
57. Olmsted, "Park"; Beveridge and Hoffman, *The Papers of Frederick Law Olmsted*, Vol. 1, 311.
58. Olmsted, Vaux, and Company, *Preliminary Report to the Commissioners for Laying out a Park in Brooklyn*. See also Olmsted, "Park."
59. Olmsted, *Preliminary Report upon the Yosemite*; Frederick Law Olmsted, *Seventh Annual Report, for the Year 1881* (Boston: Rockwell and Churchill, 1882), 517; Olmsted, Vaux, and Company, *Preliminary Report to the Commissioners for Laying out a Park in Brooklyn*; Charles E. Beveridge, "Frederick Law Olmsted's Theory of Landscape Design," *Nineteenth Century* 3 (1977); Beveridge and Hoffman, *The Papers of Frederick Law Olmsted*, Vol. 1.
60. Rosenzweig and Blackmar, *The Park and the People*, 3; Beveridge and Hoffman, *The Papers of Frederick Law Olmsted*, Vol. 1, 107–8.

61. Olmsted, Vaux, and Company, *Report on the Proposed City Park*, 1868, Frederick Law Olmsted Papers, Library of Congress.
62. Cranz, *The Politics of Park Design*; Downing, "A Talk about Public Parks and Gardens"; Olmsted, *Preliminary Report upon the Yosemite*; *Public Parks and the Enlargement of Towns*; Olmsted, Vaux, and Company, *Preliminary Report to the Commissioners for Laying out a Park in Brooklyn*; Olmsted, Vaux, and Company, *Report on the Proposed City Park*; Schuyler and Censer, *The Papers of Frederick Law Olmsted*, Vol. 6, 424.
63. Beveridge and Hoffman, *The Papers of Frederick Law Olmsted*, Vol. 1, 108, 310; Olmsted, "Park."
64. Olmsted, Vaux, and Company, *Preliminary Report to the Commissioners for Laying out a Park in Brooklyn*.
65. Beveridge and Schuyler, *The Papers of Frederick Law Olmstead*, Vol. 3, 18, 130, 154; Frederick Law Olmsted, *The Greensward Plan*, 1858, Frederick Law Olmsted Papers, Library of Congress; Beveridge, "Frederick Law Olmsted's Theory of Landscape Design," 38–43.
66. Olmsted, Vaux, and Company, *Preliminary Report to the Commissioners for Laying out a Park in Brooklyn*.
67. Cranz, *The Politics of Park Design*; Downing, "A Talk about Public Parks and Gardens"; Olmsted, *Preliminary Report upon the Yosemite*; *Public Parks and the Enlargement of Towns*; Olmsted, Vaux, and Company, *Preliminary Report to the Commissioners for Laying out a Park in Brooklyn*; *Report on the Proposed City Park*; Beveridge and Schuyler, *The Papers of Frederick Law Olmstead*, Vol. 3, 18.
68. Olmsted, Vaux, and Company, *Preliminary Report to the Commissioners for Laying out a Park in Brooklyn*, 58.
69. Ibid.
70. See also Cranz, *The Politics of Park Design*.
71. At the Pacific end of the Panama Canal Olmsted boarded the more spacious steamer *Constitution* to take him up the coast to San Francisco. Beveridge and Schuyler, *The Papers of Frederick Law Olmstead*, Vol. 3, 18–19; Ranney et al., *The Papers of Frederick Law Olmsted*, Vol. 5, 10–11.
72. Beveridge and Schuyler, *The Papers of Frederick Law Olmstead*, Vol. 3, 18–19, 144–45, 151, 175; Frederick Law Olmsted to Mary Perkins Olmsted, September 25, 1863, and Frederick Law Olmsted to Ignaz Pilat, September 26, 1863, both letters in Frederick Law Olmsted Papers, Library of Congress; Charles E. Beveridge, Charles Capen McLaughlin, and David Schuyler, *The Papers of Frederick Law Olmsted*, Vol. 2: *Slavery and the South, 1852–1858* (Baltimore: Johns Hopkins University Press, 1981); Ranney et al., *The Papers of Frederick Law Olmsted*, Vol. 5, 80–92.
73. Olmsted to Ignaz Pilat, September 26, 1863; Ranney et al., *The Papers of Frederick Law Olmsted*, Vol. 5, 85.
74. Olmsted, Vaux, and Company, *Preliminary Report to the Commissioners for Laying out a Park in Brooklyn*.
75. Olmsted to Ignaz Pilat, September 26, 1863; Ranney et al., *The Papers of Frederick Law Olmsted*, Vol. 5, 85.

76. Muir, *A Thousand-Mile Walk to the Gulf*, 187–88.

77. This is in sharp contrast to the wilderness Transcendentalism described in Muir's Yosemite writings. See also Beveridge and Hoffman, *The Papers of Frederick Law Olmsted*, Vol. 1, 47.

78. Ibid.; Cranz, *The Politics of Park Design*; Downing, "A Talk about Public Parks and Gardens"; Olmsted, *Public Parks and the Enlargement of Towns*; Schuyler, *The New Urban Landscape*; Beveridge and Rocheleau, *Frederick Law Olmsted*, 36–38.

79. Beveridge and Rocheleau, *Frederick Law Olmsted*, 32, 38.

80. Olmsted, "Park."

81. Downing, "A Talk about Public Parks and Gardens," 144–45; Schuyler, *The New Urban Landscape*.

82. Olmsted, Vaux, and Company, *Preliminary Report to the Commissioners for Laying out a Park in Brooklyn*; Beveridge and Hoffman, *The Papers of Frederick Law Olmsted*, Vol. 1, 95, 109.

83. Rybczynski, *A Clearing in the Distance*, 168.

84. Frederick Law Olmsted, "Rules and Conditions of Service of the Central Park Keepers," submitted to the Board of Commissioners of Central Park, March 12, 1859, Frederick Law Olmsted Papers, Manuscript Division, Library of Congress.

85. Olmstead, *The Greensward Plan*; Olmsted, Vaux, and Company, *A Review of Recent Changes, and Changes Which Have Been Projected, in the Plans for the Central Park: by the Landscape Architects*, submitted to the Board of Commissioners of the Department of Public Parks, May 1, 1872, Frederick Law Olmsted Papers, Manuscript Division, Library of Congress.

86. Olmstead, *The Greensward Plan*.

87. Schuyler, *The New Urban Landscape*; [Clarence Cook] "Mr. Hunt's Designs for the Gates of the Central Park," *New-York Daily Tribune*, August 2, 1865.

88. Olmsted, *Public Parks and the Enlargement of Towns*.

89. Beveridge and Schuyler, *The Papers of Frederick Law Olmstead*, Vol. 3; Olmstead, *The Greensward Plan*.

90. Olmsted, Vaux and Company, *Preliminary Report to the Commissioners for Laying Out a Park in Brooklyn*.

91. Beveridge and Schuyler, *The Papers of Frederick Law Olmstead*, Vol. 3, 16–17; Chadwick, *The Park and the Town*; Rybczynski, *A Clearing in the Distance*, 161–64; Cohen and Augustyn, *Manhattan in Maps*, 134.

92. Downing, "The New York Park," 150.

93. Bunce, "New York and Brooklyn," 555–56.

94. Rosenzweig and Blackmar, *The Park and the People*, 60, 63.

95. Cohen and Augustyn, *Manhattan in Maps*, 130–31, 134.

96. Board of Commissioners of the Department of Public Parks, *Description of the Central Park*, second *Annual Report*, 59–68. There were eleven commissioners of Central Park. Of these, five were wealthy businessmen and four were lawyers. Six of the commissioners were Republican, four Democrat, and one Know-Nothing. See Scobey, *Empire City*, 228.

97. Rosenzweig and Blackmar, *The Park and the People*, 60, 62, 64, 67.
98. Ibid., 60, 63–64.
99. Cranz, *The Politics of Park Design*; Beveridge and Hoffman, *The Papers of Frederick Law Olmsted*, Vol. 1.
100. Olmsted, Vaux, and Company, *Report on the Proposed City Park*.
101. Rosenzweig and Blackmar, *The Park and the People*, 63.
102. Ellen Gamerman, "New Economy Headed for Harlem: Some Fear the Neighborhood's Identity Is at Stake," *Baltimore Sun*, February 19, 2001, 1; Craig Unger, "Can Harlem Be Born Again?," *New York Magazine*, November 19, 1984, 28–36; Elisabeth Bumiller, "Deal Allows Clinton to Lease Space He Wants in Harlem," *New York Times*, February 17, 2001, 6; Susan Saulny, "Harlem's Executive Perks? Park View, Golden Arches," *New York Times*, February 13, 2001, 4; Allison Fass, "Home Prices Rise in Harlem," *New York Times*, November 9, 2000, 6.
103. Rosenzweig and Blackmar, *The Park and the People*, 65–77.
104. Ibid., 65–71, 89, 175–76.
105. Ibid., 65–77, 90–91; Cohen and Augustyn, *Manhattan in Maps*, 137; *Plan of Drainage for the Grounds of the Central Park* (1855), Egbert Ludovicus Viele, cartographer, hand-colored manuscript, New York Municipal Archives; Spann, *The New Metropolis*, 70, 106–7.
106. Blackmar, *Manhattan for Rent*, 173–75; Anbinder, *Five Points*, 7–13.
107. Spann, *The New Metropolis*, 26–27.
108. *Monthly Record of the Five Points House of Industry* 2 (June 1858): 34–35.
109. Blackmar, *Manhattan for Rent*, 175; Anbinder, *Five Points*, 7–13; "Our Charitable Institutions. The New-York Magdalen Asylum—Incentives Offered the Penitent—Daily Routine of Life Led by the Inmates," *New York Times*, January 14, 1862, 5.
110. Blackmar, *Manhattan for Rent*, 175–76, 209; Leonard Curry, *The Free Black in Urban America, 1800–1850* (Chicago: University of Chicago Press, 1981); Paul Gilje, *The Road to Mobocracy: Popular Disorder in New York City, 1763–1834* (Chapel Hill: University of North Carolina Press, 1988), 145–70; James Ford, *Slums and Housing, with Special Reference to New York City* (Cambridge, Mass.: Harvard University Press, 1936).
111. Philip Kasinitz, "Gentrification and Homelessness: The Single Room Occupant and the Inner City Revival," *Urban and Social Change Review* 17 (1984): 9–14; Phillip Clay, *Neighborhood Renewal* (Lexington, Mass.: Lexington Books, 1979). Clay studied fifty-seven gentrified communities in thirty-nine cities. He found that before gentrification about half the neighborhoods were predominantly black. However, after gentrification, 80 percent of the neighborhoods were predominantly white; only 2 percent were predominantly black.
112. Spann, *The New Metropolis*, 71, 74; Rosenzweig and Blackmar, *The Park and the People*, 55–56; Burrows and Wallace, *Gotham*, 833–34; John A. C. Gray and Charles W. Elliott, "New-York City. Central Park. Shall the Contract System be Adopted?" *New York Times*, August 26, 1858, 8; Robert J. Dillon, "Report of the Minority," *New York Times*, August 26, 1858, 2.
113. Spann, *The New Metropolis*, 71, 74; Burrows and Wallace, *Gotham*, 849.

114. Burrows and Wallace, *Gotham*, 845–846.
115. Spann, *The New Metropolis*, 71, 74; Bridges, *A City in the Republic*, 116; Burrows and Wallace, *Gotham*, 848.
116. Thanks to his father's continued financial support, by the time Olmsted began working at Central Park he had already apprenticed as a surveyor and tried his hand at several other careers, including sailing, farming, publishing, and being a newspaper correspondent. Rybczynski, *A Clearing in the Distance*, 19–152; Piven and Cloward, *Poor People's Movements*; Beveridge and Schuyler, *The Papers of Frederick Law Olmstead*, Vol. 3; Board of Commissioners of the Central Park, *Seventh Annual Report*, September 1864; Calvert Vaux to Frederick Law Olmsted, September 17, 1864, Frederick Law Olmsted Papers, Library of Congress; "Meetings of the Unemployed," *New York Times*, November 6, 1857, 2; "The Central Park," *New York Times*, November 21, 1857, 4.
117. Beveridge and Schuyler, *The Papers of Frederick Law Olmstead*, Vol. 3; Frederick Law Olmsted, "Influence," undated draft manuscript, Frederick Law Olmsted Papers, Library of Congress.
118. Frederick Law Olmsted, *Passages from the Life of an Unpractical Man*, 1857, Frederick Law Olmsted to John Olmsted, October 9, 1857, and Frederick Law Olmsted to Charles Loring Brace, December 8, 1860, all in Frederick Law Olmsted Papers, Library of Congress; Roper, *FLO*; "Meetings of the Unemployed"; "The Central Park," *New York Times*; McLaughlin, Beveridge, and Schuyler, *The Papers of Frederick Law Olmsted*, Vol. 2; Beveridge and Hoffman, *The Papers of Frederick Law Olmsted*, Vol. 1; Beveridge and Schuyler, *The Papers of Frederick Law Olmstead*, Vol. 3.
119. Burrows and Wallace, *Gotham*, 849–850.
120. Piven and Cloward, *Poor People's Movements*; Scobey, *Empire City*, 152.
121. Burrows and Wallace, *Gotham*, 1036–37; Hogan, *Moments in New Haven Labor History*, 14–15, 20–21.
122. Burrows and Wallace, *Gotham*, 605–6, 636–37, 763, 1036–38, 1202; Lisa Weilbacker, "7th Regiment," in *The Encyclopedia of New York City*, edited by Kenneth T. Jackson, 1061–62; Homberger, *Mrs. Astor's New York*, 242.
123. Cranz, *The Politics of Park Design*; Beveridge and Schuyler, *The Papers of Frederick Law Olmstead*, Vol. 3; Beveridge and Hoffman, *The Papers of Frederick Law Olmsted*, Vol. 1.
124. Rosenzweig and Blackmar, *The Park and the People*, 175–76; Bennett, *The Shaping of Black America*, 237–41; Burrows and Wallace, *Gotham*, 555–56, 744.
125. Rybczynski, *A Clearing in the Distance*, 105–8.
126. Prior to working on Central Park, Olmsted had limited contact with the working class. His first extended contact came in 1843 during his year-long voyage to China, where he worked on the *Ronaldson* as a deckhand. He was disgusted by what he saw as the idleness, coarse behavior, and crude language of the crew. Frederick Law Olmsted to John and Mary Olmsted, August 6, 1843; Frederick Law Olmsted to Andrew H. Green, November 3 and 10, 1860; Frederick Law Olmsted to the Board of Commissioners of the Central Park, November 13, 1860; all letters in Frederick Law Olmsted Papers, Library of Congress. Frederick Law Olmsted, "General Order for the Orga-

nization and Routine of Duty of the Keepers' Service of the Central Park," Document No. 43, *Minutes*, March 31, 1873, and Frederick Law Olmsted, "Report of the Landscape Architect on the Recent Changes in the Keepers' Service," Document No. 47, *Minutes*, July 17, 1873, both in New York Department of Public Parks; Olmsted, *Seventh Annual Report*; Beveridge and Hoffman, *The Papers of Frederick Law Olmsted*, Vol. 1; Rybzynski, *A Clearing in the Distance*, 54.

127. Olmsted, *Public Parks and the Enlargement of Towns*; Olmsted and Kimball, *Forty Years of Landscape Architecture*, 534; Rosenzweig and Blackmar, *The Park and the People*, 266.

128. Rosenzweig and Blackmar, *The Park and the People*, 267–75; Rogers, *Rebuilding Central Park*, 75.

129. As the Civil War approached construction slowed on Central Park. After leaving Central Park in 1861 Olmsted took a position as the U.S. Sanitary Commission's secretary and chief executive officer in Washington. He left the Commission in 1863 and went to work in the Yosemite region as director of the Mariposa Mining Company. The company went bankrupt in 1865. Olmsted took over as the head of the first Yosemite Commission for the Yosemite Reserve in 1864 soon after it was created. He drafted the first management plan for the park in 1865. During the time he was away from Central Park he kept in contact with Vaux, who was still affiliated with the park and who encouraged Olmsted to return. Olmsted finally returned to New York in October 1865. Jane Turner Censer, *The Papers of Frederick Law Olmsted*, Vol. 4: *Defending the Union: The Civil War and the U.S. Sanitary Commission, 1861–1863* (Baltimore: Johns Hopkins University Press, 1986), 4, 59; Ranney et al., *The Papers of Frederick Law Olmsted*, Vol. 5, 1, 29. On his Staten Island farm Olmsted required his six employees to provide daily reports of their activities. He also kept a detailed inventory of tools, and other items. At Central Park he devised an elaborate system for recording the hours the laborers worked. Beveridge and Schuyler, *The Papers of Frederick Law Olmstead*, Vol. 3, 2.

130. Frederick Law Olmsted, "Memoranda," March 19, 1874, and April 27, July 8, 1875, all in Frederick Law Olmsted Papers, Library of Congress; Andrew H. Green, "The City Improvements: Fitzjohn's Porter's Incompetence," *New York Times*, November 17, 1875; Rosenzweig and Blackmar, *The Park and the People*, 277, 325–27.

131. The estate, formerly owned by John Fremont, lies just west of what is now Yosemite National Park. Beveridge and Rocheleau, *Frederick Law Olmsted*, 25.

132. To compensate the workers for lost wages, Olmsted did lower the rates charged at the company's four boardinghouses from $1 to $0.85 per day. Ranney et al., *The Papers of Frederick Law Olmsted*, Vol. 5, 15–16, 192–205; Frederick Law Olmsted to George W. Farlee, March 1, 1864; Frederick Law Olmsted to James Hoy, March 2 and 5, 1864; all in Frederick Law Olmsted Papers, Library of Congress; Roper, *FLO*, 260–61.

133. Ranney et al., *The Papers of Frederick Law Olmsted*, Vol. 5, 15–16, 192–205; Morrison G. Wong, *Chinese Americans* (Thousand Oaks, Calif.: Sage, 1995), 73–74; L. Cheng and E. Bonacich, *Labor Migration under Capitalism: Asian Workers in the United States before World War II* (Los Angeles: University of California Press, 1984); B. L. Sung, *The Story of the Chinese in America* (New York: Collier, 1971), 44–45.

134. Olmsted, "Memoranda," March 19, 1874, April 27 and July 8, 1875; Green, "The City Improvements"; Rosenzweig and Blackmar, *The Park and the People*, 277, 325–27.
135. Anbinder, *Five Points*, 399; Burrows and Wallace, *Gotham*, 1127; Mary McDonald, "Chinatown," in *The Encyclopedia of New York City*, edited by Kenneth T. Jackson, 215.
136. Melvyn Dubofsky, *Industrialism and the American Worker* (3rd. ed.; Wheeling, W.Va.: Harlan Davidson, 1996), 94–95, 102–3, 118–19; New York Board of Commissioners of the Department of Public Parks, *Minutes*, 1871–98, March 10, 29, 1880; March 15, April 5, 1882; March 7, May 25, 1883; July 2, 1884; March 31, 1886, New York: Department of Public Parks; microfilm in New York Public Library; Rosenzweig and Blackmar, *The Park and the People*, 298.
137. Scobey, *Empire City*, 212.
138. See New York City, Department of Parks, *Six Years of Park Progress* (New York: Department of Parks, 1940), 3; C. P. Keyser, "Parks and Cities," *Parks and Recreation* 20 (February 1937): 264; Cranz, *The Politics of Park Design*.
139. Kelly, *Leisure*.
140. Olmsted, "General Order for the Organization and Routine of Duty of the Keepers' Service of the Central Park."
141. Olmsted, *Public Parks and the Enlargement of Towns*.
142. Board of Commissioners of the Department of Public Parks, *Description of the Central Park, Second Annual Report*; Blodgett, "Frederick Law Olmstead"; Downing, "The New York Park"; Olmsted, Vaux, and Company, *Report Accompanying a Design for Washington Park*, submitted to the president of the Board of Commissioners of Prospect Park, Brooklyn, 1867.
143. Cranz, *The Politics of Park Design*.
144. Rosenzweig, "Middle-Class Parks and Working-Class Play"; Peiss, *Cheap Amusements*.
145. Beveridge and Schuyler, *The Papers of Frederick Law Olmstead*, Vol. 3; McLaughlin, Beveridge, and Schuyler, *The Papers of Frederick Law Olmsted*, Vol. 2, 197; Olmsted to Charles Loring Brace, December 8, 1860; Olmsted, "Park"; Frederick Law Olmsted, "Central Park Changes," *New-York Daily Tribune*, June 3, 1873, 5; Chadwick, *The Park and the Town*; Beveridge and Hoffman, *The Papers of Frederick Law Olmsted*, Vol. 1; Rogers, *Rebuilding Central Park*, 76.
146. Board of Commissioners of the Department of Public Parks, *Description of the Central Park, Second Annual Report*, 59–68. The park had cricket grounds and accommodated horseback riding and carriages, activities attracting middle-class users. Recreational facilities that attracted working-class users, such as baseball fields, were not provided. Brooklyn Park Commissioners, *Eleventh Annual Report*, 1871, 37–59, Humanities Collection, New York Public Library; Olmsted, Vaux, and Company, *Report Accompanying a Design for Washington Park*.
147. Olmsted, "Central Park Changes."
148. Other landscape architects designed more elaborate gates for other parks. In fact, when a design competition was held in 1863 for the main Central Park entrance, designers drew very elaborate entrances. However, Olmsted, Vaux, and their supporters

opposed such ornate entranceways, believing they detracted from the rest of the park. Schuyler, *The New Urban Landscape.*

149. Beveridge and Schuyler, *The Papers of Frederick Law Olmstead*, Vol. 3; Cranz, *The Politics of Park Design.*
150. Olmsted, "General Order for the Organization and Routine of Duty of the Keepers' Service of the Central Park"; "Report of the Landscape Architect on the Recent Changes in the Keepers' Service."
151. Cranz, *The Politics of Park Design.*
152. Rosenzweig and Blackmar, *The Park and the People*, 314.
153. "The Central Park and Other City Improvements," *New York Herald*, September 6, 1857, 4; Olmsted, *Public Parks and the Enlargement of Towns*; Beveridge and Hoffman, *The Papers of Frederick Law Olmsted*, Vol. 1.
154. Schuyler, *The New Urban Landscape*, 94.
155. While Olmsted was away from Central Park, the keepers' position became a patronage job. When he regained control of the keepers in 1872 there were 149 of them. Beveridge and Hoffman, *The Papers of Frederick Law Olmsted*, Vol. 1; Olmsted to Charles Loring Brace, December 8, 1860.
156. Olmsted to Andrew H. Green, November 3 and 10, 1860; Olmsted to the Board of Commissioners of the Central Park, November 13, 1860; "General Order for the Organization and Routine of Duty of the Keepers' Service of the Central Park"; "Report of the Landscape Architect on the Recent Changes in the Keepers' Service"; Beveridge and Hoffman, *The Papers of Frederick Law Olmsted*, Vol. 1.
157. Schuyler, *The New Urban Landscape*, 94.
158. Beveridge and Hoffman, *The Papers of Frederick Law Olmsted*, Vol. 1.
159. Rogers, *Rebuilding Central Park*, 83.
160. Olmsted to Charles Loring Brace, December 8, 1860; Frederick Law Olmsted to the Board of Commissioners of the Central Park, April 1861, Frederick Law Olmsted Papers, Library of Congress; Olmsted, *Public Parks and the Enlargement of Towns*; "Skating in Central Park," *New York Herald*, December 16, 1860, 8; "A Word to the Central Park Commissioners," *New York Herald*, December 28, 1860, 4; Board of Commissioners of the Central Park, *Seventh Annual Report*; Vaux to Frederick Law Olmsted, September 17, 1864; Chadwick, *The Park and the Town*; Beveridge and Hoffman, *The Papers of Frederick Law Olmsted*, Vol. 1; Rogers, *Rebuilding Central Park*, 23; Olmsted and Kimball, *Forty Years of Landscape Architecture*, 534–36.
161. "Skating in Central Park"; "A Word to the Central Park Commissioners"; Cranz, *The Politics of Park Design*; "Central Park: II," *Scribners Monthly* 6 (October 1873): 680; Beveridge and Hoffman, *The Papers of Frederick Law Olmsted*, Vol. 1; Homberger, *Mrs. Astor's New York*, 80–81.
162. Schuyler and Censer, *The Papers of Frederick Law Olmsted*, Vol. 6, 2–3, 37–42; Jerome Mushkat, "Tweed Ring," in *The Encyclopedia of New York City*, edited by Kenneth T. Jackson, 1206–7.
163. Olmsted to Andrew H. Green, November 10, 1860.
164. Board of Commissioners of the Central Park, *Seventh Annual Report*; Vaux to Frederick Law Olmsted, September 17, 1864; Frederick Law Olmsted, *Minutes*, Board of

Commissioners of the Department of Public Parks, October 23, 1872, Document 41, Frederick Law Olmsted Papers, Library of Congress; Schuyler and Censer, *The Papers of Frederick Law Olmsted*, Vol. 6, 42, 569–70.

165. Beveridge and Hoffman, *The Papers of Frederick Law Olmsted*, Vol. 1; Olmstead, "Central Park Changes"; Olmsted, "Report of the Landscape Architect on the Recent Changes in the Keepers' Service"; Olmsted, "Report of the Landscape Architect on the Recent Changes in the Keepers' Service," 1873; Rogers, *Rebuilding Central Park*, 8; Rybczynski, *A Clearing in the Distance*, 332–37.

166. See Michel Foucault, *Discipline and Punish: The Birth of the Prison* (New York: Vintage, 1979); Michel Foucault, *The Birth of the Clinic: An Archeology of Medical Perception* (London: Tavistock, 1973); Martin Jay, "In the Empire of the Gaze: Foucault and the Denigration of Vision in Twentieth-Century French Thought," in *Foucault: A Critical Reader*, edited by David Couzens Hoy (Cambridge: Basil Blackwell, 1986), 175–204.

167. D. E. Taylor, "Central Park as a Model for Social Control," 420–77.

168. Olmsted to Andrew H. Green, November 10, 1860; Olmsted to the Board of Commissioners of the Central Park, April 1861.

169. Board of Commissioners of the Central Park, *Seventh Annual Report*; Vaux to Frederick Law Olmsted, September 17, 1864.

170. Chadwick, *The Park and the Town*; Beveridge and Hoffman, *The Papers of Frederick Law Olmsted*, Vol. 1.

171. Board of Commissioners of the Central Park, *Seventh Annual Report*; Vaux to Frederick Law Olmsted, September 17, 1864.

172. Board of Commissioners of the Central Park, *Seventh Annual Report*; Vaux to Frederick Law Olmsted, September 17, 1864; Ranney et al., *The Papers of Frederick Law Olmsted*, Vol. 5; Schuyler and Censer, *The Papers of Frederick Law Olmsted*, Vol. 6.

173. Roper, *FLO*; Schuyler and Censer, *The Papers of Frederick Law Olmsted*, Vol. 6; Beveridge and Hoffman, *The Papers of Frederick Law Olmsted*, Vol. 1; Olmsted, Vaux, and Company, *Report Accompanying a Design for Washington Park*; Olmsted, Vaux, and Company, *Architect's Report to the Board of Commissioners of the Newark Park*, October 5, 1867, Frederick Law Olmsted Papers, Library of Congress.

174. Brooklyn Park Commissioners, *Eleventh Annual Report*, 37–51; Olmsted, Vaux, and Company, *To the Chairman of the Committee on Plans of the Park Commission of Philadelphia*, December 4, 1867, Frederick Law Olmsted Papers, Library of Congress.

175. Beveridge and Hoffman, *The Papers of Frederick Law Olmsted*, Vol. 1; Olmsted, *Public Parks and the Enlargement of Towns*; Board of Commissioners of Central Park, *Tenth Annual Report*, 1867, 36–37, Humanities Collection, New York Public Library; Rosenzweig and Blackmar, *The Park and the People*.

176. Though Olmsted toured the campus in 1865 he did not submit his completed report until 1866. Olmsted, Vaux, and Company, *Report upon a Projected Improvement of the Estate of the College of California*.

177. Schuyler and Censer, *The Papers of Frederick Law Olmsted*, Vol. 6, 24; "The New Park Project," *Buffalo Commercial Advertiser*, August 26, 1868, 3; Beveridge and Hoffman, *The Papers of Frederick Law Olmsted*, Vol. 1.

178. Olmsted, Vaux, and Company, *Report on the Proposed City Park.*
179. Rosenzweig and Blackmar, *The Park and the People*, 312.
180. "Hints for the Park Commissioners," *New York Times*, August 25, 1871, 4.
181. Rosenzweig and Blackmar, *The Park and the People*, 312.
182. "Keep off the Grass Order Now Ignored," *New York Times*, July 8, 1901, 10.
183. Olmsted, Vaux, and Company, *To the Board of Park Commissioners of the Borough of New Britain, Connecticut*, March 23, 1870, Frederick Law Olmsted Papers, Library of Congress.
184. Olmsted, Vaux, and Company, *Report Accompanying Plan for Laying Out the South Park*, 1871, Chicago South Park Commission, Frederick Law Olmsted Papers, Library of Congress.
185. Beveridge and Hoffman, *The Papers of Frederick Law Olmsted*, Vol. 1.
186. Cranz, *The Politics of Park Design.*
187. Ross, *Social Control.*
188. Olmsted, *Public Parks and the Enlargement of Towns*; Rosenzweig and Blackmar, *The Park and the People*, 18; Rogers, *Rebuilding Central Park*, 12; Cohen and Augustyn, *Manhattan in Maps*, 134.

9. Social Class, Activism, and Park Use

1. Piven and Cloward, *Poor People's Movements.*
2. It cost the city two-and-a-half times the Olmsted-Vaux estimate to build the smaller Branch Brook Park on the same site thirty years later. Frederick Law Olmsted to Charles K. Hamilton, July 22, 1870, Frederick Law Olmsted Papers, Library of Congress; Roper, *FLO*; Beveridge and Schuyler, *The Papers of Frederick Law Olmstead*, Vol. 3; Board of Commissioners of the Central Park, *Seventh Annual Report*; Vaux to Frederick Law Olmsted, 1864; Chadwick, *The Park and the Town*; Cranz, *The Politics of Park Design.*
3. Rosenzweig and Blackmar, *The Park and the People*, 393.
4. Kelly, *Leisure*; Jerry G. Dickason, "The Origin of the Playground: The Role of Boston Women's Clubs, 1885–1890," *Leisure Sciences* 8, no. 1 (1983): 83–98; Piven and Cloward, *Poor People's Movements.*
5. Herbert J. Gans, *People, Plans, and Politics: Essays on Poverty, Racism, and Other National Urban Problems* (New York: Columbia University Press, 1993); Davis, *Spearheads for Reform*, 12–13, 15–17, 27; William Dwight Porter Bliss, "The Church and Social Reform Workers," *Outlook*, January 20, 1906, 122–25.
6. Dickason, "The Origin of the Playground"; W. Domhoff, *Feminine Half of the Upper Class* (New York: Random House, 1970); Joseph Lee, *Play and Education* (2nd ed.; New York: Macmillan, 1929); Mann, *Yankee Reformers*, vii.
7. See Cranz, *The Politics of Park Design*, for a discussion of the reform park movement.
8. For example see Rosenzweig and Blackmar, *The Park and the People*; Davis, *Spearheads for Reform*, 3–25.
9. For an in-depth study of this neighborhood in the late 1930s, see William Foote Whyte, *Street Corner Society: The Social Structure of an Italian Slum* (4th ed.; Chi-

cago: University of Chicago Press, 1993). Whyte includes detailed descriptions of the neighborhood and the lives of the neighborhood children at the turn of the century. Dickason, "The Origin of the Playground"; J. C. Croly, *The History of the Woman's Club Movement in America* (New York: Henry G. Allen, 1898); Mann, *Yankee Reformers*, 4; Cranz, *The Politics of Park Design*.

10. W. S. Heywood and A. B. Heywood, *Fifty-Second Annual Report of the Benevolent Fraternity of Churches* (Cambridge, Mass.: John Wilson and Son University Press, 1886); Massachusetts Emergency Hygiene Association, *Second Annual Report*, 1886, Boston Public Library.
11. Veblen, *The Theory of the Leisure Class*; Davis, *Spearheads for Reform*, 17.
12. Massachusetts Emergency Hygiene Association, *Second Annual Report*, 12.
13. Boyer, *Urban Masses*, 242.
14. Massachusetts Emergency Hygiene Association, *Fifth Annual Report*, 1889, Boston Public Library, 31; Rosenzweig and Blackmar, *The Park and the People*, 393.
15. Massachusetts Emergency Hygiene Association, *Fifth Annual Report*, 29–30.
16. See Whyte, *Street Corner Society*, 6–7.
17. Rosenzweig and Blackmar, *The Park and the People*, 223; Homberger, *Mrs. Astor's New York*, 221.
18. Hewitt, a nativist, received 90,552 votes; George received 68,110; and Roosevelt, 60,435. Some George supporters believed the election was rigged. Republicans were terrified of the George campaign platform as he gained popularity among the working class. Some political analysts think that Republican voters, believing that Roosevelt did not have a strong enough platform to beat George in the election and that Hewitt stood a better chance, switched their votes and cast ballots for Hewitt instead of Roosevelt. Henry George, *Poverty and Progress: An Inquiry into the cause of Industrial Depression and of Increase of Want with the Increase of Wealth* (1879; New York: Modern Library, 1938); Louis Freeland Post and Frederic Cyrus Leubuscher, *An Account of the George-Hewitt Campaign for the Mayor of New York Municipal Election of 1886* (New York: J. W. Lovell, 1886), 27, 123; C. Lowell Harriss, "George, Henry," in *The Encyclopedia of New York City*, edited by Kenneth T. Jackson, 461–62; Burrows and Wallace, *Gotham*, 1106.
19. Coombs claims that the Common wasn't actually set aside until 1684, when the colony that eventually became Worcester was more firmly established. Zelotes W. Coombs, "Worcester and Worcester Common," n.d., City of Worcester, Mass., and Worcester Department of Public Works and Parks, "City Parks," 2006, both at http://www.ci.worcester.ma.us/dpw/parks_rec; Roy Rosenzweig, "Middle-Class Parks and Working-Class Play: The Struggle over Recreational Space in Worcester, 1870–1910," in *The New England Working Class and the New Labor History*, edited by Herbert G. Gutman and Donald H. Bell (Urbana: University of Illinois Press, 1987), 214–25.
20. Worcester Department of Public Works and Parks, "City Parks."
21. Ibid.; Coombs, "Worcester and Worcester Common," n.d.
22. Worcester Department of Public Works and Parks, "City Parks"; Rosenzweig, "Middle-Class Parks and Working-Class Play," 214–25; Bruce Cohen, "Worcester,

Mass.," in *The Encyclopedia of New England*, edited by Burt Feintuch and David H. Watters, 254–55.

23. See Olmsted, Vaux. and Company, *To the Board of Park Commissioners of the Borough of New Britain, Connecticut*, for examples of parks designed for active recreation; *Buffalo Commercial Advertiser*, August 26, 1868, 3.
24. Rosenzweig, "Middle-Class Parks and Working-Class Play," 217; John Brinckerhoff Jackson, *American Space: The Centennial Years, 1865–1876* (New York: Norton, 1972), 214–15.
25. Friedrich Engels, *The Condition of the Working Class in England in 1844* (London: Allen and Unwin, 1892), 118–19; H. R. Wilensky, "Work, Careers, and Social Integration," *International Social Science Journal* 12 (1960): 543–60; William R. Burch, "The Social Circles of Leisure," *Journal of Leisure Research* 1, no. 2 (1969): 143; George Bammel and Lei-Lane Burrus-Bammel, *Leisure and Human Behavior* (Dubuque, Iowa: W. C. Brown, 1996), 202–5.
26. Bammel and Burrus-Bammel, *Leisure and Human Behavior*, 204–5; Wilensky, "Work, Careers, and Social Integration," 544.
27. Olmsted, *Public Parks and the Enlargement of Towns*; Beveridge and Hoffman, *The Papers of Frederick Law Olmsted*, Vol. 1; Rosenzweig and Blackmar, *The Park and the People*, 1992.
28. "The Central Park and Other City Improvements," 4. See also Rosenzweig and Blackmar, *The Park and the People*, 48–49; Blackmar, *Manhattan for Rent*, 175–76.
29. Frederick Law Olmsted to James T. Fields, October 21, 1860, Frederick Law Olmsted Papers, Library of Congress; Olmsted, *Public Parks and the Enlargement of Towns*; Frederick Law Olmsted, "A Consideration of the Justifying Value of a Park," talk presented to the American Social Science Association, 1880, Frederick Law Olmsted Papers, Library of Congress.
30. Peiss, *Cheap Amusements*, 13–14.
31. Rosenzweig, "Middle-Class Parks and Working-Class Play," 226.
32. "Editor's Table," *Appleton's Journal*, September 11, 1875, 340–41.
33. "Central Park Visitors Charged for Seats," *New York Times*, June 23, 1901, 23; "Shade for Park Chairs," *New York Times*, July 3, 1901, 6; "Fun with the Easy Chairs in the Parks," *New York Times*, July 5, 1901, 5; "Park Chair Issues Now in the Courts," *New York Times*, July 11, 1901, 10.
34. "Central Park Visitors Charged for Seats," 23.
35. Ibid.
36. Ibid.
37. "Sat in Park Chair and Was Arrested," *New York Times*, July 2, 1901, 1.
38. "Shade for Park Chairs," 6; "Mr. Murphy's Order about Park Chars," *New York Times*, July 4, 1901, 12; "Fun with the Easy Chairs in the Parks," 5.
39. "Shade for Park Chairs," 6.
40. "Mr. Murphy's Order about Park Chars," 12.
41. "A Plea for the Park Chairs," *New York Times*, July 5, 1901, 6.
42. Holly, J. Arthur, "Park Chairs Indorsed," *New York Times*, July 5, 1901, 6.

43. B. M., "A Plea for the 'Pay Chairs,'" *New York Times*, July 9, 1901, 10.
44. "Shade for Park Chairs," 6.
45. Tecumseh Swift, "Opposed to the Chair," *New York Times*, July 7, 1901, 5.
46. "Keep Off the Grass Order Ignored," *New York Times*, July 8, 1901, 10.
47. "'Pay Chair' Men Have Lively Time," *New York Times*, July 9, 1901, 14.
48. "Park Chair Men Have Lively Time; Squad of Police Called to Tackle Crowd in Madison Square," *New York Times*, July 10, 1901, 14; "More Free Chairs for Parks," *New York Times*, July 10, 1901, 14.
49. "Park Chair Issues Now in the Courts," 10; "Park Chair Licenses will be Revoked; Commissioner Clausen Decides to Give Up the Fight," *New York Times*, July 10, 1901, 1; "More Chair Riots; Capt. Flood Tells a Prisoner of the Law Protects the Seats—The Madison Square Crowds," *New York Times*, July 10, 1901, 1; "To Take their Own Chairs; Park Frequenters to Adopt a New System—They Secure Advice from Lawyers," *New York Times*, July 10, 1901, 1; "In Injunction Routs Mr. Spate's Forces; Philadelphian Occupies the Last Chair in Madison Square," *New York Times*, July 12, 1901, 3; "The Park Pay Chairs," *New York Times*, July 10, 1901, 6.
50. "Hints for the Park Commissioners," *New York Times*, August 25, 1871, 4.
51. Rosenzweig, "Middle-Class Parks and Working-Class Play," 217–18, 220; Rosenzweig, *Eight Hours for What We Will: Workers and Leisure in an Industrial City, 1870–1920* (Cambridge: Cambridge University Press, 1983), 132.
52. *Worcester Daily Times*, January 6, 12, 25, February 17, March 25, April 30, May 25, June 23, 1885.
53. *Worcester Evening Star*, August 5, 7, 1879; *Worcester Daily Times*, March 18, 1881, June 25, July 25, August 1, 14, 1879.
54. Rosenzweig, *Eight Hours for What We Will*; Worcester Department of Public Works and Parks, "City Parks."
55. *Worcester Sunday Times*, December 28, 1884.
56. Worcester Department of Public Works and Parks, "City Parks."
57. Meg Kerr, "The Blackstone River," paper prepared for Kingston, Rhode Island Sea Grant, 1990; Blackstone Canal Conservancy, "Blackstone Canal: Historical Background," 2004, Hopedale, Mass.
58. Rosenzweig, "Middle-Class Parks and Working-Class Play," 222–23; Rosenzweig, *Eight Hours for What We Will*, 135.
59. Cass Gilbert and Frederick Law Olmsted (Jr.), *Report of the New Haven Civic Improvement Commission* (New Haven: Civic Improvement Commission, 1910), 35. See also Beveridge and Schuyler, *The Papers of Frederick Law Olmstead*, Vol. 3, for similar arguments about the virtues of urban parks.
60. Worcester Department of Public Works and Parks, "City Parks."
61. Rosenzweig, "Middle-Class Parks and Working-Class Play," 222–23.
62. *Worcester Daily Times*, July 5, 1887.
63. Worcester Department of Public Works and Parks, "City Parks."
64. Rosenzweig, *Eight Hours for What We Will*, 136; Worcester Department of Public Works and Parks, "City Parks."

65. Rosenzweig, *Eight Hours for What We Will*, 137; Rosenzweig, "Middle-Class Parks and Working-Class Play," 222–23.
66. Rosenzweig and Blackmar, *The Park and the People*, 304.
67. "Devery Gives Away a Lot of Money," *New York Times*, July 17, 1902, 1.
68. Sletcher, *New Haven*, 111.
69. Ibid., 109–11.
70. Campanella, "Elm Street," 223.
71. Boyer, *Urban Masses*, 236.
72. "Public Parks of Leading Cities," part 1, *New York Times*, August 4, 1895, 20.
73. "Public Parks of Leading Cities," part 2, *New York Times*, September 1, 1895, 28.
74. "Public Parks of Leading Cities," part 1, 20.
75. Ibid.
76. City Homes Association, *Tenement Conditions in Chicago*, 168–72; "Public Parks of Leading Cities," part 1, 20.
77. "Public Parks of Leading Cities," part 1, 20.
78. "Public Parks of Leading Cities," part 2, 28.
79. "Public Parks of Leading Cities," part 1, 20.
80. "Public Parks of Leading Cities," part 2, 28.
81. Ibid.
82. Beveridge and Hoffman, *The Papers of Frederick Law Olmsted*, Vol. 1; "Public Parks of Leading Cities," part 1, 20.
83. "Public Parks of Leading Cities," part 1, 20.
84. Ibid.
85. Ibid.
86. Ibid.
87. *Worcester Evening Gazette*, May 1, 1877.
88. U.S. Department of Commerce, Historical Statistics of the United States (Washington, D.C.: Government Printing Office, 1975), 164–68; Rosenzweig and Blackmar, *The Park and the People*, 232; Nye. *Consuming Power*, 161; Rogers, *Rebuilding Central Park*, 233–34.
89. Tristram Lozaw, "Amusement and Theme Parks," in *The Encyclopedia of New England*, edited by Burt Feintuch and David H. Watters, 1406–7.
90. Anbinder, *Five Points*, 178, 188–90, 195–97.
91. Rosenzweig, *Eight Hours for What We Will*, 173, 179, 181; Peiss, *Cheap Amusements*, 115–38.
92. Hogan, *Moments in New Haven Labor History*, 32.
93. The Davis family had originally offered the city land for the water park more than twenty years earlier. Worcester Department of Public Works and Parks, "City Parks." Lincoln's quotes from "Lake Park," City of Worcester, Massachusetts Public Works and Parks, n.d., http://www.ci.worcester.ma.us/dpw/parks_rec.
94. Rosenzweig, *Eight Hours for What We Will*, 176; *Worcester Telegram*, July 7, 1896; *Worcester Evening Gazette*, July 15, 1907.
95. Rosenzweig, *Eight Hours for What We Will*, 144–45.

96. Rosenzweig and Blackmar, *The Park and the People*, 233–37; *New-York Tribune*, July 6, 1859; *New-York Tribune*, July 4, 1866; Joy Kestenbaum, "Jones Wood," in *The Encyclopedia of New York City*, edited by Kenneth T. Jackson, 626.
97. *Buffalo Commercial Advertiser*, August 26, 1868, 3; Olmsted, Vaux, and Company, *To the Board of Park Commissioners of New Britain, Connecticut*, March 23, 1870.
98. Gilbert and Olmsted, *Report of the New Haven Civic Improvement Commission*, 35. See also D. H. Perkins, ed., *Report of the Special Park Commission upon a Metropolitan Park System* (Chicago: W. J. Hartman, 1905).
99. See Gary Alan Fine, *With the Boys: Little League Baseball and Preadolescent Culture* (Chicago: University of Chicago Press, 1987); Allen Guttman, A *Whole New Ballgame: An Interpretation of American Sports* (Chapel Hill: University of North Carolina Press, 1988).
100. Rosenzweig, *Eight Hours for What We Will*, 144; Boyer, *Urban Masses*, 240.
101. Lee, *Play and Education*; Gans, *People, Plans, and Politics*, 125; Rosenzweig, *Eight Hours for What We Will*, 145–46.
102. Rosenzweig, *Eight Hours for What We Will*, 150–51; *Worcester Sunday Telegram*, August 4, 1912. See also Parsons, *The Social System*; Gibbs, *Norms, Deviance, and Social Control*, for discussions of social control.
103. Boyer, *Urban Masses*, 220.
104. Dubofsky, *Industrialism and the American Worker*, 114; Nye, *Consuming Power*, 163; Lewis Erenberg, *Steppin' Out: New York Nightlife and the Transformation of American Culture, 1890–1930* (Westport, Conn.: Greenwood Press, 1981), 146–71; David Nasaw, *Going Out: The Rise and Fall of Public Amusements* (New York: Basic Books, 1993); Riis, *How the Other Half Lives*, 176–77.
105. Burrows and Wallace, *Gotham*, 798.
106. Karen J. Blair, "Colony Club," and "Women's Clubs," both in *The Encyclopedia of New York City*, edited by Kenneth T. Jackson, 255, 1270; David Von Drehle, *Triangle: The Fire That Changed America* (New York: Grove Press, 2003), 24.
107. Elisabeth Israels Perry, "Women's City Club of New York," in *The Encyclopedia of New York City*, edited by Kenneth T. Jackson, 1269.
108. Peiss, *Cheap Amusements*, 163; Nye, *Consuming Power*, 163.
109. Burrows and Wallace, *Gotham*, 1132–34; Homberger, *Mrs. Astor's New York*, 168.
110. Burrows and Wallace, *Gotham*, 1135–36.
111. Ibid.
112. Peiss, *Cheap Amusements*, 164–67, 178–79.
113. *Steel Labor*, July 1951, 12, October 1951, 5; *Steelworker News*, July 12, 1945, 6, February 27, 1949, 1; Hurley, *Environmental Inequalities*, 92–93; Robert Bruno, *Steelworker Alley: How Class Works in Youngstown* (Ithaca: Cornell University Press, 1999). For discussions of framing, see D. A. Snow et al., "Frame Alignment Processes, Micro-Mobilization, and Movement Participation"; D. A. Snow and Benford, "Master Frames and Cycles of Protest"; and Gamson, "The Social Psychology of Collective Action."
114. Jack Metzgar, *Striking Steel: Solidarity Remembered* (Philadelphia: Temple University Press, 2000), 32.

115. Hurley, *Environmental Inequalities*, 95; "Indiana Dunes: History and Culture," National Park Service, 2008, http://www.nps.gov/indu/historyculture.

116. Blacks and Latinos living in Gary had access to neither Wolf Lake nor Miller's beaches.

117. Hurley, *Environmental Inequalities*, 93.

118. Ibid., 101–2.

119. Ibid., 102.

120. Ibid., 102–3.

121. Ibid., 103.

122. Drake and Cayton, *Black Metropolis*, 104.

123. Tuttle, *Race Riot*, 3–10, 35–37, 199, 234–35; *Chicago Herald-Examiner*, August 3, 1919; Drake and Cayton, *Black Metropolis*, 64–66.

124. Drake and Cayton, *Black Metropolis*, 46–47, 69–71.

125. Ibid., 105.

126. Ibid., 103–5.

127. Ibid.

128. Ibid., 106.

129. Hurley, *Environmental Inequalities*, 119–20.

130. Ibid., 120–21.

131. Ibid., 22.

10. Efforts to Finance Urban Parks

1. Eric Lipton, "City Rejects the First Two Offers for Clinton's Harlem Office Site," *New York Times*, February 15, 2001, 8; Josh Getlin, "Deal Permits Clinton to Get Harlem Office Space," *Los Angeles Times*, February 17, 2001, 14; Bumiller, "Deal Allows Clinton to Lease Space He Wants in Harlem"; Joanne Wasserman and Eric Herman, "Mayor Clears the Way and Bill's Movin' on Up," *New York Daily News*, February 17, 2001, 4.

2. Rosenzweig and Blackmar, *The Park and the People*, 515.

3. Central Park Conservancy, *Annual Report*, 2001, 11.

4. Rosenzweig and Blackmar, *The Park and the People*, 510, 515; Rogers, *Rebuilding Central Park*, 17, 153.

5. Kathy Madden, Philip Myrick, Katherine Brower, Shirley Secunda, and Andrew Schwartz, *Public Parks, Private Partners: How Partnerships Are Revitalizing Urban Parks* (New York: Project for Public Spaces, 2000), 14–15.

6. Central Park Conservancy, *Annual Report*, 2002, 5, and *Annual Report*, 2005, 10; Blaine Harden, "Neighbors Give Central Park a Wealthy Glow," *New York Times*, November 22, 1999, 1; Madden et al., *Public Parks, Private Partners*, 76.

7. An organization created by Rogers to raise funds for Central Park and to help with repairs and maintenance of the park. Rosenzweig and Blackmar, *The Park and the People*, 510; Madden et al., *Public Parks, Private Partners*, 75.

8. Barbara Stewart, "College President to Lead Central Park Conservancy," *New York Times*, December 1, 2000, 6; Rogers, *Rebuilding Central Park*, 153; Central Park Con-

servancy, "About the Central Park Conservancy," 2006, at http://www.centralparknyc.org.

9. Central Park Conservancy, *Annual Report*, 2002, 2.
10. Central Park Conservancy, *Annual Report*, 2001, cover.
11. Harden, "Neighbors Give Central Park a Wealthy Glow," 1.
12. Ibid.
13. Ibid.
14. Central Park Conservancy, "The Gift of a Lifetime," 2006, at http://www.centralparknyc.org.
15. Central Park Conservancy, *Annual Report*, 2005, 12.
16. Theresa Odendahl, *Charity Begins at Home: Generosity and Self Interest among the Philanthropic Elite* (New York: Basic Books, 1990), 150–51; Harden, "Neighbors Give Central Park a Wealthy Glow," 1; Francie Ostrower, *Why the Wealthy Give: The Culture of Elite Philanthropy* (Princeton: Princeton University Press, 1995).
17. The Wallace–Reader's Digest Funds committed $25 million to create and restore urban parks and greenways over a six-year period beginning in 1994. Veblen, *The Theory of the Leisure Class*; Harden, "Neighbors Give Central Park a Wealthy Glow," 1; Rosenzweig and Blackmar, *The Park and the People*, 382, 512–15; Rogers, *Rebuilding Central Park*, 154; Madden et al., *Public Parks, Private Partners*, 3.
18. Central Park Conservancy, "Central Park: Then and Now," 2003, at http://www.centralparknyc.org.
19. Ibid.
20. The first Women's Committee luncheon, held in 1982, had fifty guests. Bill Cunningham, "Abloom in the Park," *New York Times*, May 6, 2001, 50.
21. Downing, "A Talk about Public Parks and Gardens," 145; Olmsted, *Public Parks and the Enlargement of Towns*; Schuyler, *The New Urban Landscape*; Central Park Conservancy, "Central Park Conservancy: Support CPC," 2006, at http://www.centralparknyc.org; Central Park Conservancy, "Central Park: Then and Now"; Harden, "Neighbors Give Central Park a Wealthy Glow," 1.
22. Central Park Conservancy, "Central Park: Then and Now"
23. Prospect Park Alliance, Brooklyn, "Help the Park," 2003, at http://www.prospectpark.org.
24. The Battery Conservancy, New York, "About the Conservancy," 2006, at http://thebattery.org; Riverside Park Fund, "Riverside Park," 2006, at http://www.riversideparkfund.org.
25. City of San Diego, "Balboa Park Endowment Fund," 2006, at http://www.sandiego.gov/endowment-program.
26. Golden Gate National Parks Conservancy, San Francisco, "Support the Parks," 2006, and "Parks For All Forever," 2006, both at http://www.parksconservancy.org.
27. National AIDS Memorial Grove, San Francisco, "Support, Ensure and Remember," 2006, at http://www.aidsmemorial.org.
28. San Francisco Parks Trust, "History and Accomplishments," 2006, at http://www.sfpt.org.

29. Friends of the Public Garden, Boston, "Accomplishments," 2006, at http://friendsofthepublicgarden.org.
30. Six parks comprise the "emerald necklace": the Back Bay Fens, Riverway, Olmsted Park, Jamaica Park, Arnold Arboretum, and Franklin Park.
31. Emerald Necklace Conservancy, Brookline, Mass., "Join Us," 2006, at http://www.emeraldnecklace.org.
32. The Friends of Buttonwood Park, New Bedford, Mass., "The Friends of Buttonwood Park," 2006, at http://www.buttonwoodpark.org.
33. South Suburban Park Foundation, Littleton, Colo., "Current Events," 2006, at http://sspf.org.
34. Parks Foundation, City of Cincinnati, "Cincinnati Parks Foundation," 2006, at http://www.cincinnati-oh.gov/parks/.
35. Forest Park Forever, St. Louis, "Park History," 2006, at http://www.forestparkforever.org.
36. Garfield Park Conservatory Alliance, Chicago, "Support Us," 2006, at http://www.garfield-conservatory.org.
37. Friends of Garfield Park, Indianapolis, "Friends of Garfield Park: Become a Friend," 2006, at http://www.garfieldparkindy,org.
38. Fairmount Park Conservancy, Philadelphia. "Support Your Park," 2006, at http://www.fairmontparkconservancy.org.
39. Maymont Foundation, Richmond, Va., "Support Maymont," 2006, at http://www.maymont.org.
40. Madden et al., *Public Parks, Private Partners*, 17–22, 52–115; City of San Diego, "Balboa Park Endowment Fund"; Golden Gate National Parks Conservancy, at http://www.parksconservancy.org; National AIDS Memorial Grove, "Support, Ensure and Remember"; Parks Foundation, "Cincinnati Parks Foundation"; Forest Park Forever, "Park History"; San Francisco Parks Trust, "History and Accomplishments"; The Greening of Detroit, "Our Mission," 2006, at http://www.greeningofdetroit.com; ParkWorks, Cleveland, "History" and "Mission," both, 2006, at http://www.parkworks.org; Greater Newark Conservancy, "Our Urban Accomplishments," 2006, at http://www.citybloom.org.
41. Madden et al., *Public Parks, Private Partners*, 9–10, 23.
42. Central Park Conservancy, *Annual Report*, 2004 (New York: Central Park Conservancy, 2004), 29; Central Park Conservancy, *Annual Report*, 2003 (New York: Central Park Conservancy, 2003), 49, 51; Central Park Conservancy, *Annual Report*, 2002, 19, 54; Central Park Conservancy, *Annual Report*, 2001, 34–35.
43. Central Park Conservancy, *Annual Report*, 2001, 13, 36.
44. Central Park Conservancy, *Annual Report*, 2005; Central Park Conservancy, *Annual Report*, 2002, 11, 54; Central Park Conservancy, "Central Park: Then and Now"; Central Park Conservancy, *Annual Report*, 2003 (New York: Central Park Conservancy, 2003).
45. Central Park Conservancy, *Annual Report*, 2002, 29.
46. Ibid., 3, 13; Central Park Conservancy, *Annual Report*, 2003.

47. Prospect Park Alliance, Brooklyn, *Annual Report*, 2002, 4, 10, 16, 19; Prospect Park Alliance, Brooklyn, "About Park and Alliance: President's Message (Summer 2003)," 2003.
48. Central Park Conservancy, "Central Park: Then and Now"; *Annual Report*, 2002, 3; Prospect Park Alliance, Brooklyn, *Annual Report*, 2005 (Brooklyn: Prospect Park Alliance, 2005); Prospect Park Alliance, *Annual Report*, 2002, 3, 20–26; Prospect Park Alliance, Brooklyn, *Annual Report*, 2003, 19–20.
49. Prospect Park Alliance, *Annual Report*, 2005, 15; P. Harnik, *Local Parks, Local Financing*, vol. 2 (San Francisco: Trust for Public Land, 1998).
50. In 2006 tickets to the Halloween Ball ranged from $1,000 to $50,000. Central Park Conservancy, "Halloween Ball 2006," 2006, http://www.centralparknyc.org; Harden, "Neighbors Give Central Park a Wealthy Glow," 1; Rosenzweig and Blackmar, *The Park and the People*, 512–16, 523–24; Cohen and Augustyn, *Manhattan in Maps*, 121; Madden et al., *Public Parks, Private Partners*, 95.
51. Charity Navigator, *Charity Navigator Rating* (Mahwah, N.J.: Charity Navigator, 2006).
52. In the wake of the World Trade Center bombing, one-fourth of the money raised was sent to New York for the police and firefighters fund. D. Bisbee, "Hub Benefit Swing into Fall to Aid Parks and N.Y.," *Boston Herald*, September 24, 2001, 39.
53. Zachary Gorchow, "Detroit Seeks to Sell off 92 Parks," *Detroit Free Press*, October 26, 2007; Susan Saulny, "Detroit Considers Selling Small Parks," *New York Times*, December 29, 2007, 11.
54. Harnik, *Local Parks, Local Financing*.
55. Ibid.; Oglebay Foundation, Wheeling, W. Va., "Park History," 2006, at http://www.oglebayfoundation.org.
56. Harnik, *Local Parks, Local Financing*.
57. Madden et al., *Public Parks, Private Partners*, 43.
58. Charity Navigator, *Charity Navigator Rating*.
59. Madden et al., *Public Parks, Private Partners*, 53–56, 88–91, 107.
60. Central Park Conservancy, *Annual Report*, 2005, and *Annual Report*, 2002, 26–27.
61. Madden et al., *Public Parks, Private Partners*, 37; Harden, "Neighbors Give Central Park a Wealthy Glow," 1.
62. E. B. Rogers, "Making Partnership Work: The Central Park Model," speech given at the Urban Park Institute conference, sponsored by Project for Public Spaces, July 2004.
63. Harnik, *Local Parks, Local Financing*.
64. P. Harnik and L. Yaffe, *Who's Going to Pay for This Park?* (San Francisco: Trust for Public Land, 2006); Harnik, *Local Parks, Local Financing*; K. Hopper, *Local Parks, Local Financing* (San Francisco: Trust for Public Land, 1998).
65. Harnik and Yaffe, *Who's Going to Pay for This Park?*
66. Ibid.
67. Great Park Conservancy, Irvine, Calif., "News from the Park: The Jewel of Orange County," 2005; Denver City Council, "A Bill for an Ordinance Approving a Proposed

Agreement between the City and County of Denver and Stapleton Development Corporation Concerning the Master Lease and Disposition Agreement," 1998; *Denver Post Corporation v. Stapleton Development Corporation*, 2000, 19P.3d 36, 2000 Colo. App., November 24, 2; Laurel Hill Adaptive Reuse Citizens Advisory Committee, *Laurel Hill Draft Park Conceptual Plan*, July 2003 (Fairfax, Va.: Fairfax County Park Authority, 2003); Fairfax County Park Authority, Virginia, "Laurel Hill," 2003.

68. New York Department of City Planning, "Fresh Kills: Landfill to Landscape," 2003, at http://www.nyc.gov; New York Department of City Planning, "Mayor Michael R. Bloomberg Kicks Off the Transformation of Fresh Kills in Staten Island," press release, September 29, 2003, at http://www.nyc.gov.
69. Trust for Public Land, San Francisco, *City Park Facts*, 2006.
70. Great Park Conservancy, "News from the Park: The Jewel of Orange County"; Denver City Council, "A Bill for an Ordinance"; *Denver Post Corporation v. Stapleton Development Corporation*; Laurel Hill Adaptive Reuse Citizens Advisory Committee, *Laurel Hill Draft Park Conceptual Plan*; Fairfax County Park Authority, "Laurel Hill"; New York Department of City Planning, "Fresh Kills: Landfill to Landscape"; New York Department of City Planning, "Mayor Michael R. Bloomberg Kicks Off the Transformation of Fresh Kills in Staten Island."
71. J. O. Pasco, "Meeting of O.C. Park Panel being Investigated," *Los Angeles Times*, January 27, 2006, 10; J. O. Pasco, "Private Control of Great Park Opposed," *Los Angeles Times*, August 24, 2003, 3; Great Park Conservancy, "News from the Park," *Denver Post Corporation v. Stapleton Corporation*.
72. For discussions of power elites, see Mills, "The Power Elite"; Prewitt and Stone, *The Ruling Elites*. For discussions of interlocking directorates, see DiMaggio, "State Expansion," 147–60; Mark S. Mizruchi, "Cohesion, Structural Equivalence, and Similarity of Behavior: An Approach to the Study of Corporate Political Power," *Sociology Theory* 8 (1990): 16–32; Walter Powell and Paul DiMaggio, *The New Institutionalism in Organizational Analysis* (Chicago: University of Chicago Press, 1991).
73. Michael Schwartz and Shuva Paul, "Resource Mobilization versus the Mobilization of People: Why Consensus Movements Cannot be Instruments of Social Change," in *Frontiers in Social Movement Theory*, edited by Aldon Morris and Carol McClurg Mueller (New Haven: Yale University Press, 1992), 205–23.
74. D. E. Taylor, "The Rise of the Environmental Justice Paradigm," 508–80.

11. Class, Race, Space, and Zoning

1. Burrows and Wallace, *Gotham*, 85; Dunn and Dunn, "The Founding, 1681–1701," 1–10; Sletcher, *New Haven*, 65; Corey, "Waste Management," 620–21.
2. See Burrows and Wallace, *Gotham*, 178; Ramirez, "Greenwich Village," 506–9; Pascu, "Rutgers, Henry," 1030; Reiger, "Chelsea," 209; Janvier, *In Old New York*, 85–88, 117.
3. Burrows and Wallace, *Gotham*, 456–57.
4. Hayden, *Building Suburbia*, 46.
5. Robert Fishman, *Bourgeois Utopias: The Rise and Fall of Suburbia* (New York: Basic

Books, 1987), 18–73; John Archer, "Country and City in the American Romantic Suburb," *Journal of the Society of Architectural Historians* 42 (May 1983): 140–47; Hayden, *Building Suburbia*, 47.

6. Scherzer, "St. John's Park," 1035; Burrows and Wallace, *Gotham*, 457–58, 473.
7. Blackmar, *Manhattan for Rent*, 164.
8. Burrows and Wallace, *Gotham*, 582–83; Archer, "Country and City in the American Romantic Suburb," 150–51; Hayden, *Building Suburbia*, 47–48.
9. Burrows and Wallace, *Gotham*, 718; Scobey, *Empire City*, 116; Alexander Jackson Davis, *Rural Residences* (New York: A. J. Davis, 1837).
10. See, for instance, Beecher, *Treatise on Domestic Economy*; Beecher and Stowe, *The American Woman's Home*.
11. Olmsted, Vaux, and Company, *Report upon a Projected Improvement of the Estate of the College of California*; Schuyler and Censer, *The Papers of Frederick Law Olmsted*, Vol. 6, 28–29; Ranney et al., *The Papers of Frederick Law Olmsted*, Vol. 5, 546–73.
12. Olmstead, *The Greensward Plan*; Olmsted, Vaux, and Company, *A Review of Recent Changes*.
13. Schuyler and Censer, *The Papers of Frederick Law Olmsted*, Vol. 6, 29.
14. Olmsted, Vaux, and Company, *Report upon a Projected Improvement of the Estate of the College of California*. Though this report is signed as an Olmsted and Vaux collaboration, Olmsted probably wrote it and drew the design of Riverside on his own since Vaux was traveling in Europe at the time Olmsted made the site visit and completed the report. See Schuyler and Censer, *The Papers of Frederick Law Olmsted*, Vol. 6, 289.
15. Olmsted, Vaux, and Company, *Report upon a Projected Improvement of the Estate of the College of California*.
16. Ibid.
17. Ibid.
18. Olmsted, Vaux, and Company, *A Review of Recent Changes*; Schuyler and Censer, *The Papers of Frederick Law Olmsted*, Vol. 6, 33–34.
19. Richman, *Brooklyn's Green-Wood Cemetery*, 15–16.
20. Blackmar, *Manhattan for Rent*, 30–35, 41–42.
21. Ibid., 33.
22. Board of Commissioners of the Department of Public Parks, *Description of the Central Park*, second *Annual Report*, 59–68; Rosenzweig and Blackmar, *The Park and the People*, 60, 62, 64, 67.
23. Max Page, *The Creative Destruction of Manhattan, 1900–1940* (Chicago: University of Chicago Press, 1999), 26–28, 44; Scobey, *Empire City*, 82.
24. George Templeton Strong, *The Diary of George Templeton Strong* (New York: Macmillan, 1952), entry of March 18, 1865, 565–66.
25. Page, *The Creative Destruction of Manhattan*, 45–46; Burrows and Wallace, *Gotham*, 960, 1076–77; "Former Rockefeller Residence on Fifth Avenue Resold by Harr Mandel," *New York Times*, February 7, 1925, 28.
26. Downing, "The New York Park," 147; Kirkland, *Holidays Abroad*, 93–94.
27. Homberger, *Mrs. Astor's New York*, 65.

28. Page, *The Creative Destruction of Manhattan*, 46–48.
29. Ibid., 49–50.
30. *Rosa Zipp v. Frances E. Barker and Samuel P. Barker*, 40 A.D. 1, 57 N.Y.S. 569, 1899 N.Y. App. Div., April.
31. *Wallack Construction Company v. Smalwich Realty Corporation and United International Corporation*, 201 A.D. 133, 194 N.Y.S. 32, 1922 N.Y. App Div., May 5.
32. *Charles C. Bull as Trustee for Adelaide B. Harris, under the Will of William V. Brady, and Others, v. Frank V. Burton and J. Howes Burton*, 117 A.D. 824, 164 N.Y.S. 997, 1917 N.Y. App. Div., May 4.
33. *Trustees of Columbia College v. Thacher*, 1881, 87 N.Y. 311, 1881 N.Y., January 17.
34. *Zipp v. Barker*, 1899.
35. *Henry Roth v. Jerome Jung*, 79 A.D. 1, 79 N.Y.S. 822, 1903 N.Y. App. Div., January.
36. *Roth v. Jung*, 1903.
37. *John McClure v. Robert J. Leaycraft*, 97 A.D. 518, 90 N.Y.S. 233, 1904 N.Y. App. Div., November; *John McClure v. Robert J. Leaycraft*, 183 N. Y. 36, 75 N.E. 961, 1905 N.Y., October 27.
38. *Carl Schefer v. Thomas R. Ball*, 53 Misc. 448, 104 N.Y.S. 1028, 1907 N.Y. Misc., March; *Carl Schefer v. Thomas R. Ball*, 192 N.Y. 589, 85 N.E. 1115, 1908 N.Y., September 29.
39. *Batchelor v. Hinkle*, 1909, 132 A.D. 620, 117 N.Y.S. 542, 1909 N.Y. App. Div., June 4; *Batchelor v. Hinkle*, 1914, 210 N.Y. 243, 104 N.E. 629, 1914 N.Y., February 24.
40. *Batchelor v. Hinkle*, 1910, 140 A.D. 621, 125 N.Y.S. 929, 1910 N.Y. App. Div., November 18.
41. *Batchelor v. Hinkle*, 1912, 149 A.D. 910, 133 N.Y.S. 501, 1912 N.Y. App. Div., February.
42. *Batchelor v. Hinkle*, 1914.
43. "Upper West Side Building Tendencies—A Reconstruction Movement Imminent," *Real Estate Record and Builders' Guide*, January 1, 1910; *Real Estate Record and Builders' Guide*, January 8, 1910, 51; "Upper Fifth Avenue's Future," *Real Estate Record and Builders' Guide*, April 12, 1924; Page, *The Creative Destruction of Manhattan*, 50–51; *Schefer v. Ball*, 1907.

12. Land Use and Zoning in American Cities

1. Sletcher, *New Haven*, 110.
2. Scobey, *Empire City*, 35, 158.
3. Fifth Avenue Association, *Report for the Year*, 1912, 3, and *Report for the Year*, 1916, cover, both in New York City Department of Records, City Hall Library; Fifth Avenue Association, *Fifty Years on Fifth, 1907–1957* (New York: International Press, 1957), 36; "5th Av. Merchants Fight Paper Waste," *New York Times*, December 4, 1916, 15; Page, *The Creative Destruction of Manhattan*, 53–54.
4. "Fifth Avenue to Be the Best of Streets," *New York Times*, January 23, 1910, 10.
5. "Foretell Greater 5th Av., Newspaper Men Guests at Association's First Autumn Luncheon" *New York Times*, October 27, 1916, 8.
6. Page, *The Creative Destruction of Manhattan*, 54; "Cry Also against Fifth Avenue Signs," *New York Times*, September 17, 1910, 20.

7. Fifth Avenue Commission, *Preliminary Report of Fifth Avenue Commission* (New York: R. L. Stillson, 1912), 2.
8. Page, *The Creative Destruction of Manhattan*, 55.
9. "Foretell Greater 5th Av.," 8.
10. Page, *The Creative Destruction of Manhattan*, 55, 57; Fifth Avenue Association, *Minutes of the Fifth Avenue Association*, April 30, 1907.
11. Fifth Avenue Association, *Minutes of the Fifth Avenue Association*, April 13, 1911.
12. Fifth Avenue Association, *Minutes of the Fifth Avenue Association*, December 3, 1912 (found in *Fifth Avenue Association Minutes*, December 1, 1909–February 9, 1915, in New York City Department of Records, City Hall Library).
13. Fifth Avenue Association, *Minutes of the Fifth Avenue Association.*
14. "Convenient Parade Ground," *New York Times*, April 2, 1913, 10.
15. "Fifth Avenue Bright for Anniversary," *New York Times*, November 13, 1913, 20.
16. Fifth Avenue Association, *Report for the Year*, 1922; "Sift through Traffic out of Fifth Avenue," *New York Times*, November 5, 1916, 6; "Traffic on Fifth Av. Grows," *New York Times*, October 26, 1916, 10; "Oppose 'Right-Hand' Turn," *New York Times*, December 17, 1914, 20; "Cruising Hacks Cumber 5th Ave.," *New York Times*, April 26, 1914, 14; "Plan a Traffic Commission Now," *New York Times*, December 12, 1913, 16; "Point to Dangers in Fifth Avenue," *New York Times*, November 17, 1913, 15; "Fifth Ave. Traffic Moves More Freely," *New York Times*, December 21, 1912, 9; Page, *The Creative Destruction of Manhattan*, 58.
17. "Many Cities Limit Building Heights," *New York Times*, June 1, 1912, 9.
18. "Plans Truck Roads for Central Park," *New York Times*, July 30, 1916, 9.
19. "Fifth Ave. Traffic Moves More Freely," 9; "New Traffic Rules Urged for Fifth Avenue," *New York Times*, December 12, 1912, 11; "To Relieve Fifth Avenue," *New York Times*, December 15, 1912, 8.
20. Page, *The Creative Destruction of Manhattan*, 59; Gregory Gilmartin, *Shaping the City: New York and the Municipal Art Society* (New York: Clarkson Potter, 1995), 194.
21. Fifth Avenue Association, *Minutes of the Fifth Avenue Association*, January 24, 1911.
22. "Avenue Merchants Censor Advertising," *New York Times*, May 27, 1914, 6.
23. Fifth Avenue Commission, *Preliminary Report of Fifth Avenue Commission*, 2.
24. Fifth Avenue Association, *Minutes of the Fifth Avenue Association*, March 28, April 5, and May 3, 1910; Fifth Avenue Association, *Report for the Year*, 1922, 36–39; "Cry Also against Fifth Avenue Signs," 20; "Fifth Avenue's Factories," *New York Times*, July 25, 1910, 6; "Ugly Electric Signs Mar Fifth Avenue," *New York Times*, July 8, 1910, 6.
25. Page, *The Creative Destruction of Manhattan*, 63; Kenneth Revell, "Regulating the Landscape: Real Estate Values, City Planning, and the 1916 Zoning Ordinance," in *The Landscape of Modernity: Essays on New York City, 1900–1940*, edited by David Ward and Olivier Zunz (New York: Russell Sage Foundation, 1992), 23; *Estelle P. Anderson v. Steinway & Sons*, 178 A.D. 507, 165 N.Y.S. 608, and 1917 N.Y. App. Div., June 8.
26. *Anderson v. Steinway*, 1917.
27. *Anderson v. Steinway*, 1917. The judge argued that the decisions in the following cases indicated that there was little support in the courts for the plaintiff's arguments: *Reinman v. City of Little Rock*, 237 U.S. 171, 35 S. Ct. 511, 59 L. Ed. 900, 1915 U.S., April 5;

Hadacheck v. Sebastian, Chief of Police of the City of Los Angeles, 239 U.S. 394, 36 S. Ct. 143, 60 L. Ed. 348, 1915 U.S., December 20.

28. *The People ex rel. George Kemp v. Albert F. D'Oench*, 111 N.Y. 359, 18 N.E. 862, 1888 N.Y., November 27.
29. *People v. D'Oench*, 1888.
30. *Francis C. Welch v. George R. Swasey & Others*, 193 Mass. 364, 79 N.E. 745, 1907 Mass., January 1.
31. *Welch v. Swasey*, 1907.
32. Ibid.
33. *Commonwealth v. Boston Advertising Company*, 1905, 188 Mass. 348; 74 N.E. 601, 1905 Mass., June 20.
34. *Welch v. Swasey*, 1907.
35. "Building Height and Its Legality," *New York Times*, March 2, 1913, XX2.
36. *Laura A. Palmer v. Frank Mann*, 1923, 120 Misc. 396, 198 N.Y.S. 548, 1923 N.Y. Misc., January.
37. *Palmer v. Mann*, 1923.
38. *Thorofare Developing Corporation v. Deegan*, 1929, 235 N.Y.S. 544, 1929 N.Y. App. Div., May 24.
39. Fifth Avenue Association, *Minutes of the Fifth Avenue Association*, April 4, 1911; Donald J. Cannon, "Triangle Shirtwaist Fire," in *The Encyclopedia of New York City*, edited by Kenneth T. Jackson, 1199; "Lower Fifth Avenue Free from Noonday Congestion Beneficial Result of New Factory Commission Laws," *New York Times*, September 6, 1914, X9; Page *The Creative Destruction of Manhattan*, 63.
40. Robert Grier Cooke, "Against the Fifth Avenue Skyscraper," *New York Times*, November 24, 1912, SM9.
41. "Develop Fifth Ave. on Lines of Beauty," *New York Times*, February 8, 1911, 6.
42. Fifth Avenue Association, *Statement of the Fifth Avenue Association on the Limitation of Building Heights, to the New York City Commission and the Testimony of the Association's Representatives at a Conference*, June 19, 1913, 18; "The Unification of Building Inspection," *New York Times*, October 25, 1914, XX6.
43. Fifth Avenue Association, *Statement of the Fifth Avenue Association on the Limitation of Building Heights*, 18; "Lower Fifth Avenue Free from Noonday Congestion Beneficial Result of New Factory Commission Laws," X9; "Decrease in Fifth Avenue Factory Workers," *New York Times*, July 11, 1915, XX1; "Factory Invasion of 5th Av. Checked," *New York Times*, April 28, 1915, 13; "Factory Invasion Stirs 5th Av. Trade," *New York Times*, January 26, 1916, 19.
44. "Lower Fifth Avenue Free from Noonday Congestion Beneficial Result of New Factory Commission Laws."
45. R. G. Cooke, "Against the Fifth Avenue Skyscraper," 9.
46. Ernest Flagg, "In Favor of the Skyscraper," *New York Times*, November 24, 1912, 9.
47. R. G. Cooke, "Against the Fifth Avenue Skyscraper," 9.
48. Gilmartin, *Shaping the City*, 191; Page *The Creative Destruction of Manhattan*, 64; "Say Loiterers Kill Fifth Ave. Business," *New York Times*, April 8, 1910, 5.
49. "Fifth Avenue's Factories," 6.

50. "Fifth Avenue Bright for Anniversary," 20.

51. Ibid.

52. *Michael C. Bouvier v. Cornelia N. Segardi et al.*, 1920, 112 Misc. 689, 183 N.Y.S. 814, 1920 N.Y. Misc., August; *Batchelor v. Hinkle*, 1914, 210 N.Y. 243, 104 N.E. 629, 1914 N.Y., February 24.

53. *Wallack Construction Company v. Smalwich Realty Corporation and United International Corporation*, 1922, 201 A.D. 133, 194 N.Y.S. 32, 1922 N.Y. App Div., May 5.

54. *Spotwood D. Bowers v. Fifth Avenue and Seventy-seventh Street Corporation*, 1925, 125 Misc. 343, 209 N.Y.S. 743, 1925 N.Y. Misc., April 25. For a similar decision, see *Forstmann et al. v. Joray Holding Company*, 1926, 244 N.Y. 22, 154 N.E. 652, 1926 N.Y., November 30.

55. *Oppenheim v. Cruise*, 1922, 118 Misc. 368, 194 N.Y.S. 183, 1922 N.Y. Misc., March.

56. *State of Maryland v. John H. Gurry*, 1913, Court of Appeals of Maryland, 121 Md. 534, 88 A. 546, 1913 Md., October 7.

57. *State of Maryland v. John H. Gurry*, 1913.

58. Ibid.

59. *State v. William Darnell*, 1914, 166 N.C. 300, 81 S. E. 338, 1914 N.C.

60. *Carey et al. v. City of Atlanta et al.*, 1915, 143 Ga. 192, 84 S.E. 456, 1915 Ga., February 12.

61. *Hopkins and Others v. City of Richmond, and Coleman v. Town of Ashland*, 1915, 117 Va. 692, 86 S.E. 139, 1915 Va., September 9.

62. *Buchanan v. Warley*, 1917, 245 U.S. 60, 38 S. Ct. 16, 62 L. Ed. 149, 1917 U.S., November 5.

63. *Tyler v. Harmon*, 1925, 158 La. 439, 104 So. 200, 1925 La., March 2; *Tyler v. Harmon*, 1926, 150 La. 943, 107 So. 704, 1926 La., March 5; *Benjamin or Ben Harmon v. Joseph W. Tyler*, 1927, 273 U.S. 668, 47 S. Ct. 471, 71 L. Ed. 831, 1927 U.S., March 14.

64. Colten, "Basin Street Blues," 251; *Land Development Company of Louisiana v. City of New Orleans*, 1926, 13 F. 2d 898. U.S. Dist., July 8; *Land Development Company of Louisiana v. City of New Orleans*, 1927, 17 F. 2d 1016, U.S. App., April 1.

65. *City of Richmond et al. v. Deans*, 1930, 37 F. 2d 712, 1930 U.S. App., January 14; *City of Richmond et al. v. Deans*, 1930, 281 U.S. 704, 50 S. Ct. 407, 74 L. Ed. 1128, 1930 U.S., May 19.

66. *Elizabeth Koehler and August Koehler v. Leonard N. Rowland et al.*, 1918, 275 Mo. 573, 205 S.W. 217, 1918 Mo., July 30.

67. *Los Angeles Investment Company v. Alfred Gary et al.*, 1919, 181 Cal. 680, 186 P. 596, 1919 Cal., December 11.

68. *Parmalee v. Morris*, 1922, 218 Mich. 625, 188 N.W. 330, 1922 Mich., June 5.

69. *Corrigan et al. v. Buckley*, 1924, 55 App. D.C. 30, 299 F. 899, 1924 U.S. App., April 21; *Corrigan et al. v. Buckley*, 1926, 271 U.S. 323, 46 S. Ct. 521, 70 L. Ed. 969, 1926, May 24.

70. *Grady v. Garland et al.*, 1937, 67 App. D.C. 73, 89 F. 2d 817, 1937 U.S. App.

71. *Olive Ida Burke v. Isaac Kleiman et al.*, 1934, 355 Ill. 390, 189 N.E. 372, 1934 Ill.; *Olive Ida Burke v. Isaac Kleiman et al.*, 1934, 277 Ill. App. 519, 1934 Ill. App., November 27.

72. *Anna M. Lee et al. v. Carl A. Hansberry et al.*, 1937, 291 Ill. App. 517, 10 N.E.2d 406, 1937 Ill., October 7.
73. *Lee v. Hansberry*, 1937; *Burke v. Kleiman*, 1934, 355 Ill. 390, 189 N.E. 372; *Burke v. Kleiman*, 1934, 277 Ill. App. 519, 1934 Ill. App.
74. *Anna M. Lee et al. v. Carl A. Hansberry et al.*, 1939, 372 Ill. 369, 24 N.E.2d 37, 1939 Ill., December 13.
75. *Hansberry et al. v. Lee et al.*, 1940, 311 U.S. 32, 61 S. Ct. 115, 85 L. Ed. 22, 1940 U.S., November 12.
76. The house was bought by a real estate company and placed in the name of Josephine Fitzgerald (also a defendant), a white individual acting as a straw party. The Shelleys then bought the house from the real estate company. *Louis Kraemer and Fern E. Kraemer v. J. D. Shelley and Ethel Lee Shelley, and Josephine Fitzgerald*, 1948, 358 Mo. 364, 214 S. W.2d 525, 1948 Mo., November 8; *Louis Kraemer and Fern E. Kraemer v. J. D. Shelley and Ethel Lee Shelley, and Josephine Fitzgerald*, 1946, 355 Mo. 814, 198 S.W.2d 679, 1946 Mo., December 9; *Shelley v. Kraemer*, 1948.
77. *Sipes v. McGhee*, 1947, 316 Mich. 614, 25 N. W.2d 638, 1947 Mich., January 7.
78. *Shelley et ux. v. Kraemer et ux.*, 1948, 334 U.S. 1, 68 S. Ct. 836, 92 L. Ed. 1161, 1948 U.S., May 3; *Kraemer v. Shelley*, 1948.
79. *Hurd et ux. v. Hodge et al.*, 1948, 334 U.S. 24, 68 S. Ct. 847, 92 L. Ed. 1187, 1948 U.S., May 3; *Hurd et al. v. Hodge et al Urciolo et al. v. Same*, 1947, 82 U.S. App. D.C. 180, 162 F.2d 233, 1947 U.S. App., May 26.
80. *Hundley et ux. v. Gorewitz et al.*, 1942, 77 U.S. App. D.C. 48, 132 F. 2d 23, 1942 U.S. App.
81. *Mays et al. v. Burgess et al.*, 1945, 79 U.S. App. D.C. 343, 147 F.2d 869, 1945 U.S. App., January 29.

13. Workplace and Community Hazards

1. Douglas Martin, "Rose Freedman, Last Survivor of Triangle Fire, Dies at 107," *New York Times*, February 17, 2001, 8.
2. "Fewer Fires Here," *New York Times*, December 23, 1911, 6; Leon Stein with an introduction by William Greider, *The Triangle Fire* (Ithaca: Cornell University Press, 2001), 116.
3. "Partners' Account of the Disaster," *New York Times*, March 26, 1911, 4; Von Drehle, *Triangle*, 3, 12, 186; Factory Investigating Commission, *Preliminary Report of the Factory Investigating Commission* (Albany: Argus, 1912), 1:28.
4. Von Drehle, *Triangle*, 36–37, 42, 88.
5. Ibid., 38–39; Robert D. Parmet, "Garments," in *The Encyclopedia of New York City*, edited by Kenneth T. Jackson, 451–53.
6. Von Drehle, *Triangle*, 43–45; Caroline Rennolds Milbank, *New York Fashion: The Evolution of American Style* (New York: Harry N. Abrams, 1989), 45–48; Factory Investigating Commission, *Preliminary Report*, 1: 28.
7. Von Drehle, *Triangle*, 46.

8. Ibid., 47.

9. Ibid., 37.

10. Ibid., 41–42, 48.

11. Ibid, 3, 37; Martin, "Rose Freedman," 8; William Shepherd, "Eyewitness at the Triangle," *Milwaukee Journal*, March 27, 1911; Donald J. Cannon, "Triangle Shirtwaist Fire," in *The Encyclopedia of New York City*, edited by Kenneth T. Jackson, 1199; Dorothy Cobble, *Women and Unions* (New York: IRL Press, 1993), 98; Leon Stein and Phillip Taft, *Workers Speak* (New York: Arno Press, 1971), 116; Chris Llewellyn, *Fragments from the Fire* (New York: Viking Penguin, 1987), 8.

12. Von Drehle, *Triange*, 37–39; Linda Eisenmann, "Labor Unions," in *Historical Dictionary of Women's Education in the United States*, edited by Linda Eisenmann (Westport: Greenwood Publishing Group, 1998), 235–37.

13. Von Drehle, *Triangle*, 48.

14. Kerry Candaele and Sean Wilentz, "Labor," in *The Encyclopedia of New York City*, edited by Kenneth T. Jackson, 646–49.

15. Jackson and Dunbar, *Empire City*, 506; Von Drehle, *Triangle*, 14–16; "Chicago Garment Workers' Strike, Sep. 22 1910–Feb. 18 1911," in Women Working, 1800—1930, (Cambridge, Mass.: Harvard University Open Collections Program, n.d.), at http://ocp.hul.harvard.edu/ww/events_chicago_strike.html.

16. Leiserson, like Harris and Blanck, was an immigrant who rose from poverty to own a large garment factory. Von Drehle, *Triangle*, 6–8, 11; "Waistmakers Ask for Damages," *New York Times*, December 21, 1909, 1; "Triangle Owner Tells of Fire," *New York Times*, December 23, 1909, 6.

17. Parmet, "Garments," 451–53.

18. Von Drehle, *Triangle*, 8; Carolyn D. McCreesh, *Women in the Campaign to Organize Garment Workers, 1880–1917* (New York: Garland, 1985), 148.

19. Working-class women took over the leadership of the WTUL after finding new financial backers who supported their position that the organization should be headed by working-class women. After the First World War the WTUL focused less on organizing women into unions and more on educating workers and securing the passage of protective workplace legislation. The WTUL disbanded in 1955 because of lack of financial support. Barbuto, *American Settlement Houses and Progressive Social Reform*, 65–66, 158–60, 178–79, 226–27, 234–36; Annelise Orleck, "New York's Women's Trade Union League," in *The Encyclopedia of New York City*, edited by Kenneth T. Jackson, 849–850; "The Triangle Shirtwaist Fire: 1911," *Women's Rights on Trial* (Farmington Hills, Mich.: Gale, 1997), 321; "Administrative History of the Women's Trade Union League," 1908–44, Women's Trade Union League of Chicago Collection, University of Illinois at Chicago Library, http://www.uic.edu/depts/lib/specialcoll; Von Drehle, *Triangle*, 15–16.

20. Annelise Orleck, *Common Sense and a Little Fire: Women and Working Class Politics in the United States 1900–1965* (Chapel Hill: University of North Carolina Press, 1995), 48; Von Drehle, *Triangle*, 8.

21. Von Drehle, *Triangle*, 11. See also Robert P. Weiss, "Private Detective Agencies and

Labor Discipline in the United States, 1855–1946," *Historical Journal* 29:1 (March, 1986): 87–107.

22. Von Drehle, *Triangle*, 15–16.
23. Ibid., 36–37, 48–49.
24. Ibid., 49.
25. Ibid., 50–51; "Girl Strikers Well Treated, Says Baker," *New York Times*, November 6, 1909, 3.
26. Von Drehle, *Triangle*, 52–53; "Arrest Strikers for Being Assaulted," *New York Times*, November 5, 1909, 1; Annelise Orleck, "Mary E(lizabeth) Dreier," in *The Encyclopedia of New York City*, edited by Kenneth T. Jackson, 344.
27. "Girl Strikers Well Treated, Says Baker," 3; Von Drehle, *Triangle*, 52–53.
28. "Letter from Max Blanck and Isaac Harris to Garment Manufacturers," in *Jewish Daily Forward*, November 4, 1909, 8; "Letter from Max Blanck and Isaac Harris to Garment Manufacturers," in *New York Herald*, November 6, 1909, 4; Richard A. Greenwald, "Labor, Liberals, and Sweatshops," in *Sweatshop USA: The American Sweatshop in Historical and Global Perspective*, edited by Daniel E. Bender and Richard A. Greenwald (New York: Routledge, 2003), 77–89.
29. "The Cooper Union Meeting," *New York Call*, November 23, 1909; Richard A. Greenwald, "Bargaining for Industrial Democracy? Labor, the State, and the New Industrial Relations in Progressive Era New York," Ph.D. dissertation, New York University, 1998.
30. "The Cooper Union Meeting"; Von Drehle, *Triangle*, 9, 57–59.
31. Von Drehle. *Triangle*, 58–59.
32. Ibid., 60–62.
33. Ibid., 62–65; "Strikers Win in Yonkers," *New York Times*, December 10, 1910, 13.
34. Evelyn Alloy, *Working Women's Music: Songs and Struggles of Women in Cotton Mills, Textile Plants, and Needle Trades* (Somerville, Mass.: New England Free Press, 1976).
35. "A History of International Women's Day: 'We Want Bread and Roses Too,'" *Womankind* (March 1972), Chicago Women's Liberation Union history website at http://www.uic.edu/orgs/cwluherstory.
36. Stein and Taft, *Workers Speak*, 63.
37. Von Drehle, *Triangle*, 68–71, 74–75.
38. Miss Morgan Aids Girl Waiststrikers," *New York Times*, December 14, 1909.
39. Von Drehle, *Triangle*, 72–74.
40. Ibid., 75.
41. Ibid., 63, 74.
42. Ibid., 76–77; "Waist Strike Pickets Parade through Shop District in Autos," *New York Call*, December 22, 1909; "Autos for Strikers in Shirtwaist War," *New York Times*, December 21, 1909, 1; "Police Mishandle Girl Strike Pickets," *New York Times*, December 10, 1910, 13; "Girl Strikers Well Treated, Says Baker," 3; Parmet, "Garments," 451–53; Candaele and Wilentz, "Labor," 646–49.
43. Von Drehle, *Triangle*, 77–78.
44. Ibid., 79–84.

45. Ibid., 84–85.
46. Ibid., 107, 117; Cannon, "Triangle Shirtwaist Fire," 1199; *New York v. Harris*, 1911.
47. Von Drehle, *Triangle*, 118–23, 236; "Place a Firetrap," *New York Times*, March 28, 1911, 1; "Partners' Account of the Disaster," 4; Factory Investigating Commission, 1911, 580–83, New York State Archives, Albany, at http://www.archives.nysed.gov.
48. Von Drehle, *Triangle*, 120–21.
49. Many Tell Now of Firetraps," *New York Times*, March 29, 1911, 3.
50. Von Drehle, *Triangle*, 194–95.
51. Shepherd, "Eyewitness at the Triangle."
52. "148 Leap to Death or Perish in Flames," *Washington Post*, March 26, 1911, 1.
53. "Thrilling Incidents in Gotham Holocaust that Wiped Out One Hundred and Fifty Lives," *Chicago Sunday Tribune*, March 28, 1911, 2.
54. Von Drehle, *Triangle*, 2, 128–31.
55. Ibid., 133–34, 147–48, 156–57; Cannon, "Triangle Shirtwaist Fire," 1199; *New York v. Harris*, 1911; Stein, *The Triangle Fire*, 39, 48.
56. Von Drehle, *Triangle*, 136–43; "148 Leap to Death or Perish in Flames," 1; "Partners' Account of the Disaster," 4; Jackson and Dunbar, *Empire City*, 506; D. Martin, "Rose Freedman," 8.
57. Stein, *The Triangle Fire*, 37.
58. Ibid., 19.
59. Von Drehle, *Triangle*, 167.
60. Ibid., 1; "Sad All-Day March to Morgue Gates," *New York Times*, March 27, 1911, 2; "300,000 in Funeral March," *New York Times*, March 29, 1911, 3; Stein, *The Triangle Fire*, 172.
61. William Mailly, "The Triangle Trade Union Relief," *American Federalist*, July 1, 1911, 544–47; "The Triangle Fire Fund," *New York Times*, May 7, 1912, 10; Stein, *The Triangle Fire*, 122–23.
62. "Fire Sufferers' Fund Swelled to $50,485," *New York Times*, March 30, 1911, 3; "The Triangle Fire Fund," 10; Frank Cullen, Florence Hackman and Donald McNeilly, *Vaudeville Old and New: An Encyclopedia of Variety Performers in America* (New York: Routledge, 2007), 421, 1195–98; Stein, *The Triangle Fire*, 123–25.
63. *New York v. Harris*, 1911; Richard O. Boyer, *Max Steuer: Magician of the Law* (New York: Greenberg, 1932); "Triangle Witnesses Deny Locked Doors," *New York Times*, December 20, 1911, 11; "Deny Locked Doors, but Girls Insist," *New York Times*, December 19, 1911, 8; "Exhibit Shot Bolt at Triangle Trial," *New York Times*, December 15, 1911, 5; "Triangle Survivors Slid down Cables," *New York Times*, December 13, 1911, 20; "Girls Fought Vainly at Triangle Doors," *New York Times*, December 12, 1911, 4; "Door Was Locked at Factory Fire," *New York Times*, December 9, 1911, 3; "Say Triangle Exits Were Never Locked," *New York Times*, December 21, 1911, 20; "To Try the Owners of Burned Factory," *New York Times*, November 1, 1911, 6; "Indicted for Fire Horror," 4; Von Drehle, *Triangle*, 222–23, 228, 250–51.
64. "Triangle Witnesses Got Increased Pay," *New York Times*, December 22, 1911, 7.
65. *New York v. Harris*, 1911; "Settle Triangle Fire Suits," *New York Times*, March 12, 1914, 1; "Triangle Witnesses Deny Locked Doors," 11; "Deny Locked Doors, but Girls Insist,"

8; "Exhibit Shot Bolt at Triangle Trial," 5; "Triangle Survivors Slid down Cables," 20; "Girls Fought Vainly at Triangle Doors," 4; "Door Was Locked at Factory Fire," 3; "Say Triangle Exits Were Never Locked," 20; "Triangle Men Put on Trial," *New York Times*, December 5, 1911, 16; "Triangle Owners Acquitted by Jury," *New York Times*, December 28, 1911, 1; Von Drehle, *Triangle*, 254–55, 263–64.

66. "Holdup Triangle Insurance," *New York Times*, October 18, 1911, 5; Stein, *The Triangle Fire*, 176.
67. "Regrets Voting for Triangle Acquittal," *New York Times*, December 29, 1911, 8.
68. Despite the cloakmakers' victory the strike of the Brooklyn Boot and Shoe Workers in November 1910 resulted in a defeat for the workers. Louis Levine [Lewis Revitzki Lorwin], *The Women's Garment Workers* (1924; New York: Arno, 1969); Gus Tyler, *Look for the Union Label: A History of the International Ladies Garment Workers Union* (Armonk, N.Y.: M. E. Sharpe, 1995); McCreesh, *Women in the Campaign to Organize Garment Workers*; "For Safer Skyscrapers," *New York Times*, June 9, 1911, 16; "Urge Dix to Help Fight Fire Peril," *New York Times*, May 9, 1911, 6; "Indicted for Fire Horror," *New York Times*, April 12, 1911, 4; "Women Urge Triangle Trial," *New York Times*, February 2, 1912, 18; Von Drehle, *Triangle*, 172, 187, 193; Candaele and Wilentz, "Labor," 646–49.
69. "Blame Shifted on All Sides for Fire Horror," *New York Times*, March 28, 1911, 1; Von Drehle, *Triangle*, 178–79.
70. The members of the Factory Investigating Commission were the Senators Robert F. Wagner and Charles M. Hamilton; the Assemblymen Alfred E. Smith, Edward D. Jackson, and Cyrus W. Phillips; the corporate representative Simon Brentano; Robert E. Dowling; Samuel Gompers of the American Federation of Labor; and Mary Dreier of the WTUL. Factory Investigating Commission, *Preliminary Report*, 1: 13–20, 38–47, 128–31, 277–78; Jackson and Dunbar, *Empire City*, 511, 516; D. Martin, "Rose Freedman," 8; "The Triangle Fire Fund," 10; Irwin Yellowitz, "Factory Investigating Commission," in *The Encyclopedia of New York City*, edited by Kenneth T. Jackson (New Haven: Yale University Press, 1995), 387; Stein, *The Triangle Fire*, 172; "Urge Dix to Help Fight Fire Peril," 6; "Need More Time to Study Triangle Fire," *New York Times*, March 1, 1912, 22.
71. "Factory Law to End 5th Ave. Congestion," *New York Times*, February 5, 1914, 10; "Lower Fifth Avenue Free from Noonday Congestion Beneficial Result of New Factory Commission Laws"; R. G. Cooke, "Against the Fifth Avenue Skyscraper," 9.
72. R. G. Cooke, "Against the Fifth Avenue Skyscraper," 9; C. Docherty, *Historical Dictionary of Organized Labor* (Lanham, Md.: Scarecrow Press, 2004), 138; Richard A. Greenwald, *The Triangle Fire, The Protocols of Peace, and Industrial Democracy in Progressive Era New York* (Philadelphia: Temple University Press).
73. Carol Willis, "Lofts," in *The Encyclopedia of New York City*, edited by Kenneth T. Jackson, 689–90; Page, *The Creative Destruction of Manhattan*, 64.
74. "Lower Fifth Avenue Free from Noonday Congestion Beneficial Result of New Factory Commission Laws."
75. "Won't Lend Money for 5th Av. Lofts," *New York Times*, January 27, 1916, 19; "Factory Invasion Stirs 5th Av. Trade."

76. "Decrease in Fifth Avenue Factory Workers"; "Factory Invasion of 5th Av. Checked."
77. Von Drehle, *Triangle*, 264–65.
78. Ibid., 265; "Fines Blanck for Locks," *New York Times*, September 27, 1913, 9.
79. "Censures Triangle Co.," *New York Times*, 1913, 5.
80. Von Drehle, *Triangle*, 265–66.
81. Elizabeth Cline, "On Anniversary of Historic Triangle Fire, Landmark Designated," Labor Research Association, New York, April 1, 2003, at http://workinglife.org.
82. "Many Now Tell of Fire Traps," 3; Von Drehle, *Triangle*, 1; Stein, *The Triangle Fire*, 172.
83. Von Drehle, *Triangle*, 160–61; Factory Investigating Commission, *Preliminary Report*, 1: 43.
84. Von Drehle, *Triangle*, 161–62; "Yesterday's Fire," *New York Times*, November 2, 1902, 8.
85. Arthur E. McFarlane, "The Business of Arson," *Collier's*, April 13, 1913; Arthur E. McFarlane, "The Triangle Fire: The Story of a 'Rotten Risk,'" *Collier's*, May 13, 1913; Von Drehle, *Triangle*, 162–63; Factory Investigating Commission, *Preliminary Report*, 1: 28.
86. Llewellyn, *Fragments from the Fire*, 22; Meredith Tax, *The Rising of the Women* (New York: Monthly Review Press, 1980), 209; Von Drehle, *Triangle*, 163–64.
87. Von Drehle, *Triangle*, 163–64.
88. "Many Leap, One Dies, at Fire," *New York Times*, April 27, 1912, 4.
89. "Other Notable Disasters," *New York Times*, March 26, 1911, 4; Nye, *Consuming Power*, 84–85.
90. Nye, *Consuming Power*, 84–85; John K. Brown, *Limbs on the Levee: Steam Boat Explosions and the Origins of Federal Public Welfare Regulation, 1817–1852* (Middlebourne, W. Va.: International Steamboat Society, 1989); Morton J. Horwitz, *The Transformation of American Law, 1780–1860* (Cambridge, Mass.: Harvard University Press, 1977); *Farwell v. Boston and Worcester Railroad*, 45 Mass. (4 Met.) 49.
91. Richard Kazis and Richard L. Grossman, *Fear at Work: Job Blackmail, Labor and the Environment* (New York: Pilgrim Press, 1982), 166; Thomas O. McGarity and Sidney A. Shapiro, *Workers at Risk: The Failed Promise of the Occupational Safety and Health Administration* (Westport, Conn.: Praeger, 1993), 3–4; *Industrial Accident Statistics* (Washington: U.S. Department of Labor, Bureau of Labor Statistics, 1915), 5.
92. Gottlieb, *Forcing the Spring*, 48, 55–56.
93. Christian Warren, *Brush with Death: A Social History of Lead Poisoning* (Baltimore: Johns Hopkins University Press, 2000), 74–75.
94. Ibid., 76; Gottlieb, *Forcing the Spring*, 48, 55–56; Kazis and Grossman, *Fear at Work*, 157–58, 166, 170, 191–92.
95. Christopher C. Sellers, *Hazards of the Job: From Industrial Disease to Environmental Health Science* (Chapel Hill: University of North Carolina Press, 1997), 13–14.
96. Warren, *Brush with Death*, 36.
97. Sellers, *Hazards of the Job*, 15–17.

98. Ibid., 69, 87–88.
99. Ibid., 94; Warren, *Brush with Death*, 46.
100. Sellers, *Hazards of the Job*, 89–91, 93, 96.
101. Ibid., 95.
102. Iid., 25.
103. Ibid., 50, 60–64, 72; Warren, *Brush with Death*, 74.
104. Sellers, *Hazards of the Job*, 111–12, 119.
105. Warren, *Brush with Death*, 78–80.
106. Jacqueline Karnell Corn, *Response to Occupational Health Hazards: A Historical Perspective* (New York: Van Nostrand Reinhold, 1992), 13.
107. Wade, *Chicago's Pride*, 229.
108. Ibid., 114–26, 233–36, 241–58; Gottlieb, *Forcing the Spring*, 68–69; James Whorton, *Before Silent Spring: Pesticides and Public Health in Pre-DDT America* (Princeton: Princeton University Press, 1974); Alan Derikson, *To Be His Own Benefactor: The Founding of the Coeur d'Alene Miners' Union Hospital* (1891; Bloomington: Indiana University Press, 1989), 1–15.
109. Gottlieb, *Forcing the Spring*, 69–70.
110. *Our Toiling Children* (Chicago: Women's Temperance Publications Association, 1889); Forest C. Ensign, *Compulsory School Attendance and Child Labor* (Iowa City: Athens Press, 1921).
111. Dorothy Rose Blumberg, *Florence Kelley: The Making of a Social Pioneer* (New York: Augustus M. Kelley, 1966), 99–101.
112. D. R. Blumberg, *Florence Kelley*, 99–101; Hogan, *Moments in New Haven Labor History*, 26–27.
113. Walter Licht, *Getting Work: Philadelphia, 1840–1950* (Philadelphia: University of Pennsylvania Press, 1992), 20–21, 28.
114. Ibid., 20–22.
115. Ibid., 22–23.
116. Ibid., 18.
117. Ibid.
118. Ibid., 19, 65.
119. D. R. Blumberg, *Florence Kelley*, 101; Wade, *Chicago's Pride*, 229.

14. Environmental Reform

1. Nye, *Consuming Power*, 84.
2. Robert Dale Grinder, *The Anti-Smoke Crusades: Early Attempts to Reform the Urban Environment, 1893–1918* (Columbia: University of Missouri Press, 1973), 27–29; Scott Hamilton Dewey, *Don't Breathe the Air: Air Pollution and U.S. Environmental Politics, 1945–1970* (College Station: Texas A&M University Press, 2000), 22.
3. Charles Cist, *Cincinnati in 1841* (Cincinnati: Charles Cist, 1841), 237.
4. Grinder, *The Anti-Smoke Crusades*, 29–30; Dewey, *Don't Breathe the Air*, 23, 26.
5. Mazzari, "Fire and Firefighting," 224–26.

6. Dewey, *Don't Breathe the Air*, 23; Kazis and Grossman, *Fear at Work*, 191–92.
7. Samuel B. Flagg, *City Smoke Ordinances and Smoke Abatement*, Bulletin 49, Department of the Interior, U.S. Bureau of Mines, 1912, 7–8.
8. Department of Smoke Inspection, *Report of the Department of Smoke Inspection, City of Chicago*, 1911, 39–40.
9. Dewey, *Don't Breathe the Air*, 24.
10. Ibid., 24–25.
11. Ibid.; Tarr, "The Search for the Ultimate Sink," 10; Melosi, *Garbage in the Cities*, 93.
12. S. B. Flagg, *City Smoke Ordinances and Smoke Abatement*, 10, 21, 24.
13. Ibid., 12–13, 16–19, 21.
14. Barry I. Castleman, *Asbestos: Medical and Legal Aspects* (New York: Aspen Publishers, 1996), 662–68; Johns-Manville, "Company History," 2003; *Clarence Borel v. Fibreboard Paper Products Corporation, et al.* 493 F.2d 1076, 1973 U.S. App., September 10 [*Borel v. Fibreboard I*]; Lynn Kalanik, Mary McNulty, and Christina Stansell, "Johns Manville Corporation," *International Directory of Company Histories* 64 (1993), at http://www.enotes.com/company-histories/johns-manville-corporation.
15. Samuel Epstein, *The Politics of Cancer* (Garden City, N.J.: Anchor Press, 1979), 80–81; National Institute for Occupational Safety and Health (NIOSH), *National Occupational Research Agenda: Priorities for the 21st Century* (Washington, D.C.: U.S. Department of Health and Human Services, NIOSH, 2002), xxiii.
16. Corn, *Response to Occupational Health Hazards*, 90–91.
17. NIOSH, *Occupational Exposure to Asbestos: Criteria for Recommended Standard* (Washington, D.C.: NIOSH, 1972), 27–31.
18. Castleman, *Asbestos*, 1–6.
19. M. Auribault, "Note sur l'hygiene et la securite des ouvriers dan les filatures et tissages d'amiante," *Bulletin De L'Inspection Du Travail* 14 (1906): 126; Murray H. Montague, "Statement before the Committee in the Minutes of Evidence," in *Report of the Departmental Committee on Compensation for Industrial Disease* (London: His Majesty's Stationery Office, 1907).
20. W. E. Cooke, "Fibrosis of the Lungs Due to the Inhalation of Asbestos Dust," *British Medical Journal* 2 (1924): 147; W. E. Cooke, "Pulmonary Asbestosis," *British Medical Journal* 2 (1927): 1024–25.
21. P. Ellman, "Pulmonary Asbestosis: Its Clinical, Radiological and Pathological Features and Associated Risk of Tuberculosis Infection," *Journal of Industrial Hygiene* 15 (July 1933): 165–83; P. Ellman, "Pneumoconiosis," *British Journal of Radiology* 14 (1934): 361; J. Donnelley, "Pulmonary Asbestos," *American Journal of Public Health*, December 20, 1933, 1275–81.
22. Castleman, *Asbestos*, 26; Lynn Kalanik, Mary McNulty, and Christina Stansell, "Johns Manville Corporation."
23. Anthony J. Lanza, "Effects of Asbestos Dust on the Lungs," *Public Health Reports* 50 (1935): 1; Anthony J. Lanza, "Asbestosis," *Journal of the American Medical Association* 106 (1936): 368; Anthony J. Lanza, *Silicosis and Asbestosis* (New York: Oxford University Press, 1938); Andrea Peackock, *Libby, Montana: Asbestos and the Deadly Silence of an American Corporation* (Neenah, Wisc.: Big Earth Publishing, 2003), 161–62;

Lynn Kalanik, Mary McNulty, and Christina Stansell, "Johns Manville Corporation"; David Rosner and Gerald E. Markowitz, *Dying for Work: Workers' Safety and Health in Twentieth-Century America* (Bloomington: University of Indiana Press, 1987), 197–99; United States Gypsum Company, letter from CMP to John J. Cuneo, October 28, 1937, at http://www.mesotheliomanews.com.

24. W. C. Dreessen, J. M. Dallavalle, T. I. Edwards, J. W. Miller, and R. R. Sayers, "A Study of Asbestosis in the Asbestos Textile Industry," *Public Health Bulletin* 241 (1938).
25. W. E. Fleischer, F. J. Viles, R. L. Gade, and P. Drinker, "A Health Survey of Pipe-Covering Operations in Constructing Naval Vessels," *Journal of Industrial Hygiene* 28 (1945): 9–16; *Borel v. Fibreboard I*, 1973.
26. *Borel v. Fibreboard I*, 1973.
27. R. Doll, "Mortality from Lung Cancer in Asbestos Workers," *British Journal of Industrial Medicine* 12 (1955): 81–97.
28. Irving J. Selikoff, Jacob Churg, and E. Cuyler Hammond, "The Occurrence of Asbestosis among Insulation Workers in the United States," *Annals of the New York Academy of Sciences* 132, no. 12 (1965): 139–55.
29. Irving J Selikoff, and E. Cuyler Hammond, "Multiple Risk Factors in Environmental Cancer," in *Persons at High Risk of Cancer*, edited by J. F. Fraumeni (New York: Academic Press, 1975), 467–83; Epstein, *The Politics of Cancer*, 81–82.
30. NIOSH, *National Occupational Research Agenda*, xxiii, 3–4.
31. *Borel v. Fibreboard I*, 1973; Lanza, "Asbestosis"; Castleman, *Asbestos*, 172, 662–68.
32. Castleman, *Asbestos*, 644–47; Early, Ludwick, Sweeney, and Strauss, Attorneys at Law, "Asbestos Industry: Significant Dates," (New York, n.d.), at http://www.elslaw.com/asbestostimeline.htm; Henry Weinstein, "Did Industry Suppress Data on Hazards of Asbestos?," *Los Angeles Times*, October 23, 1978, 1.
33. Castleman, *Asbestos*, 621–24; United States Gypsum Company, letter from CMP to John J. Cuneo; Henry Weinstein, "Did Industry Suppress Data on Hazards of Asbestos?"
34. *Borel v. Fibreboard I*, 1973.
35. Ibid.
36. Ibid.
37. Ibid.
38. Ibid.
39. Ibid.
40. Ibid.
41. Ibid.
42. *Clarence Borel v. Fibreboard Paper Products Corporation, et al.*, 493 F.2d 1076, 1974 U.S. App., May 13 [*Borel v. Fibreboard II*].
43. Ibid.
44. See Castleman, *Asbestos*, for a detailed discussion of documents and correspondence between the asbestos companies that indicates that they were aware of the dangers of asbestos and sought to keep that information from workers and the public. See also David Wessel, "Asbestos: Who Knew What When?," *Boston Globe*, October 26, 1982.

45. See *Barnett v. Owens-Corning Fiberglas Corporation, et al.*, 1978, State of South Carolina, County of Greenville, Court of Common Pleas, August 23. See also Castleman, *Asbestos*, for a more detailed discussion of the asbestos Pentagon Papers.
46. Frances H. Miller, "Biological Monitoring: The Employer's Dilemma," *American Journal of Law and Medicine* 9 (1984): 387–426; Johns-Manville, "Company History."
47. RAND Institute for Civil Justice, *Asbestos Litigation Costs, Compensation, and Alternatives* (Santa Monica, Calif.: RAND, 2005); "RAND Corporation Press Release on Asbestos," September 25, 2002; Jeff Dircksen, "Gordian Knot: How the Senate's Asbestos 'Reform' Bill Entangles Taxpayers," National Taxpayers Union Policy Paper 118, Alexandria, Va., December 8, 2005; Lisa Girion, "Firms Hit Hard as Asbestos Claims Rise: Courts Recent Jury Awards Underscore Commercial Disaster's Continuing Toll," *Los Angeles Times*, December 17, 2001, 1; David Voreacos, "Big Increase in Claims Imperils Asbestos Fund," *Chicago Sun-Times*, December 17, 2001, 48; Insurance Information Institute, "Asbestos Liability," press release, 2006, at http://www.iii.org/media/hot topics/insurance/asbestos.
48. "Johns-Manville Trust Files Third Quarter Statement; Claims up 55% over Last Year," *Asbestos Litigation Reporter*, November 8, 2001, 8. For instance, there were 55,100 new claims filed during the first six months of 2001, compared to 20,800 new claims filed during the same period in 2000. See "Manville Trust Provides Second-Quarter Financials to Court," *Asbestos Litigation Reporter*, November 8, 2001, 7.
49. Jonathan D. Glater, "Suits on Silica Being Compared to Asbestos Cases," *New York Times*, September 6, 2003, C1; Girion, "Firms Hit Hard as Asbestos Claims Rise" 1; Voreacos, "Big Increase in Claims Imperils Asbestos Fund," 48; "Johns-Manville Trust Files Third Quarter Statement," 8; "What's Inside," *Asbestos Litigation Reporter*, November 8, 2001, 1; "Tillinghast-Towers Perrin Estimates $200B Price Tag for U.S. Asbestos Claims," *Insurance Journal*, June 19, 2001; Tillinghast-Towers Perrin, "Tillinghast-Towers Perrin Estimates Claims Associated with U.S. Asbestos Exposure Will Ultimately Cost $200 Billion," *Business Wire*, June 12, 2001, 1.
50. RAND, *Asbestos Litigation Costs, Compensation, and Alternatives.*
51. Insurance Information Institute, "Asbestos Liability."
52. See *Anchem Products, Inc., et al., v. George Windsor et al.*, 1997, 521 U.S. 591, 117 S. Ct. 2231, 138 L. ed. 2d 689, June 25; *Esteban Ortiz, et al., v. Fibreboard Corporation, et al.*, 1999, 527 U.S. 815, 119 S. Ct. 2295, 144 L. Ed. 2d 715, June 23.
53. Dircksen, "Gordian Knot."
54. Congressional Budget Office, *S. 852: Fairness in Asbestos Injury Resolution Act of 2005*, a report by the Senate Committee on the Judiciary on June 16, 2005.
55. Government Accountability Office, *Federal Compensation Programs: Perspectives on Four Programs*, 2005, 1–5, 34.
56. Shannon D. Murray, "USG to Create a $4B Asbestos Trust," *Daily Deal/The Deal*, January 31, 2006; Association of Trial Lawyers of America, "Senate Halts $20 Billion Corporate Bailout Bill for Asbestos Companies Flawed Bill Would Have Left Many Victims with Nothing," press release, February 14, 2006; Dircksen, "Gordian Knot"; "The Coalition for Asbestos Reform Challenges Supporters of FAIR Act," *PR Newswire US*, February 2, 2006.

57. Insurance Information Institute, "Asbestos Liability."

58. Hurley, *Environmental Inequalities*, 15.

59. David Bensman and Roberta Lynch, *Rusted Dreams: Hard Times in a Steel Community* (New York: McGraw-Hill, 1987), 12, 13–14.

60. Bruno, *Steelworker Alley*, 138, 139.

61. Rade B. Vukmir, *The Mill* (Lanham, Md.: University Press of America, 1999), 34.

62. John P. Hoerr, *And the Wolf Finally Came: The Decline of the American Steel Industry* (Pittsburgh: University of Pittsburgh Press, 1988), 4.

63. Fee et al., *The Baltimore Book*, 175, 178.

64. Ibid., 183.

65. Cliff Brown, "The Role of Employers in Split Labor Markets: An Event-Structure Analysis of Racial Conflict and AFL Organizing, 1917–1919," *Social Forces* 79, no. 2 (2000): 656; David Brady and Michael Wallace, "Deindustrialization and Poverty: Manufacturing Decline and AFDC Recipiency in Lake County, Indiana 1964–93," *Sociological Forum* 16, no. 2 (2001): 329.

66. *United States Steel Corporation v. Russell E. Train, II*, 1977, 556 F.2d 822, 1977 U.S. App., May 13.

67. Bensman and Lynch, *Rusted Dreams*, 45.

68. U.S. Bureau of the Census, *1940 Census of Population, Part II* (Washington, D.C.: Government Printing Office, 1940), 674–844; Hurley, *Environmental Inequalities*, 22–23; Edward Osann, "Indiana Sand Dunes and Steel Don't Mix," *Environmental Action*, April 27, 1974, 10–13; Pacific Environmental Services, *Background Report: AP-42 Section 12.2. Coke Production*, prepared for the U.S. Environmental Protection Agency, 1992, 6; Brady and Wallace, "Deindustrialization and Poverty," 329.

69. Hurley, *Environmental Inequalities*, 23.

70. Pacific Environmental Services, *Background Report*, 11–12, 30.

71. Metzgar, *Striking Steel*, 32.

72. Vukmir, *The Mill*, 36.

73. Bensman and Lynch, *Rusted Dreams*, 16; William Hurd, "Making Steel and Killing Men," *Everybody's Magazine*, November 1907.

74. Mark Reutter, *Sparrows Point: Making Steel—The Rise and Ruin of American Industrial Might* (New York: Summit Books, 1988), 179. For descriptions of hot and cold conditions in the mills, see also Charles Rumford Walker, *Steeltown: An Industrial Case History of the Conflict between Progress and Security* (New York: Harper and Brothers, 1950), 57.

75. Fee et al., *The Baltimore Book*, 178.

76. Bruno, *Steelworker Alley*, 63, 76–77.

77. Bensman and Lynch, *Rusted Dreams*, 20.

78. Ibid.

79. Reutter, *Sparrows Point*, 179, 187–88. See also Vukmir, *The Mill*, 68–69, for details on injuries and deaths in the mills.

80. Between 1882 and 1930, 1,663 blacks were lynched in the Cotton Belt alone, and 1,299 were executed. About 90 percent of the people lynched and executed in the South were black. See Stewart Tolnay and E. M. Beck, "Rethinking the Role of Racial Vio-

lence in the Great Migration," in *Black Exodus: The Great Migration from the American South*, edited by Alferdteen Harrison (Jackson: University of Mississippi Press, 1991), 27. Though work by Johnson and Fligstein call into question the relationship between lynchings and black out-migration from areas of lynching, Tolnay and Beck found such a relationship existed. See Charles Johnson, "How Much Is the Migration a Flight from Persecution?," *Opportunity* 1 (1923): 272–74; Neil Fligstein, *Going North: Migration of Blacks and Whites from the South, 1900–1950* (New York: Academic Press, 1981).

The boll weevil outbreak which began in Texas in 1898 spread eastward. Thousands of agricultural workers lost their jobs in the wake of the devastation. The outbreak ended the single-crop dependency of the South. Carole Marks, "The Social and Economic Life of Southern Blacks during the Migration," in *Black Exodus: The Great Migration from the American South*, edited by Alferdteen Harrison (Jackson: University of Mississippi Press, 1991), 37.

81. Bennett, *The Shaping of Black America*, 269; U.S. Bureau of the Census, *14th Census of the United States, 1920* (Washington, D.C.: Department of Commerce, 1920); Drake and Cayton, *Black Metropolis*, 58, 62; U.S. Bureau of the Census, *Historical Statistics of the United States, Colonial Times to 1970*; Tolnay and Beck, "Rethinking the Role of Racial Violence in the Great Migration," 20–21, 25; Marks, "The Social and Economic Life of Southern Blacks during the Migration," 36.
82. Cliff Brown, "Racial Conflict and Split Labor Markets," *Social Science History* 23, no. 3 (1998): 322; William J. Collins, "When the Tide Turned: Immigration and the Delay of the Great Black Migration," *Journal of Economic History* 57, no. 3 (March 1997): 607–8.
83. Collins, "When the Tide Turned," 607; Hurley, *Environmental Inequalities*, 112–13; A. Morris, *The Origins of the Civil Rights Movement*; Tuttle, *Race Riot*, 80, 163–64, 215; Dubofsky, *Industrialism and the American Worker*, 12; Marks, "The Social and Economic Life of Southern Blacks during the Migration," 46–47.
84. Hurley, *Environmental Inequalities*, 112–13.
85. Brown, "The Role of Employers in Split Labor Markets," 655; Joshua L. Rosenbloom, "Strikebreaking and the Labor Market in the United States, 1881–1894," *Journal of Economic History* 58, no. 1 (March 1998): 188.
86. Warren C. Whatley, "African-American Strikebreaking from the Civil War to the New Deal," *Social Science History* 17, no. 4 (1993): 529–36, 539.
87. Ibid., 526, 537–38.
88. John H. Keiser, "Black Strikebreakers and Racism in Illinois, 1865–1900," *Journal of Illinois State Historical Society* 65 (1972): 326.
89. Ray Marshall, *The Negro and Organized Labor* (New York: Wiley, 1965), 19.
90. Sterling D. Spero and Abram L. Harris, *The Black Worker: The Negro and the Labor Movement* (Port Washington, N.Y.: Kennikat, 1931), 144–45.
91. Whatley, "African-American Strikebreaking from the Civil War to the New Deal," 527, 546; Brown, "Racial Conflict and Split Labor Markets," 320, 325–27, 334; Interchurch World Movement, *Report on the Steel Strike of 1919* (New York: Harcourt, Brace and World, 1920), 177.

92. Whatley, "African-American Strikebreaking from the Civil War to the New Deal," 527, 538–39; Brown, "The Role of Employers in Split Labor Markets," 653–54, 665–73.
93. Whatley, "African-American Strikebreaking from the Civil War to the New Deal," 540–42, 548–49.
94. Ibid., 542–43; Brown, "The Role of Employers in Split Labor Markets," 653–54, 665–73.
95. Martin Oppenheimer, "The Student Movement as a Response to Alienation," *Journal of Human Relations* 16 (1968): 1–16; Martin Oppenheimer, "The Genesis of the Southern Negro Student Movement (Sit-in Movement): A Study of Contemporary Negro Protest," Ph.D. dissertation, University of Pennsylvania, 1963; Doug McAdam, "Tactical Innovation and the Pace of Insurgency," *American Sociological Review* 48, no. 6 (1983): 735–754.
96. Brown, "The Role of Employers in Split Labor Markets," 672.
97. Whatley, "African-American Strikebreaking from the Civil War to the New Deal," 545.
98. Ibid., 322.
99. Ibid., 545.
100. Raymond Wolters, *Negroes and the Great Depression: The Problem of Economic Recovery* (Westport, Conn.: Greenwood, 1970), 184.
101. Thomas N. Maloney, "Degrees of Inequality," *Social Science History* 19, no. 1 (1995): 544–45.
102. Brown, "Racial Conflict and Split Labor Markets," 323.
103. Hurley, *Environmental Inequalities*, 113–15.
104. Fee et al., *The Baltimore Book*, 177. For a discussion of ethnic stratification in the mills, see Brown, "Racial Conflict and Split Labor Markets," 324.
105. Fee et al., *The Baltimore Book*, 178.
106. Sterling D. Spero and Abram L. Harris, *The Black Worker: The Negro and the Labor Movement* (New York: Atheneum, 1969), 177; Brown, "Racial Conflict and Split Labor Markets," 32.
107. "Northern" refers to the northeast and north central regions: Connecticut, Massachusetts, Maine, New Hampshire, Rhode Island, Vermont, Pennsylvania, New York, New Jersey, Illinois, Indiana, Michigan, Ohio, Wisconsin, Iowa, Kansas, Missouri, Minnesota, North Dakota, Nebraska, and South Dakota. Maloney, "Degrees of Inequality," 35.
108. Hurley, *Environmental Inequalities*, 113–14; Carol Redmond et al., "Long Term Mortality Study of Steelworkers: VI, Mortality from Malignant Neoplasms among Coke Oven Workers," *Journal of Occupational Medicine* 14 (August 1972): 621–29.
109. Hurley, *Environmental Inequalities*, 116–17.
110. Ibid., 25–37, 54, 123.
111. Fee et al., *The Baltimore Book*, 175, 177; S. H. Olson, *Baltimore*, 214.
112. Fee et al., *The Baltimore Book*, 177; S. H. Olson, *Baltimore*, 214–15.
113. Research Triangle Institute, *National Emission Standards for Hazardous Air Pollutants (NESHAP) for Integrated Iron and Steel Plants—Background Information for Proposed*

Standards, prepared for the U.S. Environmental Protection Agency, January 2001, 2–1, 2–3.

114. Patricia Beeson, Lara Shore-Sheppard, and Kathryn Shaw, "Industrial Change and Wage Inequality: Evidence from the Steel Industry," *Industrial and Labor Relations Review* 54, no. 2 (2001): 466–83.
115. James B. Lane, *City of the Century: A History of Gary, Indiana* (Bloomington: University of Indiana Press, 1978), 194–96; Philip W. Nyden, *Steelworkers Rank-and-File: The Political Economy of a Union Reform Movement* (New York: Praeger, 1984), 17–32; Bruno, *Steelworker Alley*, 100–102; Reutter, *Sparrows Point*, 260–62; Maloney, "Degrees of Inequality," 45; Fee et al., *The Baltimore Book*, 186–87.
116. Bensman and Lynch, *Rusted Dreams*, 128; Reutter, *Sparrows Point*, 260–61; Drake and Cayton, *Black Metropolis*, 320.
117. Ruth Needleman, *Black Freedom Fighters in Steel: The Struggle for Democratic Unionism* (Ithaca: Cornell University Press, 2003), 43.
118. Brown, "Racial Conflict and Split Labor Markets," 332–35.
119. George Lipsitz, *Class and Culture in Post War America: A Rainbow at Midnight* (New York: Praeger, 1981), 14–36, 231–32.
120. Bensman and Lynch, *Rusted Dreams*, 41–42.
121. *U.S. Steel Corporation v. National Labor Relations Board*, 1983, 711 F.2d 772 U.S. App., June 30.
122. Lipsitz, *Class and Culture in Post War America*, 14–36, 231–32.
123. James Robinson, *Toil and Toxics: Workplace Struggles and Political Strategies for Occupational Health* (Berkeley: University of California Press, 1991), 1–13.
124. Bruno, *Steelworker Alley*, 120.
125. Warner Bloomberg, "Gary's Industrial Workers as Full Citizens: They Mean to Use Their New-Won Status and Power," *Commentary* 18 (September 1954): 202–10.
126. Hurley, *Environmental Inequalities*, 84; Bruno, *Steelworker Alley*, 76–77.
127. *Department of Revenue v. United States Steel Corporation*, 1981, 435 N.E.2d 659, September 15.
128. National Cancer Institute, *Atlas of Cancer Mortality in the United States: 1950–1994* (Bethesda, Md.: National Cancer Institute, 1999), 64–73; J. Mason and D. Guimany, "Filipinos and Unionization of the Alaskan Canned Salmon Industry," *Amerasia Journal* 8 (1981): 1–30; Epstein, *The Politics of Cancer*, 20–26.
129. Hurley, *Environmental Inequalities*, 94–95.
130. See Bureau of National Affairs, *Basic Patterns in Union Contracts, 1957–1987* (12th ed.; Washington, D.C.: Bureau of National Affairs, 1989); James C. Robinson, *Toil and Toxics: Workplace Struggles and Political Strateggies for Occupational Health* (Berkeley: University of California Press, 1991), 53–55.
131. B. I. Castleman and G. E. Ziem, "Corporate Influence on Threshold Limit Values," *American Journal of Industrial Medicine* 13 (1988): 531–59.
132. Charles Noble, *Liberalism at Work: The Rise and Fall of OSHA* (Philadelphia: Temple University Press, 1986), 52; Hurley, *Environmental Inequalities*, 97–98.
133. Hurley, *Environmental Inequalities*, 98.
134. Ibid., 100.

135. Ibid., 100–101.

136. Ibid., 86–87.

137. Bureau of Labor Statistics, *Characteristics of Major Collective Bargaining Agreements.*

138. Hurley, *Environmental Inequalities*, 103; Bill Kovach, "Gary Groups Join to Fight Dirty Air," *New York Times*, December 7, 1970, 51.

139. Hurley, *Environmental Inequalities*, 103–6.

140. Kovach, "Gary Groups Join to Fight Dirty Air," 51.

141. Ibid.

142. Ibid.

143. Hurley, *Environmental Inequalities*, 103–6.

144. Ibid., 106–7; *Gary-Post Tribune*, July 11, 1971.

145. Osann, "Indiana Sand Dunes and Steel Don't Mix," 10.

146. Hurley, *Environmental Inequalities*, 107; *Citizen's Action Coalition v. Northern Indiana Public Service Company*, 1985, 485 N.E.2d. 610, November 19.

147. *Gary-Post Tribune*, February 11, 1972; Hurley, *Environmental Inequalities*, 108.

148. Bensman and Lynch, *Rusted Dreams*, 187–92.

149. Hurley, *Environmental Inequalities*, 119–20.

150. Ibid., 111–12.

151. Ibid., 112–13.

152. Ibid., 133; *Gary Info*, December 28, 1972, 4.

153. Hurley, *Environmental Inequalities*, 133–34.

154. Ibid., 134.

155. Whatley, "African-American Strikebreaking from the Civil War to the New Deal," 527, 538–39.

156. For discussions of framing and the salience of frames, see Rokeach, *The Nature of Human Values*; D. A. Snow et al., "Frame Alignment Processes, Micro-Mobilization, and Movement Participation"; D. A. Snow and Benford, "Master Frames and Cycles of Protest"; Stryker, "Identity Salience," 558–64; McAdam, *Political Process*; McAdam and Paulsen, "Specifying the Relationship between Social Ties and Activism," 145–58.

157. D. A. Snow et al., "Frame Alignment Processes, Micro-Mobilization, and Movement Participation"; D. A. Snow and Benford, "Master Frames and Cycles of Protest"; Stryker, "Identity Salience"; McAdam, *Political Process*; McAdam and Paulsen, "Specifying the Relationship between Social Ties and Activism."

158. McAdam, *Political Process.*

159. Gary T. Marx, "External Efforts to Damage or Facilitate Social Movements: Some Patterns, Explanations, Outcomes, and Complications," in *Social Movements: Perspectives and Issues*, edited by Steven M. Buechler and F. Kurt Cylke Jr. (Mountain View, Calif.: Mayfield, 1997), 360–84; McAdam, "Tactical Innovation," 735–54.

160. For discussions about expanding salient identities, see McAdam, *Political Process*, 129–31; Friedman and McAdam. "Collective Identity and Activism," 161–62.

161. For discussion of resource mobilization, see Oberschall, *Social Conflict*; McCarthy and Zald, "Resource Mobilization"; Tilly, "Getting It Together in Burgundy."

162. D. E. Taylor, "The Rise of the Environmental Justice Paradigm," 504–80.

163. Opinion Research Corporation, 1965–1970.
164. Harris Poll, 1967; Harris Poll, 1970.
165. State of the Nation Poll, 1972–1976.

Conclusion

1. Louis Blumberg and Robert Gottlieb, *War on Waste: Can America Win Its Battle with Garbage?* (Covelo, Calif.: Island Press, 1989).
2. See, for instance, United Church of Christ, *Toxic Wastes and Race in the United States: A National Report on the Racial and Socioeconomic Characteristics of Communities with Hazardous Waste Sites* (New York: United Church of Christ, 1987); Paul Mohai, "The Demographics of Dumping Revisited: Examining the Impact of Alternative Methodologies in Environmental Justice Research," *Virginia Environmental Law Journal* 14 (1995): 615–53.
3. Conner Bailey, Charles Faupel, and James Gundlach, "Environmental Politics in Alabama's Blackbelt," in *Confronting Environmental Racism: Voices from the Grassroots*, edited by Robert Bullard (Boston: South End Press, 1993), 107–22; Bill D. Moyers, *Global Dumping Ground: The International Traffic in Hazardous Wastes* (Washington, D.C.: Seven Locks Press, 1990).
4. Popkin et al., *The Hidden War*; U.S. Department of Housing and Urban Development, "Hope VI," 2006, at http://www.hud.gov.
5. NIOSH, *National Occupational Research Agenda*, 4; Bureau of Labor Statistics, *News: National Census of Fatal Occupational Injuries in 2004*, press release, August 25, 2005, 1; Bureau of Labor Statistics, *News: Workplace Injuries and Illnesses in 2004*, press release, November 17, 2005, 1.
6. D. E. Taylor, "The Rise of the Environmental Justice Paradigm," 508–80.
7. Cambridge Reports, "Trends and Forecasts," 1989, press release, 4–6; Cambridge Reports, "The Rise of the Green Consumer," 1989, press release.
8. Roper, "Research Supplement," June 1989, 1.
9. Cambridge Reports, "Trends and Forecasts," 4–6; Cambridge Reports, "The Rise of the Green Consumer."
10. "Environment a Worry," CBS News Poll, March 1–2, 1998.
11. Cambridge Reports, "Trends and Forecasts," 4–6; Cambridge Reports, "The Rise of the Green Consumer."
12. Mark Crawford, introduction to *Toxic Waste Sites: An Encyclopedia of Endangered America* (Santa Barbara: ABC-CLIO, 1997), vii–xii; U.S. Environmental Protection Agency, "National Priorities List: NPL Site Totals by Status and Milestone," February 24, 2006, at http://www.epa.gov/superfund/sites/npl; U.S. Environmental Protection Agency, "Superfund National Accomplishments Summary, Fiscal Year 2003," November 3, 2003, 1–2, at http://www.epa.gov/superfund; Dave Ryan, "Superfund: 40 High-Priority Superfund Sites Cleaned Up," 2003, press release, U.S. Environmental Protection Agency.

INDEX

The symbol *b* stands for box; *f* for figure; *n* for note; and *t* for table.

Dorceta E. Taylor is an associate professor of environmental sociology at the University of Michigan, Ann Arbor, where she has a joint appointment in the School of Natural Resources and Environment and the Center for Afroamerican and African Studies. She is the author of *Race, Class, Gender, and American Environmentalism* (2002) and the editor of *Advances in Environmental Justice: Research Theory and Methodology*, a special issue of the *American Behavioral Scientist* in 2000.

Library of Congress Cataloging-in-Publication Data
Taylor, Dorceta E.
The environment and the people in American cities, 1600s–1900s :
disorder, inequality, and social change / Dorceta E. Taylor.
p. cm.
Includes bibliographical references and index.
ISBN 978-0-8223-4436-0 (cloth : alk. paper)
ISBN 978-0-8223-4451-3 (pbk. : alk. paper)
1. Cities and towns—United States—Growth—History. 2. Cities and towns—Growth—Environmental aspects—United States. 3. Cities and towns—Growth—Social aspects—United States. 4. City planning—United States. I. Title.
HT371.T39 2009
307.760973—dc22 2009030092